Lecture Notes in Computer Science   1764

Edited by G. Goos, J. Hartmanis and J. van Leeuwen

W0106157

Springer

Berlin
Heidelberg
New York
Barcelona
Hong Kong
London
Milan
Paris
Singapore
Tokyo

Hartmut Ehrig Gregor Engels
Hans-Jörg Kreowski Grzegorz Rozenberg (Eds.)

Theory and Application of Graph Tranformations

6th International Workshop, TAGT'98
Paderborn, Germany, November 16-20, 1998
Selected Papers

Springer

Series Editors

Gerhard Goos, Karlsruhe University, Germany
Juris Hartmanis, Cornell University, NY, USA
Jan van Leeuwen, Utrecht University, The Netherlands

Volume Editors

Hartmut Ehrig
Technical University of Berlin, Sekr. FR 6 - 1, Computer Science Department (FB13)
Franklinstr. 28/29, 10587 Berlin, Germany
E-mail: ehrig@cs.tu-berlin.de

Gregor Engels
University of Paderborn, Department of Computer Science
Warburger Str. 100, 33098 Paderborn, Germany
E-mail: engels@upb.de

Hans-Jörg Kreowski
University of Bremen, Department of Computer Science, P. O. 330440
28334 Bremen, Germany
E-mail: kreo@informatik.unibremen.de

Grzegorz Rozenberg
Leiden University
Leiden Institute of Advanced Computer Science (LIACS)
Niels-Bohr-Weg 1, 2333 CA Leiden, The Netherlands
E-mail: rozenberg@wi.leidenuniv.nl

Cataloging-in-Publication Data applied for

Die Deutsche Bibliothek - CIP-Einheitsaufnahme
Theory and application of graph transformations. - 6. 1998 (2000)-. -
Berlin ; Heidelberg ; New York ; Barcelona; Hong Kong ; London ; Milan
; Paris ; Singapore ; Tokyo : Springer, 2000
  (Lecture notes in computer science ; ...)
  Früher u.d.T.: Graph-grammars and their application to computer science
6. Paderborn, Germany, November 1998
Selected papers. - 2000
  (Lecture notes in computer science ; Vol. 1764)
  ISBN 3-540-67203-6
CR Subject Classification (1991): F.4.2-3, I.1.1, I.2.4, D.2, F. 3, I. 5.1, J.3., J.5.1

ISSN 0302-9743
ISBN 3-540-67203-6 Springer-Verlag Berlin Heidelberg New York

Springer-Verlag is a company in the BertelsmannSpringer publishing group.
© Springer-Verlag Berlin Heidelberg 2000

Typesetting: Camera-ready by author, data conversion by DA-TeX Gerd Blumenstein
Printed on acid-free paper   SPIN 10719685   06/3142   5 4 3 2 1 0

Preface

The area of graph transformation originated in the late 1960s under the name "graph grammars" – the main motivation came from practical considerations concerning pattern recognition and compiler construction. Since then, the list of areas which have interacted with the development of graph transformation has grown impressively. The areas include: software specification and development, VLSI layout schemes, database design, modeling of concurrent systems, massively parallel computer architectures, logic programming, computer animation, developmental biology, music composition, distributed systems, specification languages, software and web engineering, and visual languages.

As a matter of fact, graph transformation is now accepted as a fundamental computation paradigm where computation includes specification, programming, and implementation. Over the last three decades the area of graph transformation has developed at a steady pace into a theoretically attractive research field, important for applications.

This volume consists of papers selected from contributions to the Sixth International Workshop on Theory and Applications of Graph Transformation that took place in Paderborn, Germany, November 16-20, 1998. The papers underwent an additional refereeing process which yielded 33 papers presented here (out of 55 papers presented at the workshop). This collection of papers provides a very broad snapshot of the state of the art of the whole field today. They are grouped into nine sections representing most active research areas.

The workshop was the sixth in a series of international workshops which take place every four years. Previous workshops were called "Graph Grammars and Their Application to Computer Science". The new name of the Sixth Workshop reflects more accurately the current situation, where both theory and application play an equally central role.

The workshop has received financial support from the European Community as a TMR Euroconference, as well as through the TMR network GETGRATS and the ESPRIT Working Group APPLIGRAPH.

November 1999
H. Ehrig, G. Engels,
H.-J. Kreowski, G. Rozenberg

Organization

TAGT'98 was organized by the Department of Mathematics and Computer Science of the University of Paderborn, Germany, at the Heinz Nixdorf Museums-Forum, Paderborn.

Organizing Committee

G. Engels	University of Paderborn, D	(chair)
H. Ehrig	Technical University of Berlin, D	
H.-J. Kreowski	University of Bremen, D	
G. Rozenberg	University of Leiden, NL	

Program Committee

G. Engels	University of Paderborn, D	(co-chair)
G. Rozenberg	University of Leiden, NL	(co-chair)
B. Courcelle	LaBRI, Bordeaux, F	
H. Ehrig	Technical University of Berlin, D	
D. Janssens	University of Antwerp, B	
H.-J. Kreowski	University of Bremen, D	
U. Montanari	University of Pisa, I	
M. Nagl	RWTH Aachen, D	
F. Parisi–Presicce	University of Rome, I	
R. Plasmeijer	University of Nijmegen, NL	
A. Rosenfeld	University of Maryland, USA	
H.J. Schneider	University of Erlangen, D	

Referees

P. Baldan	J. Hage	F. Rossi
R. Bardohl	A. Habel	A. Schürr
A. Corradini	R. Heckel	M. Simeoni
G. Costagliola	D. Janssens	G. Taentzer
F. Drewes	B. Hoffmann	N. Verlinden
J. Engelfriet	M. Koch	J. Wadsack
G. Ferrari	M. Llabrés–Segura	E. Wanke
I. Fischer	M. Minas	B. Westfechtel
M. Gajewsky	U. Nickel	A. Zündorf
M. Große–Rhode	J. Padberg	
St. Gruner	D. Plump	

Table of Contents

Modularity and Refinement

Software Engineering

Some Remarks on the
Generative Power of Collage Grammars
and Chain-Code Grammars

Frank Drewes*

Department of Computer Science, University of Bremen
P.O. Box 33 04 40, D–28334 Bremen, Germany
drewes@informatik.uni-bremen.de

Abstract. Collage grammars and context-free chain-code grammars are compared with respect to their generative power. It is shown that the generated classes of line-drawing languages are incomparable, but that chain-code grammars can simulate those collage grammars which use only similarity transformations.

1 Introduction

Inspired by the comparison of chain-code and collage grammars in [DHT96], in this paper some further observations concerning the generative power of these two types of picture generating grammars are pointed out.

A context-free chain-code grammar [MRW82] is a type-2 Chomsky grammar generating a language of words over the alphabet $\{u, d, l, r, \uparrow, \downarrow\}$. Such a word is then interpreted as a sequence of instructions to a plotter-like device in order to produce a drawing consisting of horizontal and vertical line segments with endpoints in $\mathbb{Z}^2$. The letters u, d, l, and r are interpreted as instructions to draw a unit line from the current position of the pen upwards, downwards, to the left, and to the right, respectively. Furthermore, $\uparrow$ lifts the pen (so that subsequent drawing instructions only affect the position of the pen, rather than actually drawing a line) and $\downarrow$ sets the pen down, again.

Collage grammars, as introduced in [HK91] (see also the survey [DK99]), are quite different as they produce pictures by transforming any sort of basic geometric objects using affine transformations. In particular, they are not at all restricted to the generation of line drawings. However, collage grammars *can* of course generate line drawings in the sense of chain-code grammars, so that it is natural to compare these two devices with respect to their capabilities in generating this sort of pictures.

Three results in this respect are presented in this paper.

---

* Partially supported by the EC TMR Network GETGRATS (General Theory of Graph Transformation Systems) through the University of Bremen.

H. Ehrig et al. (Eds.): Graph Transformation, LNCS 1764, pp. 1–14, 2000.
© Springer-Verlag Berlin Heidelberg 2000

(1) Linear collage grammars can generate languages of line drawings that cannot be generated by context-free chain-code grammars.
(2) Conversely, linear context-free chain-code grammars can generate languages which cannot be generated by collage grammars.[1] Thus, the two classes of languages are incomparable.
(3) In contrast to (1), every language of line drawings which can be generated by a collage grammar using only similarity transformations, can as well be generated by a context-free chain-code grammar.

The results (1) and (2) are in fact slightly stronger because they remain valid if equality of line drawings is only required up to translation (i.e., if one is interested in the figures being generated, but not in their exact positions).

In order to prove (2), a necessary criterion for context-freeness of collage languages is proved in Section 5 of this paper, extending a result which was shown in [DKL97].

2   Basic Notions and Notation

It is assumed that the reader is familiar with the basic notions of affine geometry. The sets of natural numbers, integers, and real numbers are denoted by $\mathbb{N}$, $\mathbb{Z}$, and $\mathbb{R}$, respectively. $\mathbb{N}_+$ denotes $\mathbb{N} \setminus \{0\}$ and $[n]$ denotes $\{1, \ldots, n\}$ for $n \in \mathbb{N}$. The identical transformation on $\mathbb{R}^2$ is denoted by id. The cardinality of a set S is denoted by $|S|$. As usual, the set of all finite words (or strings) over an alphabet A is denoted by A^* and λ denotes the empty word. The composition of functions f and g (first f, then g) is denoted by $g \circ f$.

A *signature* Σ is a finite set whose elements are called *symbols*, such that for every $f \in \Sigma$ a natural number called its *rank* is specified. The fact that a symbol f has rank n may be indicated by writing $f^{(n)}$ instead of f. The set T_Σ of *terms over* Σ is defined as usual, i.e., it is the smallest set such that $f[t_1, \ldots, t_n] \in T_\Sigma$ for every $f^{(n)} \in \Sigma$ ($n \in \mathbb{N}$) and all $t_1, \ldots, t_n \in T_\Sigma$ (where, for $n = 0$, $f[]$ may be identified with f). The *size* of a term t is denoted by $|t|$ and is also defined in the usual way: $|f[t_1, \ldots, t_n]| = 1 + \sum_{i=1}^{n} |t_i|$ for all terms $f[t_1, \ldots, t_n]$.

A *regular tree grammar* (cf. [GS97]) is a tuple $g = (N, \Sigma, P, S)$ such that N is a finite set of *nonterminals* considered as symbols of rank 0, Σ is a signature disjoint with N, P is a set of term rewrite rules of the form $A \to t$ where $A \in N$ and $t \in T_{\Sigma \cup N}$, and $S \in N$ is the *start symbol*. The rules in P are also called *productions*. The *regular tree language* generated by g is given by $L(g) = \{t \in T_\Sigma \mid S \to_P^* t\}$, where $\to_P^*$ denotes the transitive and reflexive closure of the term rewrite relation $\to_P$ determined by P.

A regular tree grammar or context-free Chomsky grammar is said to be *linear* if every right-hand side contains at most one nonterminal symbol.

The following, well-known normal-form result for regular tree grammars is sometimes useful.

---
[1] This fact was already claimed in [DHT96], but a proof was missing until now.

Lemma 1. *Let g be a regular tree grammar. Then there is a regular tree grammar $g' = (N, \Sigma, P, S)$ such that*

1. *$L(g') = L(g)$ and*
2. *every production in P has the form $A \to f[A_1, \ldots, A_n]$ for some $n \in \mathbb{N}$, where $A, A_1, \ldots, A_n \in N$ and $f^{(n)} \in \Sigma$.*

3   Context-Free Chain-Code Picture Languages

In this section the notion of context-free chain-code grammars [MRW82] is recalled (with the addition of the symbols $\uparrow$ and $\downarrow$, which appeared later in the literature).

For every point $p = (x, y) \in \mathbb{Z}^2$, we denote by $\mathtt{u}(p)$, $\mathtt{d}(p)$, $\mathtt{l}(p)$, and $\mathtt{r}(p)$ the points $(x, y+1)$, $(x, y-1)$, $(x-1, y)$, and $(x+1, y)$, respectively. Furthermore, for every $a \in \{\mathtt{u}, \mathtt{d}, \mathtt{l}, \mathtt{r}\}$, $a\text{-}line(p)$ denotes the subset of $\mathbb{R}^2$ given by the straight line segment between p and $a(p)$.

A *line drawing* is a finite set D such that every $d \in D$ has the form $\mathtt{u}\text{-}line(p)$ or $\mathtt{r}\text{-}line(p)$ for some $p \in \mathbb{Z}^2$. The set of all line drawings is denoted by $\mathbb{D}$.

A *picture description* is a word over the alphabet $A_{cc} = \{\mathtt{u}, \mathtt{d}, \mathtt{l}, \mathtt{r}, \uparrow, \downarrow\}$. Every word $w \in A_{cc}^*$ determines a *drawn picture* $dpic(w) \in \mathbb{D} \times \mathbb{Z}^2 \times \{\uparrow, \downarrow\}$, as follows.

(i) $dpic(\lambda) = (\emptyset, (0,0), \downarrow)$

(ii) For every picture description $v \in A_{cc}^*$ with $dpic(v) = (D, p, s)$ and every $a \in A_{cc}$, if $a \in \{\uparrow, \downarrow\}$ then $dpic(va) = (D, p, a)$. Otherwise,

$$dpic(va) = \begin{cases} (D \cup \{a\text{-}line(p)\}, a(p), s) & \text{if } s = \downarrow \\ (D, a(p), s) & \text{if } s = \uparrow. \end{cases}$$

The line drawing $drawing(w)$ described by $w \in A_{cc}^*$ is the first component of $dpic(w)$, i.e., $drawing(w) = D$ if $dpic(w) = (D, p, s)$.

A (context-free) *chain-code grammar* is a context-free Chomsky grammar ccg whose alphabet of terminal symbols is A_{cc}. The *chain-code picture language* generated by ccg is the set $\mathcal{L}(ccg) = \{drawing(w) \mid w \in L(ccg)\}$.[2] The set of all chain-code picture languages generated by context-free chain-code grammars is denoted by *CFCC*. A language of line drawings which is generated by a linear grammar is called linear.

Example 1 (chain-code grammar). As an example, let ccg be the chain-code grammar whose nonterminals are the symbols S (the initial symbol), R, and L, and whose productions are

$$S \to \mathtt{ruR1Ld}$$

$$L \to \uparrow \mathtt{u} \downarrow L \mathtt{1} L \mathtt{d} \uparrow \mathtt{r} \downarrow, \; L \to \lambda,$$

$$R \to \uparrow \mathtt{u} \downarrow R \mathtt{r} R \mathtt{d} \uparrow \mathtt{1} \downarrow, \; L \to \lambda.$$

Some of the line drawings in $\mathcal{L}(ccg)$ are shown in Figure 1.

---

[2] For a Chomsky grammar g, $L(g)$ denotes the language generated by g.

Fig. 1. Some of the line drawings generated by the chain-code grammar in Example 1

4   Collage Grammars in $\mathbb{R}^2$

In this section the basic definitions concerning collage grammars are recalled. For technical convenience, we shall define collage grammars in the way introduced in [Dre96a,Dre96b] (which is also employed in [DKL97]) rather than using the original definitions from [HK91].

A *collage* is a finite set of *parts*, every part being a subset of $\mathbb{R}^2$. (Thus, in particular, line drawings are collages.) A *collage signature* is a signature Σ consisting of collages (viewed as symbols of rank 0) and symbols of the form $\ll f_1 \cdots f_k \gg$ of rank $k \in \mathbb{N}_+$, where $f_1, \ldots, f_k$ are affine transformations[3] on $\mathbb{R}^2$. A term $t \in \mathrm{T}_\Sigma$ denotes a collage $val(t)$ which is determined as follows. If t is a collage C then $val(t) = C$. Otherwise, if $t = \ll f_1 \cdots f_k \gg [t_1, \ldots, t_k]$ then $val(t) = f_1(val(t_1)) \cup \cdots \cup f_k(val(t_k))$ (where the f_i are canonically extended to collages).

A (context-free) *collage grammar* is a regular tree grammar $cg = (N, \Sigma, P, S)$ such that Σ is a collage signature. The *collage language generated by* cg is $\mathcal{L}(cg) = \{val(t) \mid t \in L(cg)\}$.

Example 2 (collage grammar). As an example, let $\Sigma = \{F^{(3)}, C_{\blacksquare}^{(0)}, C_{\square}^{(0)}\}$, where $C_{\blacksquare}$ is the collage containing the solid unit square as its only part (i.e., $C_{\blacksquare} = \{\{(x, y) \in \mathbb{R}^2 \mid 0 \leq x, y \leq 1\}\}$), $C_{\square}$ is the collage whose only part is the polygon given by the boundary of the unit square, and $F = \ll id\, f\, g \gg$. Here, f and g are as indicated in Figure 2 (where the large square is the unit square). Now, let

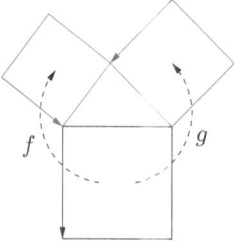

Fig. 2. The two transformations used in Example 2

---

[3] Recall that an affine transformation is a linear mapping composed with a translation by a constant vector.

$cg = (\{S\}, \Sigma, P, S)$, where

$$P = \{S \to F[C_\blacksquare, F[C_\blacksquare, S, C_\square], F[C_\blacksquare, C_\square, S]], \quad S \to C_\blacksquare\}.$$

Then the language $\mathcal{L}(cg)$ generated by cg consists of collages like those shown in Figure 3.

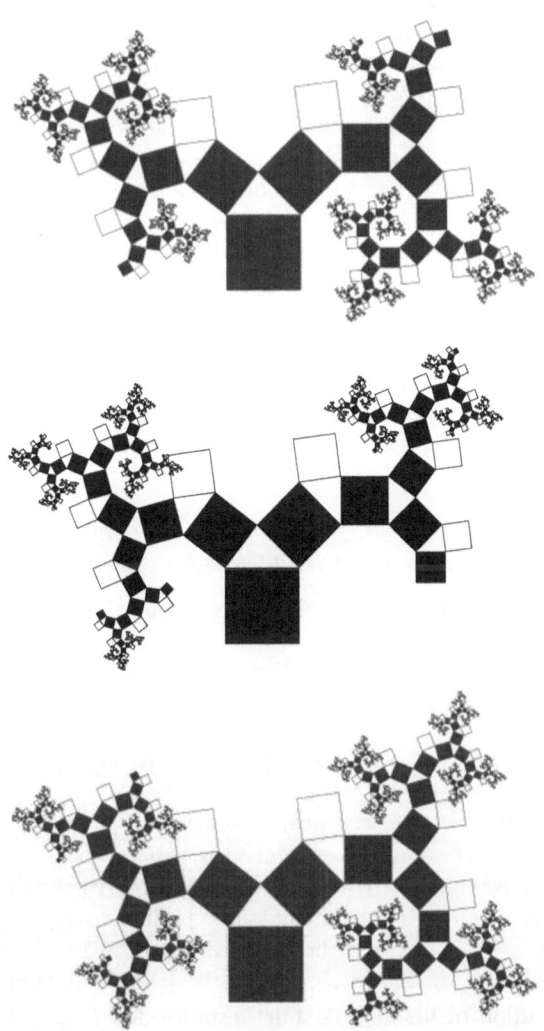

Fig. 3. Three of the collages generated by the collage grammar in Example 2

The set of all languages generated by collage grammars is denoted by *CFCL*. A collage language generated by a linear collage grammar is called linear. Furthermore, if S is a set of affine transformations then $CFCL_S$ denotes the set of all collage languages L such that $L = \mathcal{L}(cg)$ for some collage grammar $cg = (N, \Sigma, P, S)$, where $f_1, \ldots, f_k \in S$ for all $\ll f_1 \cdots f_k \gg \in \Sigma$. Thus, $CFCL_S$ is the set of all collage languages that can be generated by collage grammars using only transformations in S.

5   Collage Grammars vs. Chain-Code Grammars

In this section collage grammars and chain-code grammars are compared with respect to their generative power. Clearly, as mentioned in [DHT96], collage grammars can generate languages which cannot be generated by chain-code grammars, simply because all languages in *CFCC* consist of line drawings whereas collages may contain arbitrary subsets of $\mathbb{R}^2$ as parts. But what about the languages $L \in CFCL$ for which $L \subseteq \mathbb{D}$? It turns out that the answer is 'no' in this case, too. In fact, this negative result can be strengthened by considering line drawings as equivalent if they are equal up to translation. For this (and also for future use), let $\mathbb{GT}$ be the set of all *grid translations* τ_{rs} $(r, s \in \mathbb{Z})$ in $\mathbb{R}^2$, where $\tau_{rs}(x, y) = (x+r, y+s)$ for all $(x, y) \in \mathbb{R}^2$. Now, for two line drawings D and D', let $D \equiv D'$ if (and only if) there is some $\tau \in \mathbb{GT}$ such that $\tau(D) = D'$. Clearly, $\equiv$ is an equivalence relation. The equivalence class of $D \in \mathbb{D}$ is denoted by $[D]$. For languages $L, L' \subseteq \mathbb{D}$, we write $L \equiv L'$ if $\{[D] \mid D \in L\} = \{[D'] \mid D' \in L'\}$. Now, the first result can be formulated as follows.

Theorem 2. *There is a linear language* $L \in CFCL$ *with* $L \subseteq \mathbb{D}$ *such that there is no language* $L' \in CFCC$ *that satisfies* $L \equiv L'$.

For the proof, consider the linear collage grammar

$$cg = (\{S, A\}, \{\ll id\ id \gg, \ll f \gg, C_0, C_1\}, P, S),$$

where

$$P = \{S \to \ll id\ id \gg [C_0, A], A \to \ll f \gg [A], A \to C_1\}.$$

Here, f maps $(x, y) \in \mathbb{R}^2$ to $(2x, y)$, $C_0 = \{\text{r-}line(0, 0)\}$, and $C_1 = \{\text{u-}line(1, 0)\}$. Obviously, $\mathcal{L}(cg) = \{\{\text{r-}line(0, 0), \text{u-}line(2^n, 0)\} \mid n \in \mathbb{N}\}$.

In order to show that $\mathcal{L}(cg)$ cannot be generated by a chain-code grammar one can make use of an observation that may be interesting in its own right. As usual, call a set $Z \subseteq \mathbb{Z}^m$ *linear* if there are $z_0, \ldots, z_n \in \mathbb{Z}^m$ for some $n \in \mathbb{N}$, such that $Z = \{z_0 + \sum_{i \in [n]} a_i \cdot z_i \mid (a_1, \ldots, a_n) \in \mathbb{N}^n\}$, and say that Z is *semi-linear* if it is a finite union of linear sets. Furthermore, for $L \subseteq \mathbb{D}$ let *grid-points(L)* denote the set of all points $p \in \mathbb{Z}^2$ for which there is a line $d \in \bigcup_{D \in L} D$ such that $p \in d$. Then the following can be shown.

Lemma 3. *The set grid-points(L) is semi-linear for every chain code-picture language* $L \in CFCC$.

Proof. Using Parikh's theorem [Par66] it follows that there is a semi-linear set $M \subseteq \mathbb{N}^4$ such that $\textit{grid-points}(L) = \{(v - w, x - y) \mid (v, w, x, y) \in M\}$. Consequently, $\textit{grid-points}(L)$ itself is semi-linear, too.   □

Now, choose $L = \mathcal{L}(cg)$, where cg is as above. Let W be the set of all words $w \in A_{cc}^*$ such that $\textit{drawing}(w) \equiv D$ for some $D \in L$. It is an easy exercise to construct a finite state transduction $\textit{fst}$ which transforms every word $w \in W$ into a word $\textit{fst}(w) = \texttt{r}{\uparrow}v{\downarrow}\texttt{u}$ such that $v \in \{\texttt{u}, \texttt{d}, \texttt{l}, \texttt{r}\}^*$ and $\textit{drawing}(\texttt{r}{\uparrow}v{\downarrow}\texttt{u}) \equiv \textit{drawing}(w)$. Clearly, $\textit{drawing}(\textit{fst}(w)) \in L$ for every $w \in W$. Furthermore, it is well known that the class of context-free languages is closed under finite state transductions. Therefore, if there was a chain-code grammar ccg satisfying $\mathcal{L}(ccg) \equiv L$ then $\textit{fst}(L(ccg))$ would be context-free. But this would mean $L \in CFCC$, which is impossible because $\textit{grid-points}(L) = \{(0,0), (1,0)\} \cup \bigcup_{n \in \mathbb{N}}\{(2^n, 0), (2^n, 1)\}$ is not semi-linear.

Now, let us consider the converse question: Can collage grammars generate all the languages in $CFCC$, or at least the linear ones? Again, the answer is 'no'.

Theorem 4. *There is a linear language $L \in CFCC$ such that there is no language $L' \in CFCL$ that satisfies $L \equiv L'$.*

In fact, the proof of the theorem reveals that this does even hold if the chain-code grammars are not allowed to make use of the symbols $\uparrow$ and $\downarrow$. Consider the linear chain-code grammar

$$ccg = (\{S\}, A_{cc}, \{S \to \texttt{ru}S\texttt{dr}, S \to \texttt{r}\}, S),$$

which generates the set of all line drawings consisting of two "stairs" of equal height, as shown in Figure 4.

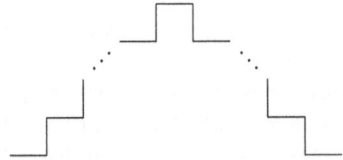

Fig. 4. The type of line drawings generated by the chain-code grammar used in the proof of Theorem 4

Denote $\mathcal{L}(ccg)$ by L. In order to show that this language cannote be generated by a collage grammar one needs a suitable criterion for context-freeness of collage languages, i.e., a criterion that allows to show that the class $CFCL$ does not contain any language L' satisfying $L' \equiv L$. Criteria of this kind have been established in [DKL97]. Unfortunately, for the present aim these criteria do not suffice. It turns out, however, that one of the results in [DKL97], namely

Theorem 1, can be generalized in a nice way (with almost the same proof as in [DKL97]). For this, view a pair $r = (L, R)$ of collages as a replacement rule in the following way. If, for a collage C, there is an affine transformation f such that $f(L) \subseteq C$ then $C \Longrightarrow_r C'$ where $C' = (C \setminus f(L)) \cup f(R)$. As usual, if S is a set of such rules let $C \Longrightarrow_S C'$ if $C \Longrightarrow_r C'$ for some $r \in S$. Then the following theorem is obtained.

Theorem 5. *For every collage language $L' \in CFCL$ there is a constant n_0 and a finite set S of pairs of collages such that the following holds. For every collage $C_0 \in L'$ there are collages $C_1, \ldots, C_n \in L'$ (for some $n \in \mathbb{N}$) such that $C_0 \Longrightarrow_S C_1 \Longrightarrow_S \cdots \Longrightarrow_S C_n$ and $|C_n| \le n_0$.*

Proof. Let $cg = (N, \Sigma, P, S)$ be a collage grammar. For the proof, let $\diamond^{(0)}$ be a symbol not in Σ. For all terms $t_0, t_1 \in T_{\Sigma \cup \{\diamond\}}$ such that t_0 contains $\diamond$ exactly once, let $t_0 \cdot t_1$ denote the term which results from t_0 by replacing the unique occurrence of $\diamond$ with t_1. (Thus, for example, $f[g[a], \diamond] \cdot h[\diamond, b] = f[g[a], h[\diamond, b]]$.) Furthermore, let $t_0^0 = \diamond$ and $t_0^{i+1} = t_0 \cdot t_0^i$ for all $i \in \mathbb{N}$. Using these notations, the well-known pumping lemma for regular tree languages can be formulated as follows.

There is some $m_0 \in \mathbb{N}$ such that every term $t \in L(cg)$ of size greater than m_0 can be written in the form $t_0 \cdot (t_1 \cdot t_2)$, such that $t_1 \ne \diamond$, $|t_1 \cdot t_2| \le m_0$, and $t_0 \cdot (t_1^i \cdot t_2) \in L(cg)$ for all $i \in \mathbb{N}$. In particular, $t_0 \cdot t_2 \in L(cg)$.

Now, define $n_0 = \max\{|val(t)| \mid t \in T_\Sigma, |t| \le m_0\}$ and let S be the set of all pairs (L, R) of collages such that $L \subseteq val(t)$ and $R = val(t')$ for some terms $t, t' \in T_\Sigma$ of size at most m_0.

Consider some collage $C_0 \in \mathcal{L}(cg)$ such that $|C_0| > n_0$. Let $t \in L(cg)$ be a minimal term such that $C_0 = val(t)$. By the choice of n_0 we have $|t| > m_0$, which means that $t = t_0 \cdot t_1 \cdot t_2$ for terms t_0, t_1, t_2 as above. If we interpret the symbol $\diamond$ as the empty collage $\emptyset$, it follows that $C_0 = val(t_0) \cup f(val(t_1 \cdot t_2))$ and $val(t_0 \cdot t_2) = val(t_0) \cup f(val(t_2))$, where f is the affine transformation obtained by composing the transformations on the path in t_0 from the root to the occurrence of $\diamond$. Since $|t_1 \cdot t_2| \le m_0$, the pair $(val(t_1 \cdot t_2) \setminus val(t_0), val(t_2))$ is in S, which yields $C_0 \Longrightarrow_S C_1$ with $C_1 = val(t_0 \cdot t_2)$. Since $|t_0 \cdot t_2| < |t|$, repeating the construction a finite number of times results in a derviation $C_0 \Longrightarrow_S C_1 \Longrightarrow_S \cdots \Longrightarrow_S C_n$, where $C_n = val(s)$ for some term $s \in L(cg)$ with $|s| \le m_0$, i.e., $|C_n| \le n_0$. ☐

Now, in order to prove Theorem 4, assume that $L' \in CFCL$ satisfies $L' \equiv L$ and let S be the corresponding set of collage-replacement rules according to Theorem 5. By the theorem, there are arbitrarily large line drawings $D \in L'$ such that $D \Longrightarrow_S D'$ for some $D' \in L'$ with $|D'| < |D|$. Consider such a pair $D, D' \in L'$ and let $D' = (D \setminus f(C)) \cup f(C')$, where $(C, C') \in S$ and f is an affine transformation. Let $d_1, \ldots, d_n, d, d'_n, \ldots, d'_1$ be the lines D consists of, ordered from left to right (thus, d_1 is the leftmost, d'_1 the rightmost, and d the topmost horizontal line). If D' contains d_1 and at least one of $d'_1, \ldots, d'_n$ then $D \subseteq D'$

and, hence, $|D'| \geq |D|$. The same holds if D' contains d'_1 and at least one of $d_1, \ldots, d_n$. Now, suppose that D is large enough to satisfy $n > |C|$. Then we have $D' \cap \{d_1, \ldots, d_n\} \neq \emptyset \neq D' \cap \{d'_1, \ldots, d'_n\}$. Therefore, from the assumption $|D'| < |D|$ it follows that $d_1, d'_1 \in f(C)$. However, since f is affine this means that the distance of $f^{-1}(d_1)$ and $f^{-1}(d'_1)$ in C, relative to the length of these lines, is the same as the distance between d_1 and d'_1. Thus, S turns out to be infinite, contradicting the assumption. This completes the proof of Theorem 4.

Theorems 2 and 4 reveal that the classes $CFCC$ and $CFCL$ are incomparable even if the latter is restricted to line-drawing languages and equality of line drawings is required only up to translation. The collage grammar used to prove Theorem 2 makes use of a non-uniform scaling, however. Intuitively, its horizontal scaling causes, in effect, an exponential translation of vertical lines. The crucial point is that a horizontal scaling of a unit vertical line results in a unit vertical line, again, because lines are one-dimensional objects. It is thus natural to wonder what happens if the collage grammars are only allowed to make use of similarity transformations[4]. The remainder of this section is devoted to the proof that, in this case, context-free chain-code grammars can indeed simulate collage grammars. Let $\mathbb{SIM}$ denote the set of all similarity transformations on $\mathbb{R}^2$. It will be shown that $CFCL_{\mathbb{SIM}}$ is nothing else than $CFCL_{\mathbb{GT}}$, and the latter is contained in $CFCC$. In order to prove that $CFCL_{\mathbb{GT}}$ contains $CFCL_{\mathbb{SIM}}$ it is convenient to prove a weaker, auxiliary statement first. For this, let $\mathbb{GR}$ ("grid rotations") denote the set of all transformations of the form f or $f \circ f'$, where f is a rotation about the origin by a multiple of $\pi/2$ and f' is the reflexion that maps $(1, 0)$ to $(-1, 0)$ and $(0, 1)$ to $(0, 1)$. Notice that $\mathbb{GR}$ is finite: it consists of eight transformations.

Now, let $\mathbb{GT}_+$ denote the set of all transformations $f \circ r$ such that $f \in \mathbb{GT}$ and $r \in \mathbb{GR}$. Obviously, $\mathbb{GT}_+$ is closed under composition. The auxiliary statement mentioned above states that $CFCL_{\mathbb{GT}}$ equals $CFCL_{\mathbb{GT}_+}$. This is proved next.

Lemma 6. $CFCL_{\mathbb{GT}} = CFCL_{\mathbb{GT}_+}$

Proof. By definition, $CFCL_{\mathbb{GT}} \subseteq CFCL_{\mathbb{GT}_+}$, so it remains to prove the converse inclusion. For this, let $cg = (N, \Sigma, P, S)$ be a collage grammar using only transformations in $\mathbb{GT}_+$, and assume without loss of generality that the productions in P have the form described in Lemma 1. Now, let $cg' = (N \times \mathbb{GR}, \Sigma', P', (S, id))$, where Σ' consists of the symbols occurring in P', and P' itself is constructed as follows.

(1) For every production $A \to C$ in P, where C is a collage, and every transformation $r \in \mathbb{GR}$, P' contains the production $(A, r) \to C'$, where $C' = r(C)$.
(2) For every production $A \to \ll f_1 \cdots f_n \gg [A_1, \ldots, A_n]$ in P and every transformation $r \in \mathbb{GR}$, since $\mathbb{GT}_+$ is closed under composition, $r \circ f_i = f'_i \circ r_i$ for some $f'_i \in \mathbb{GT}$ and $r_i \in \mathbb{GR}$. Now, let P' contain the production $(A, r) \to \ll f'_1 \cdots f'_n \gg [(r_1, A_1), \ldots, (r_n, A_n)]$.

---

[4] Recall that a similarity transformation is a transformation composed of a uniform scaling, a rotation (possibly involving a reflexion), and a translation.

For all $r \in \mathbb{GR}$, $A \in N$, and $t \in T_\Sigma$ with $A \to_P^n t$ it follows by induction on n that there is a derivation $(A, r) \to_{P'}^n t'$ with $t' \in T_{\Sigma'}$ and $val(t') = r(val(t))$. For $n = 0$, $A \to_P C$ immediately implies $(A, r) \to_{P'} r(C)$ (by (1)). For $n > 0$, the given derivation must have the form $A \to_P \ll f_1 \cdots f_n \gg [A_1, \ldots, A_n] \to_P^{n-1} \ll f_1 \cdots f_n \gg [t_1, \ldots, t_n] = t$. Using the same notation as in (2),

$$r(val(t)) = r(\ll f_1 \cdots f_n \gg (val(t_1), \ldots, val(t_n)))$$
$$= \ll r \circ f_1 \cdots r \circ f_n \gg (val(t_1), \ldots, val(t_n))$$
$$= \ll f_1' \cdots f_n' \gg (r_1(val(t_1)), \ldots, r_n(val(t_n))).$$

Furthermore, making use of the induction hypothesis,

$$(A, r) \to_{P'} \quad \ll f_1' \cdots f_n' \gg [(r_1, A_1), \ldots, (r_n, A_n)]$$
$$\to_{P'}^{n-1} \ll f_1' \cdots f_n' \gg [t_1', \ldots, t_n']$$
$$= \quad t'$$

where $val(t_i') = r_i(val(t_i))$ for all $i \in [n]$, and thus $r(val(t)) = val(t')$. The converse statement can be verified in the same way, which yields $L(cg') = L(cg)$, as claimed. □

Using the previous lemma, the first simulation result mentioned above can be proved. Intuitively, since all the parts in a collage of the language L are required to be unit lines, a collage grammar cannot significantly make use of uniform scalings or rotations, and of translations other than those in $\mathbb{GT}$, because every deviation must be remembered in the (finitely many) nonterminals in order to avoid producing "wrong" parts. This is why similarity transformations to not provide more power than the transformations in $\mathbb{GT}$ as long as only line drawings are generated.

Theorem 7. *Let $L \in CFCL_{\mathrm{SIM}}$ be a set of line drawings. Then it holds that $L \in CFCL_{\mathbb{GT}}$.*

Proof. By Lemma 6 it suffices to prove the statement of the lemma for $CFCL_{\mathbb{GT}_+}$ instead of $CFCL_{\mathbb{GT}}$. For this, let $\mathbb{T}$ be the set of all transformations of the form $f \circ g \circ h$, where f is a translation by a vector (a, b), $0 \le a, b < 1$, g is a rotation by an angle α, $0 \le \alpha < \pi/2$, and h is a (uniform) scaling. Then the following hold.

(1) For every collage $C \ne \emptyset$ there is at most one $f \in \mathbb{T}$ such that $f(C) \in \mathbb{D}$.
    (If $f(C) \in \mathbb{D}$ and $C \ne \emptyset$ then C contains a line segment. By the definition of $\mathbb{T}$ there is only one $f \in \mathbb{T}$ such that f, applied to this part, yields a part of the form $\mathbf{r}$-$line(p)$ or $\mathbf{u}$-$line(p)$ for some $p \in \mathbb{Z}^2$.)
(2) Every similarity transformation f can be decomposed as $f = g \circ h$ such that $g \in \mathbb{GT}_+$ and $h \in \mathbb{T}$.
    (We have $f = \tau_{rs} \circ \rho \circ \sigma$ for some $r, s \in \mathbb{R}$, a rotation (and possibly reflexion) ρ, and a scaling σ. This can be re-written as $\tau_{r_0 s_0} \circ \tau_{r_1 s_1} \circ \rho_0 \circ \rho_1 \circ \sigma$, where $\tau_{rs} =$

$\tau_{r_0 s_0} \circ \tau_{r_1 s_1}$, $\rho = \rho_0 \circ \rho_1$, $\tau_{r_0 s_0} \circ \rho_0 \in \mathbb{GT}_+$, $0 \le r_1, s_1 < 1$, and the rotation angle α of ρ_1 satisfies $0 \le \alpha < \pi/2$. Hence, if we set $(r_1', s_1') = \rho_0^{-1}(r_1, s_1)$ then $0 \le r_1', s_1' < 1$ because the rotation angle of ρ_0 is a multiple of $\pi/2$. As a consequence, we get $f = \tau_{r_0 s_0} \circ \rho_0 \circ \tau_{r_1' s_1'} \circ \rho_1 \circ \sigma$, where $\tau_{r_1' s_1'} \circ \rho_1 \circ \sigma \in \mathbb{T}$.)

(3) For all $g \in \mathbb{GT}_+$ and all collages C, $g(C) \in \mathbb{D}$ implies $C \in \mathbb{D}$.

(This is a direct consequence of the fact that, by definition, $\mathbb{D}$ is closed under $\mathbb{GT}_+$ and $\mathbb{GT}_+$ is closed under inversion.)

(4) For every similarity transformation f and every $D \in \mathbb{D} \setminus \{\emptyset\}$, $f(D) \in \mathbb{D}$ implies $f \in \mathbb{GT}_+$.

(This is a consequence of (1)–(3): By (2) $f(D) = g(h(D))$ for some $g \in \mathbb{GT}_+$ and $h \in \mathbb{T}$. By (3) $h(D)$ turns out to be an element of $\mathbb{D}$, too, so (1) implies that h is the identity, i.e., $f = g \in \mathbb{GT}_+$.)

Let $L = \mathcal{L}(cg) \subseteq \mathbb{D}$ for some collage grammar $cg = (N, \Sigma, P, S)$ using similarity transformations only. Without loss of generality it can be assumed that cg does not contain useless nonterminals (in the usual sense of formal language theory), that every $A \in N$ derives at least one nonempty collage, and that the productions in P have the form described in Lemma 1.

For every $A \in N$, denote by cg_A the collage grammar (N, Σ, P, A). Since A is not useless, there is some derivation $S \to^* t$ such that A occurs in t. By the definition of val, the composition of the transformations on the path in t leading from the root to this occurrence yields a similarity transformation $f \in \mathbb{SIM}$ such that, for all collages $C \in \mathcal{L}(cg_A)$, $f(C) \subseteq C_0$ for some collage $C_0 \in L$. Since, by assumption, $L \subseteq \mathbb{D}$, it follows that $f(C) \in \mathbb{D}$ for all $C \in \mathcal{L}(cg_A)$. By (2), $f = g \circ h$ for some $g \in \mathbb{GT}_+$ and $h \in \mathbb{T}$, and by (3) $h(C) \in \mathbb{D}$. By (1) this implies that h is uniquely determined (recall that $\mathcal{L}(cg_A)$ contains at least one nonempty collage). In other words, for every $A \in N$, there is a unique transformation $h_A \in \mathbb{T}$ such that $h_A(C) \in \mathbb{D}$ for all $C \in \mathcal{L}(cg_A)$.

Define $cg' = (N, \Sigma', P', S)$ as follows.

- For every rule $A \to C$ in P, where C is a collage, P' contains the rule $A \to h_A(C)$.
- For every rule $A \to \ll f_1 \cdots f_n \gg [A_1, \ldots, A_n]$ in P, P' contains the rule $A \to \ll f_1' \cdots f_n' \gg [A_1, \ldots, A_n]$, where $f_i' = h_A \circ f_i \circ h_{A_i}^{-1}$ for $i = 1, \ldots, n$.

The signature Σ' consists of all $h_A(C)$ and all $\ll f_1' \cdots f_n' \gg$ which occur in these rules.

By induction, it follows that $\mathcal{L}(cg_A') = h_A(\mathcal{L}(cg_A))$ for all $A \in N$ (where $cg_A' = (N, \Sigma', P', A)$). In particular, $\mathcal{L}(cg_A') \subseteq \mathbb{D}$ and $\mathcal{L}(cg') = h_S(\mathcal{L}(cg)) = \mathcal{L}(cg)$. The latter holds because h_S is the identical transformation (by (1) and the fact that $L \subseteq \mathbb{D}$). It remains to show that each of the transformations $f_i' = h_A \circ f_i \circ h_{A_i}^{-1}$ occurring in the rules in P' is an element of $\mathbb{GT}_+$. To see this, consider some $C \in \mathcal{L}(cg_{A_i}') \setminus \{\emptyset\}$. Since none of the nonterminals A_j in the right-hand side of $A \to \ll f_1' \cdots f_n' \gg [A_1, \ldots, A_n]$ is useless, it follows that $A \to_{P'}^* t$ for some $t \in \mathbb{T}_{\Sigma'}$ such that $val(t) = C' \supseteq h_A(f_i(h_{A_i}^{-1}(C)))$. However, as we saw above, both C' and C are in $\mathbb{D}$. Hence, (4) yields $h_A \circ f_i \circ h_{A_i}^{-1} \in \mathbb{GT}_+$. $\square$

It is now possible to prove that chain-code grammars can simulate collage grammars using only similarity transformations, provided that only line drawings are generated by the latter.

Theorem 8. *If $L \in CFCL_{\mathrm{SIM}}$ is a language of line drawings, then $L \in CFCC$.*

Proof. Using Theorem 7, this can easily be verified. Let $L \in CFCL_{\mathrm{GT}}$ be a language of line drawings, where $cg = (N, \Sigma, P, S)$ is the corresponding collage grammar (i.e., cg is assumed to use only grid translations). Then it may be assumed without loss of generality that P contains only productions of the form $A \to \ll \tau_{r_1 s_1} \cdots \tau_{r_n s_n} \gg [A_1, \ldots, A_n]$, where $A_1, \ldots, A_n \in N$, and $A \to \{d\}$, where $d = \mathbf{u}\text{-}line(0,0)$ or $d = \mathbf{r}\text{-}line(0,0)$. Now, turn cg into a chain-code grammar $ccg = (N \cup \{S_0\}, A_{cc}, P', S_0)$, where S_0 is a new nonterminal, as follows.

- For every production $A \to \ll \tau_{r_1 s_1} \cdots \tau_{r_n s_n} \gg [A_1, \ldots, A_n]$ in P, choose any two words $v_i, w_i \in \{\mathbf{u}, \mathbf{d}, \mathbf{l}, \mathbf{r}\}^*$ such that $dpic(\uparrow v_i) = (\emptyset, (r_i, s_i), \uparrow)$ and $dpic(\uparrow w_i) = (\emptyset, (-r_i, -s_i), \uparrow)$. Then P' contains the production

$$A \to v_1 A_1 w_1 v_2 A_2 w_2 \cdots v_n A_n w_n.$$

- For every production $A \to \{d\}$ in P, P' contains the production $A \to \downarrow \mathbf{ud} \uparrow$ if $d = \mathbf{u}\text{-}line(0,0)$ and $A \to \downarrow \mathbf{rl} \uparrow$ if $d = \mathbf{r}\text{-}line(0,0)$.
- In addition, P' contains the production $S_0 \to \uparrow S$.

It follows by an obvious induction that, for every $A \in N$ and every $D \in \mathbb{D}$, there is a derivation $A \to_P^* t$ for some term $t \in \mathrm{T}_\Sigma$ with $val(t) = D$, if and only if $A \to_{P'}^* w$ for some $w \in A_{cc}^*$ such that $dpic(\uparrow w) = (D, (0,0), \uparrow)$. Consequently, $\mathcal{L}(ccg) = \mathcal{L}(cg)$, which proves Theorem 8.   □

The construction of productions in the proof above can in fact be simplified if cg is a linear collage grammar. In this case it can be assumed that all productions have the form $A \to D$ or $A \to \ll id\, \tau_{r_1 s_1} \gg [D, A_1]$, where D is a line drawing and A_1 a nonterminal. Thus, the corresponding productions of a chain-code grammar would be $A \to \downarrow v_0$ and $A \to \downarrow v_0 v_1 A_1$, where $dpic(v_0) = (D, (0,0), \uparrow)$ and v_1 is as in the construction above. In particular, no w_1 to the right of A_1 is needed. Thus, the resulting grammar is regular. Together with the fact that every regular chain-code grammar can easily be transformed into an equivalent collage grammar that uses only grid translations the following corollary is obtained, which is a slight extension of Theorems 10 and 13 in [DHT96].

Corollary 9. *For every language L of line drawings the following statements are equivalent.*

(i) L can be generated by a regular chain-code grammar;
(ii) L can be generated by a linear collage grammar using only similarity transformations;
(iii) L can be generated by a linear collage grammar using only grid translations.

6   Conclusion

In the previous section it was shown that collage grammars and context-free chain-code grammars yield incomparable classes of line drawings, but that the former can be simulated by the latter if only similarity transformations are used (and, of course, the generated language is a language of line drawings). In order to prove the latter, it was shown that similarity transformations to not provide any additional power compared to the more restricted grid transformations if only line drawings are generated.

Despite the simulation result one may say that incomparability is the main characteristic of the relation between chain-code grammars and collage grammars. As Theorems 2 and 4 show, in either case not even the assumption of linearity makes a simulation possible. Intuitively, the reason for this is that both types of grammars are based on quite different concepts. While collage grammars employ a completely local generation mechanism, where the generation of a part has no effect on the rest of the generated collage, the main principle of chain-code grammars is the concatenation of line drawings (see also [Dre98]). Thus, the latter can insert new lines by shifting already generated parts to the side, as was done in the example used to prove Theorem 4. On the other hand, chain-code grammars do not provide any means of scaling, rotation, shearing, etc.—which are essential in the definition of collage grammars.

One may argue that the point of view taken in this paper is somewhat unfair against collage grammars. Most of their nice capabilities cannot be used if they are restricted to the generation of pictures consisting of unit lines. In fact, as a matter of experience (but certainly—to some extend—also as a matter of taste) collage grammars often turn out to be more appropriate and manageable than chain-code grammars because of their flexibility and their strictly local behaviour. Nevertheless, one should also notice that there are quite natural picture languages (like the one used to prove Theorem 4!) that can be generated by chain-code grammars but not by collage grammars.

Acknowledgement

I am very grateful to one of the referees, whose detailed remarks provided a number of useful suggestions.

References

DHT96.   Jürgen Dassow, Annegret Habel, and Stefan Taubenberger. Chain-code pictures and collages generated by hyperedge replacement. In H. Ehrig, H.-J. Kreowski, and G. Rozenberg, editors, *Graph Grammars and Their Application to Computer Science*, number 1073 in Lecture Notes in Computer Science, pages 412–427, 1996. 1, 2, 6, 12

DK99.   Frank Drewes and Hans-Jörg Kreowski. Picture generation by collage grammars. In H. Ehrig, G. Engels, H.-J. Kreowski, and G. Rozenberg, editors,

Handbook of Graph Grammars and Computing by Graph Transformation, volume 2, chapter 11, pages 397–457. World Scientific, 1999. 1

DKL97.   Frank Drewes, Hans-Jörg Kreowski, and Denis Lapoire. Criteria to disprove context-freeness of collage languages. In B.S. Chlebus and L. Czaja, editors, *Proc. Fundamentals of Computation Theory XI*, volume 1279 of *Lecture Notes in Computer Science*, pages 169–178, 1997. 2, 4, 7, 8

Dre96a.   Frank Drewes. Computation by tree transductions. Doctoral dissertation, University of Bremen, Germany, 1996. 4

Dre96b.   Frank Drewes. Language theoretic and algorithmic properties of d-dimensional collages and patterns in a grid. *Journal of Computer and System Sciences*, 53:33–60, 1996. 4

Dre98.   Frank Drewes. Tree-based picture generation. Report 7/98, Univ. Bremen, 1998. 13

GS97.   Ferenc Gécseg and Magnus Steinby. Tree languages. In G. Rozenberg and A. Salomaa, editors, *Handbook of Formal Languages. Vol. III: Beyond Words*, chapter 1, pages 1–68. Springer, 1997. 2

HK91.   Annegret Habel and Hans-Jörg Kreowski. Collage grammars. In H. Ehrig, H.-J. Kreowski, and G. Rozenberg, editors, *Proc. Fourth Intl. Workshop on Graph Grammars and Their Application to Comp. Sci.*, volume 532 of *Lecture Notes in Computer Science*, pages 411–429. Springer, 1991. 1, 4

MRW82.   Hermann A. Maurer, Grzegorz Rozenberg, and Emo Welzl. Using string languages to describe picture languages. *Information and Control*, 54:155–185, 1982. 1, 3

Par66.   Rohit J. Parikh. On context-free languages. *Journal of the Association for Computing Machinery*, 13:570–581, 1966. 7

Tree Languages Generated by Context-Free Graph Grammars*

Joost Engelfriet and Sebastian Maneth

Leiden University, LIACS
PO Box 9512, 2300 RA Leiden, The Netherlands
{engelfri,maneth}@liacs.nl

Abstract. A characterization is given of the class of tree languages which can be generated by context-free hyperedge replacement (HR) graph grammars, in terms of macro tree transducers (MTTs). This characterization yields a normal form for tree generating HR graph grammars. Moreover, two natural, structured ways of generating trees with HR graph grammars are considered and an inclusion diagram of the corresponding classes of tree languages is proved. Finally, the MSO definable tree transductions are characterized in terms of MTTs.

1 Introduction

A tree t (over a ranked alphabet) can conveniently be represented by a hypergraph g in the following well-known way. Each node u of t is represented in g by the same node u and a hyperedge e_u; if u has children $u_1, \ldots, u_k$ in t, then e_u is incident with $u_1, \ldots, u_k, u$ in g (in that order). Figure 1(a) shows the hypergraph g which represents the tree $t = \delta(\gamma(\beta), \alpha, \sigma(\beta, \alpha))$. We call such a hypergraph a *tree graph*, as opposed to the well-known *term graphs* which are tree graphs with sharing of subtrees (see, e.g., [Plu98]). We want to characterize the class $TR(HR)$ of tree languages which can be generated in this way by context-free hyperedge replacement (HR) grammars, in terms of macro tree transducers (MTTs) which are a well-known model for syntax-directed semantics that combines features of the top-down tree transducer and the context-free tree grammar [EV85, FV98].

The class of tree languages obtained by unfolding the term graph languages which can be generated by HR grammars, is characterized in [EH92] as the class of output tree languages of attribute grammars. More recently it was proved in [Dre99] that $TR(HR)$ can be obtained by evaluating the output tree languages of finite-copying top-down tree transducers [ERS80] in an algebra of hypergraphs in which each operation is a substitution into a tree graph. We show that the evaluation in such an algebra corresponds to a very restricted kind of MTT, viz. one which is linear and nondeleting in both the input variables and the parameters (Theorem 4). Composing this class with the class of finite-copying top-down tree transducers, we obtain our characterization of $TR(HR)$ as the class

---

* This work was supported by the EC TMR Network GETGRATS.

H. Ehrig et al. (Eds.): Graph Transformation, LNCS 1764, pp. 15–29, 2000.

of output tree languages of finite-copying MTTs which are linear and nondeleting in the parameters (Theorem 5). This characterization can be used to prove a normal form for tree generating HR grammars (Theorem 7). The normal form requires that the right-hand side of every production is a "linked forest"; a linked forest is obtained from a forest graph, i.e., a sequence of tree graphs, by combining ("linking") nonterminal hyperedges into new hyperedges.

There are two natural (known) ways of restricting the right-hand sides of the productions of an HR grammar in order to generate tree graphs in a structured way. First (see [Rao97]), every right-hand side is a "leaf-linked forest", i.e., a forest graph in which nonterminal hyperedges may only appear at leaves, but each may link several leaves (of possibly different tree graphs). In terms of our normal form this means that each nonterminal hyperedge was obtained by linking only leaves (labeled by nonterminals). Second (see [Dre99]), every right-hand side is a tree graph. In terms of our normal form this means that each forest graph consists of exactly one tree graph, without links. It turns out that these are *the* two different ways of generating tree graphs with HR grammars: the first one captures the top-down tree transducer aspect (the different trees of a forest correspond to the translations by different states) of MTTs and the second one captures the context-free tree grammar aspect of MTTs. We show that the corresponding classes of tree languages are incomparable and hence are proper subclasses of $TR(HR)$. In fact, when allowing sharing of common subtrees, i.e., when moving to term graphs, the corresponding classes of tree languages remain incomparable. Figure 3 shows a Hasse diagram of these classes of tree languages.

Adding regular look-ahead to finite-copying MTTs which are linear and non-deleting in the parameters, they compute precisely the tree transductions that are definable in monadic second order logic (Theorem 11).

2   Trees and HR Grammars

For $m \geq 0$ let $[m] = \{1, \ldots, m\}$, in particular $[0] = \varnothing$. A set Σ together with a mapping rank: $\Sigma \to \mathbb{N}$ is called a *ranked set*. For $k \geq 0$, $\Sigma^{(k)}$ is the set $\{\sigma \in \Sigma \mid \text{rank}(\sigma) = k\}$; we also write $\sigma^{(k)}$ to denote that $\text{rank}(\sigma) = k$. For a set A, $\langle \Sigma, A \rangle$ is the ranked set $\{\langle \sigma, a \rangle \mid \sigma \in \Sigma, a \in A\}$ with $\text{rank}(\langle \sigma, a \rangle) = \text{rank}(\sigma)$. By $\text{inc}(\Sigma)$ $(\text{dec}(\Sigma))$ we denote the ranked set obtained from Σ by increasing (decreasing) the rank of each symbol by one, and zero(Σ) is obtained by changing each rank into zero. The set of all trees over Σ is denoted T_Σ. For a set A, $T_\Sigma(A)$ is the set of all trees over $\Sigma \cup A$, where all elements in A have rank zero. We fix the set X of *input variables* $x_1, x_2, \ldots$ and the set Y of *parameters* $y_1, y_2, \ldots$. For $k \geq 0$, $X_k = \{x_1, \ldots, x_k\}$ and $Y_k = \{y_1, \ldots, y_k\}$.

We assume the reader to be familiar with *hypergraphs* and HR grammars (see, e.g., [BC87, Hab92, DKH97, Eng97]). For a ranked set Δ, the set of all hypergraphs over Δ is denoted HGR(Δ). A hypergraph g over Δ consists of finite sets of nodes and hyperedges; every hyperedge e of *type k* is incident with a sequence of k (distinct) nodes ("the nodes of e") and is labeled by a symbol in $\Delta^{(k)}$; g contains a sequence of (distinct) *external nodes*, the number of which is the *type* of g. Hypergraphs are drawn in the usual way; if the order of the nodes of a hyperedge e is not obvious, then we will label the tentacle to the i-th node of e by the number i (see the A-labeled hyperedges in Figs. 2 and 4(a)). We represent a tree $t \in T_\Sigma$ by a hypergraph $g = \text{gr}(t) \in \text{HGR}(\text{inc}(\Sigma))$ of type 1, as discussed in the Introduction, with the root of t being the external node of g (see Fig. 1(a)). We also represent a simple tree $t \in T_\Sigma(Y_k)$ with $t \ne y_1$ by a hypergraph $g \in \text{HGR}(\text{inc}(\Sigma))$ of type $k + 1$ (where "simple" means that each parameter in Y_k occurs exactly once in t). Then the node u of t with label y_i is the i-th external node of g, and there is no hyperedge e_u in g; the root of t is the last external node of g. Consider, e.g., the simple tree $t = \delta(\gamma(y_3), y_1, \sigma(\beta, y_2))$; the tree graph $h = \text{gr}(t)$ is depicted in Fig. 1(b).

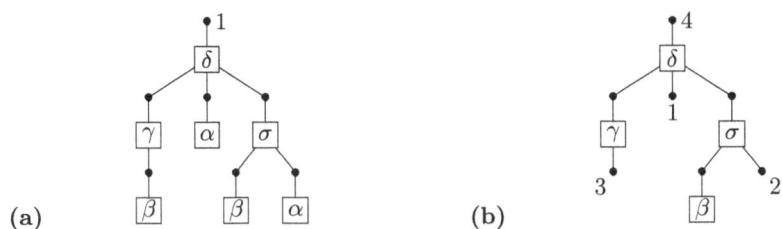

(a)   (b)

Fig. 1. The tree graphs g and h

For a tree graph g over Δ we denote by $\text{tr}(g)$ the (simple) tree it represents, which is over $\text{dec}(\Delta) \cup Y_{k-1}$ if k is the type of g. For a set L of tree graphs we denote by $\text{tr}(L)$ the tree language obtained by applying tr to the graphs in L.

A (context-free) HR grammar is a tuple $G = (N, \Delta, S, P)$, where N is a ranked alphabet of *nonterminals*, Δ is a ranked alphabet of *terminals*, S is the *initial nonterminal*, and P is a finite set of *productions* of the form $A \to g$, where $A \in N^{(k)}$, $k \ge 0$, and g is a hypergraph over $N \cup \Delta$ of type k. The language $L(G)$ generated by G is defined in the usual way. The class of all hypergraph languages generated by HR grammars is denoted by HR. The HR grammar G is *tree-generating*, if every hypergraph in $L(G)$ is a tree graph.

For a class of hypergraph languages $\mathcal{L}$ we denote by $TR(\mathcal{L})$ the class of all tree languages obtained by applying tr to the tree graph languages (of type 1) in $\mathcal{L}$. Thus $TR(HR)$ denotes the class of all tree languages which can be generated by HR grammars. If we apply tr to all tree graphs contained in *any* hypergraph language in HR, then we still obtain $TR(HR)$. This is true because the graph property of being a tree graph can be expressed in monadic second order logic (MSO) and HR is closed under intersection with MSO properties [Cou90]. Note also that both well-known types of context-free graph grammars, i.e., HR and node replacement graph grammars, generate the same class $TR(HR)$ of tree languages. This is true because any context-free node replacement graph language of bounded degree can also be generated by an HR grammar, and tree graph languages are of bounded degree (cf. Theorem 4.26 in [Eng97] and the discussion in Section 6 of [Eng97]).

3   Macro Tree Transducers

A macro tree transducer is a syntax-directed translation device in which the translation of an input tree may depend on its context. The context information is handled by parameters. We will consider *total deterministic* MTTs only.

Definition 1. A *macro tree transducer* (for short, MTT) is a tuple $M = (Q, \Sigma, \Delta, q_0, R)$, where Q is a ranked alphabet of *states*, Σ and Δ are ranked alphabets of *input* and *output symbols*, respectively, $q_0 \in Q^{(0)}$ is the *initial state*, and R is a finite set of *rules*. For every $q \in Q^{(m)}$ and $\sigma \in \Sigma^{(k)}$ with $m, k \geq 0$ there is exactly one rule of the form $\langle q, \sigma(x_1, \ldots, x_k)\rangle(y_1, \ldots, y_m) \to \zeta$ in R, where $\zeta \in T_{\langle Q, X_k \rangle \cup \Delta}(Y_m)$; the right-hand side ζ will be denoted by $\mathrm{rhs}_M(q, \sigma)$. If every state has rank zero, then M is a *top-down tree transducer* (for short, T).

Throughout this paper we will assume, for technical reasons, that macro tree transducers do not contain "erasing rules", i.e., rules that are of the form $\langle q, \sigma(x_1, \ldots, x_k)\rangle(y_1) \to y_1$. This will be called the "nonerasing assumption". The main results of this paper do not depend on this assumption.

The rules of M are used as term rewriting rules in the usual way and the derivation relation of M (on $T_{\langle Q, T_\Sigma \rangle \cup \Delta}$) is denoted by $\Rightarrow_M$. The *transduction realized by M*, denoted τ_M, is the total function $\{(s, t) \in T_\Sigma \times T_\Delta \mid \langle q_0, s\rangle \Rightarrow_M^* t\}$. The class of all transductions which can be realized by MTTs (Ts) is denoted by MTT (T). For $q \in Q^{(m)}$ and $s \in T_\Sigma$ we denote by $M_q(s)$ the tree $t \in T_\Delta(Y_m)$ such that $\langle q, s\rangle(y_1, \ldots, y_m) \Rightarrow_M^* t$ (where $\Rightarrow_M$ is extended to $T_{\langle Q, T_\Sigma \rangle \cup \Delta}(Y)$ in the obvious way); in particular, $M_{q_0}(s) = \tau_M(s)$.

If, for every $q \in Q^{(m)}$ and $\sigma \in \Sigma^{(k)}$, each $x \in X_k$ occurs exactly once in $\mathrm{rhs}_M(q, \sigma)$, then M is *simple in the input* (for short, si). If, for every $q \in Q^{(m)}$ and $\sigma \in \Sigma^{(k)}$, each $y \in Y_m$ occurs exactly once in $\mathrm{rhs}_M(q, \sigma)$, then M is *simple in the parameters* (for short, sp). Note that we use 'simple' to abbreviate the more usual 'linear and nondeleting'.

If we disregard the input of an MTT, then we obtain a *context-free tree (CFT) grammar*. A CFT grammar is a tuple $G = (N, \Delta, S, P)$, where N and Δ are ranked alphabets of *nonterminals* and *terminals*, $S \in N^{(0)}$ is the *initial nonterminal*, and P is a finite set of *productions* of the form $A(y_1, \ldots, y_m) \to \zeta$ with $A \in N^{(m)}$ and $\zeta \in T_{N \cup \Delta}(Y_m)$. The derivation relation of G is denoted $\Rightarrow_G$ and the generated tree language is denoted $L(G)$. If each $y \in Y_m$ occurs exactly once in each ζ, then G is *simple*. If there is no production $A(y_1) \to y_1$ in P, then G is *nonerasing*. The class of tree languages generated by simple CFT grammars is denoted CFT_{sp}. If $N = N^{(0)}$, then G is a *regular tree grammar*. The class of regular tree languages is denoted $REGT$. For more details on tree grammars, CFT grammars, and MTTs see, e.g., [GS84, GS97], [ES77], and [EV85, FV98], respectively.

For a class Φ of tree transductions and a class $\mathcal{L}$ of tree languages $\Phi(\mathcal{L}) = \{\varphi(L) \mid \varphi \in \Phi, L \in \mathcal{L}\}$, and $OUT(\Phi)$ is the class of ranges of transductions in Φ. For classes Φ and Φ' of tree transductions their composition is $\Phi \circ \Phi' = \{\varphi \circ \varphi' \mid \varphi \in \Phi, \varphi' \in \Phi'\}$ and $(\varphi \circ \varphi')(x) = \varphi'(\varphi(x))$; note that the order of φ and φ' is nonstandard.

Finite-Copying. The notion of finite-copying can be defined for arbitrary MTTs [EM98], but for convenience we restrict ourselves to MTTs which are simple in the parameters, i.e., sp MTTs. The definition is analogous to the one for top-down tree transducers (cf. [ERS80,AU71]); note that, trivially, all Ts are sp. Consider a derivation of an MTT M which is sp. A subtree of an input tree s may be processed by M arbitrarily many times, depending on s and the rules of M. The *state sequence of s at node u*, denoted by $\text{sts}_M(s,u)$, contains all states which process the subtree s/u of s rooted at u. Formally, the state sequence of s at its root is q_0, and if $\text{sts}_M(s,u) = q_1 \cdots q_n$, u has label σ, and $u \cdot i$ is the i-th child of u, then $\text{sts}_M(s,u\cdot i) = \pi_i(\text{rhs}_M(q_1,\sigma)\cdots\text{rhs}_M(q_n,\sigma))$, where π_i changes every $\langle q, x_i \rangle$ into q and deletes everything else. If for every $s \in T_\Sigma$ and every node u of s, $|\text{sts}_M(s,u)| \le k$ for a fixed $k \ge 0$, then M is called *k-copying* or *finite-copying* (for short, fc). The number k is a *copying bound for M*. The class of transductions realized by MTTs (Ts) which are $w \in \{\text{si, sp, fc}\}^*$ is denoted by MTT_w (T_w).

As an example, a 2-copying MTT_{sp} that translates the monadic tree $\gamma^n(\alpha)$ into $\sigma(a^n(b^n(e)), a^n(b^n(e)))$ has rules $\langle q_0, \gamma(x_1)\rangle \to \sigma(\langle q, x_1\rangle(e), \langle q, x_1\rangle(e))$, $\langle q_0, \alpha\rangle \to \sigma(e,e)$, $\langle q, \gamma(x_1)\rangle(y_1) \to a(\langle q, x_1\rangle(b(y_1)))$, and $\langle q, \alpha\rangle(y_1) \to a(b(y_1))$.

A finite-copying MTT M can be decomposed into a finite-copying top-down tree transducer followed by an MTT that is simple in the input: the top-down tree transducer simply generates $|\text{sts}_M(s,u)|$ copies of s/u.

Lemma 2. $MTT_{\text{fc,sp}} = T_{\text{fc}} \circ MTT_{\text{si,sp}}$.

Proof. ($\subseteq$) Let $M = (Q, \Sigma, \Delta, q_0, R)$ be an $MTT_{\text{fc,sp}}$. Define the T_{fc} $M' = (\text{zero}(Q), \Sigma, \Gamma, q_0, R')$ and the $MTT_{\text{si,sp}}$ $M'' = (Q, \Gamma, \Delta, q_0, R'')$ as follows. For every $q \in Q^{(m)}$ and $\sigma \in \Sigma^{(k)}$ let $\langle q, \sigma(x_1,\ldots,x_k)\rangle \to \sigma_n(\langle q_1, x_{i_1}\rangle, \ldots, \langle q_n, x_{i_n}\rangle)$ be in R' and let σ_n be in $\Gamma^{(n)}$, where $\langle q_1, x_{i_1}\rangle, \ldots, \langle q_n, x_{i_n}\rangle$ are all occurrences of elements of $\langle Q, X\rangle$ in $\text{rhs}_M(q, \sigma)$; also, let $\langle q, \sigma_n(x_1,\ldots,x_n)\rangle(y_1,\ldots,y_m) \to \zeta$ be in R'', where ζ is obtained from $\text{rhs}_M(q,\sigma)$ by changing $\langle q_j, x_{i_j}\rangle$ into $\langle q_j, x_j\rangle$. Appropriate dummy rules are added to make M'' total. It can be shown that the state sequences of M' are precisely those of M, and that for $q \in Q$ and $s \in T_\Sigma$, $M''_q(M'_q(s)) = M_q(s)$. Hence $\tau_{M'} \in T_{\text{fc}}$ and $\tau_M = \tau_{M'} \circ \tau_{M''}$.

($\supseteq$) In this direction we use a standard product construction. Let $M = (Q, \Sigma, \Gamma, q_0, R)$ be a T_{fc} and let $M' = (Q', \Gamma, \Delta, q'_0, R')$ be an $MTT_{\text{si,sp}}$. We now construct the MTT $N = (\langle Q', Q\rangle, \Sigma, \Delta, \langle q'_0, q_0\rangle, P)$ which realizes the translation $\tau_M \circ \tau_{M'}$ as follows. For $\langle q', q\rangle \in \langle Q', Q\rangle^{(m)}$, $\sigma \in \Sigma^{(k)}$, and $m, k \ge 0$, let the rule $\langle\langle q', q\rangle, \sigma(x_1,\ldots,x_k)\rangle(y_1,\ldots,y_m) \to \zeta$ be in P, where $\zeta = M''_{q'}(\text{rhs}_M(q,\sigma))$ and M'' is the extension of M' to input symbols in $\langle Q, X_k\rangle$, i.e., $M'' = (Q', \Gamma \cup \langle Q, X_k\rangle, \Delta \cup \langle\langle Q', Q\rangle, X_k\rangle, q'_0, R' \cup R'')$ and R'' consists of all rules $\langle \bar{q}', \langle \bar{q}, x_i\rangle\rangle(y_1, \ldots, y_n) \to \langle\langle \bar{q}', \bar{q}\rangle, x_i\rangle(y_1,\ldots,y_n)$ for $\langle \bar{q}, x_i\rangle \in \langle Q, X_k\rangle$, $\bar{q}' \in Q'^{(n)}$, and $n \ge 0$. It is easy to see that N is simple in the parameters (because M' and, hence, M'' are sp). Moreover, if $\text{sts}_M(s,u) = q_1\cdots q_n$, then $\text{sts}_N(s,u) = \langle q'_1, q_1\rangle\cdots\langle q'_n, q_n\rangle$ for certain states q'_i of M'. This can be shown by induction on the definition of $\text{sts}_M(s,u)$; note that since M' is si, each occurrence of $\langle \bar{q}, x_i\rangle$ in $\text{rhs}_M(q_j,\sigma)$ is translated by M'' into a unique occurrence of some $\langle\langle \bar{q}', \bar{q}\rangle, x_i\rangle$ in $\text{rhs}_N(\langle q'_j, q_j\rangle, \sigma)$

$= M''_{q'_j}(\text{rhs}_M(q_j, \sigma))$. Thus, N has the same copying bound as M. It can be shown (by induction on s) that for every $\langle q', q \rangle \in \langle Q', Q \rangle$ and $s \in T_\Sigma$, $N_{\langle q', q \rangle}(s) = M'_{q'}(M_q(s))$. $\qquad\square$

Note that the construction in the proof of Lemma 2 preserves the nonerasing assumption (but the lemma does not depend on it).

4   Tree Graph Operations

To state the characterization of $TR(HR)$ by Drewes in [Dre99], we have to recall the characterization of HR in terms of operations on hypergraphs (cf. [BC87,Eng94,Dre96]). A hypergraph that contains variables $z_1, \ldots, z_k$ can be seen as a k-ary operation on hypergraphs. Let $Z_k = \{z_i^j \mid 1 \leq i \leq k, j \geq 0\}$ with $\text{rank}(z_i^j) = j$. If $g \in \text{HGR}(\Delta \cup Z_k)$ and for every $1 \leq i \leq k$ there is exactly one j_i such that $z_i^{j_i}$ appears in g, then g is a k-ary *hypergraph operation over* Δ (on hypergraphs of type $j_1, \ldots, j_k$). For a ranked alphabet Σ and a mapping f, (Σ, f) is an *alphabet of hypergraph operations (over* Δ) if $f(\sigma)$ is a k-ary hypergraph operation over Δ for every $\sigma \in \Sigma^{(k)}$. With (Σ, f) we associate the (partial) *valuation function* $\text{val}_{\Sigma, f} : T_\Sigma \rightarrow \text{HGR}(\Delta)$. For $\sigma \in \Sigma^{(k)}$ and $s_1, \ldots, s_k \in T_\Sigma$, $\text{val}_{\Sigma, f}(\sigma(s_1, \ldots, s_k)) = f(\sigma)[z_i^j / \text{val}_{\Sigma, f}(s_i) \mid 1 \leq i \leq k, j \geq 0]$, where $g[z/h]$ is the hypergraph obtained from g by replacing its unique z-labeled hyperedge e by the hypergraph h (provided $\text{type}(h) = \text{type}(e)$). With these definitions, HR is the class of all hypergraph languages $\text{val}_{\Sigma, f}(L)$ for some alphabet of hypergraph operations (Σ, f) and a regular tree language L over Σ (assuming that all hypergraphs in a hypergraph language have the same type).

An easy way to guarantee that an HR grammar generates a tree language, is to require all its right-hand sides to be tree graphs (because tree graphs are closed under hypergraph substitution). But it is well known that the corresponding class of tree languages is a proper subclass of $TR(HR)$ (cf. [Dre99]). By the (proof of the) above characterization it is the class of all $\text{val}_{\Sigma, f}(L)$ where $f(\sigma)$ is a tree graph for every $\sigma \in \Sigma$ and L is in $REGT$. In [Dre99] Drewes shows that, to obtain $TR(HR)$, $REGT$ should be replaced by $T_{\text{fc}}(REGT)$.

Let VAL_{tr} be the set of all $\text{val}_{\Sigma, f}$ such that (Σ, f) is an *alphabet of tree graph operations*, which means that $f(\sigma)$ is a tree graph for every $\sigma \in \Sigma$. The following theorem is shown by Drewes for "hypertrees", which are a slight generalization of tree graphs; it is easy to see that it also holds for tree graphs.

Proposition 3. (Theorem 8.1 in [Dre99]) $TR(HR) = TR(VAL_{\text{tr}}(T_{\text{fc}}(REGT)))$.

We now show that the class of tree transductions $VAL_{\text{tr}} \circ TR$ is closely related to $MTT_{\text{si,sp}}$: they produce the same output tree languages when applied to a class of input tree languages with weak closure properties.

Theorem 4. Let $\mathcal{L}$ be a class of tree languages which is closed under intersection with regular tree languages and under T_{si}. Then $TR(VAL_{\text{tr}}(\mathcal{L})) = MTT_{\text{si,sp}}(\mathcal{L})$.

Proof. ($\subseteq$) If a tree graph operation h is applied to tree graphs $g_1, \ldots, g_k$, then the hyperedge of h labeled by variable $z_i^{j_i}$ is replaced by g_i. For the corresponding trees this means that in $\mathrm{tr}(h)$ the symbol $z_i^{j_i}$ of rank $j_i - 1$ is replaced by $\mathrm{tr}(g_i)$, which contains parameters $y_1, \ldots, y_{j_i-1}$. This is a simple case of term rewriting which, as it turns out, can be carried out by an $\mathrm{MTT_{si,sp}}$. Since MTTs are total, whereas a function val in VAL_{tr} is in general partial, the input for the corresponding MTT should be restricted to the domain of val, which is a regular tree language (because well-typedness is a regular property).

Let (Σ, f) be an alphabet of tree graph operations over Δ, and $L \subseteq T_\Sigma$ a tree language in $\mathcal{L}$ (with $\mathrm{val}_{\Sigma,f}(L)$ of type 1). Let $L' = L \cap \mathrm{dom}(\mathrm{val}_{\Sigma,f})$. Since $\mathcal{L}$ is closed under intersection with $REGT$ languages, $L' \in \mathcal{L}$. Let $N = \max\{\mathrm{type}(f(\sigma)) - 1 \mid \sigma \in \Sigma\}$. Define $M = (Q, \Sigma, \mathrm{dec}(\Delta), 0, R)$ with $Q = \{m^{(m)} \mid 0 \le m \le N\}$ and R consists of all rules $\langle m, \sigma(x_1, \ldots, x_k)\rangle(y_1, \ldots, y_m) \to \zeta$, where $\sigma \in \Sigma^{(k)}$, $m = \mathrm{type}(f(\sigma)) - 1$, and ζ is obtained from $\mathrm{tr}(f(\sigma))$ by changing every z_i^j into $\langle j - 1, x_i \rangle$. Since for every i there is exactly one z_i^j in $f(\sigma)$, M is si. Since $\mathrm{tr}(f(\sigma))$ is a simple tree in $T_{\mathrm{dec}(\Delta)}(Y_m)$, M is sp. For every $s \in L'$, $M_m(s) = \mathrm{tr}(\mathrm{val}_{\Sigma,f}(s))$, where $m = \mathrm{type}(\mathrm{val}_{\Sigma,f}(s)) - 1$. Hence $\tau_M(L') = \mathrm{tr}(\mathrm{val}_{\Sigma,f}(L))$.

($\supseteq$) The MTT in the above proof needs different states merely to provide the correct number of parameters, i.e., $\mathrm{val}_{\Sigma,f}$ has no state information. Thus, to realize $\tau_M \in MTT_{\mathrm{si,sp}}$ by some $\mathrm{val}_{\Sigma,f}$ we must first add to the input tree the information by which states of M its subtrees are processed. This can be done by a simple top-down tree transducer which changes the label σ of a node u in s into $\langle \sigma, q \rangle$, where q is the unique state with $q = \mathrm{sts}_M(s, u)$.

Let $M = (Q, \Sigma, \Delta, q_0, R)$ be an $\mathrm{MTT_{si,sp}}$ and let $L \subseteq T_\Sigma$ be a tree language in $\mathcal{L}$. Define the $\mathrm{T_{si}}$ $M' = (\mathrm{zero}(Q), \Sigma, \Gamma, q_0, R')$ with $\Gamma = \langle \Sigma, Q \rangle$ and R' consists of all rules $\langle q, \sigma(x_1, \ldots, x_k)\rangle \to \langle \sigma, q\rangle(\langle q_1, x_1\rangle, \ldots, \langle q_k, x_k\rangle)$ where $q \in Q$, $\sigma \in \Sigma^{(k)}$, and $\langle q_1, x_1\rangle, \ldots, \langle q_k, x_k\rangle$ are the elements of $\langle Q, X\rangle$ in $\mathrm{rhs}_M(q, \sigma)$. Define $f(\langle \sigma, q\rangle) = \mathrm{gr}(\zeta)$, where ζ is obtained from $\mathrm{rhs}_M(q, \sigma)$ by changing every $\langle q_i, x_i \rangle \in \langle Q, X\rangle^{(j)}$ into z_i^{j+1}. Since M is sp and, by the nonerasing assumption, $\zeta \neq y_1$, $\mathrm{gr}(\zeta)$ is defined, and since M is si, it is a k-ary tree graph operation over $\mathrm{inc}(\Delta)$. For $q \in Q$ and $s \in T_\Sigma$, $M_q(s) = \mathrm{tr}(\mathrm{val}_{\Gamma,f}(M'_q(s)))$. Hence $\tau_M(L) = \mathrm{tr}(\mathrm{val}_{\Gamma,f}(\tau_{M'}(L)))$ with $\tau_{M'}(L) \in \mathcal{L}$ because $\mathcal{L}$ is closed under T_{si}. $\square$

Note that the proof of Theorem 4 depends on the nonerasing assumption. This is natural because external nodes are distinct by the definition of hypergraphs and hence there is no tree graph representation of y_1 (cf. the definition of gr above Fig. 1).

5   Tree Languages Generated by HR Grammars

In the previous section we have shown how the valuation functions induced by tree graph operations are related to MTTs. By Proposition 3 the class $TR(HR)$ can be expressed in terms of tree graph operations. Thus we obtain the following characterization of $TR(HR)$.

Theorem 5. $TR(HR) = MTT_{\mathrm{fc,sp}}(REGT) = OUT(MTT_{\mathrm{fc,sp}})$.

Proof. By Proposition 3, $TR(HR)$ is equal to $TR(VAL_{\mathrm{tr}}(T_{\mathrm{fc}}(REGT)))$ which equals $MTT_{\mathrm{fc,sp}}(REGT)$ by Theorem 4 and Lemma 2. Theorem 4 is applicable because $T_{\mathrm{fc}}(REGT)$ is closed under both (i) intersection with $REGT$ and (ii) T_{si}. In fact, if $\tau \in T_{\mathrm{fc}}$ and $L, L' \in REGT$, then $\tau(L) \cap L'$ equals $\tau(L \cap \tau^{-1}(L'))$ which is in $T_{\mathrm{fc}}(REGT)$ because $REGT$ is closed under inverse top-down tree transductions (cf. Corollary IV.3.17 of [GS84]); hence (i) holds. It is straightforward to show that T_{fc} is closed under composition and hence (ii) holds (every T_{si} is 1-copying); in fact, this can be shown by following the product construction in the proof of Theorem 2 in [Rou70] and observing that the length of the state sequences of the composed top-down tree transducer is bound by the product of the copying bounds of the two given T_{fc}s (cf. also Theorem 5.4 of [ERS80], where the closure under composition is proved for T_{fc}s with regular look-ahead).

Since every $REGT$ language is the range of a 1-copying top-down tree transducer [Man98], and T_{fc} is closed under composition, the second equality follows from Lemma 2. □

This characterization is also valid without the nonerasing assumption on MTTs: this can be shown using the proof of Lemma 7.11 in [EM98] and the fact that regular look-ahead can be simulated by bottom-up relabelings (which preserve regular tree languages, see, e.g., [GS97, GS84]). Recall that Lemma 2 also holds without the assumption.

Tree Generating Normal Form. From Theorem 5 we obtain a normal form for tree generating HR grammars (similar to the one for string generating HR grammars in Theorem 5.18 of [Hab89]). In fact, an $MTT_{\mathrm{fc,sp}}$ is very close to an HR grammar (generating its range): states that process the same subtree of the input tree correspond to one nonterminal hyperedge, and so the nonterminals of the HR grammar are the state sequences of the MTT. The formal construction is essentially the one of Lemma 5.1 in [Dre99] (which is part of the proof of Proposition 3). It shows that every tree language in $TR(HR)$ can be generated by an HR grammar of which all right-hand sides of productions are "linked forests" (generalizing tree graphs).

Let us now give a formal definition of linked forests. We first need some auxiliary notions. A forest is a sequence of trees $t_1, \ldots, t_k$. It can be represented by the *forest graph* $\mathrm{gr}(t_1, \ldots, t_k) = \mathrm{gr}(t_1) \oplus \cdots \oplus \mathrm{gr}(t_k)$, where $\oplus$ is disjoint union, concatenating the sequences of external nodes (as defined in [BC87]). Intuitively, each hyperedge e of a linked forest is the combination of n hyperedges $e_1, \ldots, e_n$ of a forest graph, not necessarily from distinct trees of the forest. To check whether a graph is a linked forest we can change every e back into the n hyperedges it was obtained from. For this we need to know the types $(\rho_1, \ldots, \rho_n)$ of the hyperedges $e_1, \ldots, e_n$, respectively. We add this information to the labels: a *linked ranked alphabet* is a ranked alphabet Σ with a mapping 'link' which associates with every $\sigma \in \Sigma^{(k)}$ a sequence $\mathrm{link}(\sigma) = (\rho_1, \ldots, \rho_n)$ of natural numbers $\rho_i \geq 1$ such that $\sum_{i=1}^{n} \rho_i = k$ (where n depends on σ). Let Σ be a linked ranked alphabet. For a hypergraph g over Σ, define the hypergraph $\mathrm{cut}(g)$ which is obtained from g by "cutting all links", as follows: replace every hyperedge e with label σ and $\text{link}(\sigma) = (\rho_1, \ldots, \rho_n)$ by distinct new edges $e_1, \ldots, e_n$ such that e_j is incident with the $((\sum_{i=1}^{j-1} \rho_i) + 1)$-th up to the $(\sum_{i=1}^{j} \rho_i)$-th node of e and has any label of rank ρ_j.

A hypergraph g over Σ is a *linked forest* (over Σ) of type $(\rho_1, \ldots, \rho_k)$ if $\text{cut}(g) = h_1 \oplus \cdots \oplus h_k$ for certain tree graphs $h_1, \ldots, h_k$ of type $\rho_1, \ldots, \rho_k$, respectively.

Definition 6. An HR grammar $G = (N, \Delta, S, P)$ is in *tree generating normal form*, if (i) $N \cup \Delta$ is a linked ranked alphabet such that $S \in N^{(1)}$ and for every $\delta \in \Delta^{(k)}$, $\text{link}(\delta) = (k)$ and (ii) for every production $A \to g$ in P, g is a linked forest over $N \cup \Delta$ of type $\text{link}(A)$.

Let G be an HR grammar in tree generating normal form. Clearly G generates tree graphs only; this is true because linked forests are closed under hypergraph substitution and a terminal linked forest of type (1) is a tree graph (and note that $\text{link}(S) = (1)$). Together with Theorem 7 this implies that the HR grammars in tree generating normal form generate the full class $TR(HR)$.

Theorem 7. For every tree generating HR grammar G there is an HR grammar G' in tree generating normal form such that $L(G') = L(G)$.

Proof. By Theorem 5 there is an $\text{MTT}_{\text{fc,sp}}$ $M = (Q, \Sigma, \Delta, q_0, R)$ with range $\text{tr}(L(G))$. We want to construct an HR grammar G' in tree generating normal form such that $\text{tr}(L(G')) = \tau_M(T_\Sigma)$. Let c be a copying bound of M. Define $G' = (N, \text{inc}(\Delta), q_0, P)$, where N consists of all strings $w = q_1 \cdots q_l$ with $q_1, \ldots, q_l \in Q$, $l \leq c$, and $\text{link}(w) = (\text{rank}(q_1) + 1, \ldots, \text{rank}(q_l) + 1)$. For every $q_1 \cdots q_l \in N$, $\sigma \in \Sigma^{(k)}$, and $k \geq 0$ let the production $q_1 \cdots q_l \to g$ be in P, where g is the hypergraph obtained as follows. First, take the forest graph $\text{gr}(\text{rhs}_M(q_1, \sigma), \ldots, \text{rhs}_M(q_l, \sigma))$; now for every $i \in [k]$, combine all hyperedges $e_1, \ldots, e_n$ that are labeled by elements of $\langle Q, \{x_i\} \rangle$ into a single hyperedge e labeled $r_1 \cdots r_n$ where $\langle r_j, x_i \rangle$ is the label of e_j, such that the sequence of nodes of e is the concatenation of the sequences of nodes of $e_1, \ldots, e_n$, respectively (thus "linking" the forest graph).

Obviously, G' is in tree generating normal form. It should be clear that the nonterminal $q_1 \cdots q_l$ generates the forest graph $\text{gr}(M_{q_1}(s), \ldots, M_{q_l}(s))$ for all $s \in T_\Sigma$. $\square$

Note that the proof of Theorem 7 shows that $OUT(MTT_{\text{fc,sp}}) \subseteq TR(HR)$ which is part of Theorem 5. Figure 2 shows an HR grammar in tree generating normal form that is obtained from the example MTT in Section 3 as in the proof of Theorem 7. It generates the tree language $\sigma(a^n(b^n(e)), a^n(b^n(e)))$, $n \geq 0$. The MTT has two state sequences, viz. q_0 and qq, corresponding to nonterminals S and A, respectively. Thus, $\text{link}(A) = (2, 2)$, i.e., the A-labeled hyperedge combines two hyperedges of type 2, and each right-hand side of a production of A is obtained from two (isomorphic) tree graphs of type 2.

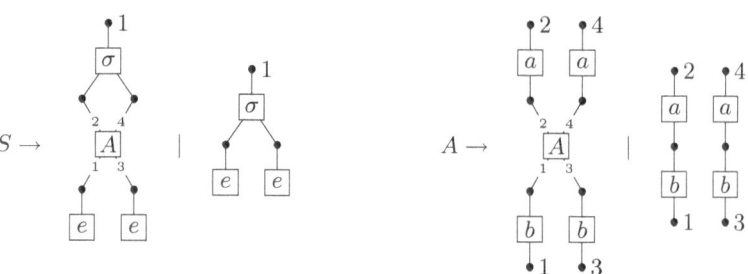

Fig. 2. A tree generating HR grammar in tree generating normal form

Special Cases. In the literature, two natural ways to generate trees by HR grammars in a structured way have been considered, viz. by restricting the right-hand sides of the productions to be (a) "leaf-linked forests", or, as already mentioned in Section 4, (b) tree graphs. These are in fact special cases of the tree generating normal form. Let G be an HR grammar in tree generating normal form. If $\text{link}(A) = (1, 1, \ldots, 1)$ for every nonterminal A of G, then the right-hand sides of the productions of G are *leaf-linked forests*. This means that all links are combinations of leaves (i.e., hyperedges of type 1). Thus, applying a production of G just corresponds to usual (parallel) tree substitution. The class of hypergraph languages generated by such grammars is denoted by HR_{fo}. If $\text{link}(A) = (\text{rank}(A))$ for every nonterminal A of G, then the right-hand sides of the productions of G are tree graphs. The corresponding class of hypergraph languages is denoted by HR_{tr}. One example of (a) and (b) is given in Fig. 4 (a) and (b), respectively.

Proposition 8. ([Rao97]) $TR(HR_{\text{fo}}) = T_{\text{fc}}(REGT)$.

Since, as observed in Section 4, $TR(HR_{\text{tr}}) = TR(VAL_{\text{tr}}(REGT))$, and since by Theorem 4 (and the known closure properties of $REGT$) this class equals $MTT_{\text{si,sp}}(REGT)$, the classes HR_{tr} and HR_{fo} correspond naturally to the "decomposition" result (Proposition 3) of Drewes, plus the one obtained from Theorem 4: $TR(HR) = TR(VAL_{\text{tr}}(T_{\text{fc}}(REGT))) = MTT_{\text{si,sp}}(T_{\text{fc}}(REGT))$. Another way of viewing these classes is to say that HR_{fo} corresponds to the top-down tree transducer aspect of MTTs (by Proposition 8), whereas HR_{tr} corresponds to the context-free tree grammar aspect of MTTs, which is stated in the next theorem.

Theorem 9. $TR(HR_{\text{tr}}) = CFT_{\text{sp}}$.

Proof. Clearly, by the argument in the beginning of the proof of Theorem 4, an HR grammar for which all productions are of the form $A \to g$ where g is a tree graph, generates the same tree language as the corresponding (non-erasing) CFT_{sp} grammar that has the same nonterminals and the productions $A(y_1, \ldots, y_m) \to \text{tr}(g)$, where $m = \text{type}(g) - 1$. As an example, the CFT grammar corresponding to HR grammar G_{b} of Fig. 4 has productions $S \to A(e)$, $A(y_1) \to a(A(b(y_1)))$, and $A(y_1) \to a(b(y_1))$.

Conversely, every nonerasing $\mathrm{CFT_{sp}}$ grammar G corresponds to an HR grammar obtained by applying gr to the right-hand sides of the productions of G. For every $\mathrm{CFT_{sp}}$ grammar G there is a nonerasing $\mathrm{CFT_{sp}}$ grammar G' such that $L(G') = L(G)$; G' is obtained from G by (i) adding for every production $A(y_1, \ldots, y_m) \to \zeta$ of G any production $A(y_1, \ldots, y_m) \to \zeta'$, where ζ' is obtained from ζ by changing some occurrences of $B(t)$ into t with $B(y_1) \Rightarrow_G^* y_1$ and (ii) deleting all productions of the form $C(y_1) \to y_1$.   □

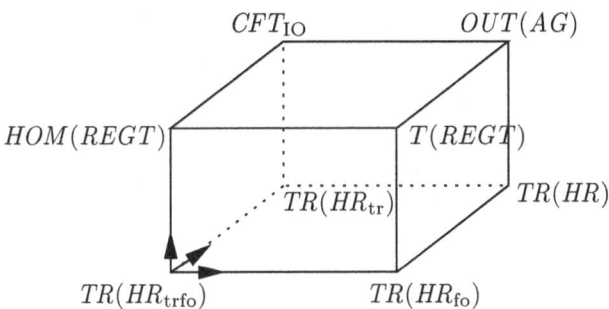

Fig. 3. Classes of tree languages generated by HR grammars

Consider an HR grammar for which the right-hand side of every production is both a leaf-linked forest *and* a tree graph, i.e., a tree graph over nonterminals A with $\mathrm{link}(A) = (1)$ only. Such a grammar is just a tree graph representation of a regular tree grammar. Let us denote the corresponding class of hypergraph languages by HR_{trfo}. Thus $TR(HR_{\mathrm{trfo}}) = REGT$. Let us now compare the classes $TR(HR_{\mathrm{trfo}})$, $TR(HR_{\mathrm{tr}})$, $TR(HR_{\mathrm{fo}})$, and $TR(HR)$ with each other and with some well-known classes of tree languages: $HOM(REGT)$, $T(REGT)$, CFT_{IO}, and $OUT(AG)$, where HOM denotes the class of tree homomorphisms, i.e., the class of transductions realized by one-state top-down tree transducers, CFT_{IO} is the class of tree languages generated by CFT grammars in IO (inside-out) derivation mode, and AG is the class of tree transductions realized by attribute grammars.

Theorem 10. Figure 3 is a Hasse diagram.

Proof. For sp CFT grammars the IO and OI (outside-in) derivation modes coincide; thus $TR(HR_{\mathrm{tr}}) = CFT_{\mathrm{sp}} \subseteq CFT_{\mathrm{IO}}$. Recall that $TR(HR) \subseteq OUT(AG)$ is known from [EH92]. The inclusion of $HOM(REGT)$ in CFT_{IO} is clear from the closure of CFT_{IO} under tree homomorphisms (Corollary 6.4 in [ES77]). In [DPSS77] it is shown that $CFT_{\mathrm{IO}} \subseteq OUT(AG)$, see also [EF81]. The inclusion of $T(REGT)$ in $OUT(AG)$ follows from the fact that top-down tree transducers coincide with attribute grammars that have synthesized attributes only (cf. Propositions 4.7 and 5.5 in [CF82]). It now suffices to prove the following three inequalities.

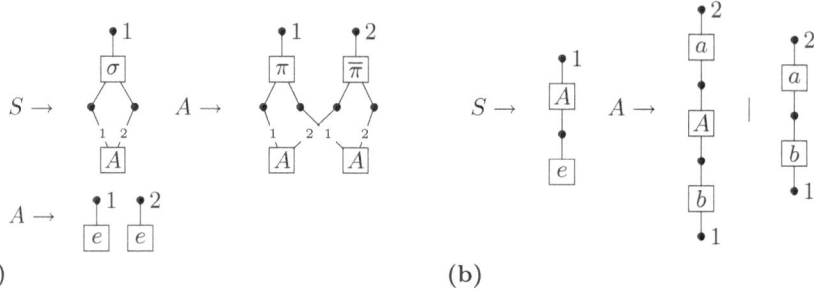

Fig. 4. Productions of G_a and G_b

(1) $TR(HR_{tr}) - T(REGT) \neq \varnothing$. Every monadic tree language in $T(REGT)$ is regular (Theorem 4 of [Rou70]). However, the HR_{tr} grammar G_b of Fig. 4(b) generates the non-regular monadic tree language $\{a^n(b^n(e)) \mid n \geq 1\}$.

(2) $TR(HR_{fo}) - CFT_{IO} \neq \varnothing$. The HR_{fo} grammar G_a of Fig. 4(a) generates all trees $\sigma(t, \bar{t})$, where t is a binary tree over $\{\pi^{(2)}, e^{(0)}\}$ and $\bar{t}$ is the same tree with each π replaced by $\bar{\pi}$. In Section 5 of [EF81] it is shown that this tree language is not in CFT_{IO}.

(3) $HOM(REGT) - TR(HR) \neq \varnothing$. Consider the one-state T with the two rules $\langle q, \gamma(x) \rangle \to \sigma(\langle q, x \rangle, \langle q, x \rangle)$ and $\langle q, \alpha \rangle \to a$. It generates full binary trees the yields of which are in $\{a^{2^n} \mid n \geq 0\}$. But the number of a's in the hypergraphs of an HR language form a semi-linear set (Theorem IV.4.3 of [Hab92]).

$\square$

We note without proof that the classes $HOM(REGT)$, $T(REGT)$, CFT_{IO}, and $OUT(AG)$ are equal to the classes $TM(HR_{trfo})$, $TM(HR_{fo})$, $TM(HR_{tr})$, and $TM(HR)$, respectively, where $TM(HR_w)$ is the class of tree languages obtained by unfolding the *term graph* languages of these HR grammars (in which sharing of subtrees is now allowed); for the last class this is shown in [EH92]. Intuitively, the reason for these equalities is that the unbounded copying facility of the formalisms mentioned above can be simulated by the use of sharing (and vice versa).

By Theorem 10 we know that the class CFT_{IO} of IO context-free tree languages is incomparable with the class $TR(HR)$ of tree languages generated by HR grammars. It is not difficult to see that also the class CFT_{OI} of tree languages generated by CFT grammars in OI derivation mode is incomparable with $TR(HR)$. In fact, $CFT_{OI} - TR(HR) \neq \varnothing$ follows from (3) in the proof of Theorem 10 and the fact that the set of full binary trees can be generated by the CFT grammar (in any derivation mode) with productions $S \to A(a)$ and $A(y) \to A(\sigma(y, y))$. The inequality $TR(HR) - CFT_{OI} \neq \varnothing$ can be proved using the tree language considered in (2) of the proof of Theorem 10 (even with $\bar{\pi}$ replaced by π; see [AD76]).

Classes of Tree Transductions. In [CE95] it is shown that the class HR of HR languages is equal to the class $MSOT(REGT)$ of output languages of graph transductions definable in monadic second order logic (cf. [Cou94]) taking regular tree languages as input. This means that $TR(HR) = TR(MSOT(REGT))$. Hence the particular MSO transductions involved translate trees into trees; moreover, they can be restricted to total functions. Let us denote this class of tree-to-tree transductions by $MSOTT$. In [BE98] $MSOTT$ is characterized in terms of attribute grammars. We now present without proof a characterization of this class in terms of MTTs (which are more powerful than attribute grammars). The (quite involved) proof can be found in [EM98]. As it turns out, the feature of regular look-ahead (cf. [EV85]) must be added to the MTTs in order to obtain a natural characterization of the class $MSOTT$. We denote regular look-ahead by a superscript R.

Theorem 11. $MSOTT = MTT^R_{\mathrm{fc,sp}}$.

Note that, by Lemma 7.11 of [EM98], Theorem 11 does not depend on the nonerasing assumption. Theorem 5 is a consequence of Theorem 11 together with the result of [CE95] that $HR = MSOT(REGT)$ and the well-known fact that the regular look-ahead can be incorporated in the regular input language. Note also that, since $MTT^R = MTT$ [EV85], every MSO tree transduction can be realized by a macro tree transducer.

Corollary 12. $MSOTT \subseteq MTT$.

Acknowledgement

We thank one of the referees for helpful comments.

References

AD76.   A. Arnold and M. Dauchet. Un théorème de duplication pour les forêts algébriques. *J. of Comp. Syst. Sci.*, 13:223–244, 1976. 26

AU71.   A. V. Aho and J. D. Ullman. Translations on a context-free grammar. *Inform. and Control*, 19:439–475, 1971.

BC87.   M. Bauderon and B. Courcelle. Graph expressions and graph rewritings. *Math. Systems Theory*, 20:83–127, 1987. 22

BE98.   R. Bloem and J. Engelfriet. A comparison of tree transductions defined by monadic second order logic and by attribute grammars. Technical Report 98-02, Leiden University, 1998. To appear in *J. of Comp. Syst. Sci.* 27

CE95.   B. Courcelle and J. Engelfriet. A logical characterization of the sets of hypergraphs defined by hyperedge replacement grammars. *Math. Systems Theory*, 28:515–552, 1995. 26, 27

CF82.   B. Courcelle and P. Franchi-Zannettacci. Attribute grammars and recursive program schemes. *Theoret. Comput. Sci.*, 17:163–191 and 235–257, 1982. 25

Cou90. B. Courcelle. The monadic second-order logic of graphs. I. recognizable sets of infinite graphs. *Information and Computation*, 85:12–75, 1990. 17

Cou94. B. Courcelle. Monadic second-order definable graph transductions: a survey. *Theoret. Comput. Sci.*, 126:53–75, 1994. 27

DKH97. F. Drewes, H.-J. Kreowski, and A. Habel. Hyperedge replacement graph grammars. In G. Rozenberg, editor, *Handbook of Graph Grammars and computing by graph transformation, Volume 1*, chapter 2, pages 95–162. World Scientific, Singapore, 1997.

DPSS77. J. Duske, R. Parchmann, M. Sedello, and J. Specht. IO-macrolanguages and attributed translations. *Inform. and Control*, 35:87–105, 1977. 25

Dre96. F. Drewes. *Computation by Tree Transductions*. PhD thesis, University of Bremen, 1996.

Dre99. F. Drewes. A characterization of the sets of hypertrees generated by hyperedge-replacement graph grammars. *Theory Comput. Systems*, 32:159–208, 1999. 15, 16, 20, 22

EF81.   J. Engelfriet and G. Filè. The formal power of one-visit attribute grammars. *Acta Informatica*, 16:275–302, 1981. 25, 26

EH92.   J. Engelfriet and L. Heyker. Context-free hypergraph grammars have the same term-generating power as attribute grammars. *Acta Informatica*, 29:161–210, 1992. 15, 25, 26

EM98.   J. Engelfriet and S. Maneth. Macro tree transducers, attribute grammars, and MSO definable tree translations. Technical Report 98-09, Leiden University, 1998. To appear in *Inform. and Comput.* 19, 22, 27

Eng94. J. Engelfriet. Graph grammars and tree transducers. In S. Tison, editor, *Proceedings of the 19th Colloquium on Trees in Algebra and Programming – CAAP 94*, volume 787 of *LNCS*, pages 15–36. Springer-Verlag, 1994.

Eng97. J. Engelfriet. Context-free graph grammars. In G. Rozenberg and A. Salomaa, editors, *Handbook of Formal Languages, Volume 3*, chapter 3. Springer-Verlag, 1997. 17

ERS80. J. Engelfriet, G. Rozenberg, and G. Slutzki. Tree transducers, L systems, and two-way machines. *J. of Comp. Syst. Sci.*, 20:150–202, 1980. 15, 22

ES77.   J. Engelfriet and E.M. Schmidt. IO and OI, Part I. *J. of Comp. Syst. Sci.*, 15:328–353, 1977. And Part II, *J. of Comp. Syst. Sci.*, 16: 67–99 (1978). 18, 25

EV85.   J. Engelfriet and H. Vogler. Macro tree transducers. *J. of Comp. Syst. Sci.*, 31:71–146, 1985. 27

FV98.   Z. Fülöp and H. Vogler. *Syntax-Directed Semantics – Formal Models based on Tree Transducers*. EATCS Monographs on Theoretical Computer Science (W. Brauer, G. Rozenberg, A. Salomaa, eds.). Springer-Verlag, 1998.

GS84.   F. Gécseg and M. Steinby. *Tree Automata*. Akadémiai Kiadó, Budapest, 1984. 22

GS97.   F. Gécseg and M. Steinby. Tree automata. In G. Rozenberg and A. Salomaa, editors, *Handbook of Formal Languages, Vol. 3*, chapter 1. Springer-Verlag, 1997.

Hab89. A. Habel. *Hyperedge Replacement: Grammars and Languages*. PhD thesis, University of Bremen, 1989. 22

Hab92. A. Habel. *Hyperedge Replacement: Grammars and Languages*, volume 643 of *LNCS*. Springer-Verlag, 1992. 26

Man98. S. Maneth. The generating power of total deterministic tree transducers. *Inform. and Comput.*, 147:111–144, 1998. 22

Plu98. D. Plump. Term graph rewriting. Technical Report CSI-R9822, Computing Science Institute Nijmegen, 1998. To appear in *Handbook of Graph Grammars and Computing by Graph Transformation, Volume 2*, World Scientific, Singapore. 15

Rao97.   J.-C. Raoult. Rational tree relations. *Bull. Belg. Math. Soc.*, 4:149–176, 1997.
          See also "Recursively defined tree transductions", Proc. RTA'93, LNCS 690,
          pages 343-357. 16, 24
Rou70.   W.C. Rounds. Mappings and grammars on trees. *Math. Systems Theory*,
          4:257–287, 1970. 22, 26

Neighborhood Expansion Grammars[*]

John L. Pfaltz

Dept. of Computer Science, University of Virginia
Charlottesville, VA 22903

Abstract. Phrase structure grammars, in which non-terminal symbols
on the left side of a production can be rewritten by the string on the right
side, together with their Chomsky hierarch classification, are familiar
to computer scientists. But, these grammars are most effective only to
generate, and parse, strings.
In this paper, we introduce a new kind of grammar in which the right
side of the production is simply appended to the intermediate structure
in such a way that the left side becomes its "neighborhood" in the new
structure.
This permits the grammatical definition of many different kinds of "n-
dimensional" discrete structures. Several examples are given.
Moreover, these grammars yield a formal theory grounded in antimatroid closure spaces. For example, we show that restricted neighborhood
expansion grammars capture the essence of finite state and context free
phrase structure grammars.

1 Overview

In this paper we consider two distinct, but intertwined, themes. The first is
neighborhood expansion grammars which constitute a considerable change from
more traditional phrase structure grammars in that intermediate symbols are
not rewritten. Instead, completely new pieces are simply added to the discrete
system being generated. Such discrete systems we model as a graph G.

The second theme is that of closure spaces and their associated lattices [16].
If the class of discrete system being generated is a closure space, we will be
able to define "good" grammars as those which homomorphically preserve the
induced lattice structure of the system with each step of the generation.

Finally, we will be able to develop a hierarchy of grammars that includes the
Chomsky hierarchy, and by which we can qualitatively measure the complexity
of discrete systems.

1.1 Neighborhood Expansion Grammars

In a phrase structure grammar, a substring, or more usually a single non-terminal
symbol A, is rewritten by some other string α. When phrase structure rewriting is

---

[*] Research supported in part by DOE grant DE-FG05-95ER25254.

H. Ehrig et al. (Eds.): Graph Transformation, LNCS 1764, pp. 30–44, 2000.

applied to more complex, non-linear structures, a subgraph H of G is rewritten as a larger subgraph H' which must then be embedded into $G' = G - H$. Describing the embedding process is not always easy [17].

Neighborhood expansion grammars do not rewrite any portion of the existing intermediate structure. Their productions have the form $N \mathrel{+}= H'$. From any intermediate structure G, we create $G' = G \cup H'$ in such a way that $N \subseteq G$ becomes the neighborhood of H' in G'. We use $\mathrel{+}=$ as our production operator, instead of the more customary $::=$ or $\models$, as a reminder of its additive quality. This process is more easily visualized by a small example.

A grammar $\mathcal{G}_{chordal}$ to generate chordal graphs can be described by a single production:

$$K_n \mathrel{+}= p \qquad n \geq 1$$

that is, any clique of order n in G can serve as the neighborhood of a new point p. Every point in K_n will be adjacent to p in G'.[1] Figure 1 illustrates a representative generation sequence. Each expanded neighborhood (in this case

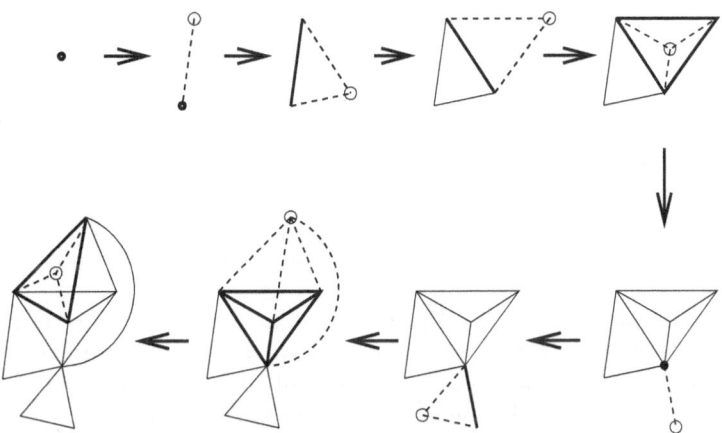

Fig. 1. A sequence of neighborhood expansions generating chordal graphs

---

[1] It has been observed that since n is unbounded, this formulation effectively specifies an infinite number of productions, thereby violating a basic property of grammars. There are 3 possible ways of meeting this objection.

(a) Instead of creating a grammar for *all* chordal graphs, one can use the same production to generate all chordal graphs with some maximal, though arbitrarily large, diameter. This bounds n.

(b) By a second property of grammars, generation must terminate after application of a finite number of productions. Since, in this case, only one node is added per production, n, for any chosen neighborhood, must necessarily be finite.

(c) Accept this kind of parameterized production as conforming to the spirit of economical description which motivates the grammar approach, and insist that n is always finite.

clique) has been made bold; and the expansion point circled. The dashed edges indicate those which define the clique as the neighborhood of the expansion point p. It is not hard to see that any graph generated in this fashion must be chordal. Because extreme points are simplicial (neighborhood is a clique), and because every chordal graph must have at least two extreme points [7,5], every chordal graph can be so generated.

1.2   Neighborhoods

We say $\bar{\eta}$ is a **complete neighborhood** operator if for all $X \subseteq \mathbf{U}$, where $\mathbf{U}$ denotes our universe,

$X \subseteq X.\bar{\eta}$,
$X \subseteq Y$ implies $X.\bar{\eta} \subseteq Y.\bar{\eta}$,
$(X \cup Y).\bar{\eta} = X.\bar{\eta} \cup Y.\bar{\eta}$.[2]

Clearly, a neighborhood operator expands the subset. Usually, we are more interested in the **deleted neighborhood**, η, defined $X.\eta = X.\bar{\eta} - X$,   because it represents this incremental addition.

Let us consider a representative neighborhood operator. If G is a graph, we define a deleted neighborhood by $X.\eta_{adj} = \{y \notin X | \exists x \in X$ adjacent to $y\}$. Readily, this is the neighborhood concept we used in $\mathcal{G}_{chordal}$. It appears in many graph-theoretic algorithms.

2   Other Expansion Grammars

We call G a **block graph** if every maximal 2-connected subgraph is an n-clique K_n [9]. A simple grammar $\mathcal{G}_{block}$ consisting of only the single rule

$$p += K_n \qquad n \geq 1$$

will generate the block graphs. Figure 2 illustrates one such derivation sequence in $\mathcal{G}_{block}$. Here, each p to be expanded is emboldened and its connection to the new K_n indicated by dashed lines. Readily, the neighborhood of each new expansion K_n is the single point. Contrasting $\mathcal{G}_{chordal}$ with $\mathcal{G}_{block}$ is inevitable. The clique K_n on the left side of the production in $\mathcal{G}_{chordal}$ is just exchanged with the new point p on the right side. If K_n, in either $\mathcal{G}_{chordal}$ or $\mathcal{G}_{block}$, is constrained to be K_1, a singleton point, then the grammar becomes $\mathcal{G}_{tree}$ which generates all, and only, undirected trees. This provides, perhaps, the clearest explanation of why block and chordal graphs are essentially tree-like. This latter result is well known, since every chordal graph is the intersection graph of the subtrees of a tree [6,12].

If we consider discrete partially ordered systems, there are two distinct neighborhood concepts:

$X.\eta_< = \{z \notin X | z < x$ and $z < y \leq x$ implies $y \in X\}$ and
$X.\eta_> = \{z \notin X | x < z$ and $x \leq y < z$ implies $y \in X\}$.

Consequently, we have at least three different grammars to generate partial orders, depending on whether one uses $\eta_<$, $\eta_>$, or $\eta_{<>}$ (which combines them

---

[2]   We prefer the suffix operator notation $X.\bar{\eta}$ to the prefix form $\bar{\eta}(X)$.

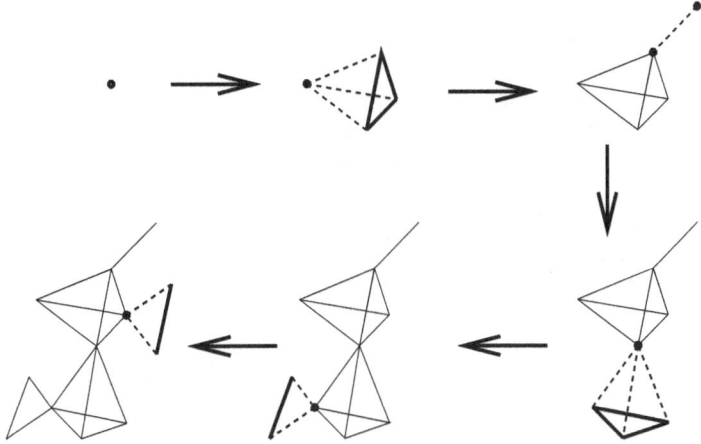

Fig. 2. A sequence of neighborhood expansions generating block graphs

both) as the neighborhood concept. For example, the grammar $\mathcal{G}_{right\_tree}$ with the single production

$$p \mathrel{+{=}} q$$

clearly generates the class of rooted trees whose orientation depends on whether p is to be $q.\eta_<$ or $q.\eta_>$. If we let the neighborhood denoted by the left hand of this production be an arbitrary subgraph N of G, as in the production

$$N \mathrel{+{=}} q$$

where $q.\eta_<$ is the neighborhood operator, we get $\mathcal{G}_{left\_root}$ generating all *left rooted* acyclic graphs. Figure 3 illustrates one such generation sequence. If in-

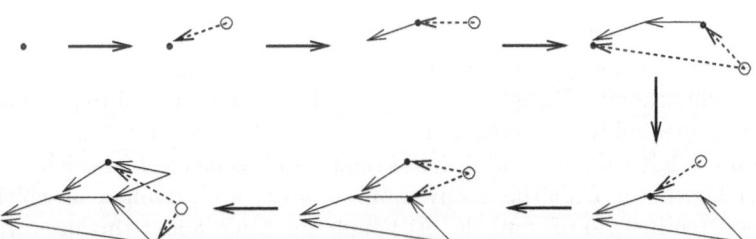

Fig. 3. A sequence of neighborhood expansions generating left rooted acyclic graphs

stead of the singleton point of Figure 3, we allow an arbitrary antichain, A_n, to be the start structure, then readily $\mathcal{G}_{left\_root}$ becomes $\mathcal{G}_{acyclic}$ which generates all acyclic graphs or partial orders.

If instead, we use the neighborhood operator $\eta_>$, the same production yields all *right rooted* acyclic graphs, which with the antichain start structure generates acyclic graphs as before.

Some interesting subfamilies of the acyclic graphs can be developed if one uses the two-sided neighborhood operator, $X.\eta_{<>} = X.\eta_< \cup X.\eta_>$. The major problem with these two-sided neighborhoods is that for an arbitrary subset H of G, it may not be evident which elements belong to which side. Such a grammar we would call **ambiguous**. An unambiguous grammar for two terminal, parallel series networks can be specified by requiring the left hand neighborhood to be a single edge. For example, $\mathcal{G}_{ttspn}$ has the two rewrite rules,

$$p_1 \longleftarrow p_2 \quad := \quad q$$

$$p_1 \longleftarrow p_2 \quad := \quad \begin{matrix} q_1 \\ q_2 \end{matrix}$$

A representative example of a sequence of structures generated by $\mathcal{G}_{ttspn}$, given a single edge as the start structure is shown in Figure 4. Again, new points of the

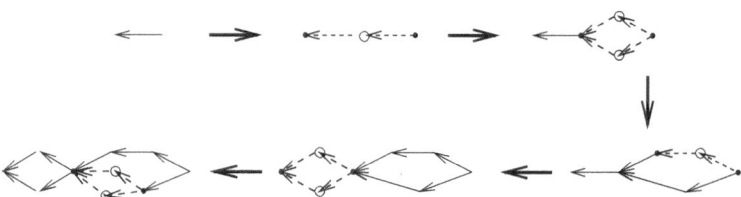

Fig. 4. A sequence of two terminal, parallel series networks generated by, $\mathcal{G}_{ttspn}$

expansion are denoted by circles with dashed lines indicating the connection to their neighborhoods. Purists may observe that neighborhood expansion grammars are expected to only enlarge the intermediate structures. $\mathcal{G}_{ttspn}$ deletes the edge on the left side after adding the right hand elements for which it is to be the neighborhood. Isn't this really a phrase structure grammar in which edges are rewritten instead of symbols? Although one could adopt this interpretation, it is more accurate to observe that if we retain the left hand edge after the expansion, the resulting structure is simply the transitive closure of a TTSPN, so their erasure is not really changing the partial order.

3   Closure Spaces

When a phase structure production is applied to a string it preserves the structure of the string. Every intermediate form is a string. We believe that all grammars should preserve the essential structure of their underlying objects. But what is the essential structure of a family of discrete objects?

We believe uniquely generated closure operators represent a basic way of describing the structure of a discrete system. Neighborhood expansions have been defined so as to homomorphically preserve this concept of structure. Any **closure operator** φ must satsify the axioms: $X \subseteq X.\varphi$, $X \subseteq Y$ implies $X.\varphi \subseteq Y.\varphi$, and $X.\varphi.\varphi = X.\varphi$, where $X, Y \subseteq \mathbf{U}$ are arbitrary subsets in a universe $\mathbf{U}$ of interest [11]. If in addition, $X.\varphi = Y.\varphi$ implies $(X \cap Y).\varphi = X.\varphi$, we say that φ is **uniquely generated**. For example, monophonic closure [5], in which $X.\varphi$ denotes all points lying on chordless paths between distinct points $p, q \in X$ is uniquely generated over chordal graphs.

A vector space, or matroid $\mathcal{M}$, is the closure (or span σ) of a set of basis vectors [18]. $\mathcal{M}$ must satisfy the **exchange axiom**: $p, q \notin X.\sigma$, and $q \in (X \cup p).\sigma$ imply that $p \in (X \cup q).\sigma$.[3] It can be shown [16] that uniquely generated closure spaces must satisfy the **anti-exchange axiom**: $p, q \notin X.\varphi$, and $q \in (X \cup p).\varphi$ imply that $p \notin (X \cup q).\varphi$. From this comes the adjective **antimatroid** closure space. Closure spaces form a kind of dual to vector spaces.

If the sets of a closure space $(\mathbf{U}, \varphi)$ are partially ordered by
$$X \leq_\varphi Y \text{ if and only if } Y \cap X.\varphi \subseteq X \subseteq Y.\varphi$$
we obtain a **closure lattice** [16], in which the sublattice of closed subsets is lower semi-modular.[4] We have asserted that this closure lattice $\mathcal{L}_{(U,\varphi)}$ describes the structure of the underlying discrete system. Figure 5(b) illustrates the closure lattice of the sixth chordal graph generated in Figure 1. This is a complex

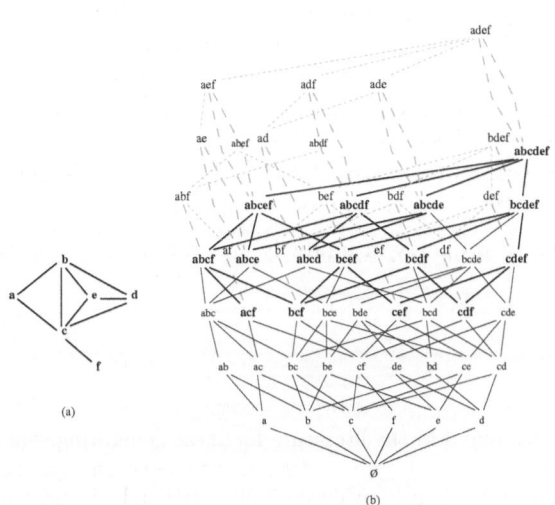

Fig. 5. A chordal graph and its closure lattice

---

3   This permits the familiar change of basis of a vector space.
4   The lattice of vector subspaces partially ordered by inclusion is upper semi-modular.

diagram; but there are significant regularities that make it amenable to analysis. For example, the sublattice of closed subsets (those which are connected by solid lines) can be shown to be a lower semi-modular lattice [13]. Consider the closed subset $\{abcf\}$ which covers the closed subsets $\{abc\}, \{acf\}, \{bcf\}$. They, in turn, cover their pairwise inf's, $\{ac\}, \{bc\}, \{cf\}$, which in turn cover their common inf, $\{c\}$, to form a distributive sublattice. This is a general property of lower semi-modularity, as shown in [3]. The closed subset $\{abcf\}$ is generated by the subset $\{abf\}$, which is above, to the left, and connected by a dashed line. Such generators, denoted $X.\kappa$, are the unique minimal subsets with a given closure.[5] The subsets contained between any closed set $X.\varphi$ and its generator $X.\kappa$ comprise a boolean algebra, $[X.\varphi, X.\kappa]$, which we sketch in with a dashed diamond. A few of the ordering relationships between non-closed subsets have been indicated by dotted lines.

Although these closure lattices grow exponentially, for example, the final chordal graph generated in Figure 1 on 9 points has 146 closed sets, one can still reason about them. We will see this presently. But, they become too large to illustrate easily.

Although there are similarities, a neighborhood operator is not in general a closure operator because (1) a neighborhood operator need not be idempotent, $X.\bar{\eta} \subseteq X.\bar{\eta}.\bar{\eta}$, and (2) a closure operator need not be union preserving, $X.\varphi \cup Y.\varphi \subseteq (X \cup Y).\varphi$. We say that η and φ are **compatible** if X closed implies $X.\bar{\eta}$ is closed.

It is not hard to show that η_{adj} and $\varphi_{geodesic}$ are compatible on block graphs. Similarly, η_{adj} and $\varphi_{monophonic}$ are compatible on chordal graphs. In [15] it is shown that at least n^n distinct operators can be defined over any n element system with $n \geq 10$. But, except for chordal graphs, η_{adj} cannot be compatible with any of these closures. Figure 6 indicates why. Assume that every point is closed, so $p.\bar{\eta}_{adj}$ is closed for all p. Consequently, the four quadrilaterals $abce, abde, bcde$, and $acde$ are closed. Because closed sets are closed under intersection, the triangles abe, ade, bce, and cde must be closed as well. This determines the upper portion of the lattice [6] which cannot be a closure lattice.

A fundamental property of uniquely generated closure lattices can be found in [16]

Many terms are found in the literature for these generating sets depending on ones approach. With convex closure in discrete geometry one speaks of *extreme points* [4]. With respect to transitive closures in relational algebras one calls them the *irreducible kernel* [1]. In [16], we called them *generators*.

[6] In Figure 6, all edges are also closed. Because closed sets are closed under intersection, the edges ae, be, ce, and de must be closed. One could assume that the remaining 4 edges are not closed, in which case ab becomes the generator of abe, etc. But, this will not change the essence of this counter example.

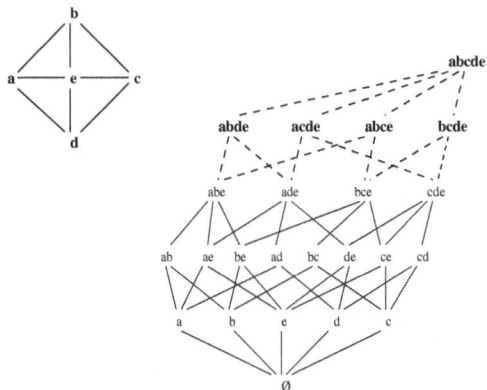

Fig. 6. Closed sets of a non-chordal graph

Lemma 1. *If φ is uniquely generated, and if $Z \neq \emptyset$ is closed, $p \in Z.\kappa$ if and only if $Z - \{p\}$ is closed, in which case $Z.\kappa - \{p\} \subseteq (Z - \{p\}).\kappa$.*

Using this property one can deduce that $\{abde\}.\kappa = \{bd\}$, $\{acde\}.\kappa = \{ac\}$, $\{abce\}.\kappa = \{ac\}$, and $\{bcde\}.\kappa = \{bd\}$. But, this is impossible. The set $\{ac\}$ cannot generate two different closed sets! Neither can $\{bd\}$.

Because Figure 6 is the simplest non-chordal graph, we have

Theorem 1. *Let the individual elements of an undirected structure G be closed. A closure operator φ can be compatible with the adjacency neighborhood operator, η_{adj} if and only if G is chordal.*

Proof. If G is chordal, then the monophonic closure operator $\varphi_{monophonic}$ satisfies the theorem. Only if follows for the example above. $\qquad\square$

Conjecture 1. Let $\mathcal{F}$ be a family of discrete undirected systems G. If there exists a neighborhood operator η and a compatible closure operator φ in which every singleton is closed, then G is essentially tree-like.

For "tree-like", we expect a definition based on intersection graphs [12].

In [16], three different closure operators are defined over partially ordered systems. They are

$$Y.\varphi_L = \{x | \exists y \in Y, x \leq y\},$$
$$Y.\varphi_R = \{z | \exists y \in Y, y \leq z\},$$
$$Y.\varphi_C = \{x | \exists y_1, y_2 \in Y, y_1 \leq x \leq y_2\}.$$

On partially ordered systems, φ_L, and φ_R are ideal closure operators, φ_C is an interval, or convex, closure operator. It is not difficult to show that $\eta_<$ is compatible with both φ_L and φ_R, as is $\eta_>$.

The induced closure lattice resulting from the closure operator φ_L on the 7 point poset of Figure 3 is shown in Figure 7.

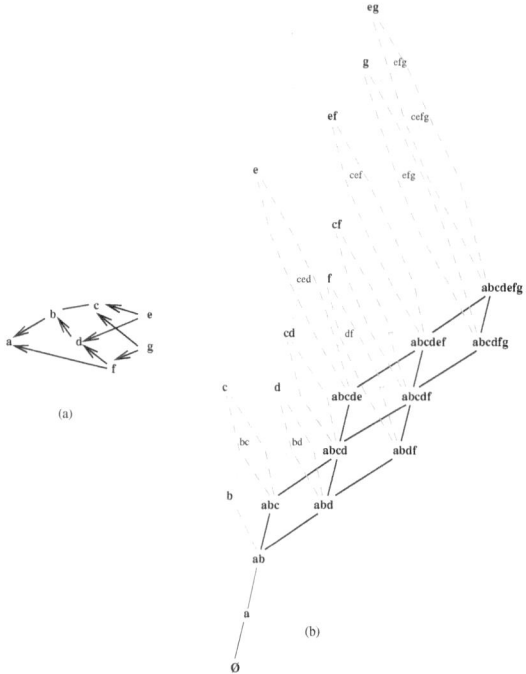

Fig. 7. A closure lattice $\mathcal{L}_{G,\varphi_L}$

As in Figure 5, the sublattice of closed subsets is denoted by solid lines; and the boolean intervals bounded by the closed subsets $X.\varphi$ and their generating sets $X.\kappa$ are denoted by dashed lines which have been oriented toward the upper left. This example is less cluttered, so the characteristic closure lattice structure is more evident. For instance, the subset $\{ef\}$ is the unique generator for the closed set $\{abcdef\}$. All 2^4 subsets in the boolean lattice they delimit (including $\{cef\}$), have $\{abcdef\}$ as their closure. The ordering between these boolean generator sub-lattices, which have been suggested by a few dotted lines, mirror the ordering of the closed sets below them.

4   Structure Preservation

Why should one adopt a process that generates a family of discrete structures by adding some completely new portion within a neighborhood rather than rewriting an existing element, which of course must have a neighborhood? A partial answer was given in sections 1 and 2. The neighborhood expansion approach can be used to generate several families of interesting discrete systems. In Section 5, we will demonstrate that the classes of regular and context free languages in the Chomsky hierarchy of the phrase structure grammars can be subsumed by neighborhood expansion classes. In this section we examine a more subtle argument based on the homomorphic preservation of discrete system structure.

Homomorphic (structure preserving) *generation* is most easily defined in terms of its inverse, homomorphic *parsing*. In this case parsing consists of simply deleting a point and all its incident edges. Figure 8 illustrates the deletion χ_d of the element d from the chordal graph of Figure 2(a). We let χ, which is suggestive of "striking out", denote the deletion operator. The subscript denotes the element, or set, being deleted. In the source lattice $\mathcal{L}_G$, pre-image equiva-

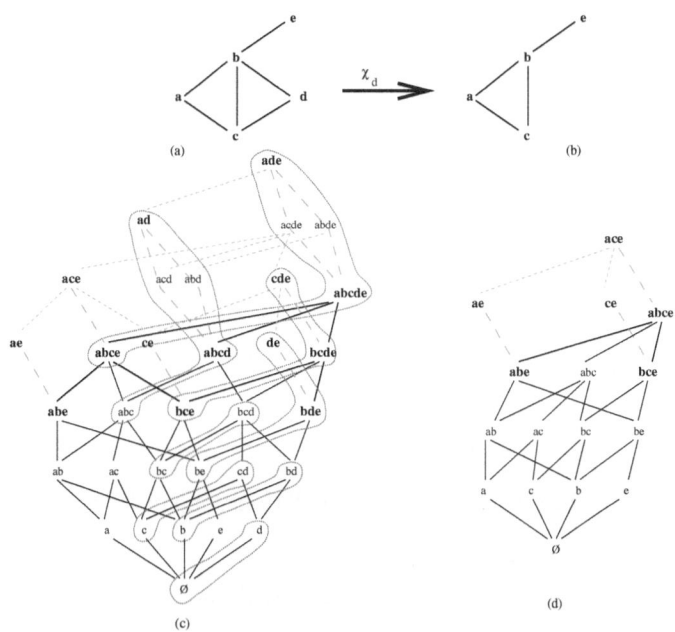

Fig. 8. A "deletion" transformation, χ_d

lence sets are indicated by dotted lines. The element d is mapped onto $\emptyset$ in the target lattice, and many sets containing d, such as $\{bcd\}$, are mapped onto the set obtained by removing d. But, as we can see from the moderately complex pre-image partition in $\mathcal{L}_G$ of Figure 8, deletion is *not* just a simple set extension of a point function.[7] For example, the set $\{ad\}$ maps onto the set $\{abc\}$. This is necessary if the the transformation is to preserve the structure of *all* the subsets, that is, if the lattice of $G.\chi_d$, denoted $\mathcal{L}_{G.\chi_d}$, is to be a semi-homomorphic contraction of $\mathcal{L}_G$. For a discrete system with n elements, its closure lattice will

---

[7] Deletion of elements make little sense within the usual definition of discrete functions because a deleted element has no image in the codomain. However, in the context of a lattice over all subsets of a discrete space, a deleted element or subspace can be simply mapped onto the empty set, $\emptyset$, in the codomain.

have 2^n elements. Consequently, reducing the size of a system greatly reduces its stuctural complexity, as is clear in Figure 8.

We now have a more formal way of defining what a homomorphic neighborhood expansion, ϵ_N, really is. Let $G^{(n)} \xrightarrow{\chi_p} G^{(n-1)}$ be any deletion in a discrete system with n points. By an expansion ϵ_N with respect to the neighborhood N we mean an inverse operator to χ_p, where N is the image in $G^{(n-1)}$ of the neighborhood N of p denoted $p.N$ in $G^{(n)}$. Thus, we always have $G^{(n)} \xrightarrow{\chi_p} G^{(n-1)} \xrightarrow{\epsilon_N} G^{(n)}$ And, since deletions always induce homomorphic mappings between the corresponding closure lattices, neighborhood expansion can be regarded as homomorphic as well.

Lemma 2. *If there exists a closure operator, φ that is compatible with the neighborhood operator, η, used by a neighborhood expansion grammar $\mathcal{G}$, then any sequence of deletions or expansions homomorphically preserves the internal structure of each intermediate system.*

The sense of this lemma is illustrated by Figure 9. Not all expansion gram-

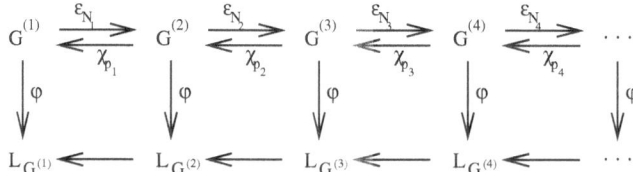

Fig. 9. A sequence of deletions/expansions

mars have compatible closure operators. But, if one exists it enforces a kind of regularity in the grammar's derivations.

5   Expansion Grammars and the Chomsky Hierarchy

How different are neighborhood expansion grammars from more customary phrase structure grammars? We will claim they are very similar, except that neighborhood expansion seems more flexible.

Theorem 2. *Let $\mathcal{L}$ is any regular language. There exists a neighborhood expansion grammar $\mathcal{G}$ in which the left side (neighborhood) of every production is a closed singleton element, such that $L(\mathcal{G}) = \mathcal{L}$.*

Proof. If $\mathcal{G}_{regular}$ is regular, or finite state, it can be cast into either a left, or right recursive form. We assume $\mathcal{G}_{regular}$ is right recursive, that is all productions are of the form $V := aV'|a$, so only the rightmost element of the string is a variable. We use $\eta_<$ and φ_R as our neighborhood and closure operators respectively. The rightmost element is closed, as are all rightmost intervals. This rightmost element provides the neighborhood, $\eta_<$, for the expansion. If $\eta_< D$ is labeled with V, it is either relabeled with a and the new expansion point labeled with V', or else $\eta_< D$ is just relabeled with the terminal symbol a, thereby ending the generation.   □

The only distinction is that expansion grammars with these properties can also generate languages of rooted trees as well.

Theorem 3. *Let $\mathcal{L}$ be any context-free language. There exists a neighborhood expansion grammar $\mathcal{G}$ in which the left side (neighborhood) of every production is closed, such that $L(\mathcal{G}) = \mathcal{L}$.*

Proof. Because a context-free grammar G_{cf} permits non-terminal symbols to be rewritten anywhere in the string, we require **convex closure** φ_C where $X.\varphi_C = \{q | p \leq q \leq r, \forall p, r \in X\}$, to emulate it. We also assume Chomsky normal form [2,8], in which every production is of the form $V := V_1 V_2$ or $V := a$. Again, it is apparent that the neighborhood labeled with V can be relabeled with V_1 and the expansion element labeled V_2.   □

By their very nature, we would assume that neighborhood expansion grammars can emulate any context-sensitive grammar, which must be non-decreasing. Neighborhood expansion grammars too must be non-decreasing, and the neighborhood dependent productions would seem to exactly capture the context sensitive quality. We believe this to be true. But, as yet, we have found no convincing proof.

It is possible to catalog neighborhood expansion grammars according to whether the left side (neighborhood) of every production is: (a) a simple closed set, (b) a closed set, (c) a singleton element, or (d) has some other well defined property which we call **neighborhood criteria**; and whether the right side (neighborhood) of every production is: (a) a closed set, (b) a singleton element, or (c) has some other well defined property which we call **replacement criteria**.

We observe that the left side of the production of $\mathcal{G}_{chordal}$ is a closed neighborhood in the intermediate structure, as in $\mathcal{G}_{block}$ as well. But, in the latter grammar every left side is also a singleton element. The language $\mathcal{L}_{block}$ is a subset of $\mathcal{L}_{chordal}$ [10]. If this production is further restricted to have only singleton elements on the right side, as in $\mathcal{G}_{tree}$, it language is $\mathcal{L}_{tree}$ which is a subset of $\mathcal{L}_{block}$.

The grammar $\mathcal{G}_{left\_root}$ with a characteristic generation as illustrated in Figure 3, satisfies none of the *neighborhood criteria* described above. If in addition to the neighborhood operator $\bar{\eta}_<$, we choose φ_L (as illustrated in Figure 7) to be its compatible closure operator, and we require each left side neighborhood to be closed, we obtain derivations such as Figure 10. Readily, requiring the left side neighborhood to be closed generates a language of transitively closed acyclic graphs.

Simply switching the compatible closure operator to be φ_R includes the set of all **basic** representations of the partial order, that is if $x < y$ and $y < z$ then the edge (x, z) is not in G [14]. If further, we require the left side neighborhood

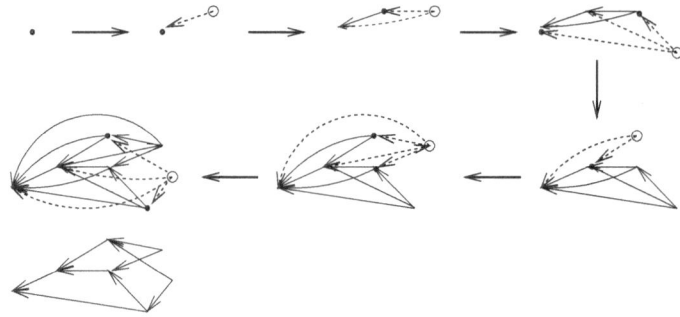

Fig. 10. Another sequence of neighborhood expansions generating left rooted acyclic graphs

to be a closed antichain, $\mathcal{G}_{left\_root}$ generates only these basic representations of a partial order.

In any case, partially ordered systems are more complex than directed, or undirected, trees with respect to the neighborhood expansion hierarchy. This accords with our algorithmic intuition. Moreover, the set of all trees is comparable to a regular language while the set of all partial orders is comparable to a context free language. Again, this seems intuitively correct. Thus we begin to see a linguistic hierarchy emerging by which we can qualitatively compare the complexity of directed and undirected discrete systems.

6 Neighborhood Operators

So far we have only considered two kinds of neighborhood operator — simple node adjacency, η_{adj}, and the comparative neighborhoods, $\eta_<, \eta_>, \eta_{<>}$ in partial orders. These are familiar, easy to work with, and make good illustrative examples. However, there exist many more ways of defining neighborhood operators and their mechanisms for binding new elements to an expanding configuration. The examples we develop is this section are suggestive of the way atoms may bond to form larger molecules, the way small procedures with some well-defined interfaces may be incorporated into a larger task, or how individual parts may be combined to create larger assemblies.

Our example is absurdly simple. The elements are equilateral triangles with edges labeled (in clockwise order) a, b, c or d, e, f as shown in Figure 11. We

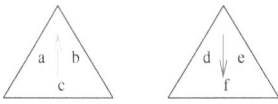

Fig. 11. Two spatial elements

have superimposed an arrow to make the orientation evident in our figures; but they are not actually part of the grammar or language. We suppose that these elements are subject to the affine transformations, translation and rotation, in the plane.

Now suppose that neighborhood operator η_1 specifies that *a can abut d, b can abut e*, and *c can abut f*. We let the grammar $\mathcal{G}_1$ consist of the single rule

$$p \ +\!= \ q$$

Now, depending on the initial orientation of the initial element a derivation must yield a uniform pattern such as Figure 12.

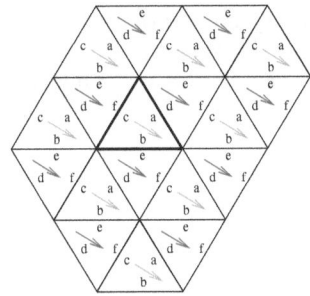

Fig. 12. A uniform field in the language $\mathcal{L}(\mathcal{G}_1)$

The rigidity of the neighborhood operator allows no flexibility in $\mathcal{G}_1$. Suppose we adjoin the possibility that *b can* also *abut d* to the operator η_1 to obtain η_2 and a grammar $\mathcal{G}_2$. Readily $\mathcal{L}(\mathcal{G}_1) \subseteq \mathcal{L}(\mathcal{G}_2)$, but $\mathcal{L}(\mathcal{G}_2)$ has many additional configurations.

If, using the new neighborhood rule, a central hexagonal "rotator" as shown in the center of Figure 13 is initially constructed, and if this is expanded in a "breadth-first" fashion, then a rigidly rotating pattern must be generated. This

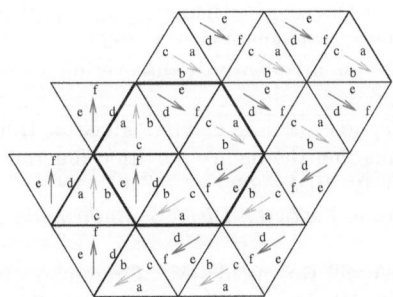

Fig. 13. A "rotating" field in the language $\mathcal{L}(\mathcal{G}_2)$ obtained by expanding the central hexagonal configuration

follows because the only edge labels that can appear on any convex exterior boundary are a, c, e, and f which uniquely determine the corresponding orientation of the expansion elements as they expand out from the central core.

This, and similar tiling examples, are not mere curiosities. They illustrate and interesting class of grammar problems for which we are primarily interested in the characteristics of the boundary of the generated structure. Such problems may serve as an abstract model of software encapsulation in which we are primarily concerned that the interface be well-formed.

References

1. Chris Brink, Wolfram Kahl, and Gunther Schmidt. *Relational Methods in Computer Science*. Springer Verlag, Wien, 1997. 36
2. Noam Chomsky. Formal Properties of Grammars. In R.D. Luce, R.R. Bush, and E. Galanter, editors, *Handbook of Mathematical Psychology*. Wiley, New York, 1963. 41
3. Paul H. Edelman. Meet-Distributive Lattices and the Anti-exchange Closure. *Algebra Universalis*, 10(3):290–299, 1980. 36
4. Paul H. Edelman and Robert E. Jamison. The Theory of Convex Geometries. *Geometriae Dedicata*, 19(3):247–270, Dec. 1985. 36
5. Martin Farber and Robert E. Jamison. Convexity in Graphs and Hypergraphs. *SIAM J. Algebra and Discrete Methods*, 7(3):433–444, July 1986. 32, 35
6. Fanica Gavril. The Intersection Graphs of Subtrees in Trees are exactly the Chordal Graphs. *J. Combinatorial Theory B*, 16:47–56, 1974. 32
7. Martin Charles Golumbic. *Algorithmic Graph Theory and Perfect Graphs*. Academic Press, New York, 1980. 32
8. John E. Hopcroft and Jeffrey D. Ullman. *Introduction to Automata Theory, Languages, and Computation*. Addison-Wesley, Reading, MA, 1979. 41
9. Robert E. Jamison-Waldner. Convexity and Block Graphs. *Congressus Numerantium*, 33:129–142, Dec. 1981. 32
10. Robert E. Jamison-Waldner. Partition Numbers for Trees and Ordered Sets. *Pacific J. of Math.*, 96(1):115–140, Sept. 1981. 41
11. Kazimierz Kuratowski. *Introduction to Set Theory and Topology*. Pergamon Press, 1972. 35
12. Terry A. McKee and Fred R. McMorris. *Topics in Intersection Graph Theory*. SIAM Monographs on Discrete Mathematics and Applications. Society for Industrial and Applied Math., Philadelphia, PA, 1999. 32, 37
13. B. Monjardet. A Use for Frequently Rediscovering a Concept. *Order*, 1:415–416, 1985. 36
14. John L. Pfaltz. *Computer Data Structures*. McGraw-Hill, Feb. 1977. 41
15. John L. Pfaltz. Evaluating the binary partition function when $N = 2^n$. *Congress Numerantium*, 109:3–12, 1995. 36
16. John L. Pfaltz. Closure Lattices. *Discrete Mathematics*, 154:217–236, 1996. 30, 35, 36, 37
17. John L. Pfaltz and Azriel Rosenfeld. Web Grammars. In *Proc. Intn'l Joint Conf on AI*, pages 609–619, Washington, DC, May 1969. 31
18. W. T. Tutte. *Introduction to the Theory of Matroids*. Amer. Elsevier, 1971. 35

Neighborhood-Preserving Node Replacements*

Konstantin Skodinis[1] and Egon Wanke[2]

[1] University of Passau, D-94032 Passau, Germany
skodinis@fmi.uni-passau.de
[2] University of Düsseldorf, D-40225 Düsseldorf, Germany
wanke@cs.uni-duesseldorf.de

Abstract We introduce a general normal form for eNCE graph grammars and use it for a characterization of the class of all neighborhood-preserving eNCE graph languages in terms of a weak notion of order-independency.

1 Introduction

Graph grammars are used for generating labeled graphs, see [3,8,7,6,5,11]. They are motivated by considerations about pattern recognition, compiler construction, and data type specification.

In this paper, we consider eNCE (edge label neighborhood controlled embedding) graph grammars. These grammars are node replacement systems and generate node and edge labeled graphs (notations adopted from [11]). An eNCE graph grammar $\mathcal{G}$ has terminal and nonterminal node labels as well as final and nonfinal edge labels. Grammar $\mathcal{G}$ consists of a finite set of productions of the form $p = A \rightarrow R$, where the left-hand side A is a nonterminal node label and the right-hand side R is a labeled graph with an embedding relation. Given a graph G, the application of p to a node u of G labeled by A is a substitution of u by R. The edges between a neighbor v of u in G and the inserted nodes of R are controlled by the embedding relation of R. These new edges are called *successor edges* of the edge between u and v in G. A graph is *terminal* if all its nodes have terminal labels. It is *final* if additionally all its edges have final labels. The language $L(\mathcal{G})$ of $\mathcal{G}$ is the set of all final graphs derivable from the start graph.

An eNCE graph grammar $\mathcal{G}$ is *order-independent*, OI-eNCE, if the order in which the productions are applied in deriving a graph with terminal nodes is irrelevant [16], see also [11]. $\mathcal{G}$ is *confluent* if the order in which productions are applied in deriving any graph is irrelevant. Here, in contrast to OI-eNCE grammars, the resulting graphs may also have nonterminal nodes. It is well-known that OI-eNCE and confluent eNCE graph grammars have the same generating power, although there are OI-eNCE graph grammars which are not confluent [11].

Graph grammars are, in general, difficult to analyze. Even simple confluent grammars like *linear* grammars define NP-hard graph languages. In this paper,

---

* The work of the first author was supported by the German Research Association (DFG) grant BR 835/7-1

H. Ehrig et al. (Eds.): Graph Transformation, LNCS 1764, pp. 45–58, 2000.
© Springer-Verlag Berlin Heidelberg 2000

we deal with the well-known neighborhood-preserving property of graph grammars, see also [18]. Intuitively, a graph grammar is neighborhood-preserving if no established neighborhood will be destroyed in any further derivation step. This property makes graph grammars and their languages more efficiently analyzable. To emphasize the importance of neighborhood-preserving derivations, we first consider the recognition problem (parsing) of graph languages. For a fixed graph grammar $\mathcal{G}$, the recognition problem is the question whether or not a given graph can be generated by $\mathcal{G}$. This question is indispensable, for example, in visual language theory [20]. The general recognition problem is PSPACE-hard [14] and still NP-hard for linear graph grammars. However, parsing can be done in polynomial time and even in parallel for confluent graph grammars generating only connected graphs of bounded degree [17,18].

Let us illustrate a basic problem that appears during the parsing process of a graph. Consider the derivation step shown in Figure 1 in which a single node u of a graph G is replaced by the right-hand side R of some graph production p. Assume that during the embedding the only edge between v and u is destroyed. In a bottom-up parsing, the derivation step is reconstructed as follows: Find an isomorphic copy of R in H and replace this subgraph with a node which has the same label as the left-hand side of the production. Then reconstruct the edges of G according to the embedding C of R. Since the edge $\{v, u\}$ is destroyed, we don't know which one of all the nodes labeled by a was the neighbor of u in G. Hence bottom-up parsing methods get in trouble if an edge can be deleted during the replacement step. This case provides an additional source of nondeterminism which makes the parsing algorithm more inefficient.

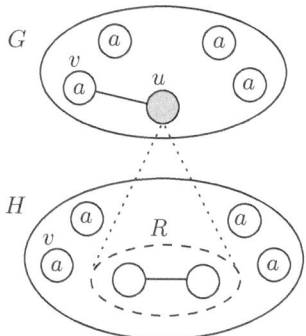

Figure1. A non neighborhood-preserving derivation step

One way to overcome the problem discussed above is to consider only grammars in which every node v adjacent to u gets always a connection to a node from R when substituting u by R. Such grammars are called neighborhood-preserving grammars.

Let $\mathcal{G}$ and $\mathcal{G}'$ be two graph grammars generating the same graph language, where $\mathcal{G}$ is neighborhood-preserving but not confluent and $\mathcal{G}'$ is confluent but not neighborhood-preserving. Using the parser for arbitrary graph grammars integrated into the "GraphEd" system of [13], experiments have shown that the parsing is often much faster for grammars like $\mathcal{G}$ than for grammars like $\mathcal{G}'$ although in the first case all different orders of applying the productions of $\mathcal{G}$ have to be tried.

Next we explain why the neighborhood-preserving property is important for the efficient analysis of graph languages with respect to a graph property. Here the problem is whether or not a given graph grammar generates some graph (or only graphs) satisfying a certain property. Such questions can often be solved more efficiently if the given graph grammar is neighborhood-preserving. For example, the question whether or not the language of a grammar contains a totally disconnected graph is undecidable for eNCE graph grammars [15] but solvable in polynomial time if the grammar is neighborhood-preserving. To see this, first remove from the graph grammar every production whose right-hand side contains adjacent nodes. Then check in polynomial time whether or not the graph language of the remaining graph grammar is empty.

Similar improvements are obtained for connectivity. If a neighborhood-preserving graph grammar contains only productions whose right-hand sides are connected then it generates only connected graphs. This does obviously not hold for grammars which are not neighborhood-preserving.

It is known from [10] that every boundary eNCE graph grammar can be transformed into an equivalent neighborhood-preserving boundary eNCE graph grammar. In an earlier version of this paper we have shown that every confluent eNCE graph grammar can be transformed into an equivalent neighborhood-preserving confluent eNCE graph grammar [19] (see also Theorems 1.3.33 and 1.3.21 in [11]).

In this paper, we give a characterization of the class of all neighborhood-preserving eNCE graph languages: We first introduce a general normal form for eNCE graph grammars, called *embedding normal form*. Intuitively, an eNCE graph grammar is in embedding normal form if all terminal graphs derivable from the right-hand sides of the productions with the same left-hand side have the "same" "embedding behavior". Then we introduce order-independent eNCE graph grammars with respect to successor edges, S-OI-eNCE, which are a proper generalization of the OI-eNCE graph grammars. An eNCE graph grammar is S-OI-eNCE if the order in which the productions are applied is irrelevant for the generation of successor edges. Here the derived graphs need not be equal or isomorphic. In fact it is not difficult to show that the class of confluent eNCE graph languages (which is the class of OI-eNCE graph languages) is properly included in the class of S-OI-eNCE graph languages. Finally, using the embedding normal form we show that every S-OI-eNCE graph grammar can be transformed into an equivalent neighborhood-preserving eNCE graph grammar, i.e., S-OI-eNCE and neighborhood-preserving eNCE graph grammars have the same generating power.

2 Preliminaries

We first recall the definitions of graphs, substitutions, and eNCE graph grammars. For technical reasons, our graphs are ordinary labeled graphs together with an embedding relations. If we talk about graphs then we sometimes mean the underlying ordinary graphs. However, it will be clear from the context whether we mean graphs with an embedding relation or without an embedding relation.

Definition 1 (Graphs). *Let Σ and Γ be two finite alphabets of labels (node and edge labels, respectively). A node and edge labeled graph $G = (V, E, \lambda, C)$ over Σ and Γ consists of*

1. *a finite set of nodes V*
2. *a node labeling $\lambda : V \to \Sigma$*
3. *a finite set of undirected labeled edges $E \subseteq \{\{u, l, v\} \mid u, v \in V, u \neq v, l \in \Gamma\}$*
4. *an embedding relation $C \subseteq \Sigma \times \Gamma \times \Gamma \times V$.*

The set of all graphs over Σ and Γ is denoted by $GRE_{\Sigma,\Gamma}$.

The components of G are also denoted as V_G, E_G, λ_G, and C_G, respectively. A node or an edge labeled by $a \in \Sigma$ or $l \in \Gamma$ is called an *a-node* or *l-edge*, respectively. We deal with undirected graphs without loops and without identically labeled multiple edges.

Definition 2 (Substitutions). *Let G and R be two (node-disjoint) graphs over Σ and Γ, and let u be a node of G. The graph $G[u/R] \in GRE_{\Sigma,\Gamma}$ is defined as follows.*

1. *Construct the union of G and R*
2. *Remove node u, the incident edges of u, and all embedding relation tuples which contain u or a node from R*
3. *Add an l'-edge between a node v from G and a node w from R if and only if $\{v, l, u\} \in E_G$ and $(\lambda_G(v), l, l', w) \in C_R$ for some edge label l*
4. *For every node w from R extend the embedding relation with all tuples (a, l, l'', w) such that $(a, l, l', u) \in C_G$ and $(a, l', l'', w) \in C_R$ for some edge label l'.*

The substitution is associative, see [4, Lemma 3.2]. That is, for all graphs G, R, R', all nodes u from G, and all nodes v from R, $G[u/R][v/R'] = G[u/R[v/R']]$.

We assume that all graphs involved in substitutions are node-disjoint. This assumption allows us to compare composed graphs in a set-theoretical sense.

The nodes in $G[u/R]$ which are from R are called *direct successor nodes* of u. If $e = \{v, l, u\}$ is an edge in G and $(\lambda_G(v), l, l', w)$ is a tuple in the embedding relation of R, then there will be an edge $e' = \{v, l', w\}$ in $G[u/R]$. We say that edge e' is a *direct successor edge* of edge e. The transitive closure of the direct successor relation is simply called *successor relation*.

Definition 3 (eNCE graph grammars). *An* eNCE *(edge label neighborhood controlled embedding) graph grammar is a system* $\mathcal{G} = (\Sigma, \Delta, \Gamma, \Omega, S, P)$ *where*

1. Σ *is the alphabet of node labels*
2. $\Delta \subseteq \Sigma$ *is the alphabet of* terminal *node labels*
3. Γ *is the alphabet of edge labels*
4. $\Omega \subseteq \Gamma$ *is the alphabet of* final *edge labels*
5. S *is the* start graph *consisting of a single node with a nonterminal label and empty embedding relation*
6. P *is a finite set of* productions. *Every production is of the form* $p = A \to R$ *where* $A \in \Sigma - \Delta$ *and* $R \in GRE_{\Sigma, \Gamma}$. A *is the left-hand side and* R *the right-hand side of* p.

Elements of $\Sigma - \Delta$ are called *nonterminal node labels* and elements of $\Gamma - \Omega$ *nonfinal edge labels*. A is the *left-hand side* and R the *right-hand side* of production $A \to R$. If a node is labeled by a terminal label, it is called *terminal node*; otherwise it is called *nonterminal node*. Similarly an edge labeled by a final label is called a *final edge* otherwise it is called a *nonfinal edge*. A graph is called *terminal* if all its nodes are terminal; it is called *final* if additionally all its edges are final.

Definition 4 (Derivations and languages). *Let* $\mathcal{G} = (\Sigma, \Delta, \Gamma, \Omega, S, P)$ *be an eNCE graph grammar,* G *be a graph over* Σ *and* Γ, u *be a node of* G, *and* $p = \lambda_G(u) \to R$ *be a production.*
$G \Longrightarrow_{u,p} H$ *is a derivation step if* $H = G[u/R]$. *A derivation is a sequence of derivation steps* $G \Rightarrow_{u_1, p_1} G_1 \Rightarrow_{u_2, p_2} G_2 \cdots \Rightarrow_{u_n, p_n} G_n$, *which is also denoted by* $G \Rightarrow^* G_n$.
A sentential form of $\mathcal{G}$ *is a graph derivable from the start graph* S. *The* language $L(\mathcal{G})$ *of an eNCE graph grammar* $\mathcal{G}$ *is*

$$L(\mathcal{G}) = \{H \in GRE_{\Delta, \Omega} \mid S \Rightarrow^* H\}.$$

Two graph grammars are called equivalent *if they define the same language.*

Observe that the embedding relations of all graphs of $L(\mathcal{G})$ are empty since the embedding relation of the start graph S is empty. Furthermore, we can assume that every production of P can be used in a derivation of a terminal graph. Useless productions can easily be removed by standard methods.

3   Embedding-Uniform eNCE Graph Grammars

Now we introduce the embedding normal form for eNCE graph grammars. First we need the notion of the *environment set* of a graph. The environment set of a graph reflects in some way its embedding behavior. Intuitively, two graphs with the same environment set have the "same" embedding behavior.

Definition 5 (Environment set). *Let* $G = (V, E, \lambda, C)$ *be a graph over* Σ *and* Γ. *The* environment set *of* G *is defined by*

$$\mathcal{E}(G) = \{(a, l, l', \lambda(u)) \mid (a, l, l', u) \in C\}.$$

The next lemma shows that the order in which nonterminal nodes of a graph G are substituted has no influence on the embedding relation and therefore on the environment set of the resulting graph. The environment set of the resulting graph only depends on the environment set of G and on the environment sets of the substituted graphs.

Lemma 1. *Let* G *be a graph over* Σ *and* Γ, *and let* $u_1, \ldots, u_k$ *be* k *pairwise distinct nodes of* G. *Let* π *be any permutation of the indices* $1, \ldots, k$. *Let* $H_1, \ldots, H_k$ *and* $J_1, \ldots, J_k$ *be graphs over* Σ *and* Γ *such that* $\mathcal{E}(H_i) = \mathcal{E}(J_i)$ *for* $i = 1, \ldots, k$. *Then*

$$\mathcal{E}(G[u_1/H_1] \cdots [u_k/H_k]) = \mathcal{E}(G[u_{\pi(1)}/J_{\pi(1)}] \cdots [u_{\pi(k)}/J_{\pi(k)}]).$$

Proof. Let $H = G[u_1/H_1] \cdots [u_k/H_k]$. It is sufficient to show that

$$\mathcal{E}(H) = \{(a, l, l', \lambda_G(v)) \mid (a, l, l', v) \in C_G \text{ and } \forall i: v \neq u_i\}$$
$$\cup \{(a, l, l', b) \mid \exists l'', i: (a, l, l'', u_i) \in C_G \text{ and } (a, l'', l', b) \in \mathcal{E}(H_i)\}$$

Let $(a, l, l', b) \in \mathcal{E}(H)$. By definition there is a b-node v from H with $(a, l, l', v) \in C_H$. If v is from G then trivially $(a, l, l', v) \in C_G$. Otherwise v is from some H_i and by the definition of substitution there exists a label l'' such that $(a, l, l'', u_i) \in C_G$ and $(a, l'', l', v) \in C_{H_i}$. That means there exists a label l'' such that $(a, l, l'', u_i) \in C_G$ and $(a, l'', l', b) \in \mathcal{E}(H_i)$. The reverse direction can be shown analogously.

Next we define embedding-uniform eNCE graph grammars.

Definition 6 (Embedding-uniform eNCE graph grammars). *Let* $\mathcal{G} = (\Sigma, \Delta, \Gamma, \Omega, P, S)$ *be an eNCE graph grammar and let* $RHS(P, A) = \{R \mid A \to R \in P\}$ *be the set of all right-hand sides of the productions with left-hand side* A.

$\mathcal{G}$ *is* embedding-uniform *if for every nonterminal node label* A *all terminal graphs derivable from the graphs of* $RHS(P, A)$ *have the same environment set, which is then denoted by* $\mathcal{E}(A)$.

In the previous definition of the embedding-uniformity the considered graphs derivable from $RHS(P, A)$ need only be terminal but not necessarily final.

By the embedding-uniformity, by the associativity of the substitution, and by the proof of Lemma 1 we get the following lemma.

Lemma 2. *Let* $\mathcal{G} = (\Sigma, \Delta, \Gamma, \Omega, S, P)$ *be an embedding-uniform eNCE graph grammar and let* $p = A \to R \in P$ *be a production.*
Then

$$\mathcal{E}(A) = \{(a, l, l', \lambda_R(v)) \mid (a, l, l', v) \in C_R, \ \lambda_R(v) \in \Delta\}$$
$$\cup \{(a, l, l', b) \mid \exists l'', u: (a, l, l'', u) \in C_R \text{ and } (a, l'', l', b) \in \mathcal{E}(\lambda_R(u)),$$
$$\text{where } \lambda_R(u) \in \Sigma - \Delta\}$$

Notice that the environment set $\mathcal{E}(A)$ of a nonterminal node label A is defined via every production with left-hand side A.

Theorem 1. *Every eNCE graph grammar $\mathcal{G} = (\Sigma, \Delta, \Gamma, \Omega, S, P)$ can be transformed into an embedding-uniform eNCE graph grammar $\mathcal{G}' = (\Sigma', \Delta, \Gamma, \Omega, S', P')$ that derives the same terminal graphs.*

Proof. Let \$ be the nonterminal label of the start graph S of $\mathcal{G}$. The set of nonterminal node labels $\Sigma' - \Delta$ consists of a new nonterminal label \$' and of all pairs (A, X) where $A \in \Sigma - \Delta$ and $X \subseteq \Sigma \times \Gamma \times \Gamma \times \Delta$. Intuitively, X will be the environment set of the terminal graphs derived from the right-hand sides of the productions with left-hand side (A, X).

The nonterminal node of the start graph S' of $\mathcal{G}'$ is labeled by the new nonterminal \$'.

The productions of $\mathcal{G}'$ are constructed as follows. Let $A \to R \in P$ be a production of $\mathcal{G}$. Then P' contains all productions $(A, X) \to R'$ such that R' is a copy of R in which every nonterminal node u_i is relabeled by a new nonterminal label $(\lambda_R(u_i), X_i)$ for some $X_i \subseteq \Sigma \times \Gamma \times \Gamma \times \Delta$, and $X = \mathcal{E}(R[u_1/H_1] \cdots [u_k/H_k])$ where H_i is some graph over Δ and Γ with $\mathcal{E}(H_i) = X_i$. Notice that $\mathcal{E}(R[u_1/H_1] \cdots [u_k/H_k])$ is uniquely defined by R and all $\mathcal{E}(H_i)$, see Lemma 1. The embedding relation $C_{R'}$ consists of all tuples $(a, l, l', v) \in C_R$ where $a \in \Delta$, and of all tuples $((B, Y), l, l', v)$ where $B \in \Sigma - \Delta$, $(B, l, l', v) \in C_R$, and $Y \subseteq \Sigma \times \Gamma \times \Gamma \times \Delta$. Thus the embedding relation of R' treats two nonterminal labels $(B, Y), (B', Y')$ in the same way if $B = B'$.

In addition for every $X \subseteq \Sigma \times \Gamma \times \Gamma \times \Delta$ grammar $\mathcal{G}'$ contains the production \$' $\to gr(\$, X)$ where $gr(\$, X)$ is the graph with a single node labeled by $(\$, X)$ and the empty embedding relation (\$ is the label of the single node of S).

By an induction on the length of the derivation it can easily be shown that the new grammar $\mathcal{G}'$ is embedding-uniform and derives the same terminal graphs as the original grammar $\mathcal{G}$. In fact, the only difference between $\mathcal{G}$ and $\mathcal{G}'$ is that every production of P is subdivided into several productions for P' such that all terminal graphs derivable from the right-hand side have the same environment set.

Next we will show that the embedding-uniformity of eNCE graph grammars can be used to characterize some important subclasses of eNCE graph languages. This basically depends on the properties shown in the following lemmas.

Lemma 3. *Let $\mathcal{G} = (\Sigma, \Delta, \Gamma, \Omega, S, P)$ be an embedding-uniform eNCE graph grammar and let H be a sentential form of $\mathcal{G}$ containing an edge $e = \{v, l, u\}$ where u is a nonterminal node. Let H' be a graph derivable from H such that*

1. *H' still contains node v and*
2. *all successor nodes of u in H' are terminal.*

Then for every $l' \in \Gamma$ and $a \in \Delta$ it holds: $(\lambda_H(v), l, l', a) \in \mathcal{E}(\lambda_H(u))$ if and only if H' has a successor edge $\{v, l', u'\}$ of e where u' is a successor node of u labeled by a.

Proof. By induction on the length of a derivation $H \Rightarrow_{u_1,p_1} H_1 \Rightarrow_{u_2,p_2}$ $H_2 \cdots \Rightarrow_{u_n,p_n} H_n = H'$. Let R_i, $1 \leq i \leq n$, be the right-hand side of p_i.

$n = 1$: Then $u_1 = u$ and the right-hand side R_1 has only terminal nodes. Since $\mathcal{G}$ is embedding-uniform Lemma 2 implies that $(\lambda_H(v), l, l', a) \in \mathcal{E}(\lambda_H(u))$ if and only if $(\lambda_H(v), l, l', u') \in C_{R_1}$ for some terminal a-node u' of R_1. That is $(\lambda_H(v), l, l', a) \in \mathcal{E}(\lambda_H(u))$ if and only if H' contains a successor edge $\{v, l', u'\}$ of e for some successor node u' of u labeled by a.

$(n-1) \rightarrow n$: If $u_1 \neq u$ then edge e is still in H_1 and the claim follows by the induction hypothesis.

If $u_1 = u$ then let $w_1, \ldots, w_k$ be the nonterminal nodes of R_1 labeled by $B_1, \ldots, B_k$, respectively. Since $\mathcal{G}$ is embedding-uniform Lemma 2 implies that $(\lambda_H(v), l, l', a) \in \mathcal{E}(\lambda_H(u))$ if and only if (i) $(\lambda_H(v), l, l', u') \in C_{R_1}$ for some terminal a-node u' of R_1 or (ii) $(\lambda_H(v), l, l'', w_j) \in C_{R_1}$ and $(\lambda_H(v), l'', l', a) \in \mathcal{E}(B_j)$ for some $l'' \in \Gamma$ and $1 \leq j \leq k$.

Case (i) holds if and only if H' contains a successor edge $\{v, l', u'\}$ where u' is a terminal node from R_1 labeled by a.

By the induction hypothesis and by the associativity of substitutions, Case (ii) holds if and only if H' has an edge $\{v, l', u'\}$ where u' is a successor a-node of some w_j.

Thus $(\lambda_H(v), l, l', a) \in \mathcal{E}(\lambda_H(u))$ if and only if H' contains a successor edge $\{v, l', u'\}$ of e for some successor node u' of u labeled by a.

Lemma 4. *Let* $\mathcal{G} = (\Sigma, \Delta, \Gamma, \Omega, S, P)$ *be an embedding-uniform eNCE graph grammar and let H be a sentential form of $\mathcal{G}$ with an edge $e = \{v, l, u\}$ where v is a terminal and u a nonterminal node. Let H' be a terminal graph derivable from H.*
Then for every $l' \in \Gamma$ and $a \in \Delta$ it holds: $(\lambda_H(v), l, l', a) \in \mathcal{E}(\lambda_H(u))$ if and only if H' has a successor edge $\{v, l', u'\}$ of e where u' is a successor node of u labeled by a.

Proof. This lemma is the sub-case of Lemma 3 in which v and H' are terminal.

Lemma 5. *Let* $\mathcal{G} = (\Sigma, \Delta, \Gamma, \Omega, S, P)$ *be an embedding-uniform eNCE graph grammar, let H be a sentential form of $\mathcal{G}$ containing an edge $e = \{v, l, u\}$ where u and v are nonterminal. Let H' be a terminal graph derivable from H by a derivation in which u and all its nonterminal successor nodes are substituted before v is substituted.*
Then for every $l' \in \Gamma$ and every $a, b \in \Delta$ it holds: $(\lambda_H(v), l, l'', a) \in \mathcal{E}(\lambda_H(u))$ and $(a, l'', l', b) \in \mathcal{E}(\lambda_H(v))$ for some $l'' \in \Gamma$ if and only if H' has a successor edge $\{v', l', u'\}$ of e where u' is a successor node of u labeled by a and v' is a successor node of v labeled by b.

Proof. Let $H \Rightarrow^* \widehat{H} \Rightarrow^* H'$ be a derivation of H' such that $\widehat{H}$ contains node v but no nonterminal successor node of u. Lemma 3 applied to the derivation $H \Rightarrow^* \widehat{H}$ and Lemma 4 applied to the derivation $\widehat{H} \Rightarrow^* H'$ yields the claim.

4   Neighborhood-Preserving Grammars

A substitution step $G[u/H]$ is *neighborhood-preserving* if every node v connected with u in G is connected with some node w from H in $G[u/H]$. An eNCE graph grammar is neighborhood-preserving if every substitution step in the derivation of a sentential form of G is neighborhood-preserving.

Neighborhood-preserving is sometimes defined in a stronger sense, see [11], which we call *edge-preserving*. A substitution step $G[u/H]$ is edge-preserving if every edge $\{v, l, u\}$ between the substituted node u and some node v from G has a direct successor edge in $G[u/H]$, i.e., if $(\lambda_G(u), l, l', w) \in C_H$ for some $l' \in \Gamma$ and $w \in V_H$.

Obviously, every edge-preserving graph grammar is neighborhood-preserving but not vice versa. However, we can show that every neighborhood-preserving eNCE graph grammar can be transformed into an equivalent edge-preserving one.

Theorem 2. *Every neighborhood-preserving eNCE graph grammar* $\mathcal{G} = (\Sigma, \Delta, \Gamma, \Omega, S, P)$ *can be transformed into an equivalent edge-preserving eNCE graph grammar* $\mathcal{G}' = (\Sigma, \Delta, \Gamma', \Omega', S, P')$.

Proof. In every right-hand side of $\mathcal{G}$ we code multiple edges by one edge such that in every sentential form of $\mathcal{G}'$ every nonterminal node is adjacent to every of its neighbors by exactly one edge.

The new set of edge labels Γ' is the old set Γ extended by the power set (set of all subsets) of Γ.

The new production set P' is defined as follows. For every production $A \rightarrow (V, E, \lambda, C)$ of P there is a production $A \rightarrow (V, E', \lambda, C')$ in P' where

$$E' = \{\{u_1, l, u_2\} \in E \mid u_1 \text{ and } u_2 \text{ terminal}\}$$
$$\cup \{\{u_1, M, u_2\} \mid M = \{l \mid \{u_1, l, u_2\} \in E\}, M \neq \emptyset, \text{ or } u_2 \text{ nonterminal}\}$$
$$C' = \{(a, M, l', v) \mid M \subseteq \Gamma, (a, l, l', v) \in C \text{ for some } l \in M, a \text{ and } v \text{ terminal}\}$$
$$\cup \{(a, M, M', v) \mid M \subseteq \Gamma, M' = \{l' \mid l \in M, \ (a, l, l', v) \in C\}, M' \neq \emptyset,$$
$$a \text{ or } v \text{ nonterminal}\}.$$

The resulting grammar $\mathcal{G}'$ is equivalent to $\mathcal{G}$ and still neighborhood-preserving. Since in every sentential form of $\mathcal{G}'$ every nonterminal node is connected to each of its neighbors by exactly one edge, $\mathcal{G}'$ is obviously edge-preserving.

Corollary 1. *Neighborhood-preserving eNCE and edge-preserving eNCE graph grammars coincide in their generating power.*

Next we define eNCE graph grammars which are order-independent with respect to successor edges.

Definition 7 (S-OI-eNCE graph grammars). *Let* $\mathcal{G}$ *be an eNCE graph grammar.*
$\mathcal{G}$ *is order-independent with respect to successor edges (S-OI-eNCE), see Figure 2, if for every sentential form G of $\mathcal{G}$, every edge e of G, and every derivation*

$G \Rightarrow_{u_1,p_1} G_1 \Rightarrow_{u_2,p_2} G_2 \cdots \Rightarrow_{u_n,p_n} G_n$ *of a terminal graph* G_n *that has a suc-cessor edge of* e, *it follows that every derivation* $G \Rightarrow_{u_{\pi(1)},p_{\pi(1)}} H_1 \Rightarrow_{u_{\pi(2)},p_{\pi(2)}}$ $H_2 \cdots \Rightarrow_{u_{\pi(n)},p_{\pi(n)}} H_n$ *defines a terminal graph* H_n *with a successor edge of* e *where* π *is any permutation of* $\{1,\ldots,n\}$.

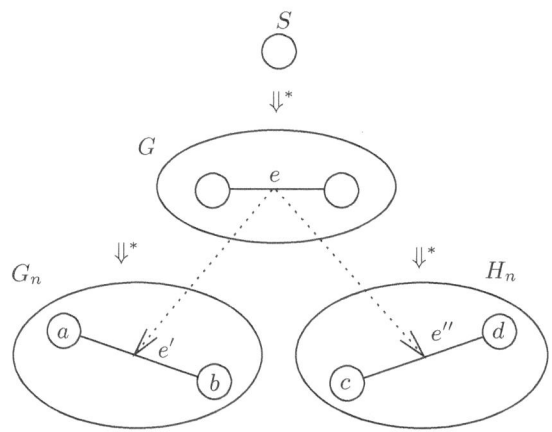

Figure2. Edge e has a successor edge in both graphs G_n and H_n.

We prove that for every S-OI-eNCE graph grammar there is an edge-pre-serving eNCE graph grammar and therefore a neighborhood-preserving eNCE graph grammar that derives the same terminal graphs.

Theorem 3. *Every S-OI-eNCE graph grammar* $\mathcal{G}$ *can be transformed into an edge-preserving eNCE graph grammar* $\mathcal{G}'$ *that derives the same terminal graphs.*

Proof. By Theorem 1 we can assume that $\mathcal{G} = (\Sigma, \Delta, \Gamma, \Omega, S, P)$ is embedding-uniform. Then $\mathcal{G}' = (\Sigma, \Delta, \Gamma, \Omega, S, P')$ where P' is defined as follows.
Let $A \to R \in P$ be a production and v and u be two nodes of R adjacent by some edge $e = \{v, l, u\}$.
We remove e from R if

1. v is terminal, u is nonterminal, and there are no $l' \in \Gamma$ and $a \in \Delta$ such that $(\lambda_R(v), l, l', a) \in \mathcal{E}(\lambda_R(u))$ or
2. v is nonterminal, u is terminal, and there are no $l' \in \Gamma$ and $a \in \Delta$ such that $(\lambda_R(u), l, l', a) \in \mathcal{E}(\lambda_R(v))$ or
3. v and u are nonterminal and there are no $l', l'' \in \Gamma$ and $a, b \in \Delta$ such that $(\lambda_R(v), l, l', a) \in \mathcal{E}(\lambda_R(u))$ and $(a, l', l'', b) \in \mathcal{E}(\lambda_R(v))$.

By the order-independency of $\mathcal{G}$ and by Lemmas 4 and 5 it follows that none of the removed edges has a successor edge in any terminal graph derivable from R.

Thus, the removal of the edges does not change anything, neither the language, nor the environment sets, nor the order-independency of the grammar.

Similarly, we remove a tuple (a, l, l', w) from C_R if

1. w is terminal, a is nonterminal, and there are no $l'' \in \Gamma$ and $b \in \Delta$ such that $(\lambda_R(w), l', l'', b) \in \mathcal{E}(a)$ or
2. w is nonterminal, a is terminal, and there are no $l'' \in \Gamma$ and $b \in \Delta$ such that $(a, l', l'', b) \in \mathcal{E}(\lambda_R(w))$ or
3. w and a are nonterminal and there are no $l'', l''' \in \Gamma$ and $b, c \in \Delta$ such that $(a, l', l'', b) \in \mathcal{E}(\lambda_R(w))$ and $(b, l'', l''', c) \in \mathcal{E}(a)$.

By the same reason as above, the removal of the tuples does not change anything, neither the language, nor the order-independency, nor the environment sets of the nonterminal symbols of the grammar.

It remains to show that $\mathcal{G}'$ is edge-preserving. Let $S = G_0 \Rightarrow_{u_1, p_1} G_1 \Rightarrow_{u_2, p_2} G_2 \cdots \Rightarrow_{u_n, p_n} G_n$ be a derivation of a terminal graph G_n such that the substitution step $G_k[u_{k+1}/R_{k+1}]$, with $0 \le k < n$, is not edge-preserving. Let $p_i = A_i \to R_i$, $1 \le i \le n$. Let $e = \{v, l', u_{k+1}\}$ be an edge of G_k that has no successor edge in $G_{k+1} = G_k[u_{k+1}/R_{k+1}]$ and therefore no successor edge in G_n.

Consider now the graph G_j, with $0 \le j < k$, such that $G_{j+1}, \ldots, G_k$ contain edge e but G_j does not contain e. That is either node v, or node u_{k+1}, or both nodes are direct successor nodes of u_{j+1}. We have to consider the following two cases: (i) Both nodes v and u_{k+1} are direct successors nodes of u_{j+1} or (ii) either node v or node u_{k+1} is a direct successor node of u_{j+1}, see Figure 3.

In the first case edge e is from the right-hand side R_{j+1} of p_{j+1}, see Figure 3 (i). This is a contradiction because of the order-independency of $\mathcal{G}'$ and Lemmas 4 and 5, one of the conditions 1.-3. above (concerning edges of the right-hand sides) must hold for e and therefore e is removed from R_{j+1}.

In the second case, we assume, without loss of generality, that u_{k+1} is a direct successor node of u_{j+1}, see Figure 3 (ii). Then v is from G_j and u_{k+1} is from the right-hand side R_{j+1} of p_{j+1}. Additionally, edge e is a successor edge of some edge $\{v, l, u_{j+1}\}$ of G_j and $(\lambda_{G_j}(v), l, l', u_{k+1}) \in C_{R_{j+1}}$ for some $l \in \Gamma$. This is also a contradiction because of the order-independency of $\mathcal{G}'$ and by Lemma 4 and 5, one of the conditions 1.-3. above (concerning embedding relations) must hold for $(\lambda_{G_j}(v), l, l', u_{k+1})$ and therefore $(\lambda_{G_j}(v), l, l', u_{k+1})$ is removed from the embedding relation of R_{j+1}. Thus $\mathcal{G}'$ is edge-preserving.

The next corollary is obtained from the fact that the transformation in the proof of the theorem above preserves the confluence property.

Corollary 2. *Every confluent eNCE graph grammar can be transformed into an equivalent confluent eNCE graph grammar which is edge-preserving.*

Furthermore, since every edge-preserving eNCE graph grammar is S-OI-eNCE we get the following theorem.

Theorem 4. *The class of neighborhood-preserving eNCE graph languages is exactly the class of S-OI-eNCE graph languages.*

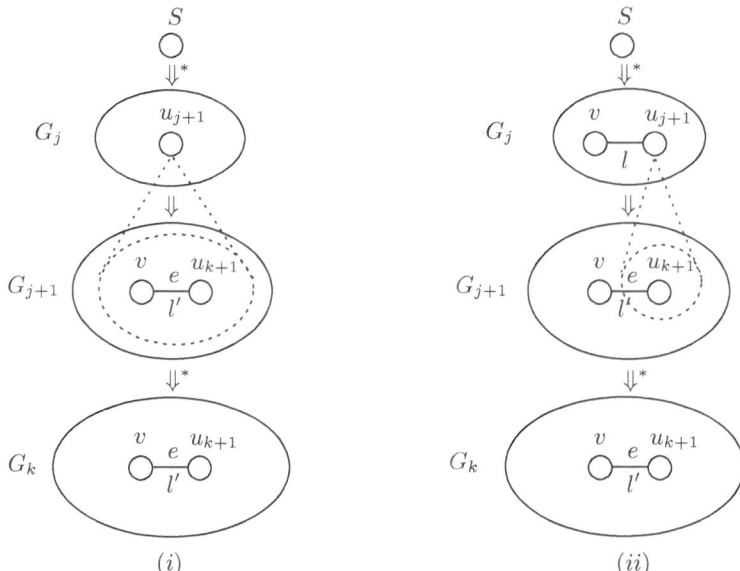

Figure3. (i) Nodes v and u_{k+1} are direct successor nodes of u_{j+1} (ii) Node u_{k+1} is a direct successor node of u_{j+1}.

In a similar way it is also possible to characterize non-blocking eNCE graph languages by transforming eNCE graph grammars that are order-independent with respect to the generation of final graphs. Here also the graphs derived by different derivation orders need not be equal or isomorphic but either all final or all nonfinal.

5 Conclusions

The main result shown in this paper is the proof that neighborhood-preserving and S-OI-eNCE graph grammars coincide in their generative power. The proof is based on the fact that every neighborhood-preserving eNCE grammar is already S-OI-eNCE and every S-OI-eNCE grammar can be transformed into an equivalent neighborhood-preserving one. It follows in particular that every confluent eNCE grammar can be transformed into an equivalent confluent eNCE one which is neighborhood-preserving.

Actually, the class of the S-OI-eNCE graph languages properly includes the confluent eNCE graph languages. To see this consider the eNCE graph grammar of Example 13 in [9] generating the set of all star graphs with 2^n leaves for $n \geq 0$. An easy inspection shows that the grammar is edge-preserving and therefore S-OI-eNCE. Furthermore, it is well known that the language of all star graphs with 2^n leaves can not be defined by some confluent grammar.

All the results in this paper can also be extended to directed graphs. However this would not be any contribution to the basic ideas used in the proofs and therefore we have restricted ourselves to undirected graphs.

The decidability of order-independency with respect to successor edges as well as the proper inclusion of the class of all S-OI-eNCE graph languages in the class of all eNCE graph languages will be discussed in the full version of this paper.

Acknowledgments

We are grateful to the referees for their fruitful suggestions and comments of an earlier version of this paper.

References

1. F.-J. Brandenburg. On partially ordered graph grammars. In *Proc. of Graph Grammars and their Application to Computer Science*, volume 291 of *Lect. Notes in Comput. Sci.*, pages 99–111. Springer-Verlag, New York/Berlin, 1987.
2. F.-J. Brandenburg. The equivalence of boundary and confluent graph grammars on graph languages of bounded degree. In *Rewriting Techniques and Applications*, volume 488 of *Lect. Notes in Comput. Sci.*, pages 312–322. Springer-Verlag, New York/Berlin, 1991.
3. V. Claus, H. Ehrig, and G. Rozenberg, editors. *Proc. of Graph Grammars and their Application to Computer Science*, volume 73 of *Lect. Notes in Comput. Sci.* Springer-Verlag, New York/Berlin, 1979. 45
4. B. Courcelle, J. Engelfriet, and G. Rozenberg. Handle-rewriting hypergraph grammars. *J. Comput. System Sci.*, 46:218–270, 1993. 48
5. J. Cuny, H. Ehrig, G. Engels, and G. Rozenberg, editors. *Proc. of Graph Grammars and their Application to Computer Science*, volume 1073 of *Lect. Notes in Comput. Sci.* Springer-Verlag, New York/Berlin, 1994. 45
6. H. Ehrig, H.J. Kreowski, and G. Rozenberg, editors. *Proc. of Graph Grammars and their Application to Computer Science*, volume 532 of *Lect. Notes in Comput. Sci.* Springer-Verlag, New York/Berlin, 1991. 45
7. H. Ehrig, M. Nagl, A. Rosenfeld, and G. Rozenberg, editors. *Proc. of Graph Grammars and their Application to Computer Science*, volume 291 of *Lect. Notes in Comput. Sci.* Springer-Verlag, New York/Berlin, 1987. 45
8. H. Ehrig, M. Nagl, and G. Rozenberg, editors. *Proc. of Graph Grammars and their Application to Computer Science*, volume 153 of *Lect. Notes in Comput. Sci.* Springer-Verlag, New York/Berlin, 1983. 45
9. J. Engelfriet, G. Leih, and E. Welzl. Boundary graph grammars with dynamic edge relabeling. *J. Comput. System Sci.*, 40:307–345, 1990. 56
10. J. Engelfriet and G. Rozenberg. A comparison of boundary graph grammars and context-free hypergraph grammars. *Inform. Comput.*, 84:163–206, 1990. 47
11. J. Engelfriet and G. Rozenberg. Node replacement graph grammars. In G. Rozenberg, editor, *Handbook of Graph Grammars and Computing by Graph Transformations*, volume 1, chapter 1, pages 1–94. World Scientific Publishing Co. Pte. Ltd., Singapore, 1997. 45, 47, 53

12. A. Habel. *Hyperedge Replacement: Grammars and Languages*, volume 643 of *Lect. Notes in Comput. Sci.* Springer-Verlag, New York/Berlin, 1992.
13. M. Himsolt. Graphed: An interactive tool for developing graph grammars. In *Proc. of Graph Grammars and their Application to Computer Science*, volume 532 of *Lect. Notes in Comput. Sci.*, pages 61–65, 1991. 47
14. D. Janssens and G. Rozenberg. Restrictions, extensions, and variations of NLC grammars. *Inform. Sci.*, 20:217–244, 1980. 46
15. D. Janssens and G. Rozenberg. Decision problems for node label controlled graph grammars. *J. Comput. System Sci.*, 22:144–177, 1981. 47
16. J. Jeffs. Order independent NCE grammars recognized in polynomial time. *Inform. Process. Lett.*, 39:161–164, 1991. 45
17. C. Lautemann. The complexity of graph languages generated by hyperedge replacement. *Acta Informat.*, 27:399–421, 1990. 46
18. G. Rozenberg and E. Welzl. Boundary NLC graph grammars - basic definitions, normal forms and complexity. *Inform. Comput.*, 69:136–167, 1986. 46
19. K. Skodinis and E. Wanke. Neighborhood-preserving confluent node replacement. Manuscript, 1995. 47
20. *Proceedings of IEEE Symposium on Visual Languages*. IEEE, Isle of Capri, Italy, 1997. 46

Complexity Issues in Switching of Graphs

Andrzej Ehrenfeucht[2], Jurriaan Hage[1], Tero Harju[3], and
Grzegorz Rozenberg[1,2]

[1] Leiden Institute of Advanced Computer Science
P.O. Box 9512, 2300 RA Leiden, The Netherlands
[2] Dept. of Computer Science, University of Colorado at Boulder
Boulder, Co 80309, U.S.A.
[3] Dept. of Mathematics, University of Turku
FIN-20014 Turku, Finland

Abstract. In the context of graph transformations we look at the operation of switching, which can be viewed as an elegant method for realizing global transformations of graphs through local transformations of the vertices. We compare the complexity of a number of problems on graphs with the complexity of these problems extended to the set of switches of a graph. Within this framework, we prove a modification of Yannakakis' result and use it to show NP-completeness for the embedding problem. Finally we prove NP-completeness for the 3-colourability problem.

1 Introduction

The operation of switching is an elegant example of a graph transformation, where the global transformation of a graph is achieved by applying local transformations to the vertices. The elegance stems from the fact that the local transformations are group actions, and hence basic techniques of group theory can be applied in developing the theory of switching.

The use of switching presented in this paper is motivated by operations for formalizing specific types of networks of processors, as introduced in [4] by Ehrenfeucht and Rozenberg (see also [3]).

For a finite undirected graph $G = (V, E)$ and a subset $\sigma \subseteq V$, the *switch* of G by σ is defined as the graph $G^\sigma = (V, E')$, which is obtained from G by removing all edges between σ and its complement $V - \sigma$ and adding as edges all nonedges between σ and $V - \sigma$. The switching class $[G]$ determined (generated) by G consists of all switches G^σ for subsets $\sigma \subseteq V$.

A switching class is an equivalence class of graphs under switching, see the survey papers by Seidel [9] and Seidel and Taylor [10]. Generalizations of this approach – where the graphs are labeled with elements of a group other than $\mathbf{Z}_2$ – can be found in Gross and Tucker [6], Ehrenfeucht and Rozenberg [4], and Zaslavsky [12]. Although, in this paper, we will restrict ourselves to graphs, many of the results on complexity hold for these generalizations.

A property $\mathcal{P}$ of graphs can be transformed into an existential property of switching classes as follows:

H. Ehrig et al. (Eds.): Graph Transformation, LNCS 1764, pp. 59–70, 2000.

$\mathcal{P}_\exists(G)$ if and only if there is a graph $H \in [G]$ such that $\mathcal{P}(H)$.

We will also refer to $\mathcal{P}_\exists$ as "the problem $\mathcal{P}$ for switching classes".

Let G be a graph on n vertices. It is easy to show that there are 2^{n-1} graphs in $[G]$, and so checking whether there exists a graph $H \in [G]$ satisfying a given property $\mathcal{P}$ requires exponential time, if each graph is to be checked separately.

However, although the hamiltonian cycle problem for graphs (Problem GT37 of Garey and Johnson [5]) is NP-complete, the hamiltonian cycle problem for switching classes can be solved in time $O(n^2)$, since one needs only check that a given graph is not complete bipartite of odd order (see [8] or [2]).

After the preliminaries we list a few problems for switching classes that are polynomial for switching classes, yet are NP-complete for ordinary graphs. We continue with giving a procedure for turning algorithms for graph properties into algorithms for the corresponding properties for switching classes.

We generalize to switching classes a result of Yannakakis [11] on graphs (GT21), which is then used to prove that the independence problem (GT20) is NP-complete for switching classes. This problem can be polynomially reduced to the embedding problem (given two graphs G and H, does there exist a graph in $[G]$ in which H can be embedded, GT48). Hence, the latter problem is also NP-complete for switching classes. It also turns out that deciding whether a switching class contains a 3-colourable graph is NP-complete (a special case of GT4). The proof of this result nicely illustrates some of the techniques that can be used to tackle such problems.

2 Preliminaries

For a (finite) set V, let $|V|$ be the cardinality of V. We shall often identify a subset $A \subseteq V$ with its characteristic function $A : V \to \mathbf{Z}_2$, where $\mathbf{Z}_2 = \{0,1\}$ is the cyclic group of order two. We use the convention that for $x \in V$, $A(x) = 1$ if and only if $x \in A$. The restriction of a function $f : V \to W$ to a subset $A \subseteq V$ is denoted by $f|_A$.

The set $E(V) = \{\{x,y\} \mid x,y \in V,\ x \neq y\}$ denotes the set of all unordered pairs of distinct elements of V. We write xy or yx for the undirected edge $\{x,y\}$. The graphs of this paper will be finite, undirected and simple, i.e., they contain no loops or multiple edges. We use $E(G)$ and $V(G)$ to denote the set of edges E and the set of vertices V, respectively, and $|V|$ and $|E|$ are called the *order*, respectively, *size* of G. Analogously to sets, a graph $G = (V,E)$ will be identified with the characteristic function $G : E(V) \to \mathbf{Z}_2$ of its set of edges so that $G(xy) = 1$ for $xy \in E$, and $G(xy) = 0$ for $xy \notin E$. Later we shall use both notations, $G = (V,E)$ and $G : E(V) \to \mathbf{Z}_2$, for graphs.

For a graph $G = (V,E)$ and $X \subseteq V$, let $G|_X$ denote the *subgraph* of G *induced* by X. Hence, $G|_X : E(X) \to \mathbf{Z}_2$. For two graphs G and H on V we define $G+H$ to be the graph such that $(G+H)(xy) = G(xy) + H(xy)$ for all $xy \in E(V)$.

A *selector* for G is a subset $\sigma \subseteq V(G)$, or alternatively a function $\sigma : V(G) \to \mathbf{Z}_2$. A *switch* of a graph G by σ is the graph G^σ such that for

all $xy \in E(V)$,
$$G^\sigma(xy) = \sigma(x) + G(xy) + \sigma(y).$$

Clearly, this definition of switching is equivalent to the one given at the beginning of the introduction. The set $[G] = \{G^\sigma \mid \sigma \subseteq V\}$ is called the *switching class* of G. We reserve lower case σ for selectors (subsets) used in switching.

A selector σ is *constant* on $X \subseteq V$ if $X \subseteq \sigma$, or $X \cap \sigma = \emptyset$. The name arises from the fact that $G|_X = G^\sigma|_X$.

Some graphs we will encounter in the sequel are K_V, the clique on the set of vertices V, and $\overline{K}_V$, the complement of K_V which is the discrete graph on V; the complete bipartite graph on A and $V - A$ is denoted by $K_{A,V-A}$. If the choice of vertices is unimportant we can write K_n, $\overline{K}_n$ and $K_{m,n-m}$ for $n = |V|$ and $m = |A|$.

We now give a few results from the literature that will be used throughout this paper.

Lemma 1. *The switching class $[\overline{K}_V]$ equals the set of all complete bipartite graphs on V.*

Proof. Given any complete bipartite graph $K_{\sigma,V-\sigma}$, we can obtain it from $\overline{K}_V$ by switching with respect to σ. Also, it is clear that for every selector σ, $\overline{K}^\sigma = K_{\sigma,V-\sigma}$. □

From the observation that applying a selector σ to G amounts to computing $G + K_{\sigma,V(G)-\sigma}$ we obtain the following result.

Lemma 2. *It holds that $G \in [H]$ if and only if $G + H \in [\overline{K}_V]$.* □

Lemma 3. *Let $G = (V, E)$ be a graph, $u \in V$ and $A \subseteq V$. If $u \notin A$, then there exists a unique graph $H \in [G]$ such that the neighbours of u in H are the vertices in A.*

Proof. The vertex u has no adjacent vertices in $G_u = G^{N(u)}$, where $N(u)$ is the set of neighbours of u in G. Switching G_u with respect to A connects u to every vertex in A (and no others) yielding H.

To show that H is unique, let H' be such that $N(u) = A$ in H'. Since H and H' are in the same switching class, $H + H'$ is a complete bipartite graph (Lemma 2), say $G_{B,V-B}$. Since u has the same neighbours in both, u is isolated in $G_{B,V-B}$. Hence, $G_{B,V-B}$ is a discrete graph and, consequently, $H = H'$. □

We define the *complement* of a switching class as follows:
$$\overline{[G]} = \{\overline{H} \mid H \in [G]\}.$$

Lemma 4. *For a graph $G = (V, E)$, $\overline{[G]} = [\overline{G}]$.*

Proof. We show that for a graph $G = (V, E)$ and $\sigma \subseteq V$: $\overline{G^\sigma} = \overline{G}^\sigma$. Indeed, let $x, y \in V$. Then $\overline{G^\sigma}(xy) = 1 - (\sigma(x) + G(xy) + \sigma(y)) = \sigma(x) + (1 - G(xy)) + \sigma(y) = \overline{G}^\sigma(xy)$, because $(1 - a) + (1 - b) = a + b$ for $a, b \in \mathbf{Z}_2$. □

3   Easy Problems for Switching Classes

We now proceed by listing a number of problems that become easy or trivial for switching classes.

Let G be a graph on n vertices. It is easy to show that there are 2^{n-1} graphs in $[G]$, and so checking whether there exists a graph $H \in [G]$ satisfying a given property $\mathcal{P}$ requires exponential time, if each graph is to be checked separately. However, although the hamiltonian cycle problem for graphs (GT37) is NP-complete, the hamiltonian problem for switching classes can be solved in time $O(n^2)$, since all one needs to check is that a given graph is not complete bipartite of odd order (see [8] or [2]).

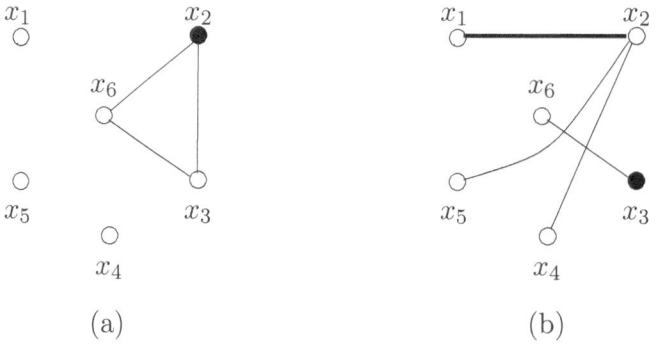

Fig. 1. The graph G and $G_1 \in [G]$ with path (x_1, x_2)

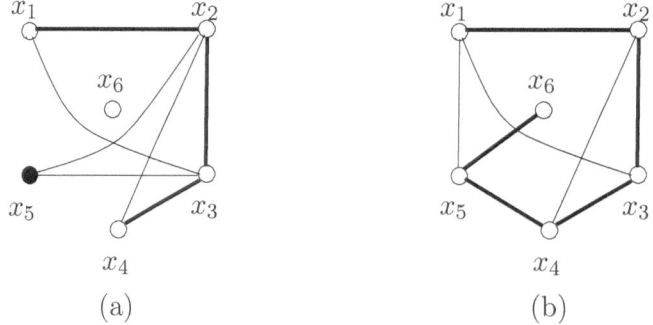

Fig. 2. The graphs G_2 and G_3

Example 1. Let G be the graph of Figure 1(a). The objective is to find a graph in $H \in [G]$ such that H has a hamiltonian path (GT39), in this case $(x_1, \ldots, x_6)$. It turns out that this is always possible, something which we shall now illustrate. The technique we use is to change the neighbourhoods of certain vertices using Lemma 3 in such a way that every time it is applied we lengthen the path, without changing the already constructed path.

Since x_2 is not adjacent to x_1 we apply the selector $\{x_2\}$ (selected vertices are marked in black) and obtain the graph G_1 of Figure 1(b). After applying the selector $\{x_3\}$ to G_3 we add both x_2x_3 and x_3x_4 to the path and obtain G_2 of Figure 2(a) with path (x_1, x_2, x_3, x_4). The last selector we have to apply is $\{x_5\}$ to obtain $H = G_3$ of Figure 2(b), which indeed has a hamiltonian path $(x_1, \ldots, x_6)$. $\diamond$

Using the same method we can obtain a spanning tree with maximum degree $\leq k$ (IND1), a spanning tree with at least k leaves (for $2 \leq k \leq n-1$) (IND2), a subgraph that is (noninduced) complete bipartite of order k (for $2 \leq k \leq n$) (a special case of GT21)). Note that these existence problems for graphs are all NP-complete, but as now made clear, trivial for switching classes.

Given any property of graphs decidable in polynomial time we can derive an algorithm for deciding the existential property for switching classes.

Theorem 1. *Let $\mathcal{P}$ be a graph property such that $\mathcal{P}(G)$, for G of order n can be decided in $\mathcal{O}(n^m)$ steps for an integer m. Let $\delta(G) \leq d(n)$ where $\delta(G)$ is the minimum degree of G. Then $\mathcal{P}_\exists$ is in $\mathcal{O}(n^{d(n)+1+max(m,2)})$.*

Proof. Let $G = (V, E)$ be the graph, $n = |V(G)|$, $\mathcal{P}$ the property, $d(n)$ the polynomial that bounds the minimum degree of G and m the constant as defined in the theorem. Algorithm 1 checks $\mathcal{P}$ for all graphs such that there is a vertex v of degree at most $d(n)$. It uses Lemma 3.

Algorithm 1.
```
P∃-DegreeBounded(G)
begin
    for all v ∈ V(G) do
        for all subsets V' of V(G) − {v} of size ≤ d(n) do
            σ = (V' − N(v)) ∪ (N(v) − V');
            if P(G^σ) then
                return true;
            else continue;
        od;
    od;
end;
```

The complexity of the algorithm is easily determined. The outer loop is executed n times, the inner loop for each value of v, $\mathcal{O}(n^{d(n)})$ times. The condition of the "if" takes $max(n^2, n^m)$ and together these yield the complexity as given in the theorem. $\square$

This leads us to the following definition which gives us the predicates such that we need only check a polynomial number of candidate graphs in each switching class.

For $\mathcal{P}$, a predicate on (or property of) graphs, $\mathcal{P}$ is of *bounded minimum degree* k if and only if for all graphs $G = (V, E)$ such that $\mathcal{P}(G)$ holds, there exists $x \in V$ such that the degree of x in G is at most k.

Corollary 1. *If $\mathcal{P}$ is a graph property of bounded minimum degree that is decidable in polynomial time, then $\mathcal{P}_\exists$ is decidable in polynomial time.*   □

Example 2. It is well known that planarity of a graph can be checked in time linear in the number of vertices. Because every planar graph has a vertex of degree at most 5, there are only a polynomial number of graphs that can possibly have this property. Since we can find these efficiently, we can check in polynomial time whether $[G]$ contains a planar graph.

Also, every acyclic graph has a vertex of degree at most 1, therefore we get a polynomial algorithm for the acyclicity of switching classes.   ◇

By complementation (Lemma 4) an analogous result can be formulated for predicates that hold for graphs that always have a vertex of degree at least $d(n)$.

4   Yannakakis' Result Modified for Switching Classes

Let $\mathcal{P}$ be a property of graphs that is preserved under isomorphisms. We say that $\mathcal{P}$ is

(i) *nontrivial*, if there exists a graph G such that $\mathcal{P}(G)$ does not hold and there are arbitrarily large graphs G such that $\mathcal{P}(G)$ does hold;

(ii) *switch-nontrivial*, if $\mathcal{P}$ is nontrivial and there exists a switching class $[G]$ such that for all $H \in [G]$, $\mathcal{P}(H)$ does not hold;

(iii) *hereditary*, if $\mathcal{P}(G|_A)$ for all $A \subseteq V(G)$ whenever $\mathcal{P}(G)$.

In the following we shall look at nontrivial hereditary properties. If there is a graph for which $\mathcal{P}(G)$ does not hold, then there are arbitrarily large graphs for which $\mathcal{P}$ does not hold, because of hereditarity.

Example 3. The following are examples of nontrivial hereditary properties of graphs that are also switch-nontrivial: G is discrete, G is complete, G is bipartite, G is complete bipartite, G is acyclic, G is planar, G has chromatic number $\chi(G) \leq k$ where k is a fixed integer, G is chordal, and G is a comparability graph.   ◇

Yannakakis proved in [11] the following general result on NP-hardness and NP-completeness.

Theorem 2. *Let $\mathcal{P}$ be a nontrivial hereditary property of graphs. Then the problem for instances (G, k) with $k \leq |V(G)|$ whether G has an induced subgraph $G|_A$ such that $|A| \geq k$ and $\mathcal{P}(G|_A)$, is NP-hard. Moreover, if $\mathcal{P}$ is in NP, then the corresponding problem is NP-complete.*   □

Example 4. If we take $\mathcal{P}$ to be the discreteness property, then Theorem 2 says that given (G, k) the problem to decide whether G has a discrete induced subgraph of order at least k is NP-complete. Note that this problem is exactly one of the standard NP-complete problems: the independence problem (GT20). The result of Yannakakis proves in one sweep that this problem is NP-complete, without having to resort to reduction. $\diamond$

We shall now establish a corresponding result for switching classes. For this, let $\mathcal{P}$ be a switch-nontrivial hereditary property. The property $\mathcal{P}_\exists$ is nontrivial, and $\mathcal{P}_\exists$ is hereditary, since

$$(G|_A)^\sigma = G^\sigma|_A \tag{1}$$

for all $A \subseteq V(G)$ and $\sigma : V(G) \to \mathbf{Z}_2$.

Theorem 3. *Let $\mathcal{P}$ be a switch-nontrivial hereditary property. Then the following problem for instances (G, k) with $k \leq |V(G)|$, is NP-hard: does the switching class $[G]$ contain a graph H that has an induced subgraph $H|_A$ with $|A| \geq k$ and $\mathcal{P}(H|_A)$? If $\mathcal{P} \in NP$ then the corresponding problem is NP-complete.*

Proof. Since $\mathcal{P}_\exists$ is a nontrivial hereditary property, we have by Theorem 2 that the problem for instances (G, k) whether G contains an induced subgraph of order at least k satisfying $\mathcal{P}_\exists$, is NP-hard. This problem is equivalent to the problem stated in the theorem, since by (1), for all subsets $A \subseteq V(G)$, $\mathcal{P}_\exists(G|_A)$ if and only if there exists a selector σ such that $\mathcal{P}((G|_A)^\sigma)$.

If $\mathcal{P}$ is in NP, then we can guess a selector σ and check whether $\mathcal{P}(G^\sigma)$ holds in nondeterministic polynomial time. Hence, $\mathcal{P}_\exists$ is in NP and NP-complete by (the second part of) Theorem 2. $\square$

5   Embedding Problems for Switching Classes

We consider now the embedding problem for switching classes. Recall that a graph H can be embedded into a graph G, denoted by $H \hookrightarrow G$, if H is isomorphic to a subgraph M of G, that is, there exists an injective function $\psi : V(H) \to V(G)$ such that

$$\{\psi(u)\psi(v) \mid uv \in E(H)\} = E(M).$$

We write $H \hookrightarrow [G]$, if $H \hookrightarrow G^\sigma$ for some selector σ. The embedding problem for graphs is known to be NP-complete, and below we show that it remains NP-complete for switching classes.

For a subset $A \subseteq V(G)$ and a selector $\sigma : V(G) \to \mathbf{Z}_2$ we have by (1) that $[G|_A] = [G]|_A$, where

$$[G]|_A = \{G^\sigma|_A \mid \sigma : V(G) \to \mathbf{Z}_2\}$$

is called the *subclass* of G induced by A.

Hence the switching class $[G]$ contains a graph H which has an independent subset A if and only if the induced subgraph $G|_A$ generates the switching class $[\overline{K}_A]$. As stated earlier in Lemma 1, $[\overline{K}_A]$ equals the set of all complete bipartite graphs on A.

An instance of the independence problem consists of a graph G and an integer $k \leq |V(G)|$, and we ask whether there exists a graph $H \in [G]$ containing an independent set A with k or more vertices. This problem is NP-complete for graphs (that is, the problem whether a graph G contains an independent subset of size $\geq k$) and, by Theorem 3, it remains NP-complete for switching classes.

Theorem 4. *The independence problem is NP-complete for switching classes. In particular, the problem whether a switching class $[G]$ has a subclass $[\overline{K}_m]$ with $m \geq k$, is NP-complete.* □

Recall that a graph $G = (V, E)$ has *clique size* $\geq k$ if there is a set $A \subseteq V$ with $|A| \geq k$ such that $G|_A$ is a clique. By Lemma 4 and Theorem 4 the following corollary holds.

Corollary 2. *For instances (G, k), where G is a graph and k an integer such that $k \leq |V(G)|$, the problem whether $[G]$ contains a graph with clique size $\geq k$, is NP-complete.* □

From the simple observation that if K_n embeds into a graph G, then it is isomorphic to a subgraph of G, we obtain

Corollary 3. *The embedding problem, $H \hookrightarrow [G]$, for switching classes is NP-complete for the instances (H, G) of graphs.* □

Since we can instantiate H with the clique on V and then use it to solve the clique problem of Corollary 2 using the same value for k, we can conclude the following.

Corollary 4. *For instances (G, H, k) for graphs G and H on the same domain V of size n and k an integer with $3 \leq k \leq n-1$, the problem whether there is a set $A \subseteq V$ with $|A| \geq k$ such that $H|_A \in [G|_A]$ is NP-complete.* □

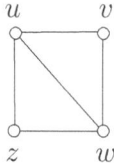

Fig. 3. A graph

Example 5. To illustrate the problem posed in Corollary 4, let $H = K_4$ and let G be the graph of Figure 3.

First of all H cannot be embedded into any graph in $[G]$. So if we take $k = 4$, then the answer to the problem posed in Corollary 4 would be "no". For $k = 3$ however the answer would be "yes", since both graphs contain a triangle.   ◇

We write $[H] \hookrightarrow [G]$, if for all $H' \in [H]$ there exists $G' \in [G]$ such that $H' \hookrightarrow G'$. It is not true that $H \hookrightarrow [G]$ implies $[H] \hookrightarrow [G]$.

By instantiating H with a clique of size k, the property $[H] \hookrightarrow [G]$ becomes the clique problem of Corollary 2. Hence,

Corollary 5. *For instances (H, G) of graphs the switching class embedding problem $[H] \hookrightarrow [G]$ is NP-complete.*   □

Note, however, that the problem to decide whether a given graph H is a subgraph of a graph in $[G]$ is easy. In this case the only difficulty lies in checking whether G can be switched so that H appears within the switch (see [7]). It occurs as a special case of the embedding problem treated here, the case where the embedding is the identity function on $V(H)$. Hence, the NP-completeness arises from the number of possible injections. Note that in Corollary 4 the problem does not lie in the number of injections, because the injection is always the identity function. The NP-completeness here arises from the number of possible subsets A. In this regard the embedding problem as treated above and the problem considered in Corollary 4 are orthogonal generalizations of the membership problem of [7].

6   3-Colourability for Switching Classes

We consider in this section the problem of 3-colourability. For a given graph $G = (V, E)$ a function $\alpha : V \to C$ for some set C is a proper colouring of G if for all $uv \in E$, $\alpha(u) \neq \alpha(v)$. The *chromatic number* of G, $\chi(G)$, is the minimum cardinality over the ranges of possible colourings of G. Note that for each $c \in C$, $\alpha^{-1}(c)$ is an independent set in G.

The 3-colourability problem for graphs (for a graph G, is $\chi(G) \leq 3$?) is NP-complete. In this section we prove that the 3-colourability problem for graphs can be reduced to the corresponding problem for switching classes, hereby proving that the latter is also NP-complete.

Theorem 5. *The problem whether a switching class $[G]$ contains a graph H with chromatic number 3, is NP-complete.*
Proof. Let $G = (V, E)$ be any graph, and let $G_9 = G + 3C_3$ be the graph which is a disjoint union of G and three disjoint triangles. Let A be the set of the nine vertices of the added triangles.

We claim that $\chi(G) \leq 3$ if and only if $[G_9]$ contains a graph H such that $\chi(H) = 3$. Since the transformation $G \mapsto G_9$ is in polynomial time, the claim follows.

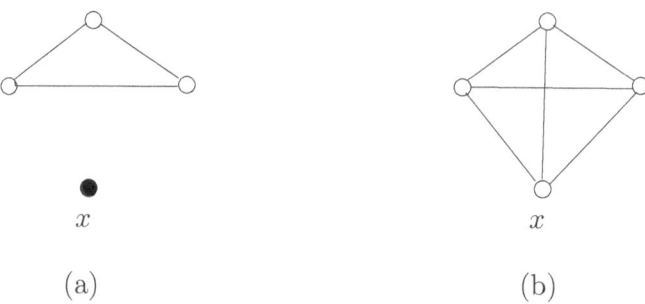

Fig. 4.

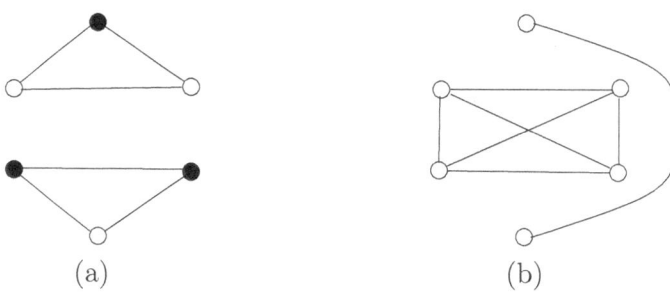

Fig. 5.

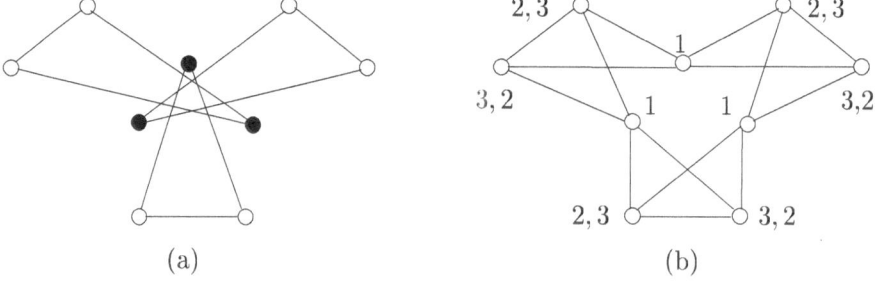

Fig. 6.

It is clear that if $\chi(G) \leq 3$ then $\chi(G_9) = 3$.

Suppose then that there exists a selector σ such that $\chi(G_9^\sigma) = 3$, and let $\alpha : V \cup A \to \{1, 2, 3\}$ be a proper 3-colouring of G_9^σ.

If σ is constant (either 0 or 1) on V, then G is a subgraph of G_9^σ, and, in this case, $\chi(G) \leq 3$.

Assume that σ is not constant on V. Since G_9^σ does not contain K_4 as a subgraph, it follows that σ is not a constant selector on any of the added triangles, see Figure 4(a) and (b). Further, each of these triangles contains equally many selections of 1 (and of 0, of course), since otherwise the subgraph $G_9^\sigma|_A$ would contain K_4 as its subgraph, see Figure 5(a) and (b).

We may assume that each of the added triangles contains exactly one vertex v with $\sigma(v) = 1$ (otherwise consider the complement of σ, $V(G_9) - \sigma$), see Figure 6(a). Let these three vertices constitute the subset $A_1 \subset A$.

In the 3-colouring α the vertices of A_1 obtain the same colour, say $\alpha(v) = 1$ for all $v \in A_1$; and in each of the added triangles the other two vertices obtain different colours, 2 and 3, since they are adjacent to each other and to a vertex of A_1 in G_9^σ, see Figure 6(b).

Each $v \in V$ with $\sigma(v) = 1$ is connected to all $u \in A - A_1$; consequently, $\alpha(v) = 1$ for these vertices. Therefore the set $B_1 = \sigma^{-1}(1) \cap V$ is an independent subset of G_9^σ. Since σ is constant on B_1, B_1 is also independent in G. The vertices in $V - B_1$ (for which $\sigma(v) = 0$) are all adjacent to the vertices in A_1 in G_9^σ, and therefore these vertices are coloured by 2 or 3. The subsets $B_2 = \alpha^{-1}(2) \cap V$ and $B_3 = \alpha^{-1}(3) \cap V$ are independent in G_9^σ. Again, since σ is constant on both B_2 and B_3, these are independent subsets of G. This shows that $\chi(G) \leq 3$.

$\square$

7   Open Problems

An interesting open problem is to get more insight into those problems which are polynomial for graphs, but become NP-complete for switching classes. For example, it is stated in [8] that the problem whether a switching class contains a k-regular graph for some k is NP-complete. Another open problem is the bipartiteness problem. It remains to be determined whether bipartiteness of switching classes is NP-complete. The current idea is that it is polynomial.

From the results in this paper one might conclude that "acyclic" problems (like hamiltonian path) become easy for switching classes, but it would also be interesting to find some more general statement. For instance hamiltonian circuit is not "acyclic", but it turned out be easy for switching classes.

Acknowledgement

We are indebted to GETGRATS for (financial) support and thank Rudy van Vliet for his helpful comments on an earlier version of this paper.

References

1. D. G. Corneil and R. A. Mathon, editors. *Geometry and Combinatorics: Selected Works of J. J. Seidel.* Academic Press, Boston, 1991. 70
2. A. Ehrenfeucht, J. Hage, T. Harju, and G. Rozenberg. Pancyclicity in switching classes. Submitted. 60, 62
3. A. Ehrenfeucht and G. Rozenberg. An introduction to dynamic labeled 2-structures. In A. M. Borzyszkowski and S. Sokolowski, editors, *Mathematical Foundations of Computer Science 1993*, volume 711 of *Lecture Notes in Computer Science*, pages 156–173, Berlin, 1993. Springer-Verlag. 59
4. A. Ehrenfeucht and G. Rozenberg. Dynamic labeled 2-structures. *Mathematical Structures in Computer Science*, 4:433–455, 1994. 59
5. M. R. Garey and D. S. Johnson. *Computers and Intractability: A Guide to the Theory of NP-Completeness.* Freeman, 1979. 60
6. J. L. Gross and T. W. Tucker. *Topological Graph Theory.* Wiley, New York, 1987. 59
7. J. Hage. The membership problem for switching classes with skew gains. Accepted for Fundamenta Informaticae. 67
8. J. Kratochvíl, J. Nešetřil, and O. Zýka. On the computational complexity of Seidel's switching. In *Combinatorics Graphs Complexity*, pages 161 – 166, Prague, 1990. Prachatice. 60, 62, 69
9. J. J. Seidel. A survey of two-graphs. In *Colloquio Internazionale sulle Teorie Combinatorie (Rome,1973)*, volume I, pages 481–511, Rome, 1996. Acc. Naz. Lincei. Reprinted in [1]. 59
10. J. J. Seidel and D. E. Taylor. Two-graphs, a second survey. In L. Lovasz and V. T. Sós, editors, *Algebraic Methods in Graph Theory (Proc. Internat. Colloq., Szeged, 1978)*, volume II, pages 689–711, Amsterdam, 1981. North-Holland. Reprinted in [1]. 59
11. M. Yannakakis. Node- and edge-deletion NP-complete problems. In *Proc. 10th Ann. ACM Symp. on theory of Computing*, pages 253 – 264, New York, 1978. ACM. 60, 64
12. T. Zaslavsky. Biased graphs. I. Bias, balance, and gains. *J. Combin. Theory, Ser. B*, 47:32–52, 1989. 59

The Power of Local Computations in Graphs with Initial Knowledge*

Emmanuel Godard[1], Yves Métivier[1], and Anca Muscholl[2]

[1] LaBRI, Université Bordeaux I, ENSERB,
351 cours de la Libératio, 33405 Talence, France
godard@labri.u-bordeaux.fr
metivier@labri.u-bordeaux.fr
[2] Institut für Informatik, Universität Stuttgart
Breitwiesenstr. 20-22, 70565 Stuttgart, Germany
muscholl@informatik.uni-stuttgart.de

Abstract. This paper deals with the delimitation of the power of local computations on graphs having some initial knowledge, e.g. the size or the topology of the network. We refine a proof technique based on coverings in order to cope with additional knowledge. Applying this method we show that several graph properties (e.g., planarity) are not recognizable even if some simple knowledge about the graph (e.g., the size) is available. Similarly, we show that election in ambiguous graphs is impossible, even if the graph topology is known by every node.

1 Introduction

Local computations on graphs, as given by graph rewriting systems (with priorities and/or forbidden contexts) [8] are a powerful model for local computations which can be executed in parallel. Rewriting systems provide a general tool for encoding distributed algorithms and for proving their correctness. This paper deals with the delimitation of the power of local computations having some initial knowledge like the size or the topology of the graph.

We focuss on two classical problems, the recognition problem and the election problem. The recognition problem asks for graph properties like acyclicity, planarity etc. Several basic properties like regularity, completeness, or acyclicity can be recognized by local computations without initial knowledge. On the other hand, planarity cannot be determined using local computations, provided that the given graph is labelled in a uniform way [9,5]. However, the presence of a distinguished vertex allows to gather information. In particular, from this fact it has been proved that it is possible to detect fixed minors [3], hence also to determine whether the graph is planar. A naturally arising question is whether some additional information encoded in the initial uniform labelling of the graph can help for deciding for example planarity.

---

* This research was partly supported by the French-German project PROCOPE and by the EC TMR Network GETGRATS (General Theory of Graph Transformation Systems) through the University of Bordeaux.

H. Ehrig et al. (Eds.): Graph Transformation, LNCS 1764, pp. 71–84, 2000.

For the election problem, algorithms are known for various families of graphs, e.g., trees and complete graphs. In these cases the algorithms do not need any additional knowledge about the graph. Further algorithms have been proposed for prime graphs [4,14], which use the size of the graph. Furthermore, it has been shown that knowing the size of the graph is also necessary for solving the election problem in prime graphs, [13].

Our proof techniques are based on coverings, which is a notion known from algebraic topology [10]. Coverings have been used for simulation, [2], and for proving impossibility results on distributed computing, [1,7]. The general proof idea is as follows. If G and H are two graphs such that G covers H, then any local computation on H induces a local computation on G. As a consequence, any graph class which is recognizable by local computations without any knowledge of the graph, must be closed under coverings. For example, the class of planar graphs is not recognizable by local computations, [9]. More generally, it has been shown that up to some few exceptions, any minor closed class of graphs (that is, a class of graphs characterized by a finite set of forbidden minors by the Graph Minor Theorem) is not closed under coverings, implying that it is not recognizable by local computations, [5]. In particular, deciding whether a fixed graph is not included as a minor in a given graph cannot be done by local computations.

With some additional knowledge about the given graph as e.g. the size, we cannot apply the above argument directly. We introduce a new construction (Proposition 15) for showing that even with some knowledge about the graph (like the number of vertices and/or the number of edges) we cannot solve the recognition problem by local computations deterministically for properties like bipartite graphs, graphs having a cut vertex or a cut edge, graphs having a given graph as minor.

In the last section we use coverings for characterizing the class of graphs where election is impossible [12]. We show that even when every vertex knows the topology of the network election cannot be solved for this graph class.

A graph G is called covering-minimal if every covering from G to some H is a bijection. The class of covering-minimal graphs plays an important role in the study of local computations. It is easy to verify (using prime rings) that the property of being covering-minimal is not recognizable without any initial knowledge about the graph. Using [12] we note that this property is recognizable if we have as initial knowledge the size of the graph. Having an odd number of vertices or having exactly one vertex with a certain label are other examples of properties which are not recognizable without initial knowledge, but are recognizable if the graph size is known. Thus recognizability under the assumption that the size is known to the algorithm is significantly more powerful than recognizability without initial knowledge.

Among models related to our model there are local computation systems as defined by Rosenstiehl et al. [16], Angluin [1] and Yamashita and Kameda [17]. In [16] a synchronous model is considered, where vertices represent (identical) deterministic finite automata. The basic computation step is to compute the next state of each processor according to its state and the states of its neighbours. In [1] an asynchronous model is considered. A basic computation step means that two adjacent vertices exchange their labels and then compute new ones. In [17] an asynchronous model is studied where a basic computation step means that a processor either changes its state and sends a message or it receives a message.

2   Basic Notions and Notation

The notation used here is essentially standard. A graph G is defined as a finite set $V(G)$ of vertices together with a set $E(G) \subseteq \binom{V}{2}$ of edges. We only consider finite, undirected and connected graphs without multiple edges and self-loops. Let $e = \{v, v'\}$ be an edge. We say that e is incident with v, and that v is a neighbour of v'. The set of neighbours of a vertex v is denoted $N_G(v)$. Let also $B_G(v) = N_G(v) \cup \{v\}$. A vertex of degree one is called a leaf. A path P from v_1 to v_i in G is a sequence $P = (v_1, e_1, v_2, e_2, \ldots, e_{i-1}, v_i)$ of vertices and edges such that for all $1 \leq j < i$, e_j is an edge incident with the vertices v_j and v_{j+1}; the integer $i - 1$ is the length of P. If $v_1 = v_i$ then P is called a cycle. If in a path (cycle) each vertex appears only once (except for $v_1 = v_k$), then the path (cycle) is called simple. A tree is a connected graph containing no simple cycle. In a tree any two vertices v and v' are connected by precisely one simple path. The distance between two vertices u, v is denoted $d(u, v)$.

G' is a subgraph of G if $V(G') \subseteq V(G)$ and $E(G') \subseteq E(G)$. The subgraph of G induced by a subset V' of $V(G)$ is the subgraph of G having V' as vertex set and containing all edges of G between vertices of V'.

Let v be a vertex, and let k be a positive integer. We denote by $B_G(v, k)$ the ball of radius k with center v, *i.e.*, the subgraph of G induced by the vertex set $V' = \{v' \in V(G) \mid d(v, v') \leq k\}$. The vertex v is called the *center* of $B_G(v, k)$.

A connected graph has at least k cycles if we can remove k edges while preserving the connectedness.

A homomorphism between two graphs G and H is a mapping $\gamma \colon V(G) \to V(H)$ such that if $\{u, v\}$ is an edge of G then $\{\gamma(u), \gamma(v)\}$ is an edge of H. Since we deal only with graphs without self-loops, this implies that $\gamma(u) \neq \gamma(v)$ whenever $\{u, v\} \in E(G)$. Note also that $\gamma(N_G(u)) \subseteq N_H(\gamma(u))$. We say that γ is a surjective homomorphism (resp. an isomorphism) if γ is surjective on vertices and edges (resp. bijective and γ^{-1} is also a homomorphism). By $G \simeq G'$ we mean that G and G' are isomorphic. A class of graphs will be any class of graphs in the set-theoretical sense containing all graphs isomorphic to some of its members.

2.1   Coverings

We say that a graph G is a *covering* of a graph H if there exists a surjective homomorphism γ from G onto H such that for every vertex v of $V(G)$ the restriction of γ to $N_G(v)$ is a bijection onto $N_H(\gamma(v))$. In particular, $\{\gamma(u), \gamma(v)\} \in E(H)$ implies $\{u, v\} \in E(G)$. The covering is proper if G and H are not isomorphic.

It is called connected if G (and thus also H) is connected. A graph G is called *covering-minimal* if every covering from G to some H is a bijection.

By a simple inductive argument we have [15]:

Lemma 1. *Suppose that G is a covering of H via γ. Let T be a subgraph of H. If T is a tree then $\gamma^{-1}(T)$ is a set of disjoint trees, each being isomorphic to T.*

The previous lemma implies:

Lemma 2. *Let H be a connected graph and let G be a covering of H via γ. Then there exists an integer q such that $card(\gamma^{-1}(v)) = q$, for every $v \in V(H)$.*

Definition 3. *Let G be a covering of H via γ and let q be such that $card(\gamma^{-1}(v)) = q$ for all $v \in V(H)$. Then the integer q is called the number of sheets of the covering. In this case we denote γ as q-sheeted covering.*

In this paper we also consider a subclass of coverings, namely *k-coverings*.

Definition 4. *Let G, H be two labelled graphs and let $\gamma : V(G) \to V(H)$ be a graph homomorphism. Let $k > 0$ be a positive integer. Then G is a k-covering of H via γ if for every vertex $v \in V(G)$, the restriction of γ on $B_G(v, k)$ is an isomorphism between $B_G(v, k)$ and $B_H(\gamma(v), k)$.*

The next two lemmas state a basic property of k-coverings (coverings, resp.), which is a main argument for the application to local computations.

Lemma 5. *[9] Let G be a k-covering of H via γ and let $v_1, v_2 \in V(G)$ be such that $v_1 \neq v_2$ and $\gamma(v_1) = \gamma(v_2)$. Then we have $B_G(v_1, k) \cap B_G(v_2, k) = \emptyset$.*

Lemma 6. *[9] Let G be a covering of H via γ and let $v_1, v_2 \in V(G)$ be such that $v_1 \neq v_2$ and $\gamma(v_1) = \gamma(v_2)$. Then we have $B_G(v_1) \cap B_G(v_2) = \emptyset$.*

Remark 7. If $q = 1$ in Def. 3 then G and H are isomorphic. Moreover, if G is a k-covering of H via γ, then G is a k'-covering of H via γ for every k' with $0 < k' \leq k$.

Lemma 8. *[13] Let $k > 0$ and consider two graphs G, G' such that G covers G' via μ. Then μ is a k-covering (resp. G is a k-covering of G') if and only if for every cycle C' in G' of length at most $2k + 1$, the inverse image $\mu^{-1}(C')$ is a disjoint union of cycles isomorphic to C'.*

3   Local Computations in Graphs

In this section we recall the definition of local computation systems and their relation with k-coverings, [9]. We consider networks of processors with arbitrary topology. A network is represented as a connected, undirected graph where vertices denote processors and edges denote direct communication links. Labels (or states) are attached to vertices and edges. Labels are modified locally, that is, on a subgraph of fixed radius k of the given graph, according to certain rules depending on the subgraph only (*k-local computations*). The relabelling is performed until no more transformation is possible, *i.e.*, until a normal form is obtained.

3.1 Labelled Graphs

Throughout the paper we will consider only connected graphs where vertices and edges are labelled with labels from a possibly infinite alphabet L. A graph labelled over L will be denoted by (G, λ), where G is a graph and $\lambda \colon V(G) \cup E(G) \to L$ is the function labelling vertices and edges. The graph G is called the underlying graph, and the mapping λ is a labelling of G. The class of labelled graphs over some fixed alphabet L will be denoted by $\mathcal{G}$. For a graph G and a label $\alpha \in L$ we let (G, Λ_α) be the graph G where every vertex and every edge is labelled by α.

Let (G, λ) and (G', λ') be two labelled graphs. Then (G, λ) is a subgraph of (G', λ'), denoted by $(G, \lambda) \subseteq (G', \lambda')$, if G is a subgraph of G' and λ is the restriction of the labelling λ' to $V(G) \cup E(G)$.

A mapping $\varphi \colon V(G) \cup E(G) \to V(G') \cup E(G')$ is a homomorphism from (G, λ) to (G', λ') if φ restricted to $V(G)$ and $V(G')$ is a graph homomorphism from G to G' which preserves the labelling, i.e., $\lambda'(\varphi(x)) = \lambda(x)$ for every $x \in V(G) \cup E(G)$. The mapping φ is an isomorphism if it is bijective.

We extend the notion of covering and k-covering to labelled graphs in an obvious way. E.g., (H, λ') is said to be covered by (G, λ) via γ if γ is an homomorphism from (G, λ) to (H, λ') whose restriction to $B_G(v)$ is an isomorphism from $(B_G(v), \lambda)$ to $(B_H(\gamma(v)), \lambda')$ for any vertex $v \in V(G)$.

An *occurrence* of (G, λ) in (G', λ') is an isomorphism φ between (G, λ) and a subgraph (H, η) of (G', λ').

3.2 Local Computations

Local computations as considered here can be described in the following general framework. Let $R \subseteq \mathcal{G} \times \mathcal{G}$ be a binary relation on the class of L-labelled graphs $\mathcal{G}$. Then R will denote a graph rewriting relation. We assume that R is closed by isomorphism, i.e., whenever $(G, \lambda)R(G', \lambda')$ and $(G_1, \lambda_1) \simeq (G, \lambda)$, then $(G_1, \lambda_1)R(G'_1, \lambda'_1)$ for some labelled graph $(G'_1, \lambda'_1) \simeq (G', \lambda')$. In the remainder of this paper R^* stands for the reflexive and transitive closure of R. The labelled graph (G, λ) is R–*irreducible* if there is no (G', λ') such that $(G, \lambda)R(G', \lambda')$. Let $(G, \lambda) \in \mathcal{G}$, then $\mathrm{Irred}_R((G, \lambda))$ denotes the set of R–irreducible graphs (or irreducible if R is fixed) which can be obtained from (G, λ) using R. The relation R is noetherian if there is no infinite chain $(G_1, \lambda_1)R(G_2, \lambda_2)R \ldots$.

Definition 9. *Let $R \subseteq \mathcal{G} \times \mathcal{G}$ be a graph rewriting relation and let $k > 0$ be an integer.*

1. *R is a relabelling relation if whenever two labelled graphs are in relation then the underlying graphs are equal (we say equal, not only isomorphic), i.e.:*

$$(G, \lambda)\, R\, (H, \lambda') \quad \Longrightarrow \quad G = H.$$

2. *Let R be a relabelling relation. Then R is called k–local, if only labels of a ball of radius k may be changed by R, i.e., $(G, \lambda)\, R\, (G, \lambda')$ implies that there*

exists a vertex $v \in V(G)$ such that

$$\lambda(x) = \lambda'(x) \text{ for every } x \notin V(B_G(v,k)) \cup E(B_G(v,k)).$$

The relation R is local *if it is $k-$local for some $k > 0$.*

The next definition states that a relabelling relation R is $k-$*locally generated* if its restriction on centered balls of radius k determines its computation on any graph. In particular, any locally generated relation is local.

Definition 10. *Let R be a relabelling relation and $k > 0$ be an integer. Then R is $k-$locally generated if the following is satisfied. For any labelled graphs (G, λ), (G, λ'), (H, η), (H, η') and any vertices $v \in V(G)$, $w \in V(H)$ such that the balls $B_G(v,k)$ and $B_H(w,k)$ are isomorphic via $\varphi \colon V(B_G(v,k)) \longrightarrow V(B_H(w,k))$ with $\varphi(v) = w$, the conditions*

1. *$\lambda(x) = \eta(\varphi(x))$ and $\lambda'(x) = \eta'(\varphi(x))$ for all $x \in V(B_G(v,k)) \cup E(B_G(v,k))$*
2. *$\lambda(x) = \lambda'(x)$, for all $x \notin V(B_G(v,k)) \cup E(B_G(v,k))$*
3. *$\eta(x) = \eta'(x)$, for all $x \notin V(B_H(w,k)) \cup E(B_H(w,k))$*

imply that $(G, \lambda) \, R \, (G, \lambda')$ if and only if $(H, \eta) \, R \, (H, \eta')$.
R is locally generated *if it is $k-$locally generated for some $k > 0$.*

3.3   Distributed Computations

The notion of relabelling defined in the previous section corresponds to *sequential computations*. But note that a $k-$locally generated relabelling relation also allows parallel rewritings, since non-overlapping k-balls may be relabelled independently. Thus we can define a distributed way of computing by saying that two consecutive relabelling steps concerning non-overlapping $k-$balls may be applied in any order. We say that such relabelling steps commute and they may be applied concurrently. More generally, any two relabelling sequences such that the latter one may be obtained from the former one by a succession of such commutations lead to the same resulting labelled graph. Hence, our notion of relabelling sequence may be regarded as a *serialization* [11] of some distributed computation. This model is clearly asynchronous. Several relabelling steps *may* be done at the same time but we do not require that all of them have to be performed. In the sequel we will essentially deal with sequential relabelling sequences but the reader should keep in mind that such sequences may be done in a distributed way.

3.4   Local Computations and Coverings

We give now a fundamental lemma which establishes the connection between k-coverings and k-locally generated relabelling relations. It states that if G is a k-covering of G' then any $k-$local computation in G' can be lifted to a $k-$local computation in G compatible with the k-covering relation. This is expressed in the following diagram:

$$
\begin{array}{ccc}
(G, \lambda_1) & \xrightarrow{\quad R^* \quad} & (G, \lambda_2) \\
\downarrow{\scriptstyle \text{k-covering}} & & \downarrow{\scriptstyle \text{k-covering}} \\
(G', \lambda'_1) & \xrightarrow{\quad R^* \quad} & (G', \lambda'_2)
\end{array}
$$

Using Lemma 5 we easily obtain:

Lemma 11. *[9] Let R be a k-locally generated relabelling relation and let (G, λ_1) be a k-covering of (G', λ'_1) via γ. Moreover, let $(G', \lambda'_1)R^*(G', \lambda'_2)$. Then a labelling λ_2 of G exists such that $(G, \lambda_1)R^*(G, \lambda_2)$ and (G, λ_2) is a k-covering of (G', λ'_2).*

4 Graph Recognizers and Initial Knowledge

4.1 The Recognition Problem

The problem addressed in this section can be informally described as follows. Let $\mathcal{F}$ be some class of (labelled) graphs. We will say that $\mathcal{F}$ can be *locally recognized* if a locally generated graph relabelling relation exists such that starting from a uniformly labelled graph (G, Λ_α) we obtain some irreducible graph satisfying a final condition if and only if $G \in \mathcal{F}$.

A final condition is a logical formula defined inductively in the following way:

1. for every label $l \in L$, l is a formula,
2. if φ and ψ are formulas then $\neg\varphi$, $\varphi \vee \psi$ and $\varphi \wedge \psi$ are formulas.

For $l \in L$, a labelled graph satisfies the formula l if $\lambda^{-1}(l) \neq \emptyset$,. By induction it satisfies $\neg\varphi$ if it does not satisfy φ, resp. it satisfies $\varphi \vee \psi$ if it satisfies φ or ψ, resp. it satisfies $\varphi \wedge \psi$ if it satisfies φ and ψ.

A set $\mathcal{K}$ of labelled graphs is defined by a *final condition*, if it is the set of graphs satisfying a final condition.

Definition 12. *A labelled graph recognizer is a pair $(R, \mathcal{K})$ where R is a graph relabelling relation and $\mathcal{K}$ is a class of labelled graphs defined by a final condition.*
A graph (G, λ) is recognized if $Irred_R(G, \lambda) \cap \mathcal{K} \neq \emptyset$.

We are interested in recognizing graphs which have a certain initial knowledge encoded in the initial labelling.

Definition 13. *A graph recognizer with initial knowledge is a triplet $(R, \mathcal{K}, \iota)$ where $(R, \mathcal{K})$ is a labelled graph recognizer, and ι is a function which associates with each graph G a label $\iota(G) \in L$. The set of graphs recognized by $(R, \mathcal{K}, \iota)$ is given as $\{G \mid (G, \Lambda_{\iota(G)}) \text{ is recognized by } (R, \mathcal{K})\}$.*
A recognizer $(R, \mathcal{K}, \iota)$ is said to be deterministic, if restricted to inputs $(G, \Lambda_{\iota(G)})$ we have the following two properties:

- R *is noetherian.*
- *Either $Irred_R(G, \Lambda_{\iota(G)}) \cap \mathcal{K} = \emptyset$ or $Irred_R(G, \Lambda_{\iota(G)}) \subseteq \mathcal{K}$.*

We can now define recognizable classes of graphs.

Definition 14. *A class $\mathcal{F}$ of graphs is said to be* (deterministically) recognizable *with initial knowledge ι if there exists a locally generated (deterministic) graph recognizer $(R, \mathcal{K}, \iota)$ recognizing exactly $\mathcal{F}$.*

If $\iota(G)$ is the size of the graph, the class of graphs is said to be *size-recognizable.*

4.2   A Generalized k-Covering Technique

It was shown in [9] that locally recognizable families of graphs must be closed under k-coverings for some k. This implies that certain families of graphs are not recognizable (without initial knowledge), because they are not closed under k-coverings. However if G is a proper covering of H then the size of G is strictly greater than the size of H, *i.e.*, size-recognizable graph families are not necessarily closed under k-coverings. Thus, it is not possible to extend the constructions of [9] directly to relabelling systems with initial knowledge like the size of the graph. We give in this section a simple relation between graph recognizers and k-coverings which allows us to cope with recognizers with initial knowledge.

Proposition 15. *Let $(R, \mathcal{K}, \iota)$ be a k-locally generated, deterministic graph recognizer with initial knowledge. Let G, G' be graphs such that $\iota(G) = \iota(G')$ and there exists H such that both G and G' are k-coverings of H. Then G is recognized by $(R, \mathcal{K}, \iota)$ if and only if G' is recognized.*

For $k > 0$ we define the relation σ_k^ι by letting $G \sigma_k^\iota G'$ if $\iota(G) = \iota(G')$ and there exists a graph H such that G and G' are both k-coverings of H. Let $\sim_k^\iota$ denote the transitive closure of σ_k^ι.

Corollary 16. *Let $\mathcal{F}$ be a class of graphs which is deterministically recognizable with the initial knowledge ι. Then $\mathcal{F}$ is closed under $\sim_k^\iota$ for some k.*

From Proposition 15 we obtain in the case where ι is the size:

Proposition 17. *The following graph classes are not deterministically size-recognizable:*

1. *bipartite graphs,*
2. *graphs having a cut edge,*
3. *graphs having a cut vertex,*
4. *Hamiltonian graphs.*

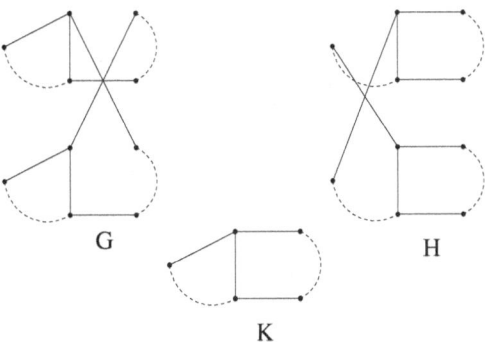

Fig. 1. Bipartite graphs are not recognizable knowing the size.

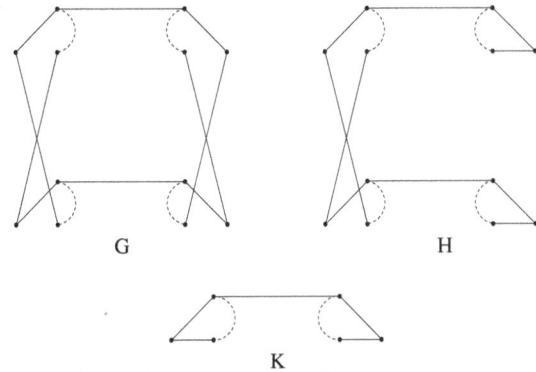

Fig. 2. Cut-edge is not recognizable knowing the size.

Proof. Let k be an integer. In the figures below the dashed lines are simple paths of $2k+1$ vertices, hence all cycles involved are at least $2k+1$ long. Thus, all coverings are also k-coverings by Lemma 8. We apply Proposition 15 to the graphs $G \sim_k^{\text{size}} H$ (both are k-coverings of K). Note that only one of the covering graphs is in the given family. Thus, no k-locally generated relabelling system can recognize these graph classes.

Figure 1 A graph is bipartite if and only if it contains no odd cycles. Thus G is bipartite, but H is not bipartite.

Figure 2. The graph G has no cut-edge and H has two cut-edges.

Figure 3. The graph G has two cut-vertices and H has no cut-vertex.

Figure 4. The graph G has the Hamilton property, but not H.

Remark 18. Using Lemma 1 we note that if G and H are both coverings of a graph K and have the same number of vertices then they have the same number of edges and thus their cycle spaces have the same dimension.

Our main result concerning minor-closed graph classes is:

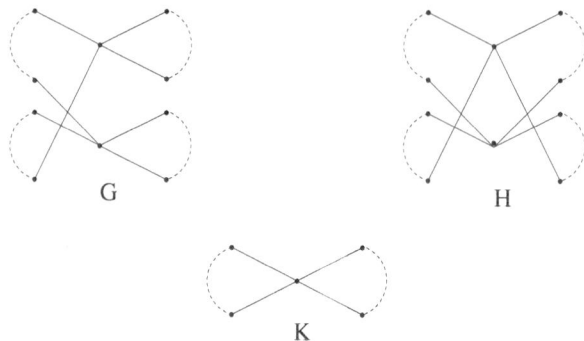

Fig. 3. Cut-vertex is not recognizable knowing the size.

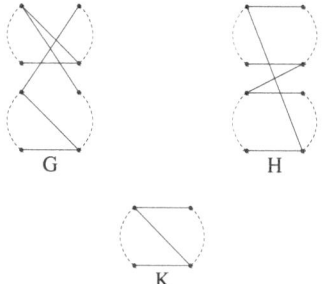

Fig. 4. The Hamilton property is not recognizable knowing the size.

Theorem 19. *Let $\mathcal{G}$ be a minor-closed class of connected graphs and let $G \notin \mathcal{G}$ be a connected graph with p vertices. Assume that there is a connected graph H such that*

1. *H has a k-covering $K' \in \mathcal{G}$ with at least $2p^3 \times |V(H)|$ vertices.*
2. *H contains three edges $e_i = \{x_i, y_i\}$, $i = 1, 2, 3$, such that $d_{H_i}(x_i, y_i) > 2k$ in the graph $H_i = (V(H), E(H) \setminus \{e_i\})$.*

Then $\mathcal{G}$ is not deterministically size-recognizable by $k-$locally generated relabelling systems.

Proof. We only sketch the proof idea. We start with the graphs H and G. Using the edges e_1, e_2, e_3 of H and Lemma 8 we are able to construct a k-covering K of H such that K and K' have the same number of sheets and K has a parallelipiped $P(l, m, n)$ as a minor. It is known that every planar graph with p vertices can be embedded into a grid of size $2p - 4$ by $p - 2$ [6]. Thus, we can choose $l \geq 2p - 4, m \geq p - 2, n \geq p$ such that G is a minor of $P(l, m, n)$, hence of K.

Since $G \notin \mathcal{G}$ and $\mathcal{G}$ is minor closed, we have that $K \notin \mathcal{G}$. This contradicts Proposition 15, which implies that K is size-recognized by $(R, \mathcal{K})$ if and only if K' is size-recognized by $(R, \mathcal{K})$.

Corollary 20. *The class of connected planar graphs is not deterministically size-recognizable.*

5   Remarks on Ambiguous Graphs, Coverings and the Election Problem

5.1   The Election Problem Knowing the Topology

Angluin showed that no election algorithm exists for a family of graphs containing a graph H and a proper covering of H, see [1]. The proof goes similarly to Lemma 11. As for graph recognizers this technique cannot be directly extended to graphs with initial knowledge. In our setting, an election algorithm denotes a noetherian relabelling relation R, which is generated by the neighborhoods of vertices. That is, we consider relations satisfying the conditions of Definition 10, with balls $B_G(v, k)$, $B_G(w, k)$ replaced by neighborhoods $B_G(v)$, $B_G(w)$, respectively. Moreover, every R-irreducible graph obtained from an uniformly labelled graph has a distinguished (elected) vertex. It is known that election is possible for rings having a prime number of vertices, see e.g. [12]. For proving that there is no algorithm when the size of a ring is composite, it is shown that it is possible to go from a symmetric situation to another symmetric situation. It follows that certain computations never terminate. The next proposition generalizes this result for general relabelling rules on graphs having some initial knowledge like the size or the topology. Let G be a graph and let v be a vertex of G. We say that v knows the topology of G if v's label encodes the incidence matrix of a graph $G' \simeq G$, but no information enables v to know which vertex of G' corresponds to v. Let G be a graph which is not covering-minimal and let H be such that G is a proper covering of H via the morphism γ. A subgraph K of G is *free modulo* γ if $\gamma^{-1}(\gamma(K))$ is a disjoint union of graphs isomorphic to K. We say that a labelling λ is $\gamma-lifted$ if

$$\gamma(x) = \gamma(y) \Longrightarrow \lambda(x) = \lambda(y).$$

We say that an algorithm *operates on subgraphs from a given set* S if every relabelling step is performed only on subgraphs of the given graph, which belong to S. We have

Proposition 21. *Let G be a graph which is not covering-minimal and let γ be a covering from G onto some graph $H \not\simeq G$. Then there is no election algorithm for G which operates on subgraphs free modulo γ, even if the topology of G is known by each vertex of G.*

The proof of the previous proposition is based on the next lemma. It expresses the fact that from any symmetric situation, an algorithm as above can reach another symmetric situation.

Lemma 22. *Let $\mathcal{A}$ be an algorithm which operates on subgraphs free modulo γ, where γ is a covering from G onto some graph $H \not\simeq G$. Let λ be a γ−lifted labelling of G. Suppose that a relabelling step of $\mathcal{A}$ modifies only labels of a subgraph of (G, λ) which is free modulo γ, yielding (G, λ'). Then there exists a γ−lifted labelling λ'' of G that can be obtained using $\mathcal{A}$ from (G, λ').*

5.2   Ambiguous Graphs

Non-ambiguous graphs have been introduced in [12], where an election algorithm is given for this family. In the next lemma we prove that a graph is non-ambiguous if and only if it is covering-minimal. For example, trees, complete graphs and T-prime graphs are non-ambiguous. From the previous section we know that there is no election algorithm for ambiguous graphs, even if every vertex knows the topology of the graph.

Let (G, λ) be a labelled graph. As usual, the labelling is called bijective if $\lambda(v) = \lambda(v')$ implies $v = v'$, for each $v, v' \in V$.

Definition 23. *Let (G, λ) be a labelled graph. The labelling λ is called locally bijective if it verifies the following two conditions:*

1. *For each vertex $v \in V$ we have that all vertices from $B_G(v)$ are differently labelled, i.e., $\lambda(v') = \lambda(v'')$ implies $v' = v''$ for all $v', v'' \in B_G(v)$.*
2. *For all vertices $v, v' \in V$, if $\lambda(v) = \lambda(v')$ then $B_G(v)$ and $B_G(v')$ have the same labels, i.e., $\lambda(N_G(v)) = \lambda(N_G(v'))$.*

A graph G is ambiguous if there exists a non-bijective labelling of G which is locally bijective.

Thus, a graph is ambiguous if there exists a labelling λ of G which is locally bijective, such that $|\lambda(V)| < |V|$.

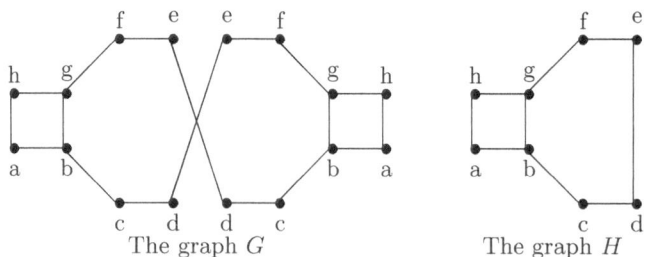

Fig. 5. The graph G is ambiguous, it is a 2-covering of the graph H.

Proposition 24. *A graph G is ambiguous if and only if it is not covering-minimal.*

By Propositions 24, 21 we have shown that there is no election algorithm for an ambiguous graph, even knowing its topology. This impossibility result is optimal, since the protocol given by [12] provides an election algorithm for covering-minimal graphs when the size is known.

6 Conclusion

For covering-minimal graphs, the knowledge of the size enables the realization of an election algorithm. But we have also shown that even knowing the size of the graph, there are properties of graphs that are still not recognizable. Hence, election and recognition of graph classes are not reducible to each other.

The remaining problem for election are to determine the classes of covering-minimal graphs that have an election algorithm with no initial knowledge. Concerning the recognition problem, the remaining problems are to characterize the recognizable and size-recognizable families of graphs, especially investigating $\sim_k^{\text{size}}$ equivalence classes. Conversely, given some family of graphs, it would be interesting to know what kind of initial information is exactly needed in order to recognize it by local computations.

References

1. D. Angluin. Local and global properties in networks of processors. In *Proceedings of the 12th Symposium on Theory of Computing*, pages 82–93, 1980. 72, 73, 81
2. H.-L. Bodlaender and J. Van Leeuwen. Simulation of large networks on smaller networks. *Information and Control*, 71:143–180, 1986. 72
3. A. Bottreau and Y. Métivier. Minor searching, normal forms of graph relabelling: two applications based on enumerations by graph relabelling. In *Foundations of Software Science and Computation Structures (FoSSaCS'98)*, volume 1378 of *Lecture notes in computer science*, pages 110–124. Springer-Verlag, 1998. 71
4. J. E. Burns and J. Pachl. Uniform self-stabilizing rings. *ACM Trans. Program. Lang. Syst.*, 11:330–344, 1989. 72
5. B. Courcelle and Y. Métivier. Coverings and minors: application to local computations in graphs. *Europ. J. Combinatorics*, 15:127–138, 1994. 71, 72
6. H. De Fraysseix, J. Pach, and R. Pollack. Small sets supporting Fáry embeddings of planar graphs. In *Proc. of the 20th Ann. ACM Symp. Theory of Computing, Chicago, 1988*, pages 426–433. ACM Press, 1988. 80
7. M. J. Fisher, N. A. Lynch, and M. Merritt. Easy impossibility proofs for distributed consensus problems. *Distrib. Comput.*, 1:26–29, 1986. 72
8. I. Litovsky, Y. Métivier, and E. Sopena. Different local controls for graph relabelling systems. *Math. Syst. Theory*, 28:41–65, 1995. 71
9. I. Litovsky, Y. Métivier, and W. Zielonka. On the recognition of families of graphs with local computations. *Information and Computation*, 118(1):110–119, 1995. 71, 72, 74, 77, 78

10. W. S. Massey. *A basic course in algebraic topology.* Springer-Verlag, 1991. Graduate texts in mathematics. 72

11. A. Mazurkiewicz. Trace theory. In W. Brauer et al., eds., *Petri nets, applications and relationship to other models of concurrency*, volume 255 of *Lecture notes in computer science*, pages 279–324. Spinger-Verlag, 1987. 76

12. A. Mazurkiewicz. Distributed enumeration. *Inf. Processing Letters*, 61:233–239, 1997. 72, 81, 82, 83

13. Yves Métivier, Anca Muscholl, and Pierre-André Wacrenier. About the local detection of termination of local computations in graphs. In D. Krizanc and P. Widmayer, editors, *Proceedings of the 4th International Colloquium on Structural Information and Communication Complexity (SIROCCO'97), Ascona, Switzerland*, number 1 in Proceedings in Informatics, pages 188–200, Canada, 1997. Carleton Scientific. 72, 74

14. Yves Métivier and Pierre-André Wacrenier. A distributed algorithm for computing a spanning tree in anonymous T-prime graphs. Submitted, 1998. 72

15. K. Reidemeister. *Einführung in die kombinatorische Topologie.* Vieweg, Brunswick, 1932. 74

16. P. Rosenstiehl, J.-R. Fiksel, and A. Holliger. Intelligent graphs. In R. Read, editor, *Graph theory and computing*, pages 219–265. Academic Press (New York), 1972. 72

17. M. Yamashita and T. Kameda. Computing on anonymous networks: Part i – characterizing the solvable cases. *IEEE Transactions on parallel and distributed systems*, 7(1):69–89, 1996. 72, 73

Double-Pullback Graph Transitions: A Rule-Based Framework with Incomplete Information[*]

Hartmut Ehrig[1], Reiko Heckel[2], Mercè Llabrés[3], Fernando Orejas[4], Julia Padberg[1], and Grzegorz Rozenberg[5]

[1] Technical University Berlin, Germany
{ehrig,padberg}@cs.tu-berlin.de
[2] University of Paderborn, Germany
reiko@uni-paderborn.de
[3] University of the Balearic Islands, Spain
merce@ipc4.uib.es
[4] Technical University of Catalonia, Spain
orejas@lsi.upc.es
[5] University of Leiden, The Netherlands
rozenber@wi.leidenuniv.nl

Abstract. Reactive systems perform their tasks through interaction with their users or with other systems (as parts of a bigger system). An essential requirement for modeling such systems is the ability to express this kind of interaction. Classical rule-based approaches like Petri nets and graph transformation are not suited for this purpose because they assume to have complete control about the state and its transformation. Therefore, in this paper we propose a general framework which extends a given rule-based approach by a loose semantics where the rules of the system (e.g., graph productions or Petri net transitions) are considered as incomplete descriptions of the transformations to be performed: they still determine the changes to the matched substructure but for the remaining part (the context) unspecified changes are possible representing the application of (unknown) rules from the environment.

The framework is applied to graph transformation systems in the double-pushout approach as well as place-transition Petri nets leading, respectively, to the concepts of *graph transition* and *open step*.

1 Introduction

Rule-based formalisms such as Petri nets and graph transformation are suitable for modeling various kinds of concurrent and distributed systems where operations affect the system's state only locally. Due to a built-in *frame condition* the part outside the redex of the rule application (also called *context*) remains

---

[*] Research partially supported by the German Research Council (DFG), the Spanish DGES project PB96-0191-CO2, the TMR network GETGRATS, and the ESPRIT Basic Research Working Group APPLIGRAPH

H. Ehrig et al. (Eds.): Graph Transformation, LNCS 1764, pp. 85–102, 2000.

unchanged. This means that the system's behavior is fully determined by the rules and no additional, unspecified effect is taken into account. This kind of modeling is adequate for closed systems where we have complete information about all possible evolutions of the system. For open systems, however, also the possible effects of operations of the environment have to be modeled.

Conceptually, we may think of the environment as another, concurrently active, rule-based system whose rules are not visible to us. Still we are able to observe the effects of their application on the system's state. That means we have to model rule-based transformations with incomplete information: the rules of the system are known, but the rules of the environment are not known although they influence the transformation of the system. In other words, such transformations consist not only of an application of a given rule, but also include a change of context caused by unknown rules in the environment. To stress this fact we will use in the sequel the term "transition" rather than "transformation": a transition is a transformation (caused by a visible rule) together with a change of context (caused by the environment):

transition = transformation + change of context

Notions of transition following this intuition have been developed for algebraic graph transformation in order to support a view-based approach to system design [8], and for Petri nets in order to study and integrate scenarios in train control systems [14]. In the first case so-called *graph transitions* are defined by means of a double pullback construction which replaces the well-known double pushout [10]. In the latter case, transitions represent *open steps* of open Petri nets which allow autonomous actions on designated open (i.e., interface) places [6].

In this paper we present a general framework for rule-based systems with incomplete information on an informal level. Two central problems are stated and solved for graph transformation and Petri nets: the characterization of transitions independently of transformations and the recovery of information about the unspecified effects.

We present in some detail the case of algebraic graph transformation focusing in particular on the *construction* of graph transitions. Given a production and a match we provide conditions that ensure the *existence* of a corresponding transition and construct certain distinguished transitions, like minimal transitions and lazy transitions (i.e., where unspecified effects are kept to minimum). Moreover, we give a construction that allows to enumerate all transitions and compare transitions with DPO derivations. These investigations shall support the operational intuition behind DPB transitions in analogy to the gluing conditions and constructions of DPO derivations in the classical approach.

2   Conceptual Framework for Rule-Based Modification with Incomplete Information

In this section we demonstrate how to extend a given technique for rule-based modification so that transformations with incomplete information can be handled. More precisely, we demonstrate how to define transitions following the principle:

transition = transformation + change of context

stated in the introduction.

Assumption 1 *Given a technique for rule-based modification of specific structures we can assume that the following notions are available:*

- *rule or production $p : L \Longrightarrow R$ with the left hand side L and the right hand side R*
- *redex $l : L \rightarrow G$ for the application of a rule p to a structure G*
- *transformation $(p, l, r) : G \Longrightarrow H$ defined by application of the rule p to the structure G at left redex l leading to the derived structure H with right redex $r : R \rightarrow H$ (cf. Figure 1)*

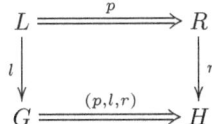

Fig. 1. A direct transformation.

In addition to the given rule-based technique we assume to have for each transformation $(p, l, r) : G \Longrightarrow H$ a class of left and right redex extensions, e.g. $e_G : G \rightarrow G'$ and $e_H : H \rightarrow H'$, such that the composition with redices is defined, i.e. $l' = e_G \circ l : L \rightarrow G'$ and $r' = e_H \circ r : R \rightarrow H'$ are redices.

Definition 1 (transitions associated to transformations). *Given a transformation $(p, l, r) : G \Longrightarrow H$ like in Figure 1 and a suitable pair of redex extensions $e_G : G \rightarrow G'$ and $e_H : H \rightarrow H'$ the transition $(p, l', r') : G' \rightsquigarrow H'$ associated to $(p, l, r) : G \Longrightarrow H$ via (e_G, e_H) is given by*

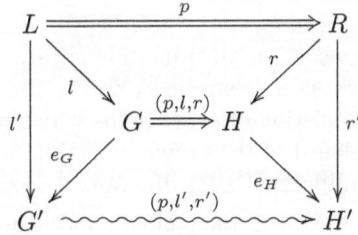

where $l' = e_G \circ l$ and $r' = e_H \circ r$.

Remark 1. The change of context is expressed by the fact that we allow different redex extensions $e_G : G \to G'$ and $e_H : H \to H'$ on the left and right hand side respectively. Without change of context we would have to require that the left context $G' - e_G(G)$ is equal to the right context $H' - e_H(H)$, or a similar condition like $G' - l'(L) = H' - r'(R)$, where we assume to have a context construction $S - S'$ for a given structure S with substructure S'. In fact, the implicit frame condition for transformations $(p, l, r) : G \Longrightarrow H$ for most rule-based techniques means that the left context $G - l(L)$ is equal to the right context $H - r(R)$.

In Definition 1 transitions are obtained from given transformations via redex extensions. In order to have a notion of transition independently of this construction we have to solve the following problems.

Problem 1 (characterization of transitions). Given a rule $p : L \Longrightarrow R$ characterize all transitions

$$(p, l', r') : G' \rightsquigarrow H'$$

independent of the notion of transformation and redex extension.

Variations on this problem include a characterization of the *existence* of transitions for a given rule p and redex l as well as the enumeration of these transitions.

In some sense reverse to Problem 1 is the problem how to recover the information which operations of the environment of the open system may have caused the change of context in a given transition.

Problem 2 (recovery of information). Characterize each transition

$$(p_0, l', r') : G' \rightsquigarrow H'$$

by a direct transformation or a transformation sequence

$$G' \Longrightarrow^* H' \text{ via } p'$$

between the same structures G' and H', where p' is a parallel and/or sequentially composed production built up over p_0 and some productions $p_1, ..., p_n$ modeling the change of the context in the transition $G' \rightsquigarrow H'$.

The conceptual framework developed in this section will be applied to graph transformation systems and Petri nets in the following sections leading to the notions of graph transitions and open Petri nets respectively.

3   Graph Transitions

Following the conceptual framework established in Section 2, in this section we introduce graph transitions. A graph transition ensures that we preserve, delete, and add at least as much as it is specified by the production, but it allows to model also addition and deletion of other items which may be caused by the environment. Solving Problem 1 of the previous section, we will characterize graph transitions by double-pullback (DPB) diagrams.[1] Before introducing formally

---

[1] This should not be confused with the pullback approach [1,2], which has been proposed as a categorical formulation of the Node Label Controlled (NLC) approach to graph transformation [7].

these concepts, we recall the basic notions of graphs and graph transformation according to the DPO approach [3].

A *graph signature* is a unary algebraic signature $\Sigma = \langle S, OP \rangle$ given by a set of sorts S and a family of sets $OP = (OP_{s,s'})_{s,s' \in S}$ of unary operation symbols. For $op \in OP_{s,s'}$ we write $op : s \to s' \in \Sigma$. *Graphs* and *graph morphisms* for a graph signature Σ are Σ-algebras $G = \langle (G_s)_{s \in S}, (op^G)_{op \in OP} \rangle$ and Σ-homomorphisms $f = (f_s)_{s \in S} : G \to G'$, respectively. The category of Σ-graphs and Σ-graph morphisms is denoted by $\mathcal{G}r(\Sigma)$. Referring to the elements of the carriers of graph G we usually skip the sort index, that is, $x \in G$ means $x \in G_s$ for a suitable sort $s \in S$ that shall be clear from the context.

Graph productions according to the DPO approach are spans of injective graph morphisms $p = (L \leftarrow K \to R)$. A *direct DPO derivation* from G to H, denoted by $(p, l, r) : G \Longrightarrow H$, is given by a diagram like in the left of Figure 3 where both (1) and (2) are pushout squares.

In order to define transitions based on DPO derivations, according to Definition 1 we have to provide a class of redex extensions for each transformation step. For this, the following concept shall be needed.

Definition 2 (weak and horizontal injectivity). *Given a morphism $h : D \to E$ and a commutative diagram (1) like below we say that h is weakly injective w.r.t. (1) when for every $x, y \in D$ such that $h(x) = h(y)$ and $x \neq y$ then $x, y \in g_1(A) - g_1 f_1(K)$ or $x, y \in g_2(B) - g_2 f_2(K)$.*

$$
\begin{array}{ccc}
K & \xrightarrow{f_1} & A \\
{\scriptstyle f_2}\downarrow & (1) & \downarrow{\scriptstyle g_1} \\
B & \xrightarrow{g_2} & D \xrightarrow{h} E
\end{array}
$$

Moreover, h is called horizontally injective *if for all $x, y \in D$ such that $h(x) = h(y)$ and $x \neq y$ it holds that $x, y \in g_1(A) - g_2 f_2(K)$.*   △

Notice that, like in the definition above, the notion of horizontal injectivity shall always be used in connection with some diagram making clear what is referred to as the horizontal dimension.

Figure 2 is an example where the morphism h on the left is weakly injective and the same morphism on the right is not weakly injective because $h(2) = h(3)$ but $2 = g_1(2)$ and $3 = g_2(3)$.

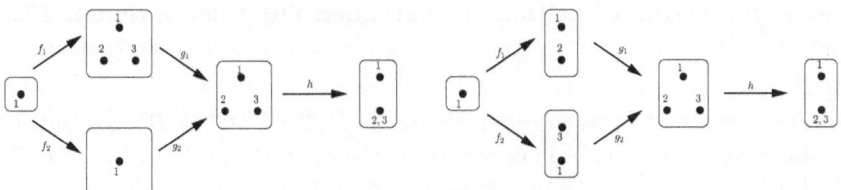

Fig. 2. Weak injectivity (example).

Now, three variants of graph transitions are defined using, respectively, weakly injective, horizontally injective, and injective morphisms as redex extensions.

Definition 3 (graph transition associated to direct derivation). *Let in Figure 3 $p = (L \leftarrow K \rightarrow R)$ be a production, $(p, l, r) : G \Longrightarrow H$ be a direct DPO derivation, and morphisms $e_G : G \rightarrow G'$ and $e_H : H \rightarrow H'$ be weakly injective (horizontally injective, injective) w.r.t. the pushout squares (1) and (2), respectively. Then, the diagram in the middle of Figure 3 represents a weakly injective (horizontally injective, injective) graph transition $(p, l', r') : G' \rightsquigarrow H'$ associated to the direct derivation $(p, l, r) : G \Longrightarrow H$.*

Notice that our presentation of transitions differs form the original one in [10,9]. There, graph transitions are defined via *double-pullback (DPB) diagrams*, i.e., diagrams like in Figure 3 on the right where both (1') and (2') are pullback squares, and (the injective version of) Definition 3 is given as a characterization of transitions by means of DPO derivations. In order to fit within the general framework introduced in the previous section, here we exchange the roles of definition and characterization.

In general, the problem is to understand which kind of redex extension (weakly injective, horizontally injective, injective) corresponds to a certain class of DPB diagrams. Consider, for example, the DPO diagram in Figure 3 on the left and assume injective graph morphisms $e_G : G \rightarrow G'$ and $e_H : H \rightarrow H'$ like in the middle diagram of Figure 3 as redex extensions. The new redices l' and r' are defined by composition of the original ones l and r with e_G and e_H, respectively. It is shown in [10] that the so-constructed outer diagram is a *faithful* DPB transition, i.e., a DPB diagram like in Figure 3 on the right where the redices $l' : L \rightarrow G'$ and $r' : R \rightarrow H'$ satisfy the identification condition w.r.t. p and $p^{-1} : (R \leftarrow K \rightarrow L)$, respectively. Vive versa, each faithful transition can be represented in this way. Thus, assuming direct DPO derivations as our notion of transformation and injective graph morphisms as redex extensions, the notion of faithful transition provides the solution to Problem 1.

Intuitively this means that we have a change of context from the left context $G' - e_G(G)$ to the right context $H' - e_H(H)$. In addition to the effect of the transformation $(p, l, r) : G \Longrightarrow H$ in the transition $(p, l', r') : G' \rightsquigarrow H'$ the part $G' - e_G(G)$ is replaced by $H' - e_H(H)$, where this replacement is given by the span of graph morphisms $G' \longleftarrow D \longrightarrow H'$.

In the theorem below, this result shall be extended to cover horizontally and weakly injective transitions.

Theorem 2 (DPB transitions as extended DPO derivations). *Given a production $p = (L \leftarrow K \rightarrow R)$, the two statements below are equivalent (cf. Figure 3).*

1. *There exists a weakly injective transition $(p, l', r') : G' \rightsquigarrow H'$ associated to a direct derivation $(p, l, r) : G \Longrightarrow H$ as shown in the middle of Figure 3.*
2. *There is a DPB diagram like in Figure 3 on the right such that pullbacks (1') and (2') form the two outer squares of the diagram in the middle.*

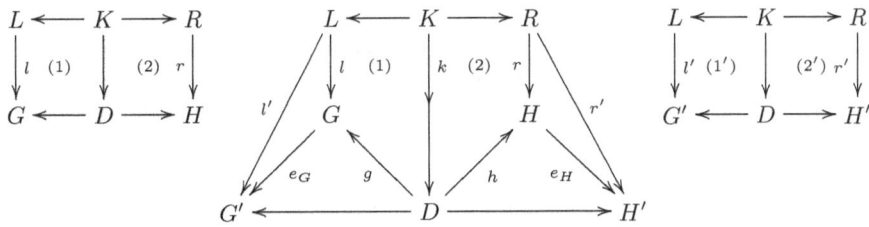

Fig. 3. DPB transitions as DPO derivations plus change of context/

The statement remains true if we restrict to horizontally injective transitions in 1. and require for the DPB diagram in 2. the injectivity of the morphisms in the bottom span. The restriction to injective transitions in 1. corresponds to the class of faithful DPB diagrams in 2.

Proof. First notice that every DPO diagram is already a DPB diagram because pushout squares over an injective graph morphism are pullback squares. Thus, it is enough to prove that we can reconstruct the diagram in the middle of Figure 3 from the one on the right. Assuming pullback squares (1') and (2') we build the pushouts (1) and (2) and by universal pushout property we obtain the morphisms e_G and e_H making commutative the diagram. Weak (resp. horizontal) injectivity of this morphisms is checked by using the pullback characterization lemma (see Appendix). □

Thus, graph transitions are characterized by replacing the double-pushout diagram of direct derivations with a double-*pullback*. In the following, if not stated otherwise, we will use this simpler definition of *DPB graph transition* in its general, injective, and faithful version, respectively.

Next, the solution to Problem 2 (recovery of information) in the case of algebraic graph transformations is given in the following theorem [10].

Theorem 3 (recovery of information). *Given a rule p_1 the following statements are equivalent:*

1. *There is a transition $(p_1, l_1, r_1) : G \rightsquigarrow H$ given by a double pullback diagram.*
2. *There are productions p_1 and p_0, such that p_0 is common subproduction of p_1 and p_2 leading to an amalgamated production $p_1 \oplus_{p_0} p_2$, and a transformation $(p_1 \oplus_{p_0} p_2, l, r) : G \Longrightarrow H$.*

In other words the change of context provided by the transition $(p_1, l_1, r_1) : G \rightsquigarrow H$ is recovered by the production p_2 which overlaps with p_1 in the common subproduction p_0.

4   Construction of DPB Transitions

In this section we first provide a necessary and sufficient condition for the existence of DPB graph transitions. Then, we give a construction of minimal graph transitions which forms the basis for constructing *all* graph transitions by suitable extensions. Finally, we define *lazy graph transitions* and show that they coincide with DPO derivations whenever the gluing condition is satisfied.

The following *weak identification condition* states that, given a production with a match, there are no conflicts between preservation and deletion of elements: If two elements from the left-hand side are identified by the match, then either both are deleted or both are preserved.

Definition 4 (weak identification condition). *Given an injective morphism* $f : K \to A$ *and a morphism* $m : A \to D$ *we say that* m *satisfies the* weak *identification condition w.r.t.* f *when* $m(a) = m(a')$ *implies* $a, a' \in f(K)$ *or* $a, a' \notin f(K)$ *for all* $a, a' \in A$.   △

The weak identification condition is necessary and sufficient for the existence of a pullback complement as shown in the following two propositions. Notice that we restrict to the case where the first of the two given morphisms (i.e., f in Figure 4) is injective. For a corresponding characterization where f need not be injective we refer to [2].

Proposition 1 (weak identification condition). *Let the square in Figure 4 on the left be a pullback with* f *an injective morphism. Then* m *satisfies the weak identification condition w.r.t.* f

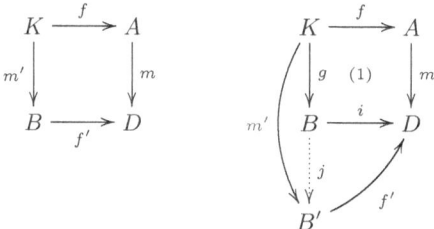

Fig. 4. Existence and minimality of pullback complements.

Assuming the weak identification condition, the next proposition gives an explicit construction of a *minimal* pullback complement. Thus, all pullback complements can be constructed by building first the minimal one and then extending it appropriately.

Proposition 2 (minimal pullback complement). *Let $f : K \rightarrow A$ in the right of Figure 4 be an injective morphism and $m : A \rightarrow D$ a morphism such that m satisfies the weak identification condition w.r.t. f.*

Let B be the image of $m \circ f$ in D. Then B together with the embedding $i : B \rightarrow D$ and the morphism $g = m \circ f : A \rightarrow D$ (considered as a morphism from K to B) is a pullback complement for f and m in $\mathcal{G}r(\Sigma)$. Moreover, it is minimal in the sense that for every pullback complement (B', f', m') of f and m there is a unique injective morphism $j : B \rightarrow B'$ such that $j \circ g = m'$ and $f' \circ j = i$.

Proof. It is clear that the square (1) in the right of Figure 4 commutes and using the pullback characterization lemma it follows that in fact it is a pullback square. If (B', f', m') is another pullback complement of f and m, we define $j : B \rightarrow B'$ as $j(b) = m'(k)$ and considering Remark 6 in the Appendix it is a well defined injective morphism. □

Figure 5 is an example of this minimal pullback complement for two different pullback complements.

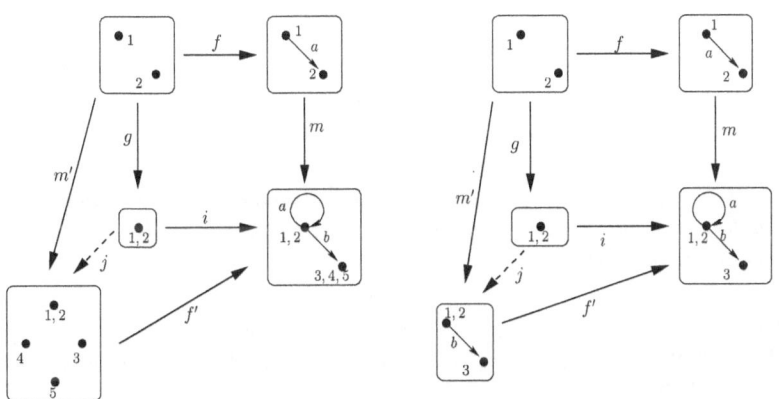

Fig. 5. Minimal pullback complement (example).

This result may be used in order to construct PB complements for a given pair of morphisms by suitable extensions of the minimal PB complement, given by factorizations $f' \circ j$ of the morphism i in the right of Figure 4. In fact, the above proposition states the *completeness* of such construction, that is, *all* PB complements can be obtained in this way.

Below we characterize those factorizations $f' \circ j = i$ that yield again PB complements as *weakly disjoint factorizations*.

Definition 5 (weakly disjoint factorization). *Given a commutative diagram like in Figure 4 on the right we say that the pair (j, f') is a weakly disjoint factorization of i w.r.t. (1) if $f'(B' - j(B)) \cap m(A) = \emptyset$.* △

Figure 6 is an example of two different factorizations (j, f') of i such that the factorization on the left is a weakly disjoint factorization and the factorization on the right is not a weakly disjoint factorization because $f'(B' - j(B)) \cap m(A) = 1, 2, 3$.

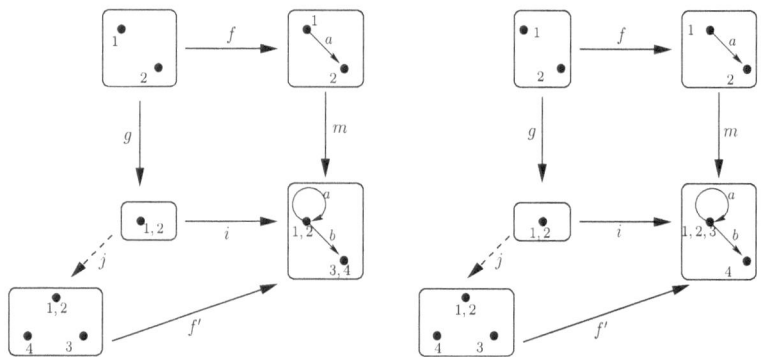

Fig. 6. Weakly disjoint factorization (example).

Construction 4 (PB complements) *Given morphisms f and m as in the right of Figure 4 such that the weak identification condition is satisfied, each PB complement m', f' of f and m can be constructed by forming first the minimal PB complement (1), choosing a weakly disjoint factorization $f' \circ j$ of i, and defining m' by $j \circ g$. Moreover, all squares constructed in this way are PB complements of f and m.*

Proof. By Lemma 5 the factorization $f' \circ j$ of i constructed in Proposition 2 is weakly disjoint, which provides the completeness of the construction. Correctness follows by Lemma 4.   □

In a similar way, all DPB graph transitions can be constructed as extensions of *minimal transitions*.

Theorem 5 (minimal DPB graph transitions). *Given a production $p = (L \leftarrow K \rightarrow R)$ and a morphism $m : L \rightarrow G$, there exists a graph H and a DPB graph transition from G to H with match m if and only if m satisfies the weak identification condition w.r.t. l.*

Moreover, in this case, a minimal *DPB graph transition with match $m : L \rightarrow G$ is constructed in the diagram below on the left, by forming first the minimal PB complement (1) and then the pushout (2). Minimality here means that for every DPB graph transition from G with the same match m to any graph H', given by the outer diagram, there exist unique morphisms $j : B \rightarrow B'$ and $h : H \rightarrow H'$ making the diagram below commute.*

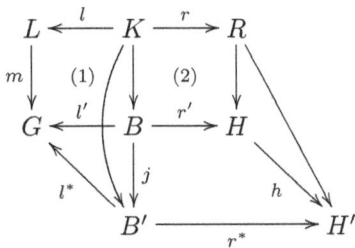 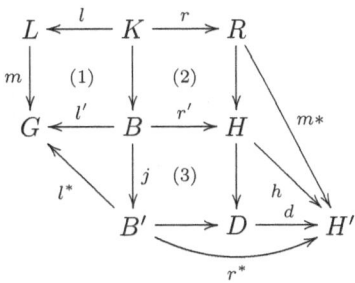

Construction 6 (DPB graph transitions) *Given production p and match m as above on the right, the outer diagram on the right above is a DPB graph transition if it is obtained by the following construction.*

1. *Construct squares (1) and (2) as minimal DPB graph transition*
2. *Choose $l^* \circ j$ as a weakly disjoint factorization of l'*
3. *Construct the pushout (3)*
4. *Choose a weakly injective morphism d and define m^* and r^* such that the diagram commutes*

Vice versa, all DPB graph transitions can be constructed in this way.

   Figure 7 is an example of a DPB construction where the left hand side is the example given in Figure 6.

Proof. By Construction 4, Lemma 2 and 3 and the fact that pushout (2+3) is also a pullback.   □

A DPB transition using a minimal pullback complement deletes as much as possible from the given graph G, that is, all items that are not in the image of the interface graph K (and are thus explicitly preserved) are deleted by the transition. Considering the operational behavior of a pushout complement (where no additional deletion takes place) we may ask for a pullback complement which deletes as *little* as possible.

Proposition 3 (maximal pullback complement). *Let $f : K \to A$ be an injective morphism and $m : A \to D$ a morphism such that m satisfies the weak identification condition w.r.t. f. Given $X = (D - m(A)) \cup mf(K)$ and*

$$D' = \{d \in X \mid \varphi^D(d) \in m(A - f(K)) \text{ for some operation } \varphi \in Op\}$$

then the graph $B = X - D'$ together with the embedding $i : B \to D$ and the morphism $g : K \to B$ consisting of $m \circ f$ is a pullback complement for f and m in $\mathcal{G}r(\Sigma)$. Moreover, it is the maximal one with an injective morphism $i : B \to D$ in the sense that for every pullback complement (B', f', m') of f and m with f' an injective morphism there is a unique injective morphism $j : B' \to B$ such that $g = j \circ m'$ and $f' = i \circ j$.

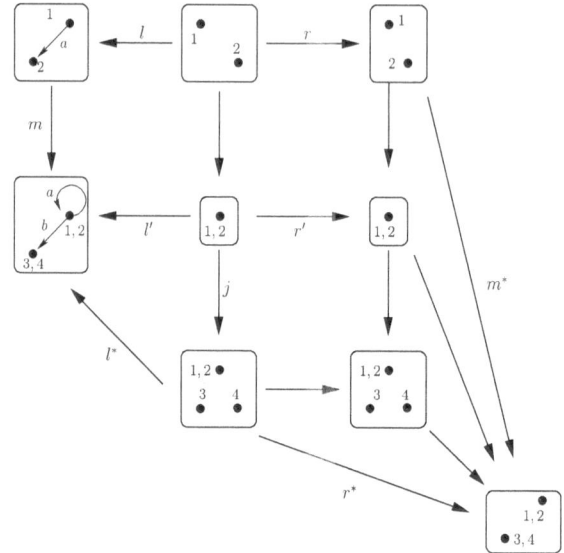

Fig. 7. Construction of DPB graph transition (example).

Proof. The weak identification condition together with the definition of D' entails that B is a graph. In a similar way like in Proposition 2 we can prove that it is in fact a pullback complement and finally, if there is another pullback complement (B', f', m') of f and m with f' an injective morphism, the pullback characterization lemma entails that $f'(B') \subset B$. The injectivity of these morphisms ends the proof. □

Remark 2. A maximal pullback complement acts like a pushout complement but for the possible deletion of dangling edges. In fact, whenever the given morphisms satisfy the gluing condition, then D' in Proposition 3 is empty such that the two constructions coincide.

Based on maximal pullback complements we now define *lazy transitions*, which produce as little unspecified effects as possible and coincide with the DPO constructions whenever the latter exist.

Definition 6 (lazy DPB transition). *A double-pullback transition is* lazy *if the left square is a maximal pullback complement and the right square is a pushout square.* △

Theorem 7 (DPO derivation as lazy DPB transition). *A lazy DPB transition whose match satisfies the gluing condition is a DPO derivation.*

There is an even stronger connection with direct derivations in the single-pushout (SPO) approach [12]. In fact, *every* lazy transition corresponds to a direct SPO derivation. However, there are more SPO derivations since the weak identification condition (usually called *conflict-freeness* in the SPO framework) does not need to be satisfied.

Example 1. Consider the transitions in Figure 8. The production in the left transition deletes two vertices 1 and 2 and generates $1'$ and $2'$. The match identifies 1 and 2, i.e., it does not satisfy the identification condition of the DPO approach [3], but the weak identification condition. Hence, there is a transition removing vertex $1 = 2$ from the given graph. Symmetrically, on the right-hand side, the transition decides to generate only one vertex $1' = 2'$ instead of two as stated in the production.

In the middle of Figure 8 there is another example of a match for a production violating the identification condition: At the same time, the production tries to preserve and to delete the vertex $1 = 2$ of the given graph. Obviously, this leads to a conflict, i.e., also the weak identification condition is violated and there is no transition using this match.

Finally, on the right-hand side, a transition is shown where the left-hand side morphism does not satisfy the dangling condition of the DPO approach [3]. Here, the transition exists and removing the dangling edge is an unspecified effect. Symmetrically, it is possible to attach edges to newly created vertices, which is shown in the right-hand side. Summarizing, transitions may not be faithful with respect to the number of deleted or generated elements and they may delete dangling edges. They are not able, however, to resolve conflicts of preservation and deletion.

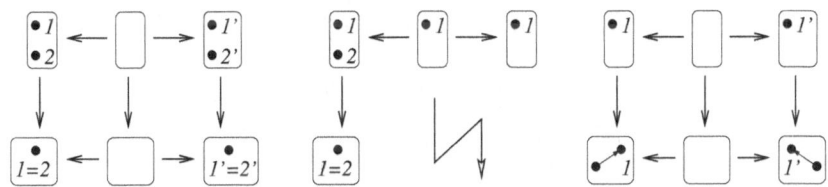

Fig. 8. Transitions that do not satisfy the gluing conditions.

5   Open Petri Nets

As a second example of our general framework we consider Petri nets and *open nets* [6,14] which are motivated by the modeling of scenarios in European Train Control Systems (ETCS) [11]. In this case systems with complete information are modeled by place-transition nets, while systems with incomplete information are modeled by open nets with distinguished open places, where autonomous actions caused by the environment are allowed. We will show that open steps defined by firing net transitions in parallel with autonomous actions on open places correspond exactly to transitions in the sense of Section 2, which are associated to transformations defined by firing net transitions.

In the context of this paper it suffices to consider the following notion of open nets based on place transition nets. According to the presentation of Petri nets by monoids [13], pre- and post-domain of net transitions are defined by functions $pre, post : T \to P^\oplus$ from the set of transitions T to the free commutative monoid $P^\oplus$ over the set of places P. That means, a net transition t like

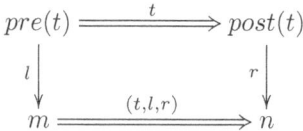

is defined by $pre(t) = 2p_1 + 3p_2$ and $post(t) = p_3 + 2p_4$.

An open net $ON = (T, P, pre, post, P_0)$ consists of a set T of net transitions, a set P of net places with subset P_0 of open places, and two functions $pre, post : T \to P^\oplus$ assigning to each net transition t the predomain $pre(t)$ and postdomain $post(t)$ of t as elements of the free commutative monoid $P^\oplus$ over P. Given a marking $m \in P^\oplus$ a net transition $t \in T$ is enabled if we have $pre(t) \leq m$, i.e., if there is a marking l such that $m = pre(t) + l$. In this case a firing step is given by $m[t\rangle n$ where $n = m - pre(t) + post(t)$.

An open step $t : m' \rightsquigarrow n'$ is then given by a firing $m[t\rangle n$ and markings $m_0, n_0 \in P_0^\oplus$ such that $m' = m + m_0, n' = n + n_0$. Hence, markings m_0 and n_0 represent tokens that are spontaneously deleted or added on open places extending in this way the effect of the net transition t.

Remark 3 (relationship to general framework). A rule in the general framework corresponds to a net transition $t : pre(t) \implies post(t)$. The existence of a left redex $l : pre(t) \to m$ means that $m = pre(t) + l$, similarly $r : post(t) \to n$ means that $n = post(t) + r$. Now a transformation $(t, l, r) : m \implies n$ is given by

$$
\begin{array}{ccc}
pre(t) & \overset{t}{=\!=\!=\!\Longrightarrow} & post(t) \\
l \downarrow & & \downarrow r \\
m & \underset{(t,l,r)}{=\!=\!=\!\Longrightarrow} & n
\end{array}
$$

where $n = m - pre(t) + post(t)$. Hence the transformation $(t, l, r) : m \implies n$ corresponds to a firing step $m[t\rangle n$.

Now we consider redex extensions $m_0, n_0 \in P_0^\oplus$, denoted by $m_0 : m \to m', n_0 : n \to n'$ with $m' = m + m_0$ and $n' = n + n_0$ leading to a transition $(t, l', r') : m' \rightsquigarrow n'$ associated to the transformation $(t, l, r) : m \implies n$.

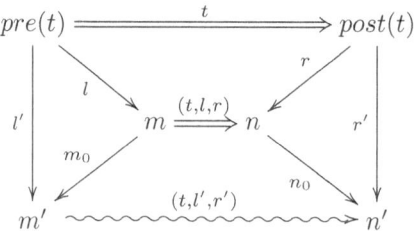

This means that a transition $(t, l', r') : m' \rightsquigarrow n'$ associated to a transformation $(t, l, r) : m \Longrightarrow n$ via (m_0, n_0) in the sense of Definition 1 corresponds exactly to an open step $t : m' \rightsquigarrow n'$ in the sense of open nets (see above).

Next we provide for open nets the solutions to Problem 1 and 2. The proofs of the following theorems are reported in [5].

Theorem 8 (characterization of transitions for open nets). *Given an open net* $ON = (T, P, pre, post, P_0)$, *markings* $m', n' \in P_0^{\oplus}$ *and a rule (a net transition)* $t : pre(t) \Longrightarrow post(t)$ *we have an open step*

$$t : m' \rightsquigarrow n'$$

if and only if we have a transition step $t : m' \rightsquigarrow n'$ *where*

(a) $pre(t) \leq m'$
(b) $post(t) \leq n'$
(c) there is $m_0 \in P_0^{\oplus}$ *such that*
   (c1) $m_0 \leq m' - pre(t)$
   (c2) $m_0 + (n' - m') - (post(t) - pre(t)) \in P_0^{\oplus}$

Remark 4. Condition (c) is equivalent to the following condition

(d) there is $n_0 \in P_0^{\oplus}$ *such that*
   (d1) $n_0 \leq n' - post(t)$
   (d2) $n_0 + (m' - n') - (pre(t) - post(t)) \in P_0^{\oplus}$

In (c2) and (d2) we don't require that each subterm is in $P_0^{\oplus}$, but rather we require that the sum / difference of all subterms is an element of $P_0^{\oplus}$.

Theorem 9 (recovery of information for open nets). *Given an open net* $ON = (T, P, pre, post, P_0)$, *a net transition* $t \in T$ *and markings* $m', n' \in P_0^{\oplus}$ *then the following statements are equivalent:*

1. *There is an open step* $t : m' \rightsquigarrow n'$
2. *There is a transition step* $t : m' \rightsquigarrow n'$ *(see Theorem 8)*
3. *There are* $m_0, n_0 \in P_0^{\oplus}$ *defining a new net transition* $t_0 \notin T$ *with* $pre(_0) = m_0, post(t_0) = n_0$ *and in the new net* $N = (T + \{t_0\}, P, pre, post)$ *there is a concurrent step* $m'[t + t_0\rangle n'$
4. *Define for each open place* $p \in P_0$ *two new net transitions* $t_p^+, t_p^- (\notin T)$ *with* $pre(t_p^+) = 0 = post(t_p^-)$ *and* $post(t_p^+) = p = pre(t_p^-)$ *leading to the new net* $ON^+ = (T + \{t_p^+, t_p^- \mid p \in Pi\}, P, pre, post)$, *called closure of the open net* ON. *Then there are* $\lambda_p, \mu_p \in \mathbb{N}(p \in P_0)$ *with finite* P_0-*support (i.e.* λ_p, μ_p *are zero for all but a finite number of* $p \in P_0$) *and there is a concurrent step in* ON^+ *as follows:*

$$m'[t + \sum_{p \in P_0} (\lambda_p t_p^+ + \mu_p t_p^-)\rangle n'$$

Remark 5. Both nets N and ON^+ are place transition nets without open places. But the net N depends on the markings $m_0, n_0 \in P_0^{\oplus}$, while ON^+ depends only on P_0. This means ON^+ is independent of the specific choice of an open step, which includes a choice of $m_0, n_o \in P_0^{\oplus}$. Although for infinite P_0 we may have infinitely many new transitions t_p^+, t_p^- the sum ,above is finite, because $\lambda_p, \mu_p \in \mathbb{N}(p \in P_0)$ are assumed to have finite P_0-support.

6   Conclusion

In this paper we have provided a general framework for defining rule-based transformations with incomplete information, called *transitions*. In addition to the changes caused by the rule applications a transition also allows a change of context. As main examples of this general framework we considered double-pullback graph transitions and open firing steps of (open) Petri nets.

The main results of this paper analyze transitions in both approaches w.r.t. the following problems:

1. Characterization of transitions independent of the notion of transformation
2. Characterization of transitions in terms of parallel transformations

Moreover, in the graph transformation context we studied in detail the construction of double-pullback transitions. In analogy to the gluing condition in the DPO-approach we presented a weak identification condition which is necessary and sufficient for the existence of a DPB-transition for a given rule and redex.

In the DPO-approach the result of a production application (whenever defined) is determined up to isomorphism by the production and its match. This is essentially different for DPB transitions where a variety of non-isomorphic results can be obtained. In order to allow an explicit construction we have characterized minimal DPB-transitions and presented a construction for all possible DPB-transitions for a given rule and redex. Moreover, in [4], we have characterized the existence and construction of DPB transitions for a given production with left and right match.

These results can be considered as a starting point for a general theory of DPB-transitions similar to that of DPO-transformations presented in [3]. In particular, rewriting properties like parallel and sequential independence of DPB-transitions can be defined in the same way like for DPO-transformations. This allows to prove a local Church-Rosser theorem for injective DPB-transitions as well as a weaker variant of the parallelism theorem without a bijective correspondence between parallel DPB-transitions and sequentially independent DPB-transition sequences, see [4].

References

1. M. Bauderon. A category-theoretical approach to vertex replacement: The generation of infinite graphs. In *5th Int. Workshop on Graph Grammars and their Application to Computer Science, Williamsburg '94, LNCS 1073*, pages 27 – 37, Springer Verlag, 1996. 88
2. M. Bauderon and H. Jacquet. Categorical product as a generic graph rewriting mechanism. *Applied Categorical Structures*, 1999. To appear. Also Tech. Rep. 1166-97, University of Bordeaux. 88, 92
3. A. Corradini, U. Montanari, F. Rossi, H. Ehrig, R. Heckel, and M. Löwe. Algebraic approaches to graph transformation, Part I: Basic concepts and double pushout approach. In Rozenberg [15], pages 163–245. 89, 97, 100

4. H. Ehrig, R. Heckel, M. Llabres, and F. Orejas. Basic properties of double pullback graph transitions. Technical Report 99-02, TU Berlin, 1999. 100, 101
5. H. Ehrig, R. Heckel, J. Padberg, and G. Rozenberg. Graph transformation and other rule-based formalisms with incomplete information. In *Prelim. Proc. 6th Int. Workshop on Theory and Application of Graph Transformation (TAGT'98)*, *Paderborn*, 1998. 99
6. H. Ehrig, A. Merten, and J. Padberg. How to transfer concepts of abstract data types to petri nets? *EATCS Bulletin, 62*, pages 106–114, 1997. 86, 97
7. J. Engelfriet and G. Rozenberg. Node replacement graph grammars. In Rozenberg [15], pages 1 – 94. 88
8. G. Engels, R. Heckel, G. Taentzer, and H. Ehrig. A combined reference model- and view-based approach to system specification. *Int. Journal of Software and Knowledge Engeneering*, 7(4):457–477, 1997. 86
9. R. Heckel. *Open Graph Transformation Systems: A New Approach to the Compositional Modelling of Concurrent and Reactive Systems*. PhD thesis, TU Berlin, 1998. 90
10. R. Heckel, H. Ehrig, U. Wolter, and A. Corradini. Double-pullback transitions and coalgebraic loose semantics for graph transformation systems. *Applied Categorical Structures*, 1999. To appear, see also TR 97-07 at `http://www.cs.tu-berlin.de/cs/ifb/TechnBerichteListe.html`. 86, 90, 91
11. A. Janhsen, K. Lemmer, B. Ptok, and E. Schnieder. Formal specifications of the european train control system. In *IFAC Transportation Systems, 8th Symposium on Transportation Systems*, 1997. 97
12. M. Löwe. Algebraic approach to single-pushout graph transformation. *TCS*, 109:181–224, 1993. 96
13. J. Meseguer and U. Montanari. Petri nets are monoids. *Information and Computation*, 88(2):105–155, 1990. 98
14. J. Padberg, L. Jansen, R. Heckel, and H. Ehrig. Interoperability in train control systems: Specification of scenarios using open nets. In *Proc. Integrated Design and Process Technology (IDPT'98), Berlin*, 1998. 86, 97
15. G. Rozenberg, editor. *Handbook of Graph Grammars and Computing by Graph Transformation, Volume 1: Foundations*. World Scientific, 1997. 100, 101

Appendix: Pullbacks and Pullback Complements

We recall the construction of pullbacks in $\mathcal{G}r(\Sigma)$ and give a set-theoretic characterization. Then, we list some properties of pullbacks and pullback complements that are needed throughout the paper. For more details see [4].

Construction 10 (pullbacks of graphs) *Let $B \xrightarrow{f} D \xleftarrow{g} A$ be graphs and graph morphisms and $X = \{(a, b) \in A \times B \mid g(a) = f(b)\}$. Then, the subgraph K of the product $A \times B$ generated by X together with the morphisms $f' = \pi_A \circ i_K$ and $f' = \pi_B \circ i_K$, is the pullback of the pair f, g.*

Remark 6. Notice that by the pullback construction, if f' is injective then f restricted to $f^{-1}(g(A))$ is injective.

Lemma 1 (pullback characterization). *Given any commutative diagram like below on the left, it is a pullback square if and only if for every $a \in A$ and $b \in B$ such that $f(b) = g(a)$ there exists a unique $k \in K$ such that $f'(k) = a$ and $g'(k) = b$.* □

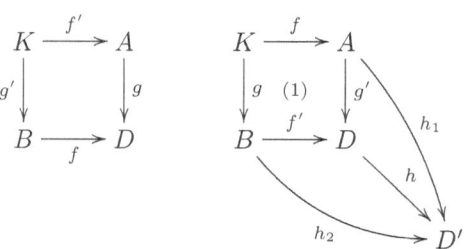

Lemma 2 (pullback extension). *Given the commutative diagram on the right above, if (1) is a pushout square, f is injective and $h : D \to D'$ is weak injective w.r.t. (1) then the outer diagram is a pullback square.* □

Lemma 3 (weak injectivity). *Given a commutative diagram like on the left below where the square in the top is a pushout square, the outer diagram is a pullback square and f,i are injective morphism then h is weak injective w.r.t. (1).* □

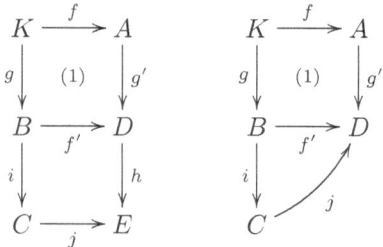

Lemma 4 (pullback complement extension). *Given a commutative diagram like on the right above where (1) is a pullback square, f and i are injective morphism and the pair (i,j) is a weakly disjoint factorization of f' w.r.t. (1) then the outer diagrams is a pullback square.* □

Lemma 5 (weak disjointness). *Given a commutative diagram like on the right above where (1) and the outer diagram are pullback squares and f is injective, then (i,j) is a weakly disjoint factorization of f' w.r.t. (1).* □

Double-Pushout Approach
with Injective Matching⋆

Annegret Habel[1], Jürgen Müller[2], and Detlef Plump[3]

[1] Universität Oldenburg, Fachbereich Informatik
Postfach 2503, 26111 Oldenburg, Germany
Annegret.Habel@informatik.uni-oldenburg.de
[2] Universitá di Pisa, Dipartimento di Informatica
Corso Italia 40, 56125 Pisa, Italy.
jmueller@di.unipi.it
[3] Universität Bremen, Fachbereich Mathematik und Informatik
Postfach 33 04 40, 28334 Bremen, Germany
det@informatik.uni-bremen.de

Abstract. We investigate and compare four variants of the double-pushout approach to graph transformation. Besides the traditional approach with arbitrary matching and injective right-hand morphisms, we consider three variations by employing injective matching and/or arbitrary right-hand morphisms in rules. For each of the three variations, we clarify whether the well-known commutativity theorems are still valid and–where this is not the case–give modified results. In particular, for the most general approach with injective matching and arbitrary right-hand morphisms, we establish sequential and parallel commutativity by appropriately strengthening sequential and parallel independence. We also show that injective matching provides additional expressiveness in two respects, viz. for generating graph languages by grammars without nonterminals and for computing graph functions by convergent graph transformation systems.

1   Introduction

In [EPS73], the first paper on double-pushout graph transformation, matching morphisms are required to be injective. But the vast majority of later papers on the double-pushout approach–including the surveys [Ehr79,CMR+97]–considers arbitrary matching morphisms. Despite this tradition, sometimes it is more natural to require that matching morphisms must be injective. For example, the set of all (directed, unlabelled) loop-free graphs can be generated from the empty

---

⋆ This research was partly supported by the ESPRIT Working Group APPLIGRAPH. Research of the second author was also supported by the EC TMR Network GET-GRATS, through the Universities of Antwerp and Pisa. Part of the work of the third author was done while he was visiting the Vrije Universiteit in Amsterdam.

H. Ehrig et al. (Eds.): Graph Transformation, LNCS 1764, pp. 103–117, 2000.

graph by the following two rules if injective matching is assumed:

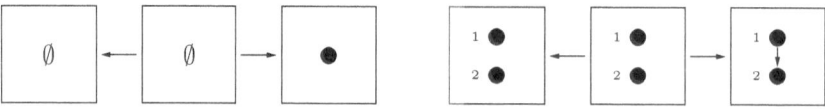

We will prove that in the traditional double-pushout approach, this graph class cannot be generated without nonterminal labels.

To give a second, non-grammatical example, consider the problem of merging in a graph all nodes labelled with some fixed label **a**. This is easily achieved–assuming injective matching–by applying as long as possible the following rule:

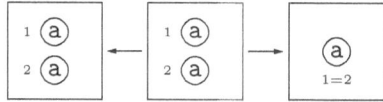

We will show that without injective matching, there is no finite, convergent graph transformation system–possibly containing identifying rules–solving this problem for arbitrary input graphs. (A graph transformation system is *convergent* if repeated rule applications to an input graph always terminate and yield a unique output graph.)

We consider two variants of the double-pushout approach in which matching morphisms are required to be injective. We denote these approaches by $\text{DPO}^{i/i}$ and $\text{DPO}^{a/i}$, where the first component of the exponent indicates whether right-hand morphisms in rules are injective or arbitrary, and where the second component refers to the requirement for matching morphisms. (So our second example belongs to $\text{DPO}^{a/i}$.)

Besides the traditional approach $\text{DPO}^{i/a}$, we will also consider $\text{DPO}^{a/a}$. Obviously, $\text{DPO}^{a/a}$ and $\text{DPO}^{a/i}$ contain $\text{DPO}^{i/a}$ and $\text{DPO}^{i/i}$, respectively, as the rules of the latter approaches are included in the former approaches. Moreover, using a quotient construction for rules, we will show that $\text{DPO}^{i/i}$ and $\text{DPO}^{a/i}$ can simulate $\text{DPO}^{i/a}$ and $\text{DPO}^{a/a}$, respectively, in a precise and strong way. Thus the relationships between the approaches can be depicted as follows, where "$\hookrightarrow$" means "is included in" and "$\rightarrow$" means "can be simulated by":

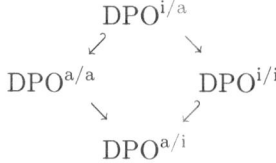

The question, then, arises to what extent the theory of $\text{DPO}^{i/a}$ carries over to the stronger approaches. We answer this question for the classical commutativity theorems, by either establishing their validity or by giving counterexamples and providing modified results.

Lack of space prevents us to give all the proofs of our results. The omitted proofs can be found in the long version of this paper, see [HMP99].

2   Preliminaries

In this section the double-pushout approach to graph transformation is briefly reviewed. For a comprehensive survey, the reader may consult [Ehr79,CMR$^+$97].

A *label alphabet* $C = \langle C_V, C_E \rangle$ is a pair of sets of *node labels* and *edge labels*. A *graph* over C is a system $G = (V, E, s, t, l, m)$ consisting of two finite sets V and E of *nodes* (or *vertices*) and *edges*, two *source* and *target functions* $s, t : E \rightarrow V$, and two *labelling functions* $l : V \rightarrow C_V$ and $m : E \rightarrow C_E$.

A *graph morphism* $g : G \rightarrow H$ between two graphs G and H consists of two functions $g_V : V_G \rightarrow V_H$ and $g_E : E_G \rightarrow E_H$ that preserve sources, targets, and labels, that is, $s_H \circ g_E = g_V \circ s_G$, $t_H \circ g_E = g_V \circ t_G$, $l_H \circ g_V = l_G$, and $m_H \circ g_E = m_G$. The graphs G and H are the *domain* and *codomain* of g, respectively. A morphism g is *injective* (*surjective*) if g_V and g_E are injective (surjective), and an *isomorphism* if it is both injective and surjective. In the latter case G and H are *isomorphic*, which is denoted by $G \cong H$.

A *rule* $p = \langle L \leftarrow K \rightarrow R \rangle$ consists of two graph morphisms with a common domain, where we throughout assume that $K \rightarrow L$ is an inclusion. Such a rule is *injective* if $K \rightarrow R$ is injective. Given two graphs G and H, G *directly derives* H *through* p, denoted by $G \Rightarrow_p H$, if the diagrams (1) and (2) below are graph pushouts (see [Ehr79] for the definition and construction of graph pushouts).

$$
\begin{array}{ccccc}
L & \longleftarrow & K & \longrightarrow & R \\
\downarrow & (1) & \downarrow & (2) & \downarrow \\
G & \longleftarrow & D & \longrightarrow & H
\end{array}
$$

The notation $G \Rightarrow_{p,g} H$ is used when $g : L \rightarrow G$ shall be made explicit.

The *application* of a rule $p = \langle L \leftarrow K \rightarrow R \rangle$ to a graph G amounts to the following steps:

(1) Find a graph morphism $g : L \rightarrow G$ and check the following *gluing condition*:
Dangling condition. No edge in $G - g(L)$ is incident to a node in $g(L) - g(K)$.
Identification condition. For all distinct items $x, y \in L$, $g(x) = g(y)$ only if $x, y \in K$. (This condition is understood to hold separately for nodes and edges.)
(2) Remove $g(L) - g(K)$ from G, yielding a graph D, a graph morphism $K \rightarrow D$ (which is the restriction of g), and the inclusion $D \rightarrow G$.
(3) Construct the pushout of $D \leftarrow K \rightarrow R$, yielding a graph H and graph morphisms $D \rightarrow H \leftarrow R$.

3   Three Variations of the Traditional Approach

A direct derivation $G \Rightarrow_{p,g} H$ is said to be in

- DPO$^{i/a}$ if p is injective and g is arbitrary (the "traditional approach"),
- DPO$^{a/a}$ if p and g are arbitrary,
- DPO$^{i/i}$ if p and g are injective,
- DPO$^{a/i}$ if p is arbitrary and g is injective.

Note that in DPO$^{i/i}$ and DPO$^{a/i}$, step (1) in the application of a rule is simpler than above because the gluing condition reduces to the dangling condition.

We now show that DPO$^{i/a}$ and DPO$^{a/a}$ can be simulated by DPO$^{i/i}$ and DPO$^{a/i}$, respectively. The idea is to replace in a graph transformation system each rule p by a finite set of rules $Q(p)$ such that every application of p corresponds to an application of a rule in $Q(p)$ obeying the injectivity condition.

Definition 1 (Quotient rule). *Given a rule* $p = \langle L \leftarrow K \rightarrow R \rangle$, *a rule* $p' = \langle L' \leftarrow K' \rightarrow R' \rangle$ *is a* quotient rule *of* p *if there is a surjective graph morphism* $K \rightarrow K'$ *such that* L' *and* R' *are the pushout objects of* $L \leftarrow K \rightarrow K'$ *and* $R \leftarrow K \rightarrow K'$, *respectively. The set of quotient rules of* p *is denoted by* $Q(p)$.

Since the number of non-isomorphic images K' of K is finite, we can without loss of generality assume that $Q(p)$ is finite. Note also that every quotient rule of an injective rule is injective.

Lemma 1 (Quotient Lemma). *For all graphs* G *and* H, *and all rules* p:

1. $G \Rightarrow_{Q(p)} H$ *implies* $G \Rightarrow_p H$.
2. $G \Rightarrow_{p,g} H$ *implies* $G \Rightarrow_{p',g'} H$ *for some* $p' \in Q(p)$ *and some injective* g'.

Proof. 1. Let $G \Rightarrow_{p'} H$ for some $p' = \langle L' \leftarrow K' \rightarrow R' \rangle$ in $Q(p)$. Then the diagrams (1), (2), (1') and (2') below are graph pushouts. By the composition property for pushouts, the composed diagrams (1)+(1') and (2)+(2') are pushouts as well. Hence $G \Rightarrow_p H$.

2. Let $G \Rightarrow_{p,g} H$ and (e, g') be an epi-mono factorization of g, that is, $e : L \rightarrow L'$ is a surjective morphism and $g' : L' \rightarrow G$ is an injective morphism such that $g = g' \circ e$. Since g satisfies the gluing condition with respect to p, e satisfies the gluing condition with respect to p as well: As e is surjective, it trivially satisfies the dangling condition; since g satisfies the identification condition, e also satisfies this condition. Therefore, a quotient rule $p' = \langle L' \leftarrow K' \rightarrow R' \rangle$ of p can be constructed by a direct derivation $L' \Rightarrow_{p,e} R'$. Moreover, g' satisfies the gluing condition with respect to p': Since g' is injective, g' satisfies the identification condition; since g satisfies the dangling condition with respect to p, g' satisfies the dangling with respect to p'. Therefore, there exists a direct derivation $G \Rightarrow_{p',g'} H'$. Then, by the composition property for pushouts, there is also a direct derivation $G \Rightarrow_{p,g} H'$. Since the result of such a step is unique up to isomorphism, we have $H' \cong H$ and hence $G \Rightarrow_{p',g'} H$.

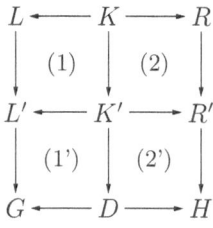
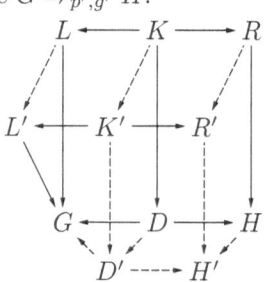

$\square$

Theorem 1 (Simulation Theorem). *Let $x \in \{a, i\}$. Then for all graphs G and H, and every rule p, $G \Rightarrow_p H$ in $DPO^{x/a}$ if and only if $G \Rightarrow_{Q(p)} H$ in $DPO^{x/i}$.*

Proof. Immediate consequence of Lemma 1.

4   Parallel Independence

In this section we consider pairs of direct derivations $H \Leftarrow_{p_1} G \Rightarrow_{p_2} \bar{H}$ and look for conditions under which there are direct derivations $H \Rightarrow_{p_2} M \Leftarrow_{p_1} \bar{H}$ or $H \Rightarrow_{Q(p_2)} M \Leftarrow_{Q(p_1)} \bar{H}$. We fomulate the notions of parallel and strong parallel independence and present three parallel commutativity results.

In the following let $p_i = \langle L_i \leftarrow K_i \rightarrow^{r_i} R_i \rangle$, for $i = 1, 2$. (In the diagrams of this section, the morphism r_1 appears to the left of K_1.)

Definition 2 (Parallel independence). *Two direct derivations $H \Leftarrow_{p_1} G \Rightarrow_{p_2} \bar{H}$ are parallelly independent if there are graph morphisms $L_1 \rightarrow D_2$ and $L_2 \rightarrow D_1$ such that $L_1 \rightarrow D_2 \rightarrow G = L_1 \rightarrow G$ and $L_2 \rightarrow D_1 \rightarrow G = L_2 \rightarrow G$.*

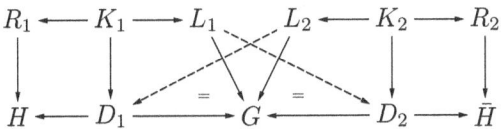

The notion of parallel independence and its following characterization are well-known, see [EK76] and [ER76], respectively.

Lemma 2 (Characterization of parallel independence). *For $x, y \in \{a, i\}$, two direct derivations $H \Leftarrow_{p_1, g_1} G \Rightarrow_{p_2, g_2} \bar{H}$ in $DPO^{x/y}$ are parallelly independent if and only if the intersection of L_1 and L_2 in G consists of common gluing items, that is, $g_1(L_1) \cap g_2(L_2) \subseteq g_1(K_1) \cap g_2(K_2)$.*

$$R_1 \longleftarrow K_1 \longrightarrow L_1 \qquad L_2 \longleftarrow K_2 \longrightarrow R_2$$
$$\downarrow \qquad \downarrow \qquad \searrow^{g_1} \quad \swarrow^{g_2} \qquad \downarrow \qquad \downarrow$$
$$H \longleftarrow D_1 \longrightarrow G \longleftarrow D_2 \longrightarrow \bar{H}$$

Theorem 2 (Parallel commutativity I). *In $DPO^{i/a}$, $DPO^{a/a}$ and $DPO^{i/i}$, for every pair of parallelly independent direct derivations $H \Leftarrow_{p_1} G \Rightarrow_{p_2} \bar{H}$ there are two direct derivations of the form $H \Rightarrow_{p_2} M \Leftarrow_{p_1} \bar{H}$.*

Proof. In [EK76,Ehr79], the statement is proved for direct derivations in $DPO^{i/a}$ and $DPO^{a/a}$. For $DPO^{i/i}$, the statement follows by inspecting the proof: If the original derivations are in $DPO^{i/i}$, then all morphisms occurring in the proof are injective. Consequently, the direct derivations $H \Rightarrow_{p_2} M \Leftarrow_{p_1} \bar{H}$ are in $DPO^{i/i}$. $\square$

The following counterexample demonstrates that the parallel commutativity property does not hold for direct derivations in $\text{DPO}^{\text{a/i}}$.

Example 1. Consider the following direct derivations.

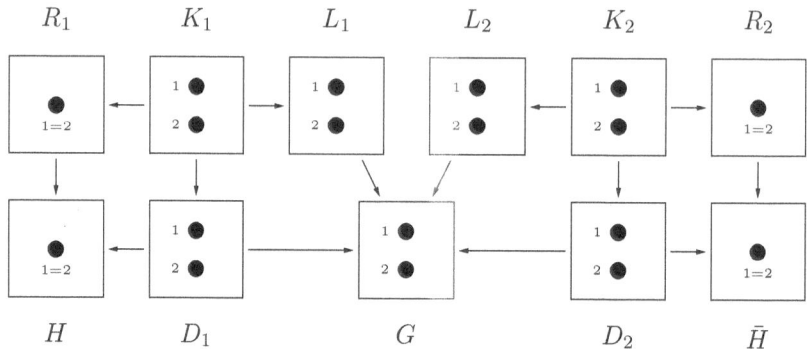

$$R_1 \qquad K_1 \qquad L_1 \qquad L_2 \qquad K_2 \qquad R_2$$

$$H \qquad D_1 \qquad G \qquad D_2 \qquad \bar{H}$$

The direct derivations are parallelly independent, but no direct derivations of the form $H \Rightarrow_{p_2} M \Leftarrow_{p_1} \bar{H}$ exist in $\text{DPO}^{\text{a/i}}$. The reason is that the composed morphisms $L_1 \rightarrow D_2 \rightarrow \bar{H}$ and $L_2 \rightarrow D_1 \rightarrow H$ are not injective.

This counterexample suggests to strengthen parallel independence as follows.

Definition 3 (Strong parallel independence). *Two direct derivations $H \Leftarrow_{p_1} G \Rightarrow_{p_2} \bar{H}$ are strongly parallelly independent if there are graph morphisms $L_1 \rightarrow D_2$ and $L_2 \rightarrow D_1$ such that (a) $L_1 \rightarrow D_2 \rightarrow G = L_1 \rightarrow G$ and $L_2 \rightarrow D_1 \rightarrow G = L_2 \rightarrow G$ and (b) $L_1 \rightarrow D_2 \rightarrow \bar{H}$ and $L_2 \rightarrow D_1 \rightarrow H$ are injective.*

Theorem 3 (Parallel commutativity II). *In $\text{DPO}^{\text{a/i}}$, for every pair of strongly parallelly independent direct derivations $H \Leftarrow_{p_1} G \Rightarrow_{p_2} \bar{H}$ there are two direct derivations of the form $H \Rightarrow_{p_2} M \Leftarrow_{p_1} \bar{H}$.*

Proof. Let $H \Leftarrow_{p_1,g_1} G \Rightarrow_{p_2,g_2} \bar{H}$ in $\text{DPO}^{\text{a/i}}$ be strongly parallelly independent. Then the two steps are in $\text{DPO}^{\text{a/a}}$ and are parallelly independent. By the proof of Theorem 2, there are direct derivations $H \Rightarrow_{p_2,g_2'} M \Leftarrow_{p_1,g_1'} \bar{H}$ in $\text{DPO}^{\text{a/a}}$ such that $g_2' = L_2 \rightarrow D_1 \rightarrow H$ and $g_1' = L_1 \rightarrow D_2 \rightarrow \bar{H}$. Since both morphisms are injective by strong parallel independence, the derivations are in $\text{DPO}^{\text{a/i}}$.   □

Note that by Theorem 2 and the fact that strong parallel independence implies parallel independence, Theorem 3 holds for $\text{DPO}^{\text{i/a}}$, $\text{DPO}^{\text{a/a}}$ and $\text{DPO}^{\text{i/i}}$ as well.

Theorem 4 (Parallel commutativity III). *In $\text{DPO}^{\text{a/i}}$, for every pair of parallelly independent direct derivations $H \Leftarrow_{p_1} G \Rightarrow_{p_2} \bar{H}$ there are two direct derivations of the form $H \Rightarrow_{Q(p_2)} M \Leftarrow_{Q(p_1)} \bar{H}$.*

Proof. Let $H \Leftarrow_{p_1} G \Rightarrow_{p_2} \bar{H}$ in $\text{DPO}^{a/i}$ be parallelly independent. Then the derivations are in $\text{DPO}^{a/a}$ and, by Theorem 2, there are direct derivations $H \Rightarrow_{p_2} M \Leftarrow_{p_1} \bar{H}$ in $\text{DPO}^{a/a}$. Hence, by the Simulation Theorem, there are direct derivations $H \Rightarrow_{Q(p_2)} M \Leftarrow_{Q(p_1)} \bar{H}$ in $\text{DPO}^{a/i}$.   □

By Theorem 2 and the fact that every rule p is contained in $Q(p)$, Theorem 4 holds for $\text{DPO}^{i/a}$, $\text{DPO}^{a/a}$ and $\text{DPO}^{i/i}$ as well.

5   Sequential Independence

We now switch from parallel independence to sequential independence, looking for conditions under which two consecutive direct derivations can be interchanged.

Definition 4 (Sequential independence). *Two direct derivations* $G \Rightarrow_{p_1} H \Rightarrow_{p_2} M$ *are sequentially independent if there are morphisms* $R_1 \rightarrow D_2$ *and* $L_2 \rightarrow D_1$ *such that* $R_1 \rightarrow D_2 \rightarrow H = R_1 \rightarrow H$ *and* $L_2 \rightarrow D_1 \rightarrow H = L_2 \rightarrow H$.

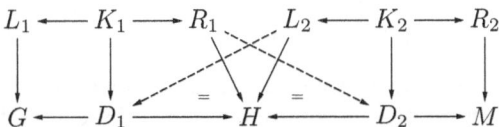

The following characterization of sequential independence can be proved analogously to Lemma 2, see [ER76].

Lemma 3 (Characterization of sequential independence). *For* $y \in \{a, i\}$, *two direct derivations* $G \Rightarrow_{p_1, g_1} H \Rightarrow_{p_2, g_2} M$ *in* $\text{DPO}^{i/y}$ *are sequentially independent if and only if the intersection of* R_1 *and* L_2 *in* H *consists of common gluing items, that is,* $h_1(R_1) \cap g_2(L_2) \subseteq h_1(r_1(K_1)) \cap g_2(K_2)$.

$$L_1 \longleftarrow K_1 \xrightarrow{\;r_1\;} R_1 \qquad L_2 \longleftarrow K_2 \longrightarrow R_2$$
$$G \longleftarrow D_1 \xrightarrow{\;\;\;\;\;\;\;} H \xleftarrow{\;\;\;\;\;\;\;} D_2 \longrightarrow M$$

This characterization may break down, however, in the presence of non-injective rules. So Lemma 3 does not hold in $\text{DPO}^{a/a}$ and $\text{DPO}^{a/i}$, as is demonstrated by the next example.

Example 2. Consider the following direct derivations.

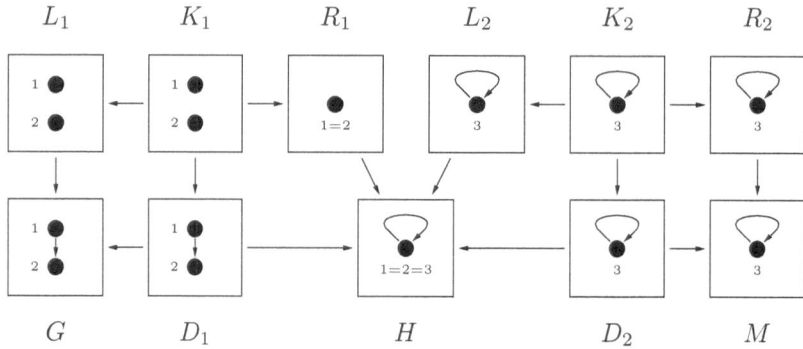

The intersection of R_1 and L_2 in H consists of common gluing items. But there does not exist a graph morphism $L_2 \rightarrow D_1$ (see also [Mül97], Example 2.5).

Theorem 5 (Sequential commutativity I). *In* $\mathrm{DPO}^{\mathrm{i/a}}$, $\mathrm{DPO}^{\mathrm{a/a}}$, $\mathrm{DPO}^{\mathrm{i/i}}$, *for every pair of sequentially independent direct derivations* $G \Rightarrow_{p_1} H \Rightarrow_{p_2} M$ *there are two sequentially independent direct derivations* $G \Rightarrow_{p_2} \bar{H} \Rightarrow_{p_1} M$.

Proof. In [EK76], the statement is proved for direct derivations in $\mathrm{DPO}^{\mathrm{i/a}}$ and $\mathrm{DPO}^{\mathrm{a/a}}$. For $\mathrm{DPO}^{\mathrm{i/i}}$, the statement follows by inspecting the proof: If the original derivations are in $\mathrm{DPO}^{\mathrm{i/i}}$, then all morphisms occurring in the proof become injective. Consequently, the direct derivations $G \Rightarrow_{p_2} \bar{H} \Rightarrow_{p_1} M$ are in $\mathrm{DPO}^{\mathrm{i/i}}$. (A self-contained proof of Theorem 5 can be found in [HMP99].) $\square$

Example 3. The following direct derivations demonstrate that in $\mathrm{DPO}^{\mathrm{a/i}}$, sequential independence does not guarantee sequential commutativity:

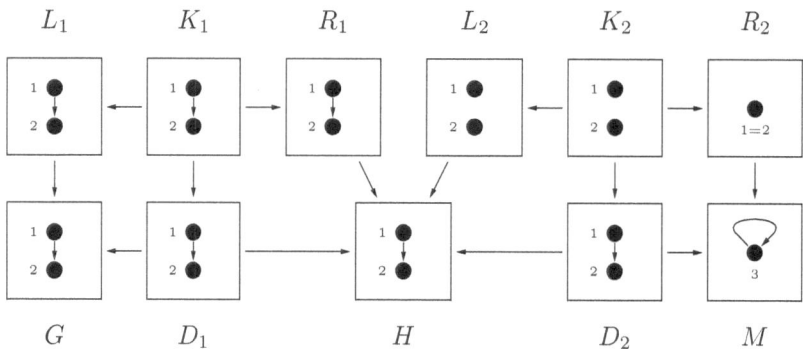

The two steps are sequentially independent. Moreover, there exists a direct derivation of the form $G \Rightarrow_{p_2} \bar{H}$ in $\mathrm{DPO}^{\mathrm{a/i}}$. But there is no step $\bar{H} \Rightarrow_{p_1} M$ in $\mathrm{DPO}^{\mathrm{a/i}}$. The reason is that the composed morphism $R_1 \rightarrow D_2 \rightarrow M$ is not injective (see also [Mül97], Example 2.1).

We now strengthen sequential independence by requiring that $R_1 \rightarrow D_2 \rightarrow M$ is injective. We need not require $L_2 \rightarrow D_1 \rightarrow G$ to be injective, though, because this follows from the injectivity of $L_2 \rightarrow H$ and $K_1 \rightarrow L_1$.

Definition 5 (Strong sequential independence). *Two direct derivations* $G \Rightarrow_{p_1} H \Rightarrow_{p_2} M$ *are strongly sequentially independent if there are graph morphisms* $R_1 \to D_2$ *and* $L_2 \to D_1$ *such that (a)* $R_1 \to D_2 \to H = R_1 \to H$ *and* $L_2 \to D_1 \to H = L_2 \to H$ *and (b)* $R_1 \to D_2 \to M$ *is injective.*

Theorem 6 (Sequential commutativity II). *In* $DPO^{a/i}$, *for every pair of strongly sequentially independent direct derivations* $G \Rightarrow_{p_1} H \Rightarrow_{p_2} M$ *there are two sequentially independent direct derivations of the form* $G \Rightarrow_{p_2} \bar{H} \Rightarrow_{p_1} M$.

Since strong sequential independence implies sequential independence, Theorem 6 also holds for $DPO^{i/a}$, $DPO^{a/a}$ and $DPO^{i/i}$. Moreover, it can be shown that the direct derivations $G \Rightarrow_{p_2} \bar{H} \Rightarrow_{p_1} M$ are even strongly sequentially independent in $DPO^{a/i}$ and $DPO^{i/i}$, but this need not hold in $DPO^{i/a}$ and $DPO^{a/a}$.

Next we combine Theorem 5 with the Simulation Theorem.

Theorem 7 (Sequential commutativity III). *In* $DPO^{a/i}$, *for every pair of sequentially independent direct derivations* $G \Rightarrow_{p_1} H \Rightarrow_{p_2} M$ *there are two direct derivations of the form* $G \Rightarrow_{p_2} \bar{H} \Rightarrow_{Q(p_1)} M$.

Proof. Let $G \Rightarrow_{p_1} H \Rightarrow_{p_2} M$ in $DPO^{a/i}$ be sequentially independent. Then the direct derivations are in $DPO^{a/a}$ and, by Theorem 5, there are sequentially independent direct derivations $G \Rightarrow_{p_2} \bar{H} \Rightarrow_{p_1} M$ in $DPO^{a/a}$. Since the original derivations are in $DPO^{a/i}$, the morphism $L_2 \to G$ is injective. By the Simulation Theorem, there is a direct derivation $\bar{H} \Rightarrow_{Q(p_1)} M$ in $DPO^{a/i}$. Therefore, there are two direct derivations $G \Rightarrow_{p_2} \bar{H} \Rightarrow_{Q(p_1)} M$ in $DPO^{a/i}$. $\square$

To see that the direct derivations $G \Rightarrow_{p_2} \bar{H} \Rightarrow_{Q(p_1)} M$ in $DPO^{a/i}$ need not be sequentially independent, consider the sequentially independent direct derivations of Example 3. There the resulting steps $G \Rightarrow_{p_2} \bar{H} \Rightarrow_{Q(p_1)} M$ are not sequentially independent.

By Theorem 5 and the fact that every rule p is contained in $Q(p)$, Theorem 7 holds for $DPO^{i/a}$, $DPO^{a/a}$ and $DPO^{i/i}$ as well.

6 Expressiveness

In this section we show that injective matching provides additional expressiveness in two respects. First we study the generative power of graph grammars without nonterminal labels and show that these grammars can generate in $DPO^{x/i}$ more languages than in $DPO^{x/a}$, for $x \in \{a, i\}$. Then we consider the computation of functions on graphs by convergent graph transformation systems and prove that in $DPO^{a/i}$ more functions can be computed than in $DPO^{a/a}$.

Given two graphs G, H and a set of rules $\mathcal{R}$, we write $G \Rightarrow_{\mathcal{R}} H$ if there is a rule p in $\mathcal{R}$ such that $G \Rightarrow_p H$. A *derivation* from G to H over $\mathcal{R}$ is a sequence of the form $G = G_0 \Rightarrow_{\mathcal{R}} G_1 \Rightarrow_{\mathcal{R}} \ldots \Rightarrow_{\mathcal{R}} G_n \cong H$, which may be denoted by $G \Rightarrow_{\mathcal{R}}^* H$.

6.1   Generative Power

We study the expressiveness of graph grammars in $\mathrm{DPO}^{x/y}$, for $x, y \in \{\mathrm{a}, \mathrm{i}\}$. It turns out that all four approaches have the same generative power for unrestricted grammars, but $\mathrm{DPO}^{x/\mathrm{i}}$ is more powerful than $\mathrm{DPO}^{x/\mathrm{a}}$ if we consider grammars without nonterminal labels. In particular, we prove that there is a grammar without nonterminal labels in $\mathrm{DPO}^{\mathrm{i}/\mathrm{i}}$ generating a language that cannot be generated by any grammar without nonterminal labels in $\mathrm{DPO}^{\mathrm{a}/\mathrm{a}}$.

Let $x, y \in \{\mathrm{a}, \mathrm{i}\}$. A *graph grammar in* $\mathrm{DPO}^{x/y}$ is a system $\mathcal{G} = \langle \mathcal{C}, \mathcal{R}, S, \mathcal{T} \rangle$, where $\mathcal{C}$ is a finite label alphabet, $\mathcal{T}$ is an alphabet of *terminal* labels with $\mathcal{T}_V \subseteq \mathcal{C}_V$ and $\mathcal{T}_E \subseteq \mathcal{C}_E$, S is a graph over $\mathcal{C}$ called the *start graph*, and $\mathcal{R}$ is a finite set of rules over $\mathcal{C}$ such that if $x = \mathrm{i}$, all rules in $\mathcal{R}$ are injective. The *graph language generated by* $\mathcal{G}$ is the set $\mathrm{L}(\mathcal{G})$ consisting of all graphs G over $\mathcal{T}$ such that there is a derivation $S \Rightarrow_{\mathcal{R}}^{*} G$ in $\mathrm{DPO}^{x/y}$. We denote by $\mathcal{L}^{x/y}$ the class of all graph languages generated by graph grammars in $\mathrm{DPO}^{x/y}$.

Theorem 8. $\mathcal{L}^{\mathrm{i}/\mathrm{a}} = \mathcal{L}^{\mathrm{i}/\mathrm{i}} = \mathcal{L}^{\mathrm{a}/\mathrm{a}} = \mathcal{L}^{\mathrm{a}/\mathrm{i}}$.

Next we put Theorem 8 into perspective by showing that injective matching provides more generative power if we restrict ourselves to grammars without nonterminal labels. To this end we denote by $\mathcal{L}_{\mathrm{T}}^{x/y}$ the class of all graph languages generated by a grammar $\langle \mathcal{C}, \mathcal{R}, S, \mathcal{T} \rangle$ in $\mathrm{DPO}^{x/y}$ with $\mathcal{T} = \mathcal{C}$. Note that for such a grammar, every graph derivable from S belongs to the generated language.

Theorem 9. *For* $x \in \{\mathrm{a}, \mathrm{i}\}$, $\mathcal{L}_{\mathrm{T}}^{x/\mathrm{a}} \subset \mathcal{L}_{\mathrm{T}}^{x/\mathrm{i}}$.

Proof. Let $\mathcal{G} = \langle \mathcal{C}, \mathcal{R}, S, \mathcal{T} \rangle$ be a grammar without nonterminal labels in $\mathrm{DPO}^{x/\mathrm{a}}$ and $\mathrm{Q}(\mathcal{R}) = \{\mathrm{Q}(p) \mid p \in \mathcal{R}\}$. Then $\mathrm{Q}(\mathcal{G}) = \langle \mathcal{C}, \mathrm{Q}(\mathcal{R}), S, \mathcal{T} \rangle$ is a grammar without nonterminal labels in $\mathrm{DPO}^{x/\mathrm{i}}$ such that, by the Simulation Theorem, $\mathrm{L}(\mathcal{G}) = \mathrm{L}(\mathrm{Q}(\mathcal{G}))$. Hence $\mathcal{L}_{\mathrm{T}}^{x/\mathrm{a}} \subseteq \mathcal{L}_{\mathrm{T}}^{x/\mathrm{i}}$.

To show that the inclusion is strict, let $\mathcal{G} = \langle \mathcal{C}, \mathcal{R}, S, \mathcal{T} \rangle$ be the grammar in $\mathrm{DPO}^{\mathrm{i}/\mathrm{i}}$ where S is the empty graph, $\mathcal{C}_V$ and $\mathcal{C}_E$ are singletons, $\mathcal{T} = \mathcal{C}$, and $\mathcal{R}$ consists of the following two rules (already shown in the introduction):

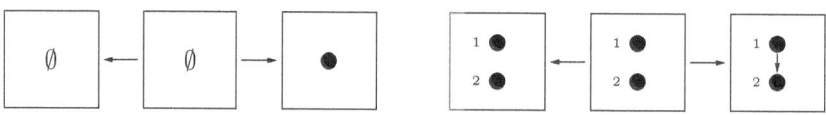

We will show that no grammar in $\mathrm{DPO}^{\mathrm{a}/\mathrm{a}}$ without nonterminal labels can generate $\mathrm{L}(\mathcal{G})$, the set of all loop-free graphs over $\mathcal{C}$. To this end, suppose the contrary and consider a grammar $\mathcal{G}' = \langle \mathcal{C}, \mathcal{R}', S', \mathcal{T} \rangle$ in $\mathrm{DPO}^{\mathrm{a}/\mathrm{a}}$ such that $\mathrm{L}(\mathcal{G}') = \mathrm{L}(\mathcal{G})$.

First we show that rules that identify nodes cannot occur in derivations of graphs in $\mathrm{L}(\mathcal{G}')$. Let $p = \langle L \leftarrow K \rightarrow^r R \rangle$ be a rule in $\mathcal{R}'$ such that $r_V(v) = r_V(v')$ for two distinct nodes v and v' in K. Then L cannot be loop-free as otherwise $L^{\oplus}$, obtained from L by adding an edge between v and v', would also be loop-free and hence belong to $\mathrm{L}(\mathcal{G}')$. The latter is impossible since $L^{\oplus} \Rightarrow_p R^{\oplus} \notin \mathrm{L}(\mathcal{G}')$, where $R^\oplus$ is obtained from R by attaching a loop to $r_V(v)$. Thus each rule in $\mathcal{R}'$ that identifies nodes must contain a loop in its left-hand side, and hence it cannot occur in any derivation of a graph in $L(\mathcal{G}')$.

The idea, now, is to show that in every derivation of a complete graph of sufficient size, a rule of $\mathcal{R}'$ must be applied that creates an edge between two existing nodes. This fact contradicts the absence of loops in $L(\mathcal{G}')$, as will be easy to show.

Let k be the maximal number of nodes occurring in S' or in a right-hand side of $\mathcal{R}'$. Consider a loop-free graph G with $k+1$ nodes such that there is an edge between each two distinct nodes. Let $S' \cong G_0 \Rightarrow_{\mathcal{R}'} G_1 \Rightarrow_{\mathcal{R}'} \ldots \Rightarrow_{\mathcal{R}'} G_n = G$ be a derivation generating G, and $i \in \{0, \ldots, n-1\}$ the largest index such that $G_i \Rightarrow_{\mathcal{R}'} G_{i+1}$ creates a node v that is not removed in the rest derivation $G_{i+1} \Rightarrow^*_{\mathcal{R}'} G_n$. Then $V_G = V_{G_n}$ is contained in $V_{G_{i+1}}$ up to isomorphism. W.l.o.g. we assume $V_G \subseteq V_{G_{i+1}}$. Since V_G contains more nodes than the right-hand side of the rule applied in $G_i \Rightarrow_{\mathcal{R}'} G_{i+1}$, there must exist a node v' in $V_G \subseteq V_{G_{i+1}}$ that is not in the image of the right-hand side. Thus, because v is created in $G_i \Rightarrow_{\mathcal{R}'} G_{i+1}$, there is no edge between v and v' in G_{i+1}. As there is an edge between v and v' in G, and $G_{i+1} \Rightarrow^*_{\mathcal{R}'} G$ does not identify nodes, there is a step $G_j \Rightarrow_{\mathcal{R}'} G_{j+1}$ with $j \geq i+1$ that creates an edge between v and v' while there is no such edge in G_j. Let $p = \langle L \leftarrow K \rightarrow R \rangle$ be the rule applied in $G_j \Rightarrow_{\mathcal{R}'} G_{j+1}$. Then there are two distinct nodes v_1 and v_2 in K such that there is an edge between (the images of) v_1 and v_2 in R but not in L.

Next observe that L is loop-free because G_j is. So $\tilde{L}$, obtained from L by identifying v_1 and v_2, is loop-free as well. On the other hand, there is a step $\tilde{L} \Rightarrow_{p,g} \tilde{R}$ where $g\colon L \rightarrow \tilde{L}$ is the surjective morphism associated with the construction of $\tilde{L}$. But then $\tilde{R}$ contains a loop, contradicting the fact that $\tilde{L}$ belongs to $L(\mathcal{G}')$.   □

6.2 DPO-Computable Functions

Graph transformation systems that transform (or "reduce") every graph into a unique irreducible graph provide a natural model for computing functions on graphs. Since the graphs resulting from double-pushout derivations are unique only up to isomorphism, we consider derivations and functions on isomorphism classes of graphs.

An *abstract graph* over a label alphabet $\mathcal{C}$ is an isomorphism class of graphs over $\mathcal{C}$. We write $[G]$ for the isomorphism class of a graph G, and $\mathcal{A}_\mathcal{C}$ for the set of all abstract graphs over $\mathcal{C}$. A *graph transformation system in* $\mathrm{DPO}^{x/y}$, for $x, y \in \{\mathrm{a}, \mathrm{i}\}$, is a pair $\langle \mathcal{C}, \mathcal{R} \rangle$ where $\mathcal{C}$ is a label alphabet and $\mathcal{R}$ a set of rules with graphs over $\mathcal{C}$ such that if $x = \mathrm{i}$, then all rules are injective. Such a system is *finite* if $\mathcal{C}_V$, $\mathcal{C}_E$ and $\mathcal{R}$ are finite sets. We will often identify $\langle \mathcal{C}, \mathcal{R} \rangle$ with $\mathcal{R}$, leaving $\mathcal{C}$ implicit. The relation $\Rightarrow_\mathcal{R}$ is lifted to $\mathcal{A}_\mathcal{C}$ by

$$[G] \Rightarrow_\mathcal{R} [H] \text{ if } G \Rightarrow_\mathcal{R} H.$$

This yields a well-defined relation since for all graphs G, G', H, H' over $\mathcal{C}$, $G' \cong G \Rightarrow_\mathcal{R} H \cong H'$ implies $G' \Rightarrow_\mathcal{R} H'$.

A graph transformation system $\mathcal{R}$ is *terminating* if there is no infinite sequence $G_1 \Rightarrow_{\mathcal{R}} G_2 \Rightarrow_{\mathcal{R}} \ldots$ of graphs in $\mathcal{A}_\mathcal{C}$. Let $\Rightarrow_{\mathcal{R}}^*$ be the transitive-reflexive closure of $\Rightarrow_{\mathcal{R}}$. Then $\mathcal{R}$ is *confluent* (or has the *Church-Rosser property*) if for all $G, H_1, H_2 \in \mathcal{A}_\mathcal{C}$, $H_1 \Leftarrow_{\mathcal{R}}^* G \Rightarrow_{\mathcal{R}}^* H_2$ implies that there is some $H \in \mathcal{A}_\mathcal{C}$ such that $H_1 \Rightarrow_{\mathcal{R}}^* H \Leftarrow_{\mathcal{R}}^* H_2$. If $\mathcal{R}$ is both terminating and confluent, then it is *convergent*. An abstract graph $G \in \mathcal{A}_\mathcal{C}$ is a *normal form* (with respect to $\mathcal{R}$) if there is no $H \in \mathcal{A}_\mathcal{C}$ such that $G \Rightarrow_{\mathcal{R}} H$. If $\mathcal{R}$ is convergent, then for every abstract graph G over $\mathcal{C}$ there is a unique normal form N such that $G \Rightarrow_{\mathcal{R}}^* N$. In this case we denote by $N_{\mathcal{R}}$ the function on $\mathcal{A}_\mathcal{C}$ that sends every abstract graph to its normal form.

Definition 6 (DPO-computable function). *Let $x, y \in \{a, i\}$. A function $f\colon \mathcal{A}_\mathcal{C} \to \mathcal{A}_\mathcal{C}$ is computable in* $\mathrm{DPO}^{x/y}$, *or* $\mathrm{DPO}^{x/y}$-*computable for short, if there exists a finite, convergent graph transformation system $\langle \mathcal{C}, \mathcal{R} \rangle$ in* $\mathrm{DPO}^{x/y}$ *such that $N_{\mathcal{R}} = f$.*

Theorem 10. *Each function that is computable in* $\mathrm{DPO}^{x/a}$ *is also computable in* $\mathrm{DPO}^{x/i}$, *for $x \in \{a, i\}$.*

Proof. Let $\langle \mathcal{C}, \mathcal{R} \rangle$ be a finite, convergent graph transformation system in $\mathrm{DPO}^{x/a}$. Then $\langle \mathcal{C}, \mathrm{Q}(\mathcal{R}) \rangle$ is a finite and convergent system in $\mathrm{DPO}^{x/i}$, as can easily be checked by means of the Simulation Theorem. Moreover, by the same result, every abstract graph has the same normal form with respect to $\mathcal{R}$ and $\mathrm{Q}(\mathcal{R})$, respectively. Thus $N_{\mathcal{R}} = N_{\mathrm{Q}(\mathcal{R})}$, which implies the proposition. □

So the double-pushout approach with injective matching is at least as powerful for computing functions as the traditional approach with arbitrary matching. The next result shows that–at least if identifying rules are present–injective matching indeed provides additional power.

Theorem 11. *There exists a function that is computable in* $\mathrm{DPO}^{a/i}$ *but not in* $\mathrm{DPO}^{a/a}$.

Proof. Let $\mathcal{C}$ be a label alphabet and $a \in \mathcal{C}_\mathrm{V}$. Consider the function $f\colon \mathcal{A}_\mathcal{C} \to \mathcal{A}_\mathcal{C}$ sending each abstract graph $[G]$ to $[G']$, where G' is obtained from G by merging all nodes labelled with a. Let $\mathcal{R}$ be the system in $\mathrm{DPO}^{a/i}$ consisting of the following single rule:

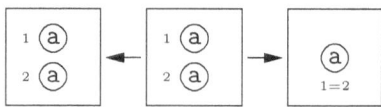

This system is terminating as each direct derivation reduces the number of nodes by one. Moreover, $\mathcal{R}$ has the following *subcommutativity property*: Whenever $H_1 \Leftarrow_{\mathcal{R}} G \Rightarrow_{\mathcal{R}} H_2$ for some $G, H_1, H_2 \in \mathcal{A}_\mathcal{C}$, then $H_1 = H_2$ or there is some $H \in \mathcal{A}_\mathcal{C}$ such that $H_1 \Rightarrow_{\mathcal{R}} H \Leftarrow_{\mathcal{R}} H_2$. This is easy to verify with the help of Theorem 3 (parallel commutativity in $\mathrm{DPO}^{a/i}$). It follows that $\mathcal{R}$ is confluent, for this is a well-known consequence of subcommutativity [Hue80]. Then it is evident that $N_\mathcal{R} = f$, that is, f is computable in $DPO^{a/i}$.

To show that f cannot be computed in $DPO^{a/a}$, suppose the contrary. Let $\mathcal{R}'$ be a finite, convergent graph transformation system in $DPO^{a/a}$ such that $N_{\mathcal{R}'} = f$. Since $\mathcal{R}'$ is finite, there is a maximal number k of edges occurring in a left-hand side of a rule in $\mathcal{R}'$. Define G to be a graph consisting of two nodes, both labelled with a, and $k+1$ edges between these nodes. (The labels and directions of these edges do not matter.) Now consider a derivation $[G] \Rightarrow^*_{\mathcal{R}'} f([G])$. This derivation contains at least one step since $[G] \neq f([G])$. Let $[G] \Rightarrow_{\mathcal{R}'} [H]$ be the first step of the derivation, with an underlying step $G \Rightarrow_{p,g} H$ on graphs. By the dangling condition, p cannot remove one of the two nodes in G. For, both nodes are incident to $k+1$ edges while p can remove at most k edges. Next consider some graph G' in $f([G])$ (consisting of a single node labelled with a and $k+1$ loops) and a surjective graph morphism $h: G \to G'$. Since p does not remove nodes and h is injective on edges, the composed morphism $h \circ g$ satisfies the gluing condition. So there is a step $G' \Rightarrow_{p, h \circ g} X$ for some X. But G' is in $f([G]) = N_{\mathcal{R}'}([G])$ and hence is a normal form. Thus our assumption that f can be computed in $DPO^{a/a}$ has led to a contradiction.   □

7 Conclusion

We have shown that injective matching makes double-pushout graph transformation more expressive. This applies to both the generative power of grammars without nonterminals and the computability of functions by convergent graph transformation systems.

The classical independence results of the double-pushout approach have been reconsidered for three variations of the traditional approach, and have been adapted where necessary. These results can be summarized as follows, where "yes" indicates a positive result and "no" means that there exists a counterexample:

	$DPO^{i/a}$	$DPO^{a/a}$	$DPO^{i/i}$	$DPO^{a/i}$
char. of parallel independence	yes	yes	yes	yes
char. of seqential independence	yes	no	yes	no
parallel commutativity I	yes	yes	yes	no
parallel commutativity II & III	yes	yes	yes	yes
sequential commutativity I	yes	yes	yes	no
sequential commutativity II & III	yes	yes	yes	yes

Corresponding results on parallelism and concurrency can be found in the long version [HMP99]. A topic for future work is to address the classical results on canonical derivations and amalgamation in the double-pushout approach.

One may also consider injective matching in the *single-pushout approach* [Löw93]. In [HHT96], parallel and sequential independence are studied for single-pushout derivations with negative application conditions. These include the dangling and the injectivity condition for matching morphisms. However, the definition of sequential independence in [HHT96] requires more information on the given direct derivations as our definition of strong sequential independence.

A further topic is to consider *high-level replacement systems*, that is, double-pushout derivations in arbitrary categories. We just mention that our independence results do not follow from the ones in [EHKP91], where non-injective rules and injective matching morphisms are not considered.

References

CMR+97.   Andrea Corradini, Ugo Montanari, Francesca Rossi, Hartmut Ehrig, Reiko Heckel, Michael Löwe.   Algebraic approaches to graph transformation. Part I: Basic concepts and double pushout approach. In G. Rozenberg, ed., Handbook of Graph Grammars and Computing by Graph Transformation, volume 1: Foundations, chapter 3, 163–245. World Scientific, 1997. 103, 105

Ehr79.   Hartmut Ehrig. Introduction to the algebraic theory of graph grammars. In Proc. Graph-Grammars and Their Application to Computer Science and Biology, Lecture Notes in Computer Science 73, 1–69, 1979. 103, 105, 107

EHKP91.   Hartmut Ehrig, Annegret Habel, Hans-Jörg Kreowski, Francesco Parisi-Presicce. Parallelism and concurrency in high level replacement systems. Mathematical Structures in Computer Science 1, 361–404, 1991. 116

EK76.   Hartmut Ehrig, Hans-Jörg Kreowski. Parallelism of manipulations in multi-dimensional information structures. In Proc. Mathematical Foundations of Computer Science, Lecture Notes in Computer Science 45, 284–293, 1976. 107, 110

EK79.   Hartmut Ehrig, Hans-Jörg Kreowski. Pushout-properties: An analysis of gluing constructions for graphs. Mathematische Nachrichten 91, 135–149, 1979.

ER76.   Hartmut Ehrig, Barry K. Rosen. Commutativity of independent transformations on complex objects. IBM Research Report RC 6251, Yorktown Heights, 1976. 107, 109

EPS73.   Hartmut Ehrig, Michael Pfender, Hans Jürgen Schneider. Graph grammars: An algebraic approach. In Proc. IEEE Conf. on Automata and Switching Theory, 167–180, 1973. 103

HHT96.   Annegret Habel, Reiko Heckel, Gabriele Taentzer. Graph grammars with negative application conditions. Fundamenta Informaticae 26, 287–313, 1996. 115, 116

HMP99.   Annegret Habel, Jürgen Müller, Detlef Plump. Double-pushout graph transformation revisited. Technical report 7/99, Fachbereich Informatik, Universität Oldenburg, 1999. 104, 110, 115

Hue80.   Gérard Huet. Confluent reductions: Abstract properties and applications to term rewriting systems. Journal of the ACM 27(4), 797–821, 1980. 115

Löw93.   Michael Löwe. Algebraic approach to single-pushout graph transformation. Theoretical Computer Science 109, 181–224, 1993. 115

Mül97.   Jürgen Müller. A non-categorical characterization of sequential independence for algebraic graph rewriting and some applications. Technical report
97-18, Technische Universität Berlin, Fachbereich Informatik, 1997.   110

Node Replacement in Hypergraphs: Translating NCE Rewriting into the Pullback Approach[*]

Hélène Jacquet[1] and Renate Klempien-Hinrichs[2]

[1] Laboratoire Bordelais de Recherche en Informatique, Université Bordeaux I
351, Cours de la Libération, 33405 Talence Cedex, France
jacquet@labri.u-bordeaux.fr
[2] Universität Bremen, Fachbereich 3,
Postfach 33 04 40, 28334 Bremen, Germany
rena@informatik.uni-bremen.de

Abstract. One of the basic concepts for the transformation of graphs is node replacement, in particular the NCE approach. As graphs do not offer a sufficiently rich model for certain applications, NCE rewriting has been generalized to directed hypergraphs where an edge may be incident to an arbitrarily long sequence of nodes. At the same time, pullback rewriting has been proposed as a generic mechanism for node replacement which allows to study this concept independently of any concrete graph (or hypergraph) model. This paper shows how node rewriting in directed hypergraphs fits in the abstract framework of the pullback approach.

1 Introduction

If one perceives a graph as a set of nodes linked by edges and wants to establish a model of graph transformation in which small bits of a graph are expanded to larger graphs, the most natural way is node rewriting. Generally speaking, there are four steps: CHOOSING a node to be rewritten, REMOVING this node together with all incident edges, ADDING a graph which replaces the node, and EMBEDDING this graph into the remainder of the first graph by additional edges. The approach of [Nag79] comes with a set of in part very liberal conditions for the insertion of an embedding edge. The NLC approach of [JR80] adds an embedding edge between a neighbour of the rewritten node and a node of the replacing graph based on the labels of these nodes (hence NLC=Node Label Controlled), whereas the NCE approach of [JR82], a special case of [Nag79], uses the identity of the node in the replacing graph (NCE=Node Controlled

---

[*] Partially supported by the EC TMR Network GETGRATS (General Theory of Graph Transformation Systems) through the University of Bordeaux I (the research for this paper has been carried out during the second author's stay in Bordeaux) and by the ESPRIT Basic Research Working Group APPLIGRAPH (Applications of Graph Transformation).

H. Ehrig et al. (Eds.): Graph Transformation, LNCS 1764, pp. 117–130, 2000.

Embedding). A recent comprehensive overview of node rewriting in graphs and particularly the NCE approach can be found in [ER97].

If, on the other hand, one perceives a graph as a set of edges which are joined at their end points, i.e. at their incident nodes, permitting an arbitrary number of nodes incident to a (hyper)edge is a natural extension. It leads to the notion of hypergraphs and the concept of hyperedge rewriting [BC87,HK87]. Other techniques of hypergraph rewriting are in particular the handle rewriting approaches in [CER93] or [KJ99]. As a first step towards integrating node and hyperedge rewriting, node rewriting in graphs is adapted to hypergraphs in [Kle96] by generalising the NCE mechanism from edges to hyperedges. The resulting concept, called hNCE rewriting, combines specific features of hyperedge rewriting and handle rewriting: Considering the context-free (sub)classes of these approaches, the generative power of confluent node rewriting includes properly that of hyperedge rewriting as well as that of separated handle rewriting [Kle], while these two are incomparable in that respect [CER93].

While (hyper)edge rewriting has grown on the solid category-theoretical foundation of the double-pushout approach [Ehr79], such a framework for node rewriting has been proposed but recently with the pullback approach of [Bau96]. With the groundwork for a general theory of this approach laid in [BJ], node rewriting can now be investigated from a more abstract perspective, too. Moreover, pullback rewriting has the practical advantage that the construction of a pullback of simple graph morphisms is easy to compute, so that it may be a suitable basis for a general implementation of node rewriting. Finally, the ideas underlying the pullback approach have proved to be stable enough to be transferred to an application such as refinement in Petri nets, see [Kle98]. It is therefore interesting to see whether the pullback approach is general enough to comprise the notion of hNCE rewriting. The work presented here gives a positive answer to this question.

In [Bau96], NLC node rewriting (and, in fact, NCE rewriting) in graphs is translated into terms of pullback rewriting. A notion of node rewriting in hypergraphs with the pullback approach is presented in [BJ97], where hypergraphs are seen as bipartite graphs or, more precisely, as graphs *structured* by a two-node loop-free graph. The associated category is a slice category of the category of simple graphs and thus inherits the property of completeness (which means in particular that all pullbacks exist). Coding hypergraphs as this kind of structured graphs is, however, not rich enough to capture hyperedges with an ordered set of tentacles as in the hypergraphs of [HK87] or [Kle96]. In this paper, we propose to code them as *structured hypergraphs*, i.e. we use a linear three-node loop-free graph to structure the graphs in our category. This allows to translate NCE node rewriting in directed hypergraphs into the pullback approach while staying in the general framework of [BJ].

The paper is organised as follows. In Section 2, the basic form of pullback rewriting in graphs is recalled, and Section 3 recalls node rewriting in directed hypergraphs. Section 4 shows how to code these hypergraphs as structured hypergraphs, i.e. in the general form used for pullback rewriting. The constructions to express node rewriting in directed hypergraphs in the pullback approach can be found in Section 5. Finally, Section 6 contains some concluding remarks.

2   Pullback Rewriting

In this section, graphs and graph morphisms are defined, as well as pullback rewriting in the associated category. For the simulation of NCE rewriting in hypergraphs, we will later consider pullback rewriting in a category of graphs *structured* by a certain graph. For now, however, we will recall the ideas underlying the basic case of the approach. We refer the reader interested in the general case of pullback rewriting in structured graphs – where the structuring graph is not explicitly given – to [BJ, Section 3]. Moreover, we do not give the general definitions of a category, a pullback, etc.; these can be found in introductory texts to category theory such as [HS79].

A(n undirected, simple) *graph* is a tuple $G = (V_G, E_G)$ consisting of a set V_G of *vertices* and a set $E_G \subseteq V_G \times V_G$ of *edges* with $\langle u, v \rangle \in E_G$ if and only if $\langle v, u \rangle \in E_G$. For all $u, v \in V_G$, we consider the edges $\langle u, v \rangle$ and $\langle v, u \rangle$ to be the same. An edge $\langle v, v \rangle \in E_G$ is a *loop*, and a vertex $v \in V_G$ which has a loop is said to be *reflexive*. A vertex $u \in V_G$ is a *neighbour* of a vertex $v \in V_G$ if there is an edge $\langle u, v \rangle$ in G. For a graph $G = (V_G, E_G)$ and a set $V \subseteq V_G$, the *subgraph of G induced by V* is $G|_V = (V, E)$ with $E = E_G \cap (V \times V)$.

In pictures of a graph, a vertex is drawn as a circle o, and an edge between two vertices as a line o—o.

Let G, G' be graphs. A *graph morphism* $f: G \to G'$ from G to G' is a pair of mappings $f = \langle f_V, f_E \rangle$ with $f_V: V_G \to V_{G'}$ and $f_E: E_G \to E_{G'}$ such that $f_E(\langle u, v \rangle) = \langle f_V(u), f_V(v) \rangle$ for all $\langle u, v \rangle \in E_G$. We also write $f(x)$ for $f_V(x)$ or $f_E(x)$. A graph morphism $f = \langle f_V, f_E \rangle$ is an *isomorphism* if both f_V and f_E are bijective. Note that a graph morphism $f: G \to G'$ is completely specified with the images of the vertices of G.

Graphs and graph morphisms form a category, which we denote by $\mathcal{G}$. This category is complete; in particular, it has all pullbacks. A pullback in $\mathcal{G}$ is constructed as follows.

CONSTRUCTING A PULLBACK OF GRAPH MORPHISMS. The *pullback* of two graph morphisms $f: B \to A$, $g: C \to A$ consists of a graph D (also called the *pullback object*) and two graph morphisms $f': D \to B$, $g': D \to C$, where $V_D = \{(u, v) \in V_B \times V_C \mid f(u) = g(v)\}$, $E_D = \{\langle (u, v), (u', v') \rangle \in V_D \times V_D \mid \langle u, u' \rangle \in E_B$ and $\langle v, v' \rangle \in E_C\}$, and for all $(u, v) \in V_D$, $f'((u, v)) = u$ and $g'((u, v)) = v$.

Example 1. Fig. 1 shows an example of a pullback in $\mathcal{G}$. The morphisms are indicated by the shades of grey of the nodes and by relative spatial arrangement; f maps, e.g., the two white vertices of B to the white vertex of A, and f' maps the upper white vertex of D to the upper white vertex of B.   ◁

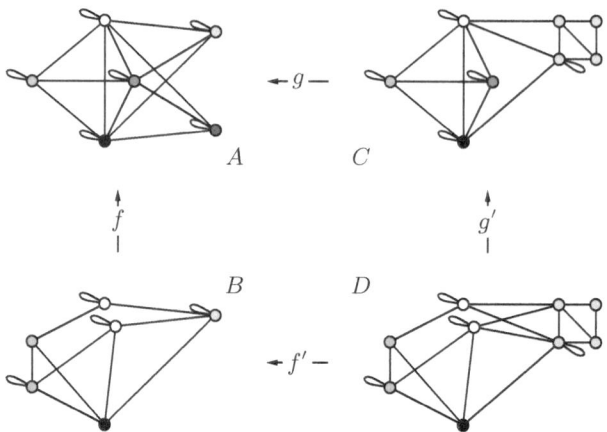

Fig. 1. A pullback in $\mathcal{G}$

The pullback of Example 1 can be interpreted as a vertex rewriting step: Imagine that the rightmost vertex u of B is to be replaced with the graph

$$C' = \raisebox{-0.5em}{\includegraphics{placeholder}}$$

such that each reflexive (white) neighbour of u will be connected to each of the left vertices in C', and each (black) neighbour which is not reflexive will be connected to the reflexive vertex of C'. Then this replacement yields the graph D of Fig. 1.

In the rest of the section, we will recall the general form of two graph morphisms so that the construction of their pullback corresponds to a vertex replacement. We start with the form of their common codomain, the generic alphabet graph A, which consists of three parts: The unknown part A^U whose inverse image is the vertex to be rewritten resp. the graph replacing it, the context part A^C which allows to reproduce the farther context of the rewritten vertex, and finally the interface part A^I which is used to specify how neighbours of the rewritten vertex turn into neighbours of vertices in the replacing graph.

The *generic alphabet graph* A consists of

- a countably infinite set A^U of reflexive vertices (we will speak of the ith unknown vertex in A),
- a countably infinite set A^I of reflexive vertices (we will speak of the interface vertices in A),
- a set A^C of one reflexive vertex (we will speak of the context vertex in A), and
- all possible edges between A^U and A^I, between A^I and A^C, and between vertices of A^I.

The graph A in Fig. 1 is a finite subgraph of the generic alphabet graph with two unknown vertices (drawn on the right) and three interface vertices (drawn in the middle). In concrete examples such as this, we will always use a "sufficiently large" subgraph of the generic alphabet graph and call it A, too.

Let G be a graph and $u \in V_G$ a reflexive vertex. An *unknown* on u is a graph morphism $\varphi_u \colon G \to A$ mapping u to a vertex in A^U (φ_u is said to have the *type* i if $\varphi_u(u)$ is the ith unknown vertex), each neighbour of u to a neighbour of $\varphi_u(u)$, and the other vertices of G to the context vertex of A.

The graph morphism f in Fig. 1 is an unknown on the right vertex in B.

A *rule* is a graph morphism $r \colon R \to A$ which maps all vertices of R in $r^{-1}(A^U)$ to the same unknown vertex v of A (r is said to have the *type* i if v is the ith unknown vertex) and whose restriction to the subgraph of R induced by $r^{-1}(A^I \cup A^C)$ and the subgraph of A induced by $A^I \cup A^C$ is an isomorphism.

The graph morphism g in Fig. 1 is a rule which maps C' – the subgraph of C induced by the vertices in $g^{-1}(A^U)$ – to one unknown vertex (and which is finite in the sense that the maximal subgraph of C disjoint from C' is finite; this subgraph would have to be adjusted accordingly if for some reason a larger subgraph of the alphabet were used).

Let $\varphi_u \colon G \to A$ be an unknown and $r \colon R \to A$ a rule of the same type. The *application* of r to φ_u is obtained by computing the pullback of φ_u and r, and we write $G \to_{(\varphi_u, r)} G'$, where G' is the graph constructed by the pullback.

The pullback of Fig. 1 corresponds to the rule application $B \to_{(f,g)} D$, where the subgraph C' of C replaces the rightmost vertex u of B as discussed above. The pullback construction reproduces C' because its image $g(C')$ in A^U is isomorphic to the inverse image $f^{-1}(g(C'))$ of this image under f (both consist of a single reflexive vertex, the terminal object of $\mathcal{G}$). Similarly, the partial isomorphy between A and C which is required for a rule guarantees that B is reproduced up to the replaced vertex u. Note that f distinguishes between neighbours of u which are going to be treated differently during the replacement of u by mapping them to distinct interface vertices of A. In other vertex-rewriting approaches, e.g. NLC or NCE grammars, such a distinction is achieved by assigning distinct labels to the vertices.

3   An NCE Approach to Node Rewriting in Hypergraphs

In this section, we define directed, labelled hypergraphs and NCE node rewriting in hypergraphs. The approach presented here differs from hNCE replacement as introduced in [Kle96] only insofar as hypergraphs may have parallel hyperedges, i.e. hyperedges with the same label and sequence of attachment nodes. Thus, they are closer to the hypergraphs of [HK87], and easier to code in a form suitable for the pullback approach (see the next section).

Throughout the paper, we will assume a finite set Σ of labels.

A (directed, labelled) *hypergraph* (over Σ) is a tuple $H = (V_H, E_H, att_H, lab_H)$ consisting of two finite and disjoint sets V_H and E_H of *nodes* and *hyperedges*, respectively, an attachment function $att_H \colon E_H \to V_H^*$ which assigns a sequence $att_H(e)$ of nodes to each hyperedge $e \in E_H$, as well as a labelling function $lab_H \colon V_H \cup E_H \to \Sigma$.

For a hyperedge $e \in E_H$ with $att_H(e) = v_1 \ldots v_k$, we write $vert_H(e, i) = v_i$ and $vset_H(e) = \{v_i \mid i \in \{1, \ldots, k\}\}$. The natural number $k \in \mathbb{N}$ is the *rank* of e, denoted $rank_H(e)$, and each $i \in \{1, \ldots, rank_H(e)\}$ is a *tentacle* of e. Thus, we have $att_H(e) = vert_H(e, 1) \ldots vert_H(e, rank_H(e))$.

In illustrations, the nodes of a hypergraph are depicted as circles ○, the hyper-edges as squares □, and the attachment relation by numbered lines, e.g. ○–1–□–2–○. The label of a node resp. hyperedge will be written next to the symbol repre-senting the node resp. hyperedge.

The left-hand side of a node rewriting production $A ::= (R, C_R)$ consists as usual of a label, and the right-hand side is a hypergraph R augmented by a set C_R of *connection instructions* of the form (ex/cr), which specify the generation of *embedding hyperedges* in a node rewriting step as follows. Suppose that the production is applied to a node u. Then a hyperedge e incident to u can be specified by the *existence* part $ex = \langle \gamma, x_1 \ldots x_m \rangle$ where γ is the label of e and $x_1 \ldots x_m$ is the sequence of labels of the nodes incident to e, with x_i changed into the special symbol $\Diamond$ if u is the ith incident node of e. Finally, the transformation of e into a hyperedge e' is specified by the *creation* part $cr = \langle \delta, y_1 \ldots y_n \rangle$ where δ is the label of e' and $y_1 \ldots y_n$ defines the node sequence of e' by letting $y_j \in V_R$ if the jth incident node of e' belongs to R, and $y_j \in \{1, \ldots, m\}$ if the jth incident node of e' is the y_jth incident node of e (which obviously may not be the rewritten node u itself, so we require $x_{y_j} \in \Sigma$ in this case).

Let $EX = (\Sigma \times (\Sigma \cup \{\Diamond\})^*) - (\Sigma \times \Sigma^*)$ be the (countably infinite) set of creation parts (over Σ) and, for a hypergraph R,

$$
\begin{aligned}
CI_R = \{(ex/cr) \mid \ & ex = \langle \gamma, x_1 \ldots x_m \rangle \in EX, \ cr = \langle \delta, y_1 \ldots y_n \rangle, \\
& \delta \in \Sigma, \ n \in \mathbb{N}, \text{ and for all } j \in \{1, \ldots, n\} : \\
& y_j \in V_R \cup \mathbb{N}, \text{ and } y_j \in \mathbb{N} \text{ implies } x_{y_j} \in \Sigma\}
\end{aligned}
$$

the set of connection instructions (over Σ). A *production* (over Σ) is a pair $(A, (R, C_R))$, usually written $A ::= (R, C_R)$, such that $A \in \Sigma$, R is a hypergraph, and the *connection relation* C_R is a finite subset of CI_R. A *rewriting system* (over Σ) is a set $\mathcal{R}$ of productions, and $N(\mathcal{R}) = \{A \in \Sigma \mid (A ::= (R, C_R)) \in \mathcal{R}\}$ contains the *nonterminal* symbols of $\mathcal{R}$.

Let G be a hypergraph and $p = (A ::= (R, C_R))$ a production (such that G and R are disjoint), and let u be a node of G with $lab_G(u) = A$. Then the *application* of p to u yields the hypergraph H, denoted $G \Rightarrow_{[u/p]} H$, if:

$V_H = (V_G \setminus \{u\}) \cup V_R$ with

- $lab_H(v) = lab_G(v)$ if $v \in V_G \setminus \{u\}$ and
- $lab_H(v) = lab_R(v)$ if $v \in V_R$;

and

$$
\begin{aligned}
E_H = \ & (E_G \setminus \{e \mid u \in vset_G(e)\}) \cup E_R \cup \\
& \{(e, (ex/cr)) \in E_G \times C_R \mid u \in vset_G(e), \ ex = \langle \gamma, x_1 \ldots x_m \rangle, \\
& \quad lab_G(e) = \gamma, \ rank_G(e) = m, \text{ and for all } i \in \{1, \ldots, m\} : \\
& \quad (vert_G(e, i) = u \wedge x_i = \Diamond) \vee (vert_G(e, i) \neq u \wedge x_i = lab_G(v_i))\}
\end{aligned}
$$

with

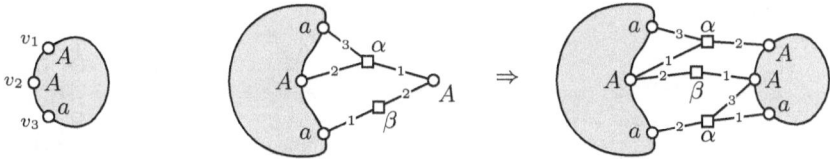

Fig. 2. R **Fig. 3.** Application of $A ::= (R, C_R)$ to a node

- $lab_H(e) = lab_G(e)$ and $att_H(e) = att_G(e)$ if $e \in E_G \setminus \{e' \mid u \in vset_G(e')\}$,
- $lab_H(e) = lab_R(e)$ and $att_H(e) = att_R(e)$ if $e \in E_R$, and
- if $e = (e', (ex/cr)) \in E_G \times C_R$ with $cr = \langle \delta, y_1 \ldots y_n \rangle$:
  $lab_H(e) = \delta$, $rank_H(e) = n$, and for all $j \in \{1, \ldots, n\}$:
  $vert_H(e, j) = y_j$ if $y_j \in V_R$ and $vert_H(e, j) = vert_G(e', y_j)$ if $y_j \in \mathbb{N}$.

Example 2. Consider a production $A ::= (R, C_R)$ where R is some hypergraph with (at least) nodes v_1, v_2, v_3 as sketched in Fig. 2, and $C_R = \{(ex_1/cr_1), (ex_1/cr_2), (ex_2/cr_3)\}$ with:

$$ex_1 = \langle \alpha, \Diamond Aa \rangle \qquad cr_1 = \langle \alpha, 2v_1 3 \rangle$$
$$cr_2 = \langle \beta, v_2 2 \rangle$$
$$ex_2 = \langle \beta, a\Diamond \rangle \qquad cr_3 = \langle \alpha, v_3 1 v_2 \rangle$$

Applying this production to the rightmost node of the left hypergraph in Fig. 3 yields the hypergraph on the right, where the upper α-labelled hyperedge gives rise to two embedding hyperedges labelled α resp. β, and the lower β-labelled hyperedge results in one α-labelled embedding hyperedge.   ◁

The notion of NCE rewriting in hypergraphs can be extended as usual to comprise derivations, grammars and their generated languages.

4   From Hypergraphs to Structured Graphs

The pullback rewriting approach has been developed in [BJ] in the framework of *structured graphs*. In order to use this framework and to obtain a suitable category for the representation of directed hypergraphs, we transform these hypergraphs to a type of structured graphs we call *structured hypergraphs*, which are subsequently *expanded* to code the labelling, too. In the following, we present the ideas for these constructions rather than a full formalization.

A directed labelled hypergraph is transformed to an expanded structured hypergraph in three steps (illustrated in Fig. 4).

STEP 1. The basic idea to represent directed hypergraphs as structured hypergraphs is to code nodes (o) and hyperedges (□) as two sorts of vertices, and introduce tentacles (•) as a third sort to describe the attachment of a node to a hyperedge.

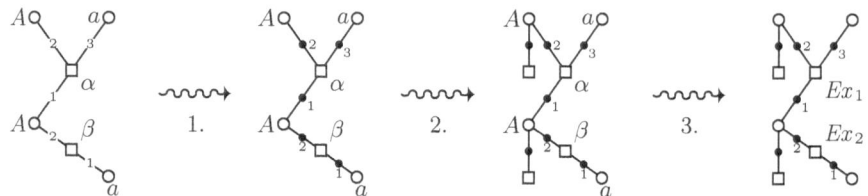

Fig. 4. Transforming a labelled hypergraph into an expanded structured hypergraph (A is a nonterminal symbol, and $ex_1 = \langle \alpha, AAa \rangle$, $ex_2 = \langle \beta, aA \rangle$)

For the rest of this paper, let us fix S as the graph 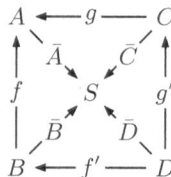, i.e. as the graph with vertices $\circ, \bullet, \square$ and edges $\langle \circ, \bullet \rangle$, $\langle \bullet, \square \rangle$, $\langle \square, \bullet \rangle$, $\langle \bullet, \circ \rangle$. A *graph structured by S* or *S-graph* is a graph morphism $\bar{G} \colon G \to S$. A vertex which an S-graph maps on $\circ$ is called a $\circ$-vertex, and analogously for $\square$-vertices and $\bullet$-vertices.

A morphism $f \colon \bar{G} \to \bar{H}$ between two S-graphs is a graph morphism $f \colon G \to H$ such that $\bar{H} \circ f = \bar{G}$. S-graphs and S-graph morphisms form a category, in fact, the *slice category* $\langle \mathcal{G} \downarrow S \rangle$. This category is complete (a property inherited from $\mathcal{G}$), so that it has in particular arbitrary pullbacks. The pullback $(\bar{D}, f', g')$ of two S-graph morphisms f and g is constructed just like the pullback (D, f', g') of the associated graph morphisms f and g, with the pullback object $\bar{D} \colon D \to S$ mapping a vertex on $\circ$, $\bullet$, or $\square$ such that f' and g' are S-graph morphisms:

$$
\begin{array}{ccc}
A & \xleftarrow{\quad g \quad} & C \\
\uparrow \quad \bar{A} & S & \bar{C} \quad \uparrow \\
f \quad & \diagdown \nearrow & \quad g' \\
\downarrow \quad \bar{B} & \bar{D} & \downarrow \\
B & \xleftarrow{\quad f' \quad} & D
\end{array}
$$

Note that not all objects of $\langle \mathcal{G} \downarrow S \rangle$ are good candidates to represent directed hypergraphs: as a $\bullet$-vertex stands for a tentacle, it must be adjacent to exactly one $\circ$-vertex and one $\square$-vertex.

STEP 2. The pullback approach rewrites a whole pattern rather than just a node: in the category of graphs, the node to be rewritten has to be reflexive; in the category of S-graphs, the pattern to be rewritten is a copy of S (which is the terminal object in $\langle \mathcal{G} \downarrow S \rangle$). Therefore, each nonterminally labelled node must be completed to such a copy by adding a new private tentacle and hyperedge.

STEP 3. Finally, to collect the information the NCE approach needs in order to classify the hyperedges, we expand the label of a hyperedge with the labels of the nodes incident to that hyperedge.

Formally, let $ex = \langle \gamma, x_1 \dots x_m \rangle$ be an existence part in EX. Then we write $ex[0] = \gamma$, $ex[1..m] = x_1 \dots x_m$, and $ex[i] = x_i$ for the ith element of $ex[1..m]$. Moreover, we associate two sets $\bullet ex$ and $ex \bullet$ of integers with ex, where $\bullet ex = \{i \mid ex[i] \neq \Diamond\}$ and $ex\bullet = \{i \mid ex[i] = \Diamond\}$, and say that $Ex = (\bullet ex, ex, ex\bullet)$ is an *expansion triple*.

We *expand* the label $lab(e)$ of a hyperedge e into the following expansion triple Ex_e: $ex_e[0]$ is the label $lab(e)$ of e, $ex_e[1..k]$ is the "attachment sequence" $lab(att(e))$ of e, $\bullet ex_e = \{1, \dots, rank(e)\}$, and $ex_e\bullet = \emptyset$. For example, the expansion triple Ex_1 corresponding to $ex_1 = \langle \alpha, AAa \rangle$ of Fig. 4 has $\bullet ex_1 = \{1, 2, 3\}$ and $ex_1\bullet = \emptyset$. (The usefulness of the symbol $\Diamond$ and the component $ex\bullet$ will become clear when we study the coding of a rewriting step.)

TO CONCLUDE. A *structured hypergraph* is a graph morphism $\bar{H}: H \to S$ such that each $\bullet$-vertex u of V_H is adjacent to exactly one $\circ$-vertex and one $\square$-vertex, and each nonterminal node is completed to a copy of S by a private tentacle and a private hyperedge. An *expanded structured hypergraph* is a structured hypergraph in which all hyperedge labels are expanded.

5   Translating a Rewriting Step

This section presents the construction of a system of morphisms which simulates the application of a production to a nonterminal node in the pullback approach. More precisely, we construct for a hypergraph G and a nonterminal node u in G an unknown morphism $\Phi_u: \bar{G} \to \bar{\mathcal{A}}$, and for a production $p = (A ::= (R, C_R))$ a rule morphism $r_p: \bar{R}_p \to \bar{\mathcal{A}}$. In these constructions, $\bar{\mathcal{A}}$ is a structured alphabet graph in $\langle \mathcal{G} \downarrow S \rangle$, $\bar{G}: G \to S$ is the expanded structured hypergraph associated with G, and $\bar{R}_p$ is constructed from R and C_R. Our unknowns and rules are special cases of those introduced in [BJ, Section 3]; in particular, our alphabet $\bar{\mathcal{A}}$ is a subgraph of the alphabet $\mathcal{A}_S$ in $\langle \mathcal{G} \downarrow S \rangle$, which is defined in that reference as the categorical product of the generic alphabet graph A (see Section 2) with the structuring graph S.

CONSTRUCTING THE ALPHABET GRAPH.   Instead of a formal definition, we explain the construction of the alphabet graph informally and step by step, as illustrated in Fig. 5.

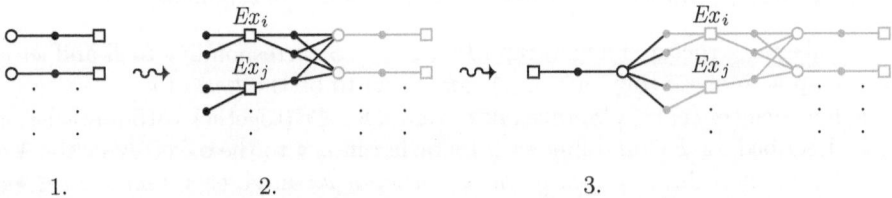

Fig. 5. Step-by-step construction of the alphabet graph

The *alphabet graph* associated with Σ is the S-graph $\bar{\mathcal{A}}: \mathcal{A} \to S$ with $\mathcal{A}$ constructed as follows:

1. Take a countably infinite set $\mathcal{A}^U$ of isomorphic copies of S (we will speak of the k-*copy of* S in $\mathcal{A}^U$, for $k \in \mathbb{N}$).
2. Take an S-graph Ex for each element $ex \in EX$, where the S-graph Ex corresponding to the expansion triple $Ex = (\bullet ex, ex, ex\bullet)$ consists of one $\square$-vertex standing for $ex[0]$ and n $\bullet$-vertices linked to it by an edge (where n is the length of the "attachment sequence" $ex[1..n]$), and put an edge between each $\circ$-vertex of $\mathcal{A}^U$ and each $\bullet$-vertex in $ex\bullet$.
3. Take an isomorphic copy $\mathcal{A}^C$ of S and put an edge between its $\circ$-vertex and each $\bullet$-vertex in $\bullet ex$ (for every Ex).

CONSTRUCTING AN UNKNOWN. Let u be a nonterminal node of an expanded structured hypergraph $\bar{G}$. An *unknown* on u is a morphism $\Phi_u \colon \bar{G} \to \bar{A}$ constructed as follows (see the left half of Fig. 6 for an example of an unknown):

1. Map u and its private tentacle and hyperedge to one k-copy of S in $\mathcal{A}^U$; k is said to be the *type* of Φ_u.
2. For each hyperedge e adjacent to u, first transform its expansion triple $Ex_e = (\bullet ex_e, ex_e, ex_e\bullet)$ to Ex'_e as follows: for all i, if $vert(e, i) = u$ then move i from $\bullet ex_e$ to $ex_e\bullet$ and replace $ex_e[i]$ by the symbol $\Diamond$. Then map e to the $\square$-vertex of the Ex in $\mathcal{A}$ with $Ex = Ex'_e$, and map the $\bullet$-vertices of Ex'_e to their corresponding $\bullet$-vertices of Ex.
   In Fig. 4 (see also Fig. 6, bottom left), choose for example the A-labelled node u which is incident to the two hyperedges. Then the expansion triples Ex_1, Ex_2 corresponding to the hyperedges with respect to u as nonterminal are $Ex_1 = (\{2, 3\}, \langle \alpha, \Diamond Aa \rangle, \{1\})$ and $Ex_2 = (\{1\}, \langle \beta, a\Diamond \rangle, \{2\})$.
3. Map the rest of G to $\mathcal{A}^C$.

CONSTRUCTING A RULE. Let $p = (A ::= (R, C_R))$ be a production. Recall that C_R is a set of connection instructions (ex/cr) with $cr \in \Sigma \times (V_R \cup N)^*$. We associate a triple $Cr = (\bullet cr, cr, cr\bullet)$ with each cr, where $\bullet cr = \{i \mid cr[i] \in \mathbb{N}\}$ and $cr\bullet = \{i \mid cr[i] \in V_R\}$.

A *pullback rule* associated with p is a morphism $r_p \colon \bar{R}_p \to \bar{A}$ constructed as follows[1] (see the upper half of Fig. 6 for an example of a pullback rule):

1. Take the structured hypergraph $\bar{R}^U \colon R^U \to S$ corresponding to R and let r_p map it to one k-copy of S in $\mathcal{A}^U$; k is said to be the *type* of r_p.
2. For each $(ex/cr) \in C_R$, take one $\square$-vertex for Cr (together with tentacles, as described for Ex) and link each tentacle i in $cr\bullet$ to the $\circ$-vertex $cr[i] \in V_R$. Let r_p map this $\square$-vertex to the $\square$-vertex of Ex in $\mathcal{A}$, each tentacle $i \in \bullet cr$ to the tentacle $cr[i]$ in $\bullet ex$ and, finally, each tentacle in $cr\bullet$ to one of the tentacles in $ex\bullet$.[2]

---

[1] In general, r_p is not unique –
[2] – because there may be more than one tentacle in $ex\bullet$. Note, moreover, that $ex\bullet$ is not empty because $ex \in EX$.

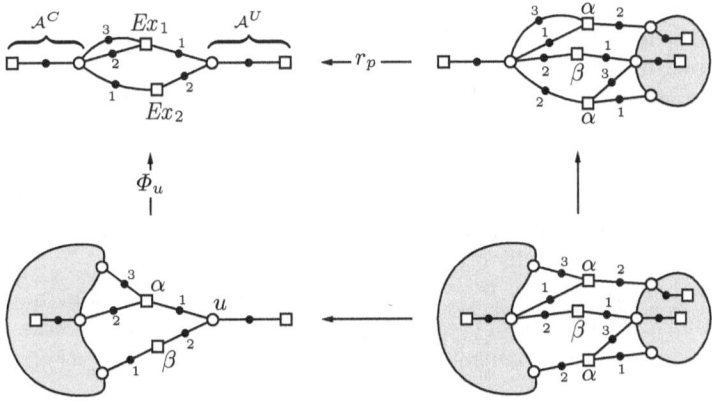

Fig. 6. Application of r_p to Φ_u (cf. Fig. 3)

3. Take one copy of S, link the ○-vertex to every •-vertex in each •cr, and let r_p map this S-copy to $\mathcal{A}^C$.

The resulting structured hypergraph is $\bar{R}_p$, which is expanded by taking the expansion of $\bar{R}^U$ and associating with each □-vertex representing a Cr the expansion corresponding to cr.

RULE APPLICATION.   Let $\Phi_u \colon \bar{G} \to \bar{A}$ be an unknown and r_p a pullback rule such that Φ_u and r_p are of the same type. Then the *application* of r_p to Φ_u is obtained by computing the pullback of (Φ_u, r_p), and we write $\bar{G} \to_{(\Phi_u, r_p)} \bar{H}$, where $\bar{H}$ is the pullback object of (Φ_u, r_p).

Example 3. Read from left to right, the lower half of Fig. 6 illustrates the application of the pullback rule r_p drawn in the upper half. We have resumed Example 2 from Section 3; note that the labels etc. are only added to make the references more obvious.   ◁

Proposition 4. *The pullback object of a rule application is a structured hypergraph.*

Proof. Let us consider the application of a rule $r_p \colon \bar{R}_p \to \bar{A}$ to an unknown $\Phi_u \colon \bar{G} \to \bar{A}$ on u, where $(\bar{H} \colon H \to S, \Phi'_u \colon \bar{H} \to \bar{G}, r'_p \colon \bar{H} \to \bar{R}_p)$ is the pullback of (Φ_u, r_p). Then $\bar{H} \colon H \to S$ is an S-graph in the category $\langle \mathcal{G} \downarrow S \rangle$ because of the pullback definition.

As S is the neutral element for the categorical product in $\langle \mathcal{G} \downarrow S \rangle$, the subgraph of $\bar{H}$ which is mapped to $\bar{A}^C$ by $\Phi_u \circ \Phi'_u$ is an isomorphic copy of the subgraph of $\bar{G}$ which is mapped to $\bar{A}^C$ by Φ_u, and is consequently a structured hypergraph. In the same way, the subgraph of $\bar{H}$ which is mapped to $\bar{A}^U$ by $r_p \circ r'_p$ is an isomorphic copy of the subgraph $\bar{R}^U$ of $\bar{R}_p$ which is mapped to $\bar{A}^U$ by r_p, and is a structured hypergraph, too.

Now consider a •-vertex x of $\bar{H}$ which is mapped to $\bar{\mathcal{A}}^I$ (by $\Phi_u \circ \Phi'_u$ and $r_p \circ r'_p$), and let x_1 denote the vertex $\Phi'_u(x) \in V_G$ and x_2 the vertex $r'_p(x) \in V_{R_p}$. As $\bar{G}$ is a structured hypergraph, x_1 is adjacent to a ○-vertex v_1 and a □-vertex w_1. For the same reason, x_2 is adjacent to a ○-vertex v_2 and a □-vertex w_2 in $\bar{R}_p$. From the construction of a rule and an unknown (and implicitly from the construction of $\bar{\mathcal{A}}$), we can deduce that v_1 and v_2 have the same image in $\bar{\mathcal{A}}$. This implies that x is adjacent to a ○-vertex in $\bar{H}$ which is mapped to v_1 by Φ'_u and to v_2 by r'_p. In the same way, w_1 and w_2 have the same image in $\bar{\mathcal{A}}$, and x must be adjacent to a □-vertex which is mapped to w_1 by Φ'_u and to w_2 by r'_p.

The fact that $\bar{G}$ and $\bar{R}_p$ are structured hypergraphs – and in particular that each of their •-vertices is adjacent to exactly one ○-vertex and exactly one □-vertex – is sufficient to ensure that each •-vertex of $\bar{H}$ is adjacent to at most one ○-vertex and at most one □-vertex.   □

Theorem 5. *Let $G \Rightarrow_{[u/p]} H$ be the application of a production p to a node u and $\bar{G} \to_{(\Phi_u, r_p)} \bar{H}$ the corresponding pullback rule application. Then $\bar{H}$ is the structured hypergraph associated with H, and there is a general construction for the expansion of $\bar{H}$ such that the result corresponds to the labelled hypergraph H.*

Proof. We have already stated in the proof of Proposition 4 that due to the neutrality property of S, an isomorphic copy of $\bar{G}$ minus the vertex u and its "incident hyperedges" is restored in $\bar{H}$, as well as an isomorphic copy of $\bar{R}^U$, the structured hypergraph corresponding to the labelled hypergraph R in the production $p = (A ::= (R, C_R))$. So now we have to verify that all the hyperedges which are created by the embedding process in an NCE derivation step are created (with the adequate tentacles) by the corresponding pullback rule application, and not more than those. The verification consists of establishing step by step that the coding of the embedding relation into the rule and unknown morphisms leads to the expected result. The rather lengthy proof does not contain particular difficulties, so we trust that all the details given in this and the previous section suffice to convince the reader of the correctness of the claim. Similarly, it is probably clear that we get the correct expansion of $\bar{H}$ as follows: Expand the part of $\bar{H}$ which is mapped on $\bar{\mathcal{A}}^C$ just like its image in $\bar{G}$ under Φ'_u, the part mapped on $\bar{\mathcal{A}}^U$ like its image in $\bar{R}^U$ under r'_p, and associate the expansion triple $(•cr \cup cr•, cr', \emptyset)$ with each □-vertex of H mapped on the □-vertex of $Cr = (•cr, cr, cr•)$ by r'_p and on the □-vertex of $Ex = (•ex, ex, ex•)$ by $r_p \circ r'_p$, where $cr'[0] = cr[0]$, $cr'[i] = ex[cr[i]]$ if $i \in •cr$, and $cr'[i] = lab_R(cr[i])$ if $i \in cr•$.   □

6   Conclusion

This paper shows how NCE rewriting in directed hypergraphs can be translated into pullback rewriting. Thus, the pullback approach is a suitable category-theoretical framework for node rewriting in directed as well as in undirected hypergraphs. The important part of the translation is that of a rewriting step, but of course, the notions of a grammar and its generated language can be developed for the pullback approach as for the NCE approach.

A topic for future research is to consider the pullback approach with respect to combining node and hyperedge rewriting. In fact, the general concept presented in [BJ], and in particular the generic form of the alphabet graph, seems to provide a natural formal basis for this notion. While it is not surprising that a technique which allows to describe node rewriting in hypergraphs can also describe hyperedge rewriting (cf. [Kle96, Theorem 1]), the combination of these two hypergraph rewriting mechanisms promises to lead to interesting questions concerning e.g. aspects of context-free rewriting.

Acknowledgement

We gratefully acknowledge the stimulating discussions with Michel Bauderon and his generosity in sharing his ideas.

The referees' remarks helped to improve the paper; we thank them for that.

Our pictures are fabricated with Frank Drewes's LATEX 2_ε-package for typesetting graphs.

References

Bau96.   Michel Bauderon. A category-theoretical approach to vertex replacement: The generation of infinite graphs. In J. Cuny, H. Ehrig, G. Engels, and G. Rozenberg, editors, *Proc. Fifth Intl. Workshop on Graph Grammars and Their Application to Comp. Sci.*, volume 1073 of *Lecture Notes in Computer Science*, pages 27–37. Springer, 1996.   118

BC87.   Michel Bauderon and Bruno Courcelle. Graph expressions and graph rewriting. *Mathematical Systems Theory*, 20:83–127, 1987.   118

BJ.   Michel Bauderon and Hélène Jacquet. Pullback as a generic graph rewriting mechanism. To appear in Applied Categorical Structures. A previous version is *Categorical product as a generic graph rewriting mechanism*, Technical Report 1166–97, LaBRI, University of Bordeaux, 1996.   118, 119, 123, 125, 129

BJ97.   Michel Bauderon and Hélène Jacquet. Node rewriting in hypergraphs. In F. D'Amore, P.G. Franciosa, and A. Marchetti-Spaccamela, editors, *Graph-Theoretic Concepts in Computer Science (Proc. WG'96)*, volume 1197 of *Lecture Notes in Computer Science*, pages 31–43. Springer, 1997.   118

CER93.   Bruno Courcelle, Joost Engelfriet, and Grzegorz Rozenberg. Handle-rewriting hypergraph grammars. *Journal of Computer and System Sciences*, 46:218–270, 1993.   118

Ehr79.   Hartmut Ehrig. Introduction to the algebraic theory of graph grammars. In V. Claus, H. Ehrig, and G. Rozenberg, editors, *Graph-Grammars and Their Application to Computer Science and Biology*, volume 73 of *Lecture Notes in Computer Science*, pages 1–69, 1979.   118

ER97.   Joost Engelfriet and Grzegorz Rozenberg. Node replacement graph grammars. In G. Rozenberg, editor, *Handbook of Graph Grammars and Computing by Graph Transformation. Vol. I: Foundations*, chapter 1, pages 1–94. World Scientific, 1997.   118

HK87.   Annegret Habel and Hans-Jörg Kreowski. May we introduce to you: Hyper-edge replacement. In H. Ehrig, M. Nagl, G. Rozenberg, and A. Rosenfeld, editors, *Proc. Third Intl. Workshop on Graph Grammars and Their Application to Comp. Sci.*, volume 291 of *Lecture Notes in Computer Science*, pages 15–26, 1987. 118, 121

HS79.   Horst Herrlich and George Strecker. *Category Theory*. Heldermann, Berlin, second edition, 1979. 119

JR80.   Dirk Janssens and Grzegorz Rozenberg. On the structure of node-label-controlled graph languages. *Information Sciences*, 20:191–216, 1980. 117

JR82.   Dirk Janssens and Grzegorz Rozenberg. Graph grammars with neighbour-hood-controlled embedding. *Theoretical Computer Science*, 21:55–74, 1982. 117

KJ99.   Changwook Kim and Tae Eui Jeong. HRNCE grammars – a hypergraph generating system with an eNCE way of rewriting. *Theoretical Computer Science*, 233:143–178, 1999. 118

Kle.   Renate Klempien-Hinrichs. The generative power of context-free node rewrit-ing in hypergraphs. Submitted for publication. 118

Kle96.   Renate Klempien-Hinrichs. Node replacement in hypergraphs: Simulation of hyperedge replacement, and decidability of confluence. In J. Cuny, H. Ehrig, G. Engels, and G. Rozenberg, editors, *Proc. Fifth Intl. Workshop on Graph Grammars and Their Application to Comp. Sci.*, volume 1073 of *Lecture Notes in Computer Science*, pages 397–411, Springer, 1996. 118, 121, 129

Kle98.   Renate Klempien-Hinrichs. Net refinement by pullback rewriting. In M. Ni-vat, editor, *Proc. Foundations of Software Science and Computation Struc-tures*, volume 1378 of *Lecture Notes in Computer Science*, pages 189–202, Springer, 1998. 118

Nag79.   Manfred Nagl. *Graph-Grammatiken: Theorie, Anwendungen, Implemen-tierungen*. Vieweg, Braunschweig, 1979. 117

Pushout Complements for Arbitrary Partial Algebras*

Mercè Llabrés and Francesc Rosselló

Departament de Matemàtiques i Informàtica, Universitat de les Illes Balears,
07071 Palma de Mallorca (Spain)
{merce,cesc}@ipc4.uib.es

Abstract. To develop a double-pushout approach to transformation in a specific category, two basic preliminary questions must be answered: *a)* when a given rule can be applied through a given occurrence?, and *b)* when the result of such an application is unique? We solve these problems in the usual category of partial algebras over an arbitrary signature.

1 Introduction

The double-pushout (DPO) approach to algebraic transformation, invented 25 years ago by H. Ehrig, M. Pfender and H. J. Schneider, has been developed mainly for objects that can be understood as unary (total or partial) algebras, but, from the very beginning, objects with non-unary structures have also been considered: for instance, relational structures in [3]. The last years have still witnessed an increasing interest in the algebraic transformation of objects that can be understood as partial algebras over arbitrary signatures. This interest is due mainly to the interaction between algebraic transformation and ADT specification. For instance, the single-pushout (SPO) transformation of partial algebras without nullary operations based on *quomorphisms* (a special type of partial homomorphisms of partial algebras; specifically, a homomorphism from a relative subalgebra of the source algebra) has been studied by Wagner and Gogolla in [8], where it has been applied to the specification of object transformation systems. Another formalism of transformation of arbitrary partial algebras, called Algebra Rewrite Systems, has been invented recently by Große-Rhode (see, for instance, [4]) and applied to the specification of state transforming dynamic systems.

In spite of this interest in the algebraic transformation of partial algebras, and although the DPO approach has a long and well-established tradition in the field of algebraic transformation, no general DPO approach to the transformation of partial algebras over an arbitrary signature has been developed so far. The goal of this paper is to clear the path towards this development, by solving for

---

* This work has been partly supported by the DGES, grant PB96-0191-C02-01. M. Llabrés has also been partly supported by the EU TMR Network GETGRATS (General Theory of Graph Transformation Systems) through the Technical University of Berlin.

H. Ehrig et al. (Eds.): Graph Transformation, LNCS 1764, pp. 131–144, 2000.
© Springer-Verlag Berlin Heidelberg 2000

arbitrary partial algebras two basic algebraic problems that are previous to any such development.

Recall that a *(DPO) production rule* in a category $\mathcal{C}$ is a pair of morphisms $P = (\mathbf{L} \xleftarrow{l} \mathbf{K} \xrightarrow{r} \mathbf{R})$. Such a production rule in $\mathcal{C}$ can be *applied* to an object $\mathbf{D}$ when there exists a morphism $m : \mathbf{L} \to \mathbf{D}$ and a diagram

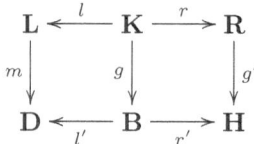

such that both squares in it are pushout squares. In other words, when there exists a *pushout complement* $(\mathbf{B}, g : \mathbf{K} \to \mathbf{B}, l' : \mathbf{B} \to \mathbf{D})$ of l and m (i.e., an object $\mathbf{B}$ and two morphisms $g : \mathbf{K} \to \mathbf{B}$ and $l' : \mathbf{B} \to \mathbf{D}$ such that $\mathbf{D}$, together with m and l', is the pushout of l and g) and a pushout of r and $g : \mathbf{K} \to \mathbf{B}$. When such a diagram exists, we say that $\mathbf{H}$ has been *derived* from $\mathbf{D}$ by the application of rule P through $m : \mathbf{L} \to \mathbf{D}$; the intermediate object $\mathbf{B}$ is called the *context object* of this derivation.

The category Alg_Σ of partial algebras of a given signature Σ with their homomorphisms as morphisms is cocomplete, and therefore it has all binary pushouts. So, to use DPO transformation in Alg_Σ (or in a cocomplete full subcategory of it) one has first to solve the following two problems:

a) To find a *gluing condition*: a necessary and sufficient condition on two homomorphisms $f : \mathbf{K} \to \mathbf{A}$ and $m : \mathbf{A} \to \mathbf{D}$ for the existence of a pushout complement of them. This condition yields a necessary and sufficient condition for the applicability of a rule through a homomorphism.

b) To find a *uniqueness condition*: a necessary and sufficient condition on a homomorphism $f : \mathbf{K} \to \mathbf{A}$ for the uniqueness (up to isomorphisms) of a pushout complement of f and a homomorphism $m : \mathbf{A} \to \mathbf{D}$, when they satisfy the gluing condition. This condition yields a general sufficient condition for the uniqueness (up to isomorphisms) of the result of the application of a rule through a given homomorphism.

For pairs of homomorphisms that do not have a unique pushout complement, it is useful to describe a *distinguished pushout complement*, characterized by means of some abstract or concrete property; this allows to distinguish (up to isomorphisms) an object among all those derived by the application of a given rule through a given homomorphism, provided that there exists at least one.

These problems were solved for partial algebras over a unary signature in [2, Sect. 2]. In this paper we solve them for partial algebras over an arbitrary signature. The solutions to these problems in the arbitrary case generalize, of course, those obtained in the unary case, but not in the 'straightforward' way one could reasonably guess at first glance. Notice furthermore that Alg_Σ is never a topos if the signature contains some operation symbol (see [7]), and thus the results obtained by Kawahara in [5] are of no use here.

This paper is organized as follows. In Sect. 2 we recall the standard construction of pushouts in Alg_Σ. Then, in Sect. 3 we show that no straightforward generalization of the gluing condition for unary partial algebras works for arbitrary partial algebras, and in Sect. 4 we find the right gluing condition for the latter. Finally, in Sect. 5 we find the uniqueness condition for arbitrary partial algebras. The lack of space does not allow us to give complete proofs of the results reported here; thus, the most difficult proofs are sketched, and the rest are simply omitted. However, the reader used to the algebraic and category theoretical reasoning should find no difficulty in filling up all the details. A full version of this paper, including detailed proofs and applications, will appear elsewhere [6]. The main results given in this paper can also be applied to determine the gluing and uniqueness conditions for total algebras over an arbitrary signature; we shall report on it elsewhere.

Due also to the lack of space, it is impossible to define in this paper all the basic concepts about partial algebras appearing in it. Thus, we refer to Appendix A of [2] for all definitions and notations on partial algebras used in this paper. In any case, there are two conventions that we want to point out here. First, given a partial algebra denoted by means of a boldface capital letter, $\mathbf{A}, \mathbf{B}$ etc., its carrier will be always denoted, unless otherwise stated, by the same capital letter, but in italic type: A, B, etc. And second, when there is no danger of confusion, and in order to lighten the notations, we shall omit all subscripts corresponding to sorts in the names of the carriers of the algebras, the components of the mappings or of the congruences etc., and denote things just as if we were dealing with one-sorted algebras.

2   Pushouts in Alg_Σ

Let $\Sigma = (S, \Omega, \eta)$ be an arbitrary signature, with set of sorts S, set of operation symbols Ω and arity function $\eta : \Omega \to S^* \times S$ with $\eta(\varphi) = (w(\varphi), \sigma(\varphi))$ for every $\varphi \in \Omega$. Let $\Omega^{(0)}$ be the set of all *nullary* operation symbols in Ω (i.e., those operation symbols in Ω of arity (λ, s) for some s) and $\Omega^{(+)} = \Omega - \Omega^{(0)}$.

Let $\Sigma^{(+)} = (S, \Omega^{(+)}, \eta|_{\Omega^{(+)}})$ be the signature obtained by removing from Ω all nullary operations. Given a partial Σ-algebra $\mathbf{A} = (A, (\varphi^{\mathbf{A}})_{\varphi \in \Omega})$, we shall always denote by $\mathbf{A}^{(+)}$ its $\Sigma^{(+)}$-*reduct*: the partial $\Sigma^{(+)}$-algebra $(A, (\varphi^{\mathbf{A}})_{\varphi \in \Omega^{(+)}})$ with its same carrier, but with only its non-nullary operations.

Let Alg_Σ be the category of partial Σ-algebras with homomorphisms as morphisms. It is well-known that Alg_Σ has all binary pushouts, whose construction we briefly recall in the following result (see [1, §4.3] for details).

Theorem 1. *Let* $\mathbf{K}, \mathbf{A},$ *and* $\mathbf{B}$ *be three partial* Σ-*algebras, and let* $f : \mathbf{K} \to \mathbf{A}$ *and* $g : \mathbf{K} \to \mathbf{B}$ *be two homomorphisms of partial* Σ-*algebras. Let*

$$\mathbf{A}^{(+)} + \mathbf{B}^{(+)} = (A \sqcup B, (\varphi^{\mathbf{A}^{(+)} + \mathbf{B}^{(+)}})_{\varphi \in \Omega^{(+)}})$$

be the partial $\Sigma^{(+)}$*-algebra with carrier the disjoint union[1]* $A \sqcup B$ *of* A *and* B, *and operations the disjoint unions of the corresponding operations on the* $\Sigma^{(+)}$*-reducts* $\mathbf{A}^{(+)}$ *and* $\mathbf{B}^{(+)}$ *of* $\mathbf{A}$ *and* $\mathbf{B}$*: i.e., for every* $\varphi \in \Omega^{(+)}$,

$$\operatorname{dom} \varphi^{\mathbf{A}^{(+)}+\mathbf{B}^{(+)}} = \operatorname{dom} \varphi^{\mathbf{A}} \sqcup \operatorname{dom} \varphi^{\mathbf{B}}$$

and if $\underline{a} \in \operatorname{dom} \varphi^{\mathbf{A}}$ *(resp.* $\underline{b} \in \operatorname{dom} \varphi^{\mathbf{B}}$*) then* $\varphi^{\mathbf{A}^{(+)}+\mathbf{B}^{(+)}}(\underline{a}) = \varphi^{\mathbf{A}}(\underline{a})$ *(resp.* $\varphi^{\mathbf{A}^{(+)}+\mathbf{B}^{(+)}}(\underline{b}) = \varphi^{\mathbf{B}}(\underline{b})$*).*

Let $\theta(f,g)$ *be the least congruence on* $\mathbf{A}^{(+)} + \mathbf{B}^{(+)}$ *containing the relation*

$$\{(f(x), g(x)) \mid x \in K\} \cup \{(\varphi_0^{\mathbf{A}}, \varphi_0^{\mathbf{B}}) \mid \varphi_0 \in \Omega^{(0)}, \ \varphi_0^{\mathbf{A}}, \varphi_0^{\mathbf{B}} \ \textit{defined}\}.$$

Let finally $\mathbf{D}$ *be the partial* Σ*-algebra whose* $\Sigma^{(+)}$*-reduct is the quotient*

$$(\mathbf{A}^{(+)} + \mathbf{B}^{(+)}) \big/ \theta(f,g)$$

and whose nullary operations are defined as follows: for every $\varphi_0 \in \Omega^{(0)}$, $\varphi_0^{\mathbf{D}}$ *is defined if and only if* $\varphi_0^{\mathbf{A}}$ *or* $\varphi_0^{\mathbf{B}}$ *are defined, in which case* $\varphi_0^{\mathbf{D}}$ *is its (or their) equivalence class modulo* $\theta(f,g)$.

Then $\mathbf{D}$*, together with the homomorphisms* $\mathbf{A} \to \mathbf{D}$ *and* $\mathbf{B} \to \mathbf{D}$ *given by the restrictions to* $\mathbf{A}$ *and* $\mathbf{B}$ *of the quotient mapping* $A \sqcup B \to (A \sqcup B) \big/ \theta(f,g)$, *is a pushout of* f *and* g *in* Alg_Σ.[2]   □

Corollary 2. *Let*

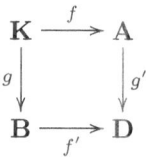

be a pushout square in Alg_Σ. *If* f *is closed and injective and* g *is injective, then* f' *is closed and injective and* g' *is injective.*

Proof. The assumptions on f and g imply the equality

$$\theta(f,g) = \{(f(x), g(x)), (g(x), f(x)) \mid x \in K\} \cup \Delta_{A \sqcup B}$$

(where $\Delta_{A \sqcup B}$ denotes the diagonal relation on $A \sqcup B$), from which the conclusion is easily deduced.   □

Example 3. Consider a one-sorted signature Σ with a nullary operation (denoted by a dashed circle in the diagrams) and a binary operation (denoted by a hyperarc in the diagrams). Fig. 3 displays a pushout square in Alg_Σ. [3]

---

[1] Formally, the disjoint union $A \sqcup B$ of A and B is defined as $A \times \{1\} \cup B \times \{2\}$, but in order to simplify the notations we shall identify A and B with their images in $A \sqcup B$.

[2] Of course, if $\Omega^{(0)} = \emptyset$ then $\Sigma^{(+)} = \Sigma$, $\mathbf{A}^{(+)} = \mathbf{A}$, $\mathbf{B}^{(+)} = \mathbf{B}$, $\theta(f,g)$ is the least congruence on $\mathbf{A} + \mathbf{B}$ containing $\{(f(x), g(x)) \mid x \in K\}$, and $\mathbf{D} = (\mathbf{A} + \mathbf{B}) \big/ \theta(f,g)$.

[3] The homomorphisms in the figures send the points of their source algebra to points with the same names, perhaps with some more subscripts, of the target algebra; for

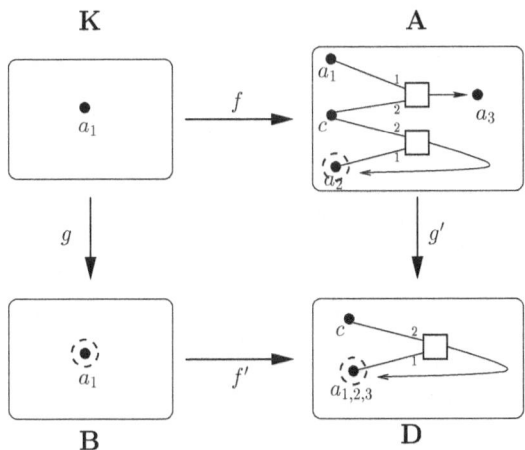

Fig. 1. An example of pushout.

3   A Failed Gluing Condition

As we have already mentioned in the introduction, we want to find a *gluing condition* in Alg_Σ: a necessary and sufficient condition for the existence of a pushout complement of two homomorphisms of partial Σ-algebras. This problem is solved for unary signatures in [2, Props. 9, 10]. We summarize in the next proposition its solution. Recall that $C_{\mathbf{A}}(X)$ denotes the least closed subset of a partial algebra $\mathbf{A}$ containing a subset X of its carrier.

Proposition 4. *Let Σ be a unary signature and let $f : \mathbf{K} \to \mathbf{A}$ and $m : \mathbf{A} \to \mathbf{D}$ be two homomorphisms of partial Σ-algebras. Then, f and m have a pushout complement in Alg_Σ if and only if they satisfy the following conditions:*

i) *(Identification condition) $m(C_{\mathbf{A}}(f(K))) \cap m(A - C_{\mathbf{A}}(f(K))) = \emptyset$ and the restriction of m to $A - C_{\mathbf{A}}(f(K))$ is injective.*

ii) *(Dangling condition) $m(A - C_{\mathbf{A}}(f(K)))$ is an initial segment of $\mathbf{D}$: i.e., for every $\varphi \in \Omega$, if $d \in \mathrm{dom}\,\varphi^{\mathbf{D}}$ and $\varphi^{\mathbf{D}}(d) \in m(A - C_{\mathbf{A}}(f(K)))$ then $d \in m(A - C_{\mathbf{A}}(f(K)))$.*

iii) *(Relative closedness condition) For every $\varphi \in \Omega$ and for every $a \in A - C_{\mathbf{A}}(f(K))$, if $m(a) \in \mathrm{dom}\,\varphi^{\mathbf{D}}$ then $a \in \mathrm{dom}\,\varphi^{\mathbf{A}}$.*

And when f and m satisfy such conditions (i) to (iii), then a pushout complement of them is given by $(\mathbf{B}, g : \mathbf{K} \to \mathbf{B}, f' : \mathbf{B} \hookrightarrow \mathbf{D})$, where $\mathbf{B}$ is the relative (actually, closed) subalgebra of $\mathbf{D}$ supported on $B = (D - m(A)) \cup m(C_{\mathbf{A}}(f(K)))$, $f' : \mathbf{B} \hookrightarrow \mathbf{D}$ is the corresponding embedding, and $g : \mathbf{K} \to \mathbf{B}$ is $m \circ f$ considered as a homomorphism from $\mathbf{K}$ to $\mathbf{B}$.   □

instance, the homomorphism g' in Fig. 3 sends a_1, a_2 and a_3 to $a_{1,2,3}$, and c to c. The order of the arguments of the binary operations in the figures is specified by the labels of the sources of the corresponding hyperarcs.

This result is no longer true if we let Σ be an arbitrary signature: for instance, f and g' in Example 3 do not satisfy the identification condition. So, it is natural to wonder whether there is a natural generalization of this proposition to arbitrary signatures.

One of the key ingredients in the proof of Proposition 4 given in [2] is that, if Σ is unary, then

$$\theta(f,g) \subseteq (C_{\mathbf{A}}(f(K)) \sqcup C_{\mathbf{B}}(g(K)))^2 \cup \Delta_{A \sqcup B},$$

and the reason why Proposition 4 is false for arbitrary signatures is essentially that this inclusion does not hold for arbitrary signatures Σ: see Example 3 again. Thus, in order to obtain a similar inclusion, we must replace $C_{\mathbf{A}}(f(K))$ and $C_{\mathbf{B}}(g(K))$ by suitable greater sets.

Given a partial Σ-algebra $\mathbf{A} = (A, (\varphi^{\mathbf{A}})_{\varphi \in \Omega})$ with $A = (A_s)_{s \in S}$, let the *algebraic preorder* $\leq_{\mathbf{A}}$ on $\bigcup_{s \in S} A_s$ be the reflexive and transitive closure of the following relation:

$a' \prec_{\mathbf{A}} a$ iff $a = \varphi^{\mathbf{A}}(\ldots, a', \ldots)$ for some operation $\varphi \in \Omega$ and some tuple $(\ldots, a', \ldots) \in \operatorname{dom} \varphi^{\mathbf{A}}$ containing a'.

Given a subset X of a partial Σ-algebra $\mathbf{A} = (A, (\varphi^{\mathbf{A}})_{\varphi \in \Omega})$, let

$$U_{\mathbf{A}}(X) = \left\{ a \in A \mid x \leq_{\mathbf{A}} a \text{ for some } x \in X \cup \{\varphi_0^{\mathbf{A}} \mid \varphi_0 \in \Omega^{(0)}, \ \varphi_0^{\mathbf{A}} \text{ defined}\} \right\}.$$

When Σ is a unary signature, $U_{\mathbf{A}}(X) = C_{\mathbf{A}}(X)$.

Lemma 5. *If $f : \mathbf{K} \to \mathbf{A}$ and $g : \mathbf{K} \to \mathbf{B}$ are two homomorphisms of partial Σ-algebras, then $\theta(f,g) \subseteq (U_{\mathbf{A}}(f(K)) \sqcup U_{\mathbf{B}}(g(K)))^2 \cup \Delta_{A \sqcup B}$.* ☐

Example 3 shows that the inclusion in this lemma may be an equality. Now, from the construction of pushouts described in Theorem 1 and the inclusion given in Lemma 5 we obtain the following two results, which generalize respectively Lemma 7 and Proposition 9 in [2] to arbitrary signatures. These results will be widely used in the sequel, mainly in the parts of the proofs that we omit.

Lemma 6. *Let*

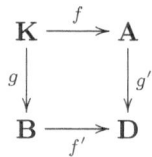

be a pushout square in Alg_{Σ}. Then:

i) $f'(B) \cup g'(A) = D$.

ii) For every $a \in A$ and $b \in B$, if $g'(a) = f'(b)$ then $a \in U_{\mathbf{A}}(f(K))$ and $b \in U_{\mathbf{B}}(g(K))$.

iii) For every $a, a' \in A$, if $g'(a) = g'(a')$ and $a \neq a'$ then $a, a' \in U_{\mathbf{A}}(f(K))$. A similar property holds for f'.

iv) For every operation $\varphi \in \Omega$, if $\underline{x} \in \operatorname{dom} \varphi^{\mathbf{D}}$ then there exists either some $\underline{a} \in \operatorname{dom} \varphi^{\mathbf{A}}$ with $g'(\underline{a}) = \underline{x}$ or some $\underline{b} \in \operatorname{dom} \varphi^{\mathbf{B}}$ with $f'(\underline{b}) = \underline{x}$. ☐

Proposition 7. *Let* $f : \mathbf{K} \to \mathbf{A}$ *and* $m : \mathbf{A} \to \mathbf{D}$ *be two homomorphisms of partial* Σ-*algebras having a pushout complement. Then:*

i) $m(U_{\mathbf{A}}(f(K))) \cap m(A - U_{\mathbf{A}}(f(K))) = \emptyset$ *and the restriction of* m *to* $A - U_{\mathbf{A}}(f(K))$ *is injective.*

ii) $m(A - U_{\mathbf{A}}(f(K)))$ *is an initial segment of* $\mathbf{D}$.

iii) For every $\varphi \in \Omega$ *and for every* $\underline{a} \notin U_{\mathbf{A}}(f(K))^{w(\varphi)}$, *if* $m(\underline{a}) \in \operatorname{dom}\varphi^{\mathbf{D}}$ *then there exists some* $\underline{a}' \in \operatorname{dom}\varphi^{\mathbf{A}}$ *such that* $m(\underline{a}) = m(\underline{a}')$; *in particular, if* $\underline{a} \in (A - U_{\mathbf{A}}(f(K)))^{w(\varphi)}$, $w(\varphi) \neq \lambda$, *and* $m(\underline{a}) \in \operatorname{dom}\varphi^{\mathbf{D}}$, *then* $\underline{a} \in \operatorname{dom}\varphi^{\mathbf{A}}$. $\quad\square$

Proposition 4 says that the converse of Proposition 7 holds when Σ is a unary signature, in which case $U_{\mathbf{A}}(f(K)) = C_{\mathbf{A}}(f(K))$. But this converse is false for arbitrary signatures Σ: conditions (i) to (iii) in Proposition 7 do not guarantee the existence of a pushout complement of f and m, as Example 8 below shows.

Example 8. Let Σ be the signature introduced in Example 3. The pair of homomorphisms described in Fig. 8 satisfy conditions (i) to (iii) in Proposition 7, but they do not have a pushout complement. Indeed, if they had a pushout complement $(\mathbf{B}, g : \mathbf{K} \to \mathbf{B}, d : \mathbf{B} \to \mathbf{D})$ then, since $m \circ f$ is injective, g would be injective, and then, since f is closed and injective, by Corollary 2 m should be injective, too.

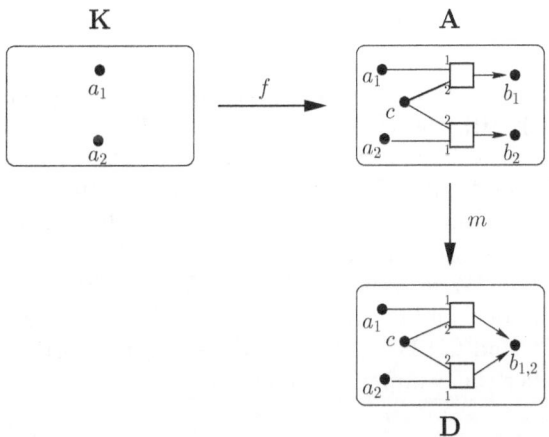

Fig. 2. These homomorphisms do not have a pushout complement.

4   The Right Gluing Condition

Example 8 above shows that a gluing condition in Alg_Σ cannot be given by properties of m relative to certain subsets of A alone (together with the relative closedness condition), but it will have to include some information on the kernel of m: specifically, the homomorphisms f and m therein have no pushout complement because m identifies b_1 and b_2 and it shouldn't. So, to find a gluing condition in Alg_Σ we follow a different approach, based on Proposition 9 below.

Following [3], we say that a pushout complement $(\mathbf{B_0}, g_0 : \mathbf{K} \to \mathbf{B_0}, d_0 : \mathbf{B_0} \hookrightarrow \mathbf{D})$ of two homomorphisms $f : \mathbf{K} \to \mathbf{A}$ and $m : \mathbf{A} \to \mathbf{D}$ of partial Σ-algebras is *natural* when $d_0 : \mathbf{B_0} \hookrightarrow \mathbf{D}$ is the embedding of a closed subalgebra.

Proposition 9. *Let $f : \mathbf{K} \to \mathbf{A}$ and $m : \mathbf{A} \to \mathbf{D}$ be two homomorphisms of partial Σ-algebras that have a pushout complement. Then:*

i) *They have one and only one natural pushout complement $(\mathbf{B_0}, g_0 : \mathbf{K} \to \mathbf{B_0}, d_0 : \mathbf{B_0} \hookrightarrow \mathbf{D})$.*

ii) *Such a natural pushout complement is terminal among all pushout complements of f and m, in the sense that if $(\mathbf{C}, m' : \mathbf{K} \to \mathbf{C}, f' : \mathbf{C} \to \mathbf{D})$ is any other pushout complement of them, then there exists an epimorphism $e_0 : \mathbf{C} \to \mathbf{B_0}$ such that $e_0 \circ m' = g_0$ and $d_0 \circ e_0 = f'$.*

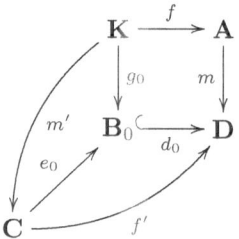

Proof. i) Let $(\mathbf{B}, g : \mathbf{K} \to \mathbf{B}, d : \mathbf{B} \to \mathbf{D})$ be a pushout complement of f and m. Let $B_0 = C_{\mathbf{D}}(d(B))$, $\mathbf{B_0}$ the closed subalgebra of $\mathbf{D}$ supported on B_0, $d_0 : \mathbf{B_0} \hookrightarrow \mathbf{D}$ the closed embedding of $\mathbf{B_0}$ into $\mathbf{D}$, $d' : \mathbf{B} \to \mathbf{B_0}$ the homomorphism d understood with target algebra $\mathbf{B_0}$, which is an epimorphism by [1, Prop. 3.6.1], and $g_0 : \mathbf{K} \to \mathbf{B_0}$ the composition $d' \circ g$. Then $(\mathbf{B_0}, g_0 : \mathbf{K} \to \mathbf{B_0}, d_0 : \mathbf{B_0} \hookrightarrow \mathbf{D})$ is also a pushout complement of f and m.

Assume now that $(\mathbf{B_1}, g_1 : \mathbf{K} \to \mathbf{B_1}, d_1 : \mathbf{B_1} \hookrightarrow \mathbf{D})$ is another natural pushout complement of f and m. Let $B' = B_0 \cap B_1$, $\mathbf{B}'$ the closed subalgebra of $\mathbf{D}$ supported on B', and $i_0 : \mathbf{B}' \hookrightarrow \mathbf{B_0}$, $i_1 : \mathbf{B}' \hookrightarrow \mathbf{B_1}$ the corresponding closed embeddings. Consider the commutative diagram

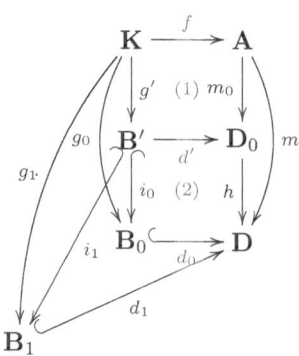

where $g' : \mathbf{K} \to \mathbf{B}'$ is $m \circ f$ understood with target algebra $\mathbf{B}'$, (1) is a pushout square and $h : \mathbf{D}_0 \to \mathbf{D}$ is the only homomorphism such that $h \circ m_0 = m$ and $h \circ d' = d_0 \circ i_0 = d_1 \circ i_1$. Since (1)+(2) is also a pushout square, (2) is a pushout square, too, and since $d_0 \circ i_0$ is a closed embedding, d' is a closed and injective homomorphism. Therefore, by Corollary 2, h is closed and injective, too. And h is also surjective because

$$D = m(A) \cup (B_0 \cap B_1) = m(A) \cup B' = h(m_0(A) \cup d'(B')) = h(D_0).$$

Hence, h is an isomorphism, and from the description on $\theta(d', i_0)$ given in the proof of Corollary 2 it easily follows that $\mathbf{B}' = \mathbf{B}_0$ and $i_0 = \mathrm{Id}_{B_0}$. Finally, by symmetry, it also happens that $\mathbf{B}' = \mathbf{B}_1$ and $i_1 = \mathrm{Id}_{B_1}$, which proves the uniqueness assertion.

ii) It is a consequence of the uniqueness of such a pushout complement and the proof of its existence given above.   □

A similar result can also be proved for homomorphisms of total algebras.

Under the light of this proposition, to find a necessary and sufficient condition for the existence of a pushout complement of $f : \mathbf{K} \to \mathbf{A}$ and $m : \mathbf{A} \to \mathbf{D}$, it is enough to detect a candidate of natural pushout complement of them. It is done in the next result.

Theorem 10. *Let* $f : \mathbf{K} \to \mathbf{A}$ *and* $m : \mathbf{A} \to \mathbf{D}$ *be two homomorphisms of partial* Σ-*algebras. Let* $B_0 = C_{\mathbf{D}}((D - m(A)) \cup mf(K))$, $\mathbf{B}_0$ *the closed subalgebra of* $\mathbf{D}$ *supported on* B_0, $g_0 : \mathbf{K} \to \mathbf{B}_0$ *the composition* $m \circ f$ *understood with target algebra* $\mathbf{B}_0$, *and* $d_0 : \mathbf{B}_0 \hookrightarrow \mathbf{D}$ *the embedding of* $\mathbf{B}_0$ *into* $\mathbf{D}$.

Then, f *and* m *have a pushout complement if and only if* $(\mathbf{B}_0, g_0 : \mathbf{K} \to \mathbf{B}_0, d_0 : \mathbf{B}_0 \hookrightarrow \mathbf{D})$ *is a pushout complement of them.*

Proof. Assume that f and m have a pushout complement, and let $(\mathbf{B}, g : \mathbf{K} \to \mathbf{B}, d : \mathbf{B} \hookrightarrow \mathbf{D})$ be a natural pushout complement of them. The joint surjectivity of pushouts (Lemma 6.(i)) implies that $B_0 \subseteq B$; let $i : \mathbf{B}_0 \hookrightarrow \mathbf{B}$ be the corresponding closed embedding. Consider now the commutative diagram

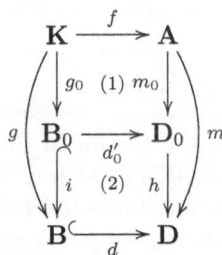

where (1) and (1)+(2) are pushout squares and $h : \mathbf{D}_0 \to \mathbf{D}$ is the only homomorphism such that $h \circ m_0 = m$ and $h \circ d'_0 = d \circ i$. An argument similar to the one used in the proof of the uniqueness assertion in Proposition 9.(i) shows that h is an isomorphism, which clearly implies the thesis.   □

By Proposition 9, if $f : \mathbf{K} \to \mathbf{A}$ and $m : \mathbf{A} \to \mathbf{D}$ have a pushout complement, then their pushout complement described in the previous theorem can be characterized both in a concrete way, as their only natural pushout complement, and in an abstract way, as a final object in a certain category of pushout complements. This abstract characterization entails that, among all derived objects of $\mathbf{D}$ through m by the application of a given production rule P having f as left-hand side homomorphism, the one obtained using the natural pushout complement of f and m as context algebra is final in a certain sense, which in particular characterizes it up to isomorphisms: see the last Remark in Sect. 2 of [2] for the details in the unary case.

Now, to obtain a *gluing condition* for homomorphisms of arbitrary partial algebras similar in spirit to the one for unary partial algebras recalled in Proposition 4, it is enough to translate Theorem 10 into a set-theoretical condition. This is done in the following proposition.

Proposition 11. *Let $f : \mathbf{K} \to \mathbf{A}$ and $m : \mathbf{A} \to \mathbf{D}$ be two homomorphisms of partial Σ-algebras. Let B_0 and $\mathbf{B}_0$ be as in Theorem 10. Let θ be the congruence on $\mathbf{A}^{(+)} + \mathbf{B}_0^{(+)}$ generated by the relation*

$$\{(a, m(a)) \mid a \in f(K)\} \cup \{(\varphi_0^{\mathbf{A}}, \varphi_0^{\mathbf{D}}) \mid \varphi_0 \in \Omega^{(0)}, \ \varphi_0^{\mathbf{A}}, \varphi_0^{\mathbf{D}} \ defined\}.$$

Then f and m have a pushout complement if and only if they satisfy the following three properties:

GC1) $\ker m \subseteq \theta$.
GC2) *If $m(a) \in B_0$ then $(a, m(a)) \in \theta$.*
GC3) *For every $\varphi \in \Omega$, if $\underline{d} \in \operatorname{dom} \varphi^{\mathbf{D}}$ and $\underline{d} \notin B_0^{w(\varphi)}$ then there exists some $\underline{a} \in \operatorname{dom} \varphi^{\mathbf{A}}$ such that $\underline{d} = m(\underline{a})$.* □

The proof is a simple although quite technical verification, and we omit it.

Conditions (GC1), (GC2) and (GC3) in this theorem are strictly stronger than conditions (i), (ii) and (iii) in Proposition 7, respectively, and they are equivalent to them (i.e., to the *identification, dangling* and *relative closedness* conditions in Proposition 4) when the signature is unary. However, in the general case it is not possible to rewrite these conditions (GC1), (GC2) and (GC3) into conditions independent of θ, as it can be done in the unary case.

5   The Uniqueness Condition

Let us search now a *uniqueness condition* in the sense explained in the introduction. Notice that, by Proposition 9, such a uniqueness condition is nothing but a necessary and sufficient condition on a homomorphism $f : \mathbf{K} \to \mathbf{A}$ that guarantees that in every pushout square

$$
\begin{array}{ccc}
\mathbf{K} & \xrightarrow{\;f\;} & \mathbf{A} \\
{\scriptstyle g}\downarrow & & \downarrow{\scriptstyle m} \\
\mathbf{B} & \xrightarrow{\;f'\;} & \mathbf{D}
\end{array}
$$

the homomorphism $f' : \mathbf{B} \to \mathbf{D}$ is closed and injective. Since

$$
\begin{array}{ccc}
\mathbf{K} & \xrightarrow{f} & \mathbf{A} \\
\scriptstyle{\mathrm{Id_K}} \downarrow & & \downarrow \scriptstyle{\mathrm{Id_A}} \\
\mathbf{K} & \xrightarrow{f} & \mathbf{A}
\end{array}
$$

is always a pushout square, the uniqueness condition must include closedness and injectivity. As a matter of fact, in the unary case closedness and injectivity are already enough by [2, Lem. 13 and Prop. 14], but in general closed and injective homomorphisms need not be inherited under pushouts, as Example 12 below shows, and thus some extra condition is needed.

Example 12. Let Σ be again the same signature as in the previous examples. Fig. 12 displays a pushout square where the homomorphism f is closed and injective, but the homomorphism f' is not injective.

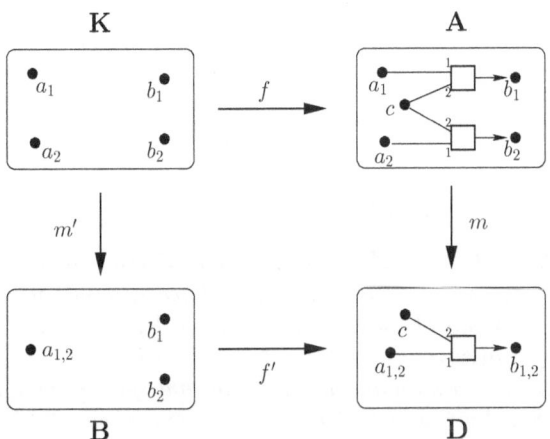

Fig. 3. Closedness and injectivity are not inherited under pushouts.

The extra condition that must be combined to closedness and injectivity to yield the uniqueness condition in Alg_Σ is given by the following notion.

Definition 13. *A closed subalgebra* $\mathbf{C}$ *of a partial* Σ-*algebra* $\mathbf{A}$ *satisfies the* minimal congruence extension property *(mce-property, for short) when for every congruence* θ *on* $\mathbf{C}$, *if* $\overline{\theta}$ *is the congruence on* $\mathbf{A}$ *generated by* θ *then:*

i) $\overline{\theta} \cap (C \times C) = \theta$.
ii) For every $\varphi \in \Omega$, *if* $(a_1, \ldots, a_n) \in \mathrm{dom}\, \varphi^{\mathbf{A}}$ *and* $(a_1, c_1), \ldots, (a_n, c_n) \in \overline{\theta}$ *with* $c_1, \ldots, c_n \in C$, *then there exist* $c_1', \ldots, c_n' \in C$ *such that* $(a_1, c_1'), \ldots, (a_n, c_n') \in \overline{\theta}$ *and* $(c_1', \ldots, c_n') \in \mathrm{dom}\, \varphi^{\mathbf{C}}$.

A *homomorphism* $f : \mathbf{K} \to \mathbf{A}$ *is an* mce-homomorphism *when it is a closed and injective homomorphism and the closed subalgebra of* $\mathbf{A}$ *supported on* $f(K)$ *satisfies the mce-property.*

With the notations of the previous definition, in the unary case we have that $\bar{\theta} = \theta \cup \Delta_A$, and then every closed subalgebra satisfies the mce-property.

Now we have the following result.

Lemma 14. *Let* $f : \mathbf{K} \to \mathbf{A}$ *be an mce-homomorphism and*

$$
\begin{array}{ccc}
\mathbf{K} & \xrightarrow{\,f\,} & \mathbf{A} \\
{\scriptstyle g}\downarrow & & \downarrow{\scriptstyle g'} \\
\mathbf{B} & \xrightarrow{\,f'\,} & \mathbf{D}
\end{array}
$$

a pushout square in Alg_Σ. *Then* $f' : \mathbf{B} \to \mathbf{D}$ *is again an mce-homomorphism, and in particular closed and injective.*

Proof. Without any loss of generality, we may assume that $f : \mathbf{K} \to \mathbf{A}$ is the embedding of a closed subalgebra satisfying the mce-property. Let θ_g be the congruence on $\mathbf{A}$ generated by $\ker g$. Using the mce-property, one shows that

$$
\theta(f,g) = \theta_g \cup \{(g(k), a), (a, g(k)) \mid k \in K, (k,a) \in \theta_g\} \cup \Delta_B
$$

from where, using the explicit construction of a pushout of f and g given in Theorem 1, the thesis is proved. $\qquad\square$

Proposition 15. i) *Let* $f : \mathbf{K} \to \mathbf{A}$ *be an mce-homomorphism, and let* $m : \mathbf{A} \to \mathbf{D}$ *be any homomorphism. If* f *and* m *have a pushout complement, then it is unique up to isomorphisms, and it is given by the closed subalgebra of* $\mathbf{D}$ *supported on* $(D - m(A)) \cup mf(K)$.

ii) *If* $f : \mathbf{K} \to \mathbf{A}$ *is a homomorphism such that, for every homomorphism* m *with source algebra* $\mathbf{A}$, f *and* m *have at most one pushout complement (up to isomorphisms), then* $f : \mathbf{K} \to \mathbf{A}$ *is an mce-homomorphism.*

Proof. i) From Proposition 9 and Lemma 14, we know that the pushout complement of f and m, if it exists, is isomorphic over $\mathbf{K}$ and $\mathbf{D}$ to the natural pushout complement $(\mathbf{B}_0, g_0 : \mathbf{K} \to \mathbf{B}_0, d_0 : \mathbf{B}_0 \hookrightarrow \mathbf{D})$ described in Proposition 9. And from the explicit description of $\theta(f, g_0)$ given in the proof of Lemma 14, the equality $B_0 = (D - m(A)) \cup mf(K)$ is easily proved.

ii) If $\mathbf{K}$ is a closed subalgebra of $\mathbf{A}$ not satisfying the mce-property, and θ is a congruence on $\mathbf{K}$ that does not satisfy conditions (i) or (ii) in Definition 13, then

$$
\begin{array}{ccc}
\mathbf{K} & \xhookrightarrow{} & \mathbf{A} \\
{\scriptstyle \pi_\theta}\downarrow & & \downarrow{\scriptstyle \pi_{\bar{\theta}}} \\
\mathbf{K}/_\theta & \xrightarrow[\bar{f}]{} & \mathbf{A}/_{\bar{\theta}}
\end{array}
$$

(where $\bar{\theta}$ is the congruence on $\mathbf{A}$ generated by θ, $\pi_\theta : \mathbf{K} \to \mathbf{K}/_\theta$ and $\pi_{\bar{\theta}} : \mathbf{A} \to \mathbf{A}/_{\bar{\theta}}$ are the corresponding quotient homomorphisms, and $\bar{f} : \mathbf{K}/_\theta \to \mathbf{A}/_{\bar{\theta}}$ is the homomorphism induced by the inclusion $\mathbf{K} \hookrightarrow \mathbf{A}$) is a pushout square and $\bar{f}$ is not a closed and injective homomorphism. □

We don't know a simple characterization of closed subalgebras satisfying the mce-property, although this kind of conditions are quite common in the literature on total algebras. We want to point out at least the following sufficient (but not necessary) condition for the mce-property.

Definition 16. *A subset X of the carrier of a partial Σ-algebra $\mathbf{A}$ is strongly convex in it when $\varphi^\mathbf{A}(a_1, \ldots, a_n) \in X$ and $x \leq_\mathbf{A} a_i$ for some $x \in X$ and some $i = 1, \ldots, n$ imply $a_1, \ldots, a_n \in X$.*

Notice that this notion of strongly convex set is strictly stronger than the usual notion of a *convex set* (if $x_0 \leq_\mathbf{A} a \leq_\mathbf{A} x_1$ with $x_0, x_1 \in X$ then $a \in X$) and strictly weaker than the notion of initial segment.

Proposition 17. *If $\mathbf{K} \hookrightarrow \mathbf{A}$ is a closed subalgebra supported on a strongly convex set then $\mathbf{K}$ satisfies the mce-property in $\mathbf{A}$.* □

Example 18. The homomorphism f' in Fig. 3 is an embedding of a closed subalgebra that satisfies the mce-property but that is not supported on a strongly convex set of $\mathbf{A}$.

6   Final Remarks and Conclusion

In this paper we have given a gluing condition and a uniqueness condition for partial algebras over an arbitrary signature, and we have distinguished a natural pushout complement of any pair of homomorphisms satisfying that gluing condition. These results generalize the corresponding results known for unary partial algebras (including graphs and hypergraphs) [2] and relational systems [3], and they shape the first basic steps in the development of the DPO transformation of arbitrary partial algebras, which we plan to carry over elsewhere.

Further steps in this development should be the study of the properties of this transformation related to parallelism and concurrency, as well as its relation with the Single Pushout (SPO) transformation of arbitrary partial algebras based on different types of partial homomorphisms, as for instance quomorphisms in [8].

On this last topic we want to advance that, since the applicability of a rule $P = (\mathbf{L} \xleftarrow{l} \mathbf{K} \xrightarrow{r} \mathbf{R})$ through a homomorphism $m : \mathbf{L} \to \mathbf{D}$ only depends on the relationship between l and m, and it has nothing to do with r, there are many cases in which the pushout of a quomorphism $p : \mathbf{L} \to \mathbf{R}$ and a homomorphism $m : \mathbf{L} \to \mathbf{D}$ does not exist but if we translate p into a DPO rule $P_p = (\mathbf{L} \xleftarrow{i} \mathbf{Dom}\, p \xrightarrow{p} \mathbf{R})$ then the derived algebra of $\mathbf{D}$ by the application of P_p through m does exist; this is the case, for instance, of both examples of non-existence of pushouts given in [8, Ex. 1]. Conversely, it is not difficult to

Mercè Llabrés and Francesc Rosselló

produce examples where the pushout of a quomorphism $p : \mathbf{L} \to \mathbf{R}$ and a homomorphism $m : \mathbf{L} \to \mathbf{D}$ exists but the derived algebra of $\mathbf{D}$ by the application of P_p through m does not exist, as well as examples where both algebras exist but they are not isomorphic. Therefore, and contrary to what happens with DPO and SPO graph transformation under the Berlin approaches, where the SPO approach subsumes (in a suitable sense) the DPO approach, DPO transformation of partial algebras is somehow transversal to SPO transformation using quomorphisms as introduced in [8].

References

1. Burmeister, P.: A Model Theoretic Oriented Approach to Partial Algebras. Mathematical Research **32**, Akademie-Verlag (1986). 133, 138
2. Burmeister, P., Rosselló, F., Torrens, J., Valiente, G.: Algebraic Transformation of Unary Partial Algebras I: Double-Pushout Approach. Theoretical Computer Science **184** (1997) 145–193. 132, 133, 135, 136, 140, 141, 143
3. Ehrig, H., Kreowski, H.-J., Maggiolo-Schettini, A., Rosen, B. K., Winkowski, J.: Deriving Structures from Structures. Lect. Notes in Comp. Science **64** (1978) 177–190. 131, 138, 143
4. Große-Rhode, M.: Specification of State Based Systems by Algebra Rewrite Systems and Refinements. Technical Report 99-04, TU Berlin (March 1999). 131
5. Kawahara, Y.: Pushout-Complements and Basic Concepts of Grammars in Toposes. Theoretical Computer Science **77** (1990) 267-289. 133
6. Llabrés, M., Rosselló, F.: Double-Pushout Transformation of Arbitrary Partial Algebras. In preparation. 133
7. Monserrat, M., Rosselló, F., Torrens, J.: When is a Category of Many–Sorted Partial Algebras Cartesian Closed? International Journal of Foundations of Computer Science **6** (1995) 51-66. 133
8. Wagner, A., Gogolla, M.: Defining Operational Behavior of Object Specifications by Attributed Graph Transformations. Fundamenta Informaticae **26** (1996) 407–431. 131, 143, 144

Unfolding of Double-Pushout Graph Grammars is a Coreflection[*]

Paolo Baldan, Andrea Corradini, and Ugo Montanari

Dipartimento di Informatica – Università di Pisa
Corso Italia, 40, 56125 Pisa, Italy
{baldan,andrea,ugo}@di.unipi.it

Abstract. In a recent paper, mimicking Winskel's construction for Petri nets, a concurrent semantics for (*double-pushout*) *DPO graph grammars* has been provided by showing that each graph grammar can be unfolded into an acyclic branching structure, that is itself a (nondeterministic occurrence) graph grammar describing all the possible computations of the original grammar.

This paper faces the problem of providing a closer correspondence with Winskel's result by showing that the unfolding construction can be described as a coreflection between the category of graph grammars and the category of occurrence graph grammars. The result is shown to hold for a suitable subclass of graph grammars, called *semi-weighted graph grammars*. Unfortunately the coreflection does not extend to the whole category of graph grammars: some ideas for solving the problem are suggested.

1 Introduction

In recent years, various concurrent semantics for graph rewriting systems have been proposed in the literature, some of which are inspired by their correspondence with Petri nets (see [5] for a tutorial introduction to the topic and for relevant references). A classical result in the theory of concurrency for Petri nets, due to Winskel [18], shows that the event structure semantics of *safe* nets can be given via a chain of coreflections starting from the category **Safe** of safe nets, through category **Occ** of occurrence nets. The event structure associated with a net is obtained by first constructing a "nondeterministic unfolding" of the net, and then by considering only its transitions and the causal and conflict relations among them. In [14,15] it is shown that essentially the same constructions work for the larger category of *semi-weighted nets*, i.e., P/T nets where the initial marking is a set and transitions can generate at most one token in each post-condition. Winskel's result has been also extended, in [2], to a more general class of nets called (semi-weighted) contextual nets or nets with read (test) arcs. Contextual nets generalize classical nets by adding the possibility of checking

---

[*] Research partially supported by MURST project Tecniche Formali per Sistemi Software, by TMR Network GETGRATS and by Esprit WG APPLIGRAPH.

H. Ehrig et al. (Eds.): Graph Transformation, LNCS 1764, pp. 145–163, 2000.

for the presence of a token in a place, without consuming it. Their capability of "preserving part" of the state in a rewriting step makes this kind of nets closer to graph grammars. Indeed, starting from these results, the paper [3] shows that a Winskel's style construction allows one to unfold each graph grammar into a nondeterministic occurrence grammar describing its behaviour. The unfolding is used to define a prime algebraic domain and an event structure semantics for the grammar.

In this paper we make a further step towards full correspondence with Winskel's result by facing the problem of characterizing the unfolding construction for DPO graph grammars just mentioned as a true coreflection.

Section 2 reviews the basics of DPO typed graph grammars and introduces the notion of grammar morphism, a slight variation of the morphisms in [6], making the class of graph grammars a category **GG**. Section 3 recalls the notion of *nondeterministic occurrence grammar* [3], which are grammars satisfying suitable acyclicity and well-foundedness requirements, representing in a unique "branching" structure several possible "acyclic" grammar computations. The full subcategory of **GG** having occurrence grammars as objects is denoted by **OGG**. By exploiting the notions of occurrence grammar and of grammar morphism, Section 4 defines *nondeterministic graph process*. As in Petri net theory, a nondeterministic process of a grammar $\mathcal{G}$ consists of a (suitable) grammar morphism from an occurrence grammar to $\mathcal{G}$. Nicely, deterministic finite processes turn out to coincide with the graph processes of [1].

Section 5 presents the unfolding construction that, when applied to a given grammar $\mathcal{G}$, yields a nondeterministic occurrence grammar $\mathcal{U}(\mathcal{G})$, which describes its behaviour. The unfolding is endowed with a morphism $\chi_{\mathcal{G}}$ into the original grammar $\mathcal{G}$, making $\mathcal{U}(\mathcal{G})$ a process of $\mathcal{G}$. Next, Section 6 faces the problem of turning the unfolding construction into a functor establishing a coreflection between the categories of graph grammars and of occurrence grammars. As in the case of Petri nets we restrict to those grammars where the initial graph and the items produced by each production are injectively typed. Such grammars, by analogy with the corresponding subclass of Petri nets, are called *semi-weighted*, and the corresponding full subcategory of **GG** is denoted by **SGG**. We show that the unfolding construction extends to a functor $\mathcal{U} : \mathbf{SGG} \to \mathbf{OGG}$ which is right adjoint to the inclusion $\mathcal{I}_O : \mathbf{OGG} \to \mathbf{SGG}$, and thus establishes a coreflection between the two categories.

In Section 7, we show that unfortunately the result cannot be extended in a trivial way to the whole category **GG** of graph grammars. Even worse, a counterexample shows that there is no way of turning the unfolding construction into a functor which is right adjoint to the inclusion $\mathcal{I} : \mathbf{OGG} \to \mathbf{GG}$. Starting from this negative result some possible ways of solving the problem are singled out.

Because of space limitations we are forced to defer to the full version the detailed comparison with the related work in the literature, comprising different notions of graph grammar morphisms [6,12,17,4] as well as the unfolding construction for SPO grammars in [17]. For the same reason also the proofs of our statements are omitted.

2 Typed Graph Grammars and Their Morphisms

This section first summarizes the basic definitions about typed graph grammars [8], a variation of classical DPO graph grammars [10,9] which uses *typed graphs*, namely graphs labelled over a structure (the *graph of types*) that is itself a graph. Next some insights are given on the relationship between typed graph grammars and Petri nets. Finally, the class of typed graph grammars is turned into a category **GG** by introducing a notion of grammar morphism.

2.1 Typed Graph Grammars

Let **Graph** be the category of (directed, unlabelled) graphs and total graph morphisms. For a graph G we will denote by N_G and E_G the sets of *nodes* and *arcs* of G, and by $s_G, t_G : E_G \to N_G$ its *source* and *target* functions. Given a graph TG, a *typed graph* G over TG is a graph $|G|$, together with a morphism $t_G : |G| \to TG$. A morphism between TG-typed graphs $f : G_1 \to G_2$ is a graph morphisms $f : |G_1| \to |G_2|$ consistent with the typing, i.e., such that $t_{G_1} = t_{G_2} \circ f$. A typed graph G is called *injective* if the typing morphism t_G is injective. The category of TG-typed graphs and typed graph morphisms is denoted by TG-**Graph** and can be sinthetically defined as the comma category (**Graph** $\downarrow TG$).

Fixed a graph TG of types, a *(TG-typed graph) production* $(L \xleftarrow{l} K \xrightarrow{r} R)$ is a pair of *injective* typed graph morphisms $l : K \to L$ and $r : K \to R$, where $|L|$, $|K|$ and $|R|$ are finite graphs. It is called *consuming* if morphism $l : K \to L$ is not surjective. The typed graphs L, K, and R are called the *left-hand side*, the *interface*, and the *right-hand side* of the production, respectively.

Definition 1 (typed graph grammar). *A (TG-typed) graph grammar $\mathcal{G}$ is a tuple $\langle TG, G_{in}, P, \pi \rangle$, where G_{in} is the initial (typed) graph, P is a set of production names, and π is a function which associates a graph production to each production name in P.*

We denote by $Elem(\mathcal{G})$ the set $N_{TG} \cup E_{TG} \cup P$. Furthermore, we will assume that for each production name q the corresponding production $\pi(q)$ is $L_q \xleftarrow{l_q} K_q \xrightarrow{r_q} R_q$, where, without loss of generality, the injective morphisms l_q and r_q are inclusions.

Since in this paper we work only with typed notions, we will usually omit the qualification "typed", and, sometimes, we will not indicate explicitly the typing morphisms. Moreover, we will consider only *consuming* grammars, namely grammars where all productions are consuming: this corresponds, in the theory of Petri nets, to the common requirement that transitions must have non-empty preconditions.

Definition 2 (direct derivation). *Given a typed graph G, a production q, and a match (i.e., a graph morphism) $g : L_q \to G$, a direct derivation δ from G to H using q (based on g) exists, written $\delta : G \Rightarrow_q H$, if and only if the diagram*

$$
\begin{array}{ccccc}
q : L_q & \xleftarrow{\ l_q\ } & K_q & \xrightarrow{\ r_q\ } & R_q \\
\ \downarrow{\scriptstyle g} & & \ \downarrow{\scriptstyle k} & & \ \downarrow{\scriptstyle h} \\
G & \xleftarrow{\ b\ } & D & \xrightarrow{\ d\ } & H
\end{array}
$$

*can be constructed, where both squares have to be pushouts in TG-**Graph**.*

Given an injective morphism $l_q : K_q \to L_q$ and a match $g : L_q \to G$ as in the above diagram, their *pushout complement* (i.e., a graph D with morphisms k and b such that the left square is a pushout) exists if and only if the *gluing condition* is satisfied. This consists of two parts:

- the *identification condition*, requiring that if two distinct nodes or arcs of L_q are mapped by g to the same image, then both must be in the image of l_q;
- the *dangling condition*, stating that no arc in $G - g(L_q)$ should be incident to a node in $g(L_q - l_q(K_q))$ (because otherwise the application of the production would leave such an arc "dangling").

A *derivation* over a grammar $\mathcal{G}$ is a sequence of direct derivations (over $\mathcal{G}$) starting from the initial graph, namely $\rho = \{G_{i-1} \Rightarrow_{q_{i-1}} G_i\}_{i \in \{1,\ldots,n\}}$, with $G_0 = G_{in}$.

2.2   Relation with Petri Nets

The notion of grammar morphism, and many definitions and constructions in this paper are better understood keeping in mind the relation between Petri nets and DPO graph grammars. The basic observation (which belongs to the folklore, see, e.g., [5]) is that a P/T Petri net is essentially a rewriting system on multisets, and that, given a set A, a multiset of A can be represented as a discrete graph typed over A. In this view a P/T net can be seen as a graph grammar acting on discrete graphs typed over the set of places, the productions being (some encoding of) the net transitions: a marking is represented by a set of nodes (tokens) labelled by the place where they are, and, for example, the unique transition t of the net in Fig. 1.(a) is represented by the graph production in the top row of Fig. 1.(b). Notice that the interface is empty since nothing is explicitly preserved by a net transition.

It is easy to check that this representation satisfies the properties one would expect: a production can be applied to a given marking if and only if the corresponding transition is enabled and, in this case, the double pushout construction produces the same marking as the firing of the transition. For instance, the firing of transition t, leading from the marking $3A + 2B$ to the marking $A + B + C + D$ in Fig. 1.(a), becomes the double pushout diagram of Fig. 1.(b).

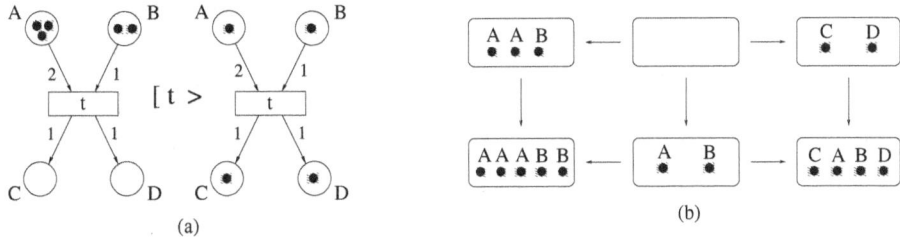

Fig. 1. Firing of a transition and corresponding DPO direct derivation.

2.3 Grammar Morphisms

The notion of grammar morphism we are going to introduce is very similar to the one originally defined in [6], which was in turn introduced as a generalization of Petri nets morphisms. Recall that a Petri net morphism [18] consists of two components: a multirelation between the sets of places, and a partial function mapping transitions of the first net into transitions of the second one. Net morphisms are required to "preserve" the pre-set and post-set of transitions, in the sense that the pre- (post-)set of the image of a transition t must be the image of the pre- (post-)set of t.

Since the items of the graph of types of a grammar can be seen as a generalization of Petri net places, the first component of a grammar morphism will be a span between the type graphs of the source and target grammars, arising as a categorical generalization of the notion of multirelation. For an extensive discussion of this idea we refer the reader to [6,4]. The following definitions will be useful.

Definition 3 (spans). *Let* **C** *be a category. A (concrete) span in* **C** *is a pair of coinitial arrows* $f = \langle f^L, f^R \rangle$ *with* $f^L : x_f \to a$ *and* $f^R : x_f \to b$. *Objects* a *and* b *are called the source an the target of the span and we will write* $f : a \leftrightarrow b$. *The span* f *will be sometimes written as* $\langle f^L, x_f, f^R \rangle$, *explicitly giving the common source object* x_f.

Consider now the equivalence $\sim$ *over the set of spans with the same source and target defined, for* $f, f' : a \leftrightarrow b$, *as* $f \sim f'$ *if there exists an isomorphism* $k : x_f \to x_{f'}$ *such that* $f'^L \circ k = f^L$ *and* $f'^R \circ k = f^R$ *(see Fig. 2.(a)). The isomorphism class of a span* f *will be denoted by* $[f]$ *and called a* semi-abstract span.

Definition 4 (category of spans). *Let* **C** *be a category with pullbacks. Then the category* **Span(C)** *has the same objects of* **C** *and semi-abstract spans on* **C** *as arrows. More precisely, a semi-abstract span* $[f]$ *is an arrow from the source to the target of* f. *The composition of two semi-abstract spans* $[f_1] : a \leftrightarrow b$ *and* $[f_2] : b \leftrightarrow c$ *is the (equivalence class) of a span* f *constructed as in Fig. 2.(b) (i.e.,* $f^L = f_1^L \circ y$ *and* $f^R = f_2^R \circ z$), *where the square is a pullback. The identity* *on an object a is the equivalence class of the span $\langle id_a, id_a \rangle$, where id_a is the identity of a in* **C**.

It can be shown that composition is well-defined, namely it does not depend on the particular choice of the representatives, and that it is associative.

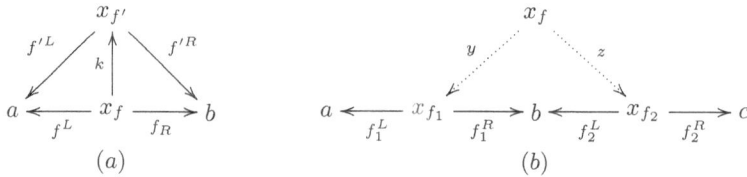

$$(a) \qquad\qquad\qquad\qquad (b)$$

Fig. 2. Equivalence and composition of spans.

Let $\mathcal{G}_1$ and $\mathcal{G}_2$ be two graph grammars and let $[f_T] : TG_1 \leftrightarrow TG_2$ be a semi-abstract span between the corresponding type graphs. Observe that $[f_T]$ induces a relation between TG_1-typed graphs and TG_2-typed graphs. In fact, let G_1 be in TG_1-**Graph**. Then we can transform G_1 as depicted in the diagram below, by first taking a pullback (in **Graph**) of the arrows $f_T^L : X_{f_T} \to TG_1$ and $t_{G_1} : |G_1| \to TG_1$, and then typing the pullback object over TG_2 by using the right part of the span $f_T^R : X_{f_T} \to TG_2$.

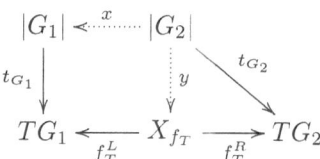

The TG_2-typed graph $G_2 = \langle |G_2|, f_T^R \circ y \rangle$ obtained with this construction, later referred to as *pullback-retyping* construction induced by $[f_T]$, is determined only up to isomorphism. This is due to the definition of pullback and to the fact that, considering semi-abstract spans, we can choose any concrete representative $f_T' \sim f_T$. Sometimes we will write $f_T\{x, y\}(G_1, G_2)$ (or simply $f_T(G_1, G_2)$ if we are not interested in morphisms x and y) to express the fact that G_1 and G_2 are related in this way by the pullback-retyping construction induced by $[f_T]$.

We are now ready to define grammar morphisms. Besides the component specifying the relation between the type graphs, a morphism from $\mathcal{G}_1$ to $\mathcal{G}_2$ includes a (partial) mapping between production names. Furthermore a third component explicitly relates the (untyped) graphs underlying corresponding productions of the two grammars, as well as the graphs underlying the initial graphs.

Definition 5 (grammar morphism). *Let $\mathcal{G}_i = \langle TG_i, G_{in_i}, P_i, \pi_i \rangle$ $(i \in \{1, 2\})$ be two graph grammars. A morphism $f : \mathcal{G}_1 \to \mathcal{G}_2$ is a triple $\langle [f_T], f_P, \iota_f \rangle$ where*

- $[f_T] : TG_1 \leftrightarrow TG_2$ *is a semi-abstract span in* **Graph**, *called the* type-span;
- $f_P : P_1 \to P_2 \cup \{\emptyset\}$ *is a total function, where* $\emptyset$ *is a new production name (not in* P_2*), with associated production* $\emptyset \leftarrow \emptyset \to \emptyset$, *referred to as the* empty production;
- ι_f *is a family* $\{\iota_f(q_1) \mid q_1 \in P_1\} \cup \{\iota_f^{in}\}$ *such that* $\iota_f^{in} : |G_{in_2}| \to |G_{in_1}|$ *and for each* $q_1 \in P_1$, *if* $f_P(q_1) = q_2$, *then* $\iota_f(q_1)$ *is triple of morphisms*

$$\langle \iota_f^L(q_1) : |L_{q_2}| \to |L_{q_1}|, \iota_f^K(q_1) : |K_{q_2}| \to |K_{q_1}|, \iota_f^R(q_1) : |R_{q_2}| \to |R_{q_1}| \rangle.$$

such that the following conditions are satisfied:

1. Preservation of the initial graph.
   There exists a morphism k *such that* $f_T\{\iota_f^{in}, k\}(G_{in_1}, G_{in_2})$, *namely such that the following diagram commutes and the square is a pullback:*

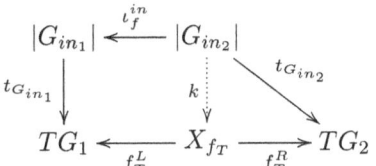

2. Preservation of productions.
   For each $q_1 \in P_1$, *with* $q_2 = f_P(q_1)$, *there exist morphisms* k^L, k^K *and* k^R *such that the diagram below commutes, and* $f_T\{\iota_f^X(q_1), k^X\}(X_{q_1}, X_{q_2})$ *for* $X \in \{L, K, R\}$.

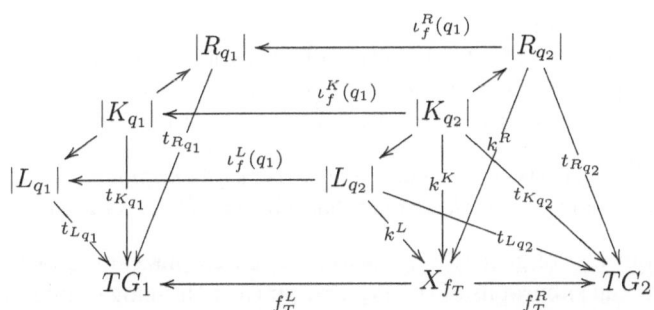

The grammar morphisms in [6] rely on the assumption of having a fixed choice of pullbacks. Consequently the pullback-retyping construction is deterministic and morphisms are required to preserve the initial graphs and the productions "on the nose". This requirement is very strict and it may imply the absence of a morphism between two grammars having isomorphic initial graph and productions. The notion of morphism just introduced is, in a sense, more liberal: we avoid a global choice of pullbacks, and, influenced by the notion of graph process in [1], we fix "locally", for each morphism f, only part of the pullback diagrams, namely the morphisms in the family ι_f.

It is worth noticing that, for technical convenience, the partial mapping on production names is represented as a total mapping by enriching the target set with a distinguished point $\emptyset$, representing "undefinedness". In this way the condition asking the preservation of productions (Condition 2) faithfully rephrases the situation of net theory where the pre- and post-set of a transition on which the morphism is undefined are necessarily mapped to the empty multiset.

As in [6] one can show that grammar morphisms are "simulations" in the sense that for every derivation ρ_1 in $\mathcal{G}_1$ there is a corresponding derivation ρ_2 in $\mathcal{G}_2$, related to ρ_1 by the pullback-retyping construction induced by the morphism. As already observed, as a consequence of the partial arbitrariness in the choice of the pullback components, such correspondence, differently from [6], is not "functional".

3   Nondeterministic Occurrence Grammars

Nondeterministic occurrence grammars, as introduced in [3], are intended to represent the computations of graph grammars in a static way, by recording the events (production applications) which can appear in all possible derivations and the dependency relations between them. Analogously to what happens for nets, occurrence grammars are "safe" grammars, where the dependency relations between productions satisfy suitable acyclicity and well-foundedness requirements. While for nets it suffices to take into account only the causality and conflict relations, for grammars the fact that a production application not only consumes and produces, but also preserves a part of the state leads to a form of asymmetric conflict between productions. Furthermore, because of the dangling condition, also the graphical structure of the state imposes some precedences between productions.

A first step towards the definition of occurrence grammar is a suitable notion of safeness [8], generalizing the usual one for P/T nets which requires that each place contains at most one token in any reachable marking.

Definition 6 ((strongly) safe grammar). *A grammar* $\mathcal{G} = \langle TG, G_{in}, P, \pi \rangle$ *is* (strongly) safe *if, for all H such that $G_{in} \Rightarrow^* H$, H is injective.*

Without loss of generality, injective typed graphs can be identified with the corresponding subgraphs of the type graph (just thinking of injective morphisms as inclusions). In particular, each TG-typed graph G reachable in a safe grammar can be identified with the subgraph $t_G(|G|)$ of the type graph TG. With the above identification, in each computation of a safe grammar starting from the initial graph a production can only be applied to the subgraph of the type graph which is the image via the typing morphism of its left-hand side. Therefore according to its typing, we can think that a production *produces*, *preserves* or *consumes* items of the type graph. Using a net-like language, we speak of *pre-set* $^\bullet q$, *context* $\underline{q}$ and *post-set* $q^\bullet$ of a production q, defined in the obvious way. Similarly, for a node or arc x in TG we write $^\bullet x$, $\underline{x}$ and $x^\bullet$ to denote the sets of productions which produce, preserve and consume x. Consider, for instance, the grammar $\mathcal{G}$ in Fig. 3, where the typing morphisms for the initial graph and the productions are represented by suitably labelling the involved graphs with items of the type graph TG. The pre-set, context and post-set of production q_1 are $^\bullet q_1 = \{A, L\}$, $\underline{q_1} = \{B\}$ and $q_1^\bullet = \{C\}$, while for the node B, $^\bullet B = \emptyset$, $\underline{B} = \{q_1\}$ and $B^\bullet = \{q_2\}$.

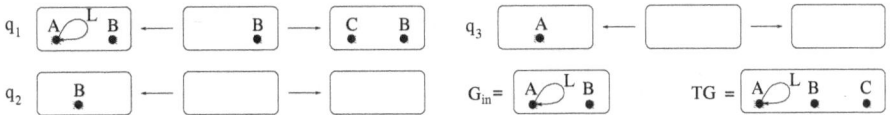

Fig. 3. The safe grammar $\mathcal{G}$.

Although the notion of causal relation is meaningful only for safe grammars, it is technically convenient to define it for general grammars. The same holds for the asymmetric conflict relation introduced below.

Definition 7 (causal relation). *The* causal relation *of a grammar* $\mathcal{G}$ *is the binary relation* $<$ *over* $Elem(\mathcal{G})$ *defined as the least transitive relation satisfying: for any node or arc* x *in the type graph* TG, *and for productions* $q_1, q_2 \in P$

1. *if* $x \in {}^\bullet q_1$ *then* $x < q_1$;
2. *if* $x \in q_1^\bullet$ *then* $q_1 < x$;
3. *if* $q_1^\bullet \cap \underline{q_2} \neq \emptyset$ *then* $q_1 < q_2$;

As usual $\leq$ *is the reflexive closure of* $<$. *Moreover, for* $x \in Elem(\mathcal{G})$ *we denote by* $\lfloor x \rfloor$ *the set of causes of* x *in* P, *namely* $\{q \in P : q \leq x\}$.

Notice that the fact that an item is preserved by q_1 and consumed by q_2, i.e., $\underline{q_1} \cap {}^\bullet q_2 \neq \emptyset$ (e.g., the node B in grammar $\mathcal{G}$ of Fig. 3), does not imply $q_1 < q_2$. Actually, since q_1 must precede q_2 in any computation where both appear, in such computations q_1 acts as a cause of q_2. However, differently from a true cause, q_1 is not necessary for q_2 to be applied. Therefore we can think of the relation between the two productions as a *weak* form of *causal dependency*. Equivalently, we can observe that the application of q_2 prevents q_1 to be applied, so that q_1 can never follow q_2 in a derivation. But the converse is not true, since q_1 *can* be applied before q_2. Thus this situation can also be interpreted naturally as an *asymmetric conflict* between the two productions (see [2,16,13]).

Definition 8 (asymmetric conflict). *The* asymmetric conflict relation *of a grammar* $\mathcal{G}$ *is the binary relation* $\nearrow$ *over the set of productions, defined by:*

1. *if* $\underline{q_1} \cap {}^\bullet q_2 \neq \emptyset$ *then* $q_1 \nearrow q_2$;
2. *if* $^\bullet q_1 \cap {}^\bullet q_2 \neq \emptyset$ *and* $q_1 \neq q_2$ *then* $q_1 \nearrow q_2$;
3. *if* $q_1 < q_2$ *then* $q_1 \nearrow q_2$.

Condition 1 is justified by the discussion above. Condition 2 essentially expresses the fact that the ordinary symmetric conflict is encoded, in this setting, as an asymmetric conflict in both directions. Finally, since $<$ represents a global order of execution, while $\nearrow$ determines an order of execution only locally to each computation, it is natural to impose $\nearrow$ to be an extension of $<$ (Condition 3).

Definition 9 ((nondeterministic) occurrence grammar). *A* (nondeterministic) occurrence grammar *is a grammar* $\mathcal{O} = \langle TG, G_{in}, P, \pi \rangle$ *such that*

1. *its causal relation $\leq$ is a partial order, and, for any $q \in P$, the set $\lfloor q \rfloor$ is finite and the asymmetric conflict $\nearrow$ is acyclic on $\lfloor q \rfloor$;*
2. *the initial graph G_{in} is the set $Min(\mathcal{O})$ of minimal elements of $\langle Elem(\mathcal{O}), \leq \rangle$ (with the graphical structure inherited from TG and typed by the inclusion);*
3. *each item x in TG is created by at most one production in P, namely $|\,{}^\bullet x\,| \leq 1$;*
4. *for each production q, the typing t_{L_q} is injective on the "consumed part" $|L_q| - l_q(|K_q|)$, and similarly t_{R_q} is injective on the "produced part" $|R_q| - r_q(|K_q|)$.*

*We denote by **OGG** the full subcategory of **GG** having occurrence grammars as objects.*

Since the initial graph of an occurrence grammar $\mathcal{O}$ is determined by $Min(\mathcal{O})$, we often do not mention it explicitly. One can show that, by the defining conditions, each occurrence grammar is *safe*.

Intuitively, conditions (1)–(3) recast in the framework of graph grammars the analogous conditions of occurrence nets (actually of occurrence contextual nets [2]). In particular, in Condition (1), acyclicity of asymmetric conflict on $\lfloor q \rfloor$ corresponds to the requirement of irreflexivity for the conflict relation in occurrence nets. Condition (4), instead, is closely related to safeness and requires that each production consumes and produces items with multiplicity one. Together with acyclicity of $\nearrow$, it disallows the presence of some productions which surely could never be applied, because they fail to satisfy the identification condition with respect to the typing morphism.

It is worth stressing that because of the dangling condition, some productions of an occurrence grammar might never be applicable, as, for example, the production q_3 of grammar $\mathcal{G}$ in Fig. 3. The reason why we did not consider the dangling condition in the definition of occurrence grammar is that checking such negative (non-monotonic) condition on a production, would require to find a possible computation which removes the potentially dangling arcs, and to verify the consistency of such computation with the production at hand. By using the Turing completeness of DPO graph grammars, it can be shown that such verification is undecidable for infinite occurrence grammars, which can be obtained as unfolding of finite grammars.

The restrictions to the behaviour imposed by the dangling condition are considered when defining the configurations of an occurrence grammar, which represent exactly all the possible deterministic runs of the grammar.

Definition 10 (configuration). *A configuration of an occurrence graph grammar $\mathcal{O} = \langle TG, P, \pi \rangle$ is a subset $C \subseteq P$ such that*

1. *if $\nearrow_C$ denotes the restriction of the asymmetric conflict relation to C, then $(\nearrow_C)^*$ is a partial order, and $\{q' \in C : q'(\nearrow_C)^* q\}$ is finite for all $q \in C$;[1]*

---

[1] As usual, for a binary relation r, with r^* we denote its transitive and reflexive closure.

2. *C is left-closed w.r.t. $\leq$, i.e., for all $q \in C$, $q' \in P$, $q' \leq q$ implies $q' \in C$;*
3. *for all $e \in TG$ and $n \in \{s(e), t(e)\}$, if $n^{\bullet} \cap C \neq \emptyset$ and ${}^{\bullet}e \subseteq C$ then $e^{\bullet} \cap C \neq \emptyset$.*

If C satisfies conditions (1) and (2), then it is called a pre-configuration.

The first two conditions are equivalent to those defining configurations of asymmetric event structures and thus of occurrence contextual nets [2]. Condition 3, instead, formalizes the dangling condition. If a configuration contains a production q consuming a node n and a production q' producing an arc e with source (or target) n, then arc e must be removed by some production in the configuration, otherwise, due to the dangling condition, q could not be executed. Similar considerations apply if the arc e is present in the initial graph, i.e., ${}^{\bullet}e = \emptyset$.

A production which does not satisfy the dangling condition in any graph reachable from the initial graph is not part of any configuration. For example, q_3 does not appear in the set of configurations of grammar $\mathcal{G}$ in Fig. 3, $Conf(\mathcal{G}) = \{\emptyset, \{q_1\}, \{q_2\}, \{q_1, q_2\}\}$.

4   Nondeterministic Graph Processes

In the theory of Petri nets the notion of occurrence net is strictly related to that of process. A (non)deterministic net process is a (non)deterministic occurrence net with a suitable morphism to the original net. Similarly, nondeterministic occurrence grammars can be used to define a notion of *nondeterministic graph processes*, generalizing the deterministic graph processes of [8,1].

A *nondeterministic graph process* is aimed at representing in a unique "branching" structure several possible computations of a grammar. The underlying occurrence grammar makes explicit the causal structure of such computations since each production can be applied at most once and each items of the type graph can be "filled" at most once. Via the morphism to the original grammar, productions and items of the type graph in the occurrence grammar can be thought of, respectively, as instances of applications of productions and instances of items generated in the original grammar by such applications. Actually, to allow for such an interpretation, some further restrictions must be imposed on the process morphism. Recall that process morphisms in Petri net theory must map places into places (rather than into multisets of places) and must be total on transitions [11]. Similarly, for graph process morphisms the left component of the type-span is required to be an isomorphism in such a way that the type-span can be thought of simply as a graph morphism. Furthermore a process morphism cannot map a production to the empty production, a requirement corresponding to totality.

Definition 11 (strong morphism). *A grammar morphism $f : \mathcal{G}_1 \to \mathcal{G}_2$ is called strong if $f_T^L : X_f \to TG_1$ is an isomorphism and $f_P(q_1) \neq \emptyset$, for any $q_1 \in P_1$.*

Hereafter we will always choose as concrete representative of the type-span of a strong grammar morphism f, a span f_T such that the left component f_T^L is the identity id_{TG_1}.

It is not difficult to verify that, if f is a strong morphism then, by Condition 1 of the definition of grammar morphism (Definition 5), $\iota_f^{in} : |G_{in_2}| \to |G_{in_1}|$ is an isomorphism. Similarly, by Condition 2, for each production $q_1 \in P_1$, $\iota_f(q_1)$ is a triple of isomorphisms, namely each production of $\mathcal{G}_1$ is mapped to a production of $\mathcal{G}_2$ with associated isomorphic (untyped) span.

Definition 12 (graph process). *Let $\mathcal{G}$ be a graph grammar. A graph process of $\mathcal{G}$ is a strong grammar morphism $\chi : \mathcal{O}_\chi \to \mathcal{G}$, where $\mathcal{O}_\chi$ is an occurrence grammar.*

We will denote by TG_χ, G_{in_χ}, P_χ and π_χ the components of the occurrence grammar $\mathcal{O}_\chi$ underlying a process χ.

Using the notions above we are naturally led to the definitions of *deterministic* occurrence grammar and process. In fact we can take an occurrence grammar $\mathcal{O}$ to be *deterministic* if the set P of its productions is a configuration of $\mathcal{O}$. Then a process χ is *deterministic* if the underlying occurrence grammar $\mathcal{O}_\chi$ is deterministic. Nicely, deterministic finite processes are exactly the (non-concatenable) graph processes of [1], which are shown there to be equivalent with the more classical trace semantics (e.g., as described in [7]).

5 Unfolding Construction

This section introduces the unfolding construction which, applied to a consuming grammar $\mathcal{G}$, produces a nondeterministic occurrence grammar $\mathcal{U}(\mathcal{G})$ describing the behaviour of $\mathcal{G}$. The unfolding is equipped with a strong grammar morphism $\chi_\mathcal{G}$ to the original grammar, making it a process of $\mathcal{G}$.

The idea consists of starting from the initial graph of the grammar, then applying in all possible ways its productions, and recording in the unfolding each occurrence of production and each new graph item generated in the rewriting process, both enriched with the corresponding causal history. According to the discussion in the previous section, during the unfolding process productions are applied without considering the dangling condition. Moreover we adopt a notion of concurrency which is "approximated", again in the sense that it does not take care of the precedences between productions induced by the dangling condition.

Definition 13 (quasi-concurrent graph). *Let $\mathcal{O} = \langle TG, P, \pi \rangle$ be an occurrence grammar. A subgraph G of TG is called* quasi-concurrent *if*

1. $\bigcup_{x \in G} \lfloor x \rfloor$ *is a pre-configuration;*
2. $\neg(x < y)$ *for all $x, y \in G$.*

Another basic ingredient of the unfolding is the *gluing* operation. It can be seen as a "partial application" of a rule to a given match, in the sense that it generates the new items as specified by the production (i.e., items of right-hand side not in the interface), but items that should have been deleted are not affected: intuitively, this is because such items may still be used by another production in the nondeterministic unfolding.

Definition 14 (gluing). *Let q be a production, G a graph and $m : L_q \to G$ a graph morphism. We define, for any symbol $*$, the gluing of G and R_q along K_q, according to m and marked by $*$, denoted by $glue_*(q, m, G)$, as the graph $\langle N, E, s, t \rangle$, where:*

$$N = N_G \cup m_*(N_{R_q}) \qquad\qquad E = E_G \cup m_*(E_{R_q})$$

with m_* defined by: $m_*(x) = m(x)$ if $x \in K_q$ and $m_*(x) = \langle x, * \rangle$ otherwise. The source and target functions and the typing are inherited from G and R_q.

The gluing operation keeps unchanged the identity of the items already in G, and records in each newly added item from R_q the given symbol $*$. Notice that the gluing, as just defined, is a concrete deterministic definition of the pushout of the arrows $G \xleftarrow{m} L_q \xhookleftarrow{l_q} K_q$ and $K_q \xhookrightarrow{r_q} R_q$.

As described below, the unfolding of a grammar is obtained as the limit of a chain of occurrence grammars, each approximating the unfolding up to a certain causal depth.

Definition 15 (depth). *Let $\mathcal{O} = \langle TG, P, \pi \rangle$ be an occurrence grammar. The function depth : $Elem(\mathcal{O}) \to \mathbb{N}$ is defined inductively as follows:*

$$
\begin{aligned}
&depth(x) = 0 && \text{for } x \in |G_{in}| = Min(\mathcal{O}); \\
&depth(q) = \max\{depth(x) \mid x \in {}^\bullet q \cup \underline{q}\} + 1 && \text{for } q \in P; \\
&depth(x) = depth(q) && \text{for } x \in q^\bullet.
\end{aligned}
$$

It is not difficult to prove that *depth* is a well-defined total function, since infinite descending chains of causality are disallowed in occurrence grammars. Moreover, given an occurrence grammar $\mathcal{O}$, the grammar containing only the items of *depth* less or equal to n, denoted by $\mathcal{O}^{[n]}$, is a well-defined occurrence grammar.

As expected an occurrence grammar $\mathcal{O}$ is the (componentwise) union of its subgrammars $\mathcal{O}^{[n]}$, of depth n. Moreover it is not difficult to see that if $g : \mathcal{O} \to \mathcal{G}$ is a grammar morphism, then for any $n \in \mathbb{N}$, g restricts to a morphism $g^{[n]} : \mathcal{O}^{[n]} \to \mathcal{G}$. In particular, if $TG^{[n]}$ denotes the type graph of $\mathcal{O}^{[n]}$, then the type-span of $g^{[n]}$ will be the equivalence class of

$$TG^{[n]} \xleftarrow{\ g_T^{L\,[n]}\ } X^{[n]} \xrightarrow{\ g_T^{R\,[n]}\ } TG_{\mathcal{G}}$$

where $X^{[n]} = \{x \in X_g \mid g_T^L(x) \in TG^{[n]}\}$. Vice versa each morphism $g : \mathcal{O} \to \mathcal{G}$ is uniquely determined by its truncations at finite depths.

We are now ready to present the unfolding construction.

Definition 16 (unfolding). *Let $\mathcal{G} = \langle TG, G_{in}, P, \pi \rangle$ be a (consuming) graph grammar. We inductively define, for each n, an occurrence grammar $\mathcal{U}(\mathcal{G})^{[n]} = \langle TG^{[n]}, P^{[n]}, \pi^{[n]} \rangle$ and a morphism $\chi^{[n]} = \langle \chi_T{}^{[n]}, \chi_P{}^{[n]}, \iota^{[n]} \rangle : \mathcal{U}(\mathcal{G})^{[n]} \to \mathcal{G}$. Then the unfolding $\mathcal{U}(\mathcal{G})$ and the folding morphism $\chi_{\mathcal{G}} : \mathcal{U}(\mathcal{G}) \to \mathcal{G}$ are the occurrence grammar and strong grammar morphism defined as the componentwise union of $\mathcal{U}(\mathcal{G})^{[n]}$ and $\chi^{[n]}$, respectively.*

Since each morphism $\chi^{[n]}$ is strong, assuming that the left component of the type-span $\chi_T{}^{[n]}$ is the identity on $TG^{[n]}$ we only need to define the right component $\chi_T^{R[n]} : TG^{[n]} \to TG$, which, by the way, makes $\langle TG^{[n]}, \chi_T^{R[n]} \rangle$ a TG-typed graph.

$(\mathbf{n} = \mathbf{0})$ *The components of the grammar $\mathcal{U}(\mathcal{G})^{[0]}$ are $TG^{[0]} = |G_{in}|$, $P^{[0]} = \pi^{[0]} = \emptyset$, while morphism $\chi^{[0]} : \mathcal{U}(\mathcal{G})^{[0]} \to \mathcal{G}$ is defined by $\chi_T^{R[0]} = t_{G_{in}}$, $\chi_P^{[0]} = \emptyset$, and $\iota^{[0]\,in} = id_{|G_{in}|}$.*

$(\mathbf{n} \to \mathbf{n} + \mathbf{1})$ *The occurrence grammar $\mathcal{U}(\mathcal{G})^{[n+1]}$ is obtained by extending $\mathcal{U}(\mathcal{G})^{[n]}$ with all the possible production applications to quasi-concurrent subgraphs of its the type graph. More precisely, let $M^{[n]}$ be the set of pairs $\langle q, m \rangle$ such that $q \in P$ is a production in $\mathcal{G}$ and $m : L_q \to \langle TG^{[n]}, \chi_T^{R[n]} \rangle$ is a match satisfying the identification condition, with $m(|L_q|)$ quasi-concurrent subgraph of $TG^{[n]}$. Then $\mathcal{U}(\mathcal{G})^{[n+1]}$ is the occurrence grammar resulting after performing the following steps for each $\langle q, m \rangle \in M^{[n]}$.*

- *Add to $P^{[n]}$ the pair $\langle q, m \rangle$ as a new production name and extend $\chi_P{}^{[n]}$ so that $\chi_P{}^{[n]}(\langle q, m \rangle) = q$. Intuitively, $\langle q, m \rangle$ represents an occurrence of q, where the match m is needed to record the "history".*

- *Extend the type graph $TG^{[n]}$ by adding to it a copy of each item generated by the application q, marked by $\langle q, m \rangle$ (in order to keep trace of the history). The morphism $\chi_T^{R[n]}$ is extended consequently. More formally, the TG-typed graph $\langle TG^{[n]}, \chi_T^{R[n]} \rangle$ is replaced by $glue_{\langle q,m \rangle}(q, m, \langle TG^{[n]}, \chi_T^{R[n]} \rangle)$.*

- *The production $\pi^{[n]}(\langle q, m \rangle)$ has the same untyped span of $\pi(q)$ and the morphisms $\iota^{[n]}(\langle q, m \rangle)$ are identities, that is $\iota(\langle q, m \rangle) = \langle id_{|L_q|}, id_{|K_q|}, id_{|R_q|} \rangle$. The typing of the left-hand side and of the interface is determined by m, and each item x of the right-hand side which is not in the interface is typed over the corresponding new item $\langle x, \langle q, m \rangle \rangle$ of the type graph.*

It is not difficult to verify that for each n, $\mathcal{U}(\mathcal{G})^{[n]}$ is a (finite depth) nondeterministic occurrence grammar, and $\mathcal{U}(\mathcal{G})^{[n]} \subseteq \mathcal{U}(\mathcal{G})^{[n+1]}$, componentwise. Therefore $\mathcal{U}(\mathcal{G})$ is a well-defined occurrence grammar. Similarly for each $n \in \mathbb{N}$ we have that $\chi^{[n]}$ is a well-defined morphism from $\mathcal{U}(\mathcal{G})^{[n]}$ to $\mathcal{G}$, which is the restriction to $\mathcal{U}(\mathcal{G})^{[n]}$ of $\chi^{[n+1]}$. This induces a unique morphism $\chi_{\mathcal{G}} : \mathcal{U}(\mathcal{G}) \to \mathcal{G}$.

It is possible to show that the unfolding construction applied to an occurrence grammar yields a grammar which is isomorphic to the original one.

6   Functorial Unfolding for Semi-weighted Grammars

The unfolding construction has been defined, up to now, only at "object level". This section makes a further step towards a full correspondence with Winskel's construction, by facing the problem of characterizing the unfolding as a coreflection between the categories of graph grammars and of occurrence grammars. As in the case of (contextual) Petri nets [15,2], we restrict to a full subcategory **SGG** of **GG** where objects satisfy conditions analogous to those defining semi-weighted P/T Petri nets. Then we show that the unfolding construction can be extended to a functor $\mathcal{U} : \mathbf{SGG} \to \mathbf{OGG}$ that is right adjoint to the inclusion functor $\mathcal{I}_O : \mathbf{OGG} \to \mathbf{SGG}$ and thus establishes a coreflection between **SGG** and **OGG**.

A graph grammar is semi-weighted if the initial graph is injective and the right-hand side of each production is injective if restricted to produced items (namely, items which are not in the interface). It is possible to show that, if we encode a Petri net N as a grammar $\mathcal{G}_N$, as sketched in Section 2, then N is a semi-weighted net if and only if $\mathcal{G}_N$ is a semi-weighted grammar.

Definition 17 (semi-weighted grammars). *A TG-typed production $L \leftarrow K \to R$ is called* semi-weighted *if t_R is injective on the "produced part" of R, namely on $|R| - r(|K|)$. A grammar $\mathcal{G}$ is called* semi-weighted *if the initial graph G_{in} is injective and for any $q \in P$ the production $\pi(q)$ is semi-weighted. We denote by* **SGG** *the full subcategory of* **GG** *having semi-weighted grammars as objects.*

The coreflection result strongly relies on the technical property which is stated in the next lemma. This is a key point where the restriction to semi-weighted grammars plays a rôle, since, as we will see, the lemma fails to hold for arbitrary grammars.

Lemma 1. *Let $\mathcal{G} = \langle TG, G_{in}, P, \pi \rangle$ be a semi-weighted grammar, let $\mathcal{O} = \langle TG', G'_{in}, P', \pi' \rangle$ be an occurrence grammar and let $f : \mathcal{O} \to \mathcal{G}$ be a grammar morphism. Then the morphism k, such that $f_T\{\iota_f^{in}, k\}(G_{in}, G_{in'})$ (see Definition 5, Condition 1) is uniquely determined. Similarly, for each $q \in P$, with $q' = f_P(q)$, the morphisms k^L, k^K and k^R such that $f_T\{\iota_f^X(q), k^X\}(X_q, X_{q'})$ for $X \in \{L, K, R\}$ (see Definition 5, Condition 2) are uniquely determined.*

A relevant property of morphisms between occurrence grammars which plays a central rôle in the proof of the coreflection is the fact that they "preserve" quasi-concurrency. This lemma can be proved along the same lines of an analogous result which hold for morphisms of contextual nets [2]. In fact, the notion of quasi-concurrency disregards the dangling condition, taking into account only causality and asymmetric conflict, whose treatment is basically the same for grammars and contextual nets.

Lemma 2 (preservation of concurrency). *Let $\mathcal{O}_1$ and $\mathcal{O}_2$ be occurrence grammars, let $f : \mathcal{O}_1 \to \mathcal{O}_2$ be a grammar morphism, and consider, for $i \in \{1,2\}$, a TG_i-typed graph G_i. If $f_T(G_1, G_2)$ and $t_{G_1}(|G_1|)$ is a quasi-concurrent subgraph of TG_1 then $t_{G_2}(|G_2|)$ is a quasi-concurrent subgraph of TG_2.*

Occurrence grammars are particular semi-weighted grammars, thus we have the inclusion functor $\mathcal{I}_O : \mathbf{OGG} \to \mathbf{SGG}$. The next theorem shows that the unfolding of a grammar $\mathcal{U}(\mathcal{G})$ and the folding morphism $\chi_{\mathcal{G}}$ are cofree over $\mathcal{G}$. Therefore $\mathcal{U}$ extends to a functor that is right adjoint of $\mathcal{I}_O$, thus establishing a coreflection between **SGG** and **OGG**.

Theorem 1 (coreflection between SGG and OGG). *Let $\mathcal{G}$ be a semi-weighted grammar, let $\mathcal{U}(\mathcal{G})$ be its unfolding and let $\chi : \mathcal{U}(\mathcal{G}) \to \mathcal{G}$ be the folding morphism as in Definition 16. Then for any occurrence grammar $\mathcal{O}$ and morphism $g : \mathcal{O} \to \mathcal{G}$ there exists a unique morphism $h : \mathcal{O} \to \mathcal{U}(\mathcal{G})$ such that the following diagram commutes:*

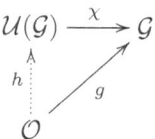

Therefore $\mathcal{I}_O \dashv \mathcal{U}$.

The proof of the existence of the morphism h uses Lemma 2 to inductively define, for each n, a morphism $h^{[n]} : \mathcal{O}^{[n]} \to \mathcal{U}(\mathcal{G})$, while uniqueness basically relies on Lemma 1.

7   Conclusions and Future Work

A natural question regards the possibility of extending the result of this paper to the whole category **GG** of graph grammars. We remark that the proof of the uniqueness of the morphism h in Theorem 1 strongly relies on Lemma 1 which in turn requires the grammar $\mathcal{G}$ to be semi-weighted. Unfortunately the problem does not reside in our proof technique: the cofreeness of the unfolding $\mathcal{U}(\mathcal{G})$ and of the folding morphism $\chi_{\mathcal{G}}$ over $\mathcal{G}$ may really fail to hold if the grammar $\mathcal{G}$ is not semi-weighted.

For instance, consider grammars $\mathcal{G}_1$ and $\mathcal{G}_2$ in Fig. 4, where typed graphs are represented by decorating their items with pairs "concrete identity:type". The grammar $\mathcal{G}_2$ is not semi-weighted since the initial graph is not injective, while $\mathcal{G}_1$ is clearly an occurrence grammar. The unfolding $\mathcal{U}(\mathcal{G}_2)$ of the grammar $\mathcal{G}_2$, according to Definition 16, is defined as follows. The initial graph and type graph of $\mathcal{U}(\mathcal{G}_2)$ coincide with $|G_{in_2}|$. Furthermore, $\mathcal{U}(\mathcal{G}_2)$ contains two productions $q'_2 = \langle q_2, m' \rangle$ and $q''_2 = \langle q_2, m'' \rangle$, which are two occurrences of q_2 corresponding to the two possible different matches $m', m'' : L_{q_2} \to G_{in_2}$ (the identity and the swap).

Now, let $g : \mathcal{G}_1 \to \mathcal{G}_2$ be a grammar morphism, with $g_P(q_1) = q_2$ and the type span g_T defined as follows: X_{g_T} is a discrete graph with two nodes x and y, $g_T^L(x) = g_T^R(y) = A$ and $g_T^L(x) = g_T^R(y) = B$ (see the bottom row of the diagram in Fig. 4). Consider the pullback-retyping diagram in Fig. 4, expressing the preservation of the initial graph for morphism g (Condition 1 of Definition 5). Notice that there are two possible different morphisms k and k' from $|G_{in_2}|$ to X_{g_T} (represented via plain and dotted arrows, respectively) making

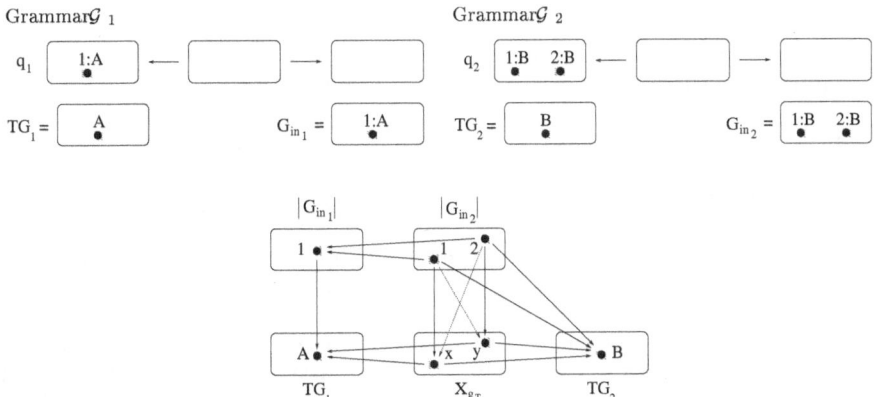

Fig. 4. The grammars $\mathcal{G}_1$ and $\mathcal{G}_2$, and the pullback-retyping diagram for their initial graphs.

the diagram commutes and the square a pullback. This provides a counterexample, showing that Lemma 1 cannot be extended to general (non semi-weighted) grammars.

Now, it is not difficult to see that, correspondingly, we can construct two different morphisms $h_i : \mathcal{G}_1 \to \mathcal{U}(\mathcal{G}_2)$ ($i \in \{1,2\}$), such that $\chi_{\mathcal{G}_2} \circ h_i = g$, the first one mapping production q_1 into q_2' and the second one mapping q_1 into q_2''. An immediate consequence of this fact is the impossibility of extending $\mathcal{U}$ to morphisms, in order to obtain a functor which is right adjoint of the inclusion $\mathcal{I} : \mathbf{OGG} \to \mathbf{GG}$.

A possible way to overcome this problem could be the choice of a different notion of grammar morphism, constraining in some way also the "k"-component of the pullback-retyping diagram. Some insights could come again from the theory of Petri nets [15], where the treatment of general P/T nets reveals similar problems which are solved there via the notions of decorated occurrence net and family morphism, at the price of obtaining a proper adjunction rather than a coreflection.

To conclude, it is worth stressing that a similar construction has been proposed by Ribeiro in her doctoral thesis [17] for the *single-pushout (SPO) approach*. She defines an unfolding functor from the category of graph grammars to a category of (abstract) *occurrence grammars*, showing that it is a right adjoint to a suitable *folding functor*. Although the basic ideas are very similar, concretely, the differences between the two settings, like the absence of the application conditions in the SPO approach, a different notion of "enabling" allowing for the concurrent application of productions related by asymmetric conflict and a different choice of grammar morphisms, makes difficult a synthetic direct comparison. For lack of space we defer to the full version a detailed analysis of the relation between the two approaches.

References

1. P. Baldan, A. Corradini, and U. Montanari. Concatenable graph processes: relating processes and derivation traces. In *Proceedings of ICALP'98*, volume 1443 of *LNCS*, pages 283–295. Springer Verlag, 1998. 146, 151, 155, 156
2. P. Baldan, A. Corradini, and U. Montanari. An event structure semantics for P/T contextual nets: Asymmetric event structures. In M. Nivat, editor, *Proceedings of FoSSaCS '98*, volume 1378, pages 63–80. Springer Verlag, 1998. 145, 153, 154, 155, 159
3. P. Baldan, A. Corradini, and U. Montanari. Unfolding and Event Structure Semantics for Graph Grammars. In W. Thomas, editor, *Proceedings of FoSSaCS '99*, volume 1578, pages 73–89. Springer Verlag, 1999. 146, 152
4. R. Banach and A. Corradini. An Opfibration Account of Typed DPO and DPB Graph Transformation: General Productions. Technical Report UMCS-96-11-2, University of Manchester, Department of Computer Science, 1996. 146, 149
5. A. Corradini. Concurrent Graph and Term Graph Rewriting. In U. Montanari and V. Sassone, editors, *Proceedings CONCUR'96*, volume 1119 of *LNCS*, pages 438–464. Springer Verlag, 1996. 145, 148
6. A. Corradini, H. Ehrig, M. Löwe, U. Montanari, and J. Padberg. The category of Typed Graph Grammars and its adjunctions with categories of derivations. In J. Cuny, H. Ehrig, G. Engels, and G. Rozenberg, editors, *Proceedings of the 5th International Workshop on Graph Grammars and their Application to Computer Science*, volume 1073 of *LNCS*. Springer Verlag, 1996. 146, 149, 151, 152
7. A. Corradini, H. Ehrig, M. Löwe, U. Montanari, and F. Rossi. An Event Structure Semantics for Graph Grammars with Parallel Productions. In J. Cuny, H. Ehrig, G. Engels, and G. Rozenberg, editors, *Proceedings of the 5th International Workshop on Graph Grammars and their Application to Computer Science*, volume 1073 of *LNCS*. Springer Verlag, 1996. 156
8. A. Corradini, U. Montanari, and F. Rossi. Graph processes. *Fundamenta Informaticae*, 26:241–265, 1996. 147, 152, 155
9. A. Corradini, U. Montanari, F. Rossi, H. Ehrig, R. Heckel, and M. Löwe. Algebraic Approaches to Graph Transformation I: Basic Concepts and Double Pushout Approach. In G. Rozenberg, editor, *Handbook of Graph Grammars and Computing by Graph Transformation. Volume 1: Foundations*. World Scientific, 1997. 147
10. H. Ehrig. Tutorial introduction to the algebraic approach of graph-grammars. In H. Ehrig, M. Nagl, G. Rozenberg, and A. Rosenfeld, editors, *Proceedings of the 3rd International Workshop on Graph-Grammars and Their Application to Computer Science*, volume 291 of *LNCS*, pages 3–14. Springer Verlag, 1987. 147
11. U. Golz and W. Reisig. The non-sequential behaviour of Petri nets. *Information and Control*, 57:125–147, 1983. 155
12. R. Heckel, A. Corradini, H. Ehrig, and M. Löwe. Horizontal and Vertical Structuring of Graph Transformation Systems. *Mathematical Structures in Computer Science*, 6(6):613–648, 1996. 146
13. R. Langerak. *Transformation and Semantics for LOTOS*. PhD thesis, Department of Computer Science, University of Twente, 1992. 153
14. J. Meseguer, U. Montanari, and V. Sassone. On the semantics of Petri nets. In *Proceedings CONCUR '92*, volume 630 of *LNCS*, pages 286–301. Springer Verlag, 1992. 145
15. J. Meseguer, U. Montanari, and V. Sassone. On the semantics of Place/Transition Petri nets. *Mathematical Structures in Computer Science*, 7:359–397, 1997. 145, 159, 161

16. G. M. Pinna and A. Poigné. On the nature of events: another perspective in concurrency. *Theoretical Computer Science*, 138:425–454, 1995. 153
17. L. Ribeiro. *Parallel Composition and Unfolding Semantics of Graph Grammars.* PhD thesis, Technische Universität Berlin, 1996. 146, 147, 161
18. G. Winskel. Event Structures. In *Petri Nets: Applications and Relationships to Other Models of Concurrency*, volume 255 of *LNCS*, pages 325–392. Springer Verlag, 1987. 145, 149

Local Views on Distributed Systems and
Their Communication

Ingrid Fischer[1][*], Manuel Koch[2], and Gabriele Taentzer[2]

[1] International Computer Science Institute, Berkeley, USA, and
IMMD2, University of Erlangen-Nuremberg, Germany
idfische@informatik.uni-erlangen.de
[2] Technical University of Berlin, Germany
{mlkoch,gabi}@cs.tu-Berlin.de

Abstract. Distributed graph transformation has been used to specify static as well as dynamic aspects of distributed systems. To support distributed designs by different developers, local views are introduced. A local view on a distributed system consists of one local system, its import and export interfaces, and connected remote interfaces. The behavior of a local system is specified by a set of graph rules that are applicable only to the local view of the local system. Local systems communicate either synchronously or asynchronously via their import and export interfaces. Asynchronous communication is modeled by sequential application of graph rules, synchronous communication by the amalgamation of graph rules. We compose a distributed system grammar from the rule sets for local systems. The operational semantics of the distributed system is given by distributed transformation sequences.

1 Introduction

Distributed graph transformation has been presented as a visual means for specifying static as well as dynamic aspects of distributed systems [8,3,6]. Whereas [8] introduces the basis concepts of distributed graph transformation based on the double-pushout approach to graph transformation, several extensions towards a specification technique customized for distributed systems have been proposed. In [3], distributed graph rules were equipped with application conditions for additional control of the rule application. Furthermore, attribution concepts have been integrated [4] to specify occurring data. In [6], both extensions have been combined to yield a specification technique suitable for describing the main aspects of distributed systems.

However, distributed graph transformation as introduced up to now assumes a global view on the whole distributed system, which is desirable in the early stage of system development to get an overview. In this first phase, network and interface structure graphs are developed and the principle network activities and

---

[*] I. Fischer was partly supported by a postdoc stipend of "Gemeinsamen Hochschulsonderprograms III von Bund und Ländern" from the DAAD.

H. Ehrig et al. (Eds.): Graph Transformation, LNCS 1764, pp. 164–178, 2000.
© Springer-Verlag Berlin Heidelberg 2000

interface services are described by distributed graph rules. After the network structure and its reconfiguration possibilities as well as the interfaces are fixed, the system developers are supposed to take a local view on network nodes and local system parts running on them.

To support this second phase of the development process as well, local views on distributed graphs are introduced in this paper. A local view contains a local system itself, its import and export interfaces and remote import and export interfaces to which the local system has connections. The behavior of a local system is specified by a set of local view rules, i.e. graph rules applicable only to the local view of the local system. Two local systems can communicate synchronously or asynchronously via their import and export interfaces. Asynchronous communication is modeled by sequential application of local view rules, whereas synchronous communication is formalized by rule amalgamation as presented in [5] for arbitrary cocomplete categories.

The synchronization of local view rules describes a kind of service request which is not already available, but can be computed. The result is put into an export interface.

A distributed system grammar contains the set of local view rules for each local system participating in the distributed system. The operational semantics of the distributed system is given by the set of all distributed transformation sequences starting at the start graph, which represents the initial state of the distributed system.

To simplify the presentation in this paper, we present concepts for labeled graphs instead of attributed graphs. All main concepts of distributed graph transformation and their local views are illustrated in a case study concerning distributed configuration management [9].

2   Distributed Configuration Management

First, we introduce a case study dealing with *Distributed Configuration Management*, as described in [9], for verifying the usability of the ideas developed later on.[1]

The size of software projects and requirements for high quality make it difficult to plan and coordinate projects. Two third of time and money spent on them go into maintenance and development after the project was finished. Furthermore, the knowledge needed to complete a greater larger project is hardly available in a single company. External project partners must contribute. Often it is also cheaper to allocate tasks to external experts. One possible scenario is to have just the core developers located in the original company. Additional personnel is situated around the world wherever the knowledge needed is available for a low price. With an ever growing demand on distributed software development, communication, coordination and quality management are needed, as all project sites must have access to a consistent actual set of project documents.

---

[1] The authors would like to thank V. Volle for numerous discussions.

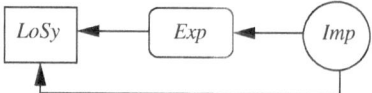

Fig. 1. The type graph $DiSy$.

When a project site changes a document leading to a new revision, it must become known in all other project sites, too. This is especially a problem when no central online archive can be used by all project partners. In this situation each project site has its own *revision archive*, a local repository where all documents and their different revisions are stored. Documents are *replicated* among revision archives to ensure that each project site has an up-to-date document set. This can be done by different means via the Internet daily, hourly, weekly or even by sending a floppy via postal mail. *Workspaces* are used when a revision archive document is to be changed or a new document has to be inserted. A workspace is connected to one archive and each archive may have an arbitrary number of workspaces. When the owner of a workspace wants to change a document he/she *checks it out* from the archive into the workspace. Then the actual change can take place or something new can be created. When this work is finished, the documents are *checked back into* the revision archive. In the following sections parts such a system will be described with distributed graph transformations.

3   Distributed Typed Graphs

We consider a distributed system consisting of local systems communicating via export and import interfaces. In export interfaces, local systems present objects accessible for remote systems whereas import interfaces contain local copies of objects from remote export interfaces. This idea of a distributed system's topology is modeled in the graph $DiSy$ shown in Figure 1. We provide a node $LoSy$ for local systems, a node Exp for export interfaces and a node Imp for import interfaces. Connections are possible between local systems and any type of interface and between import and export interfaces.

The nodes of the graph $DiSy$ are abstract in the sense that they show only the main components of a distributed system. In order to instantiate the abstract types for a concrete application, an application specific graph NTG is chosen. This graph has to possess the structure of our distributed system, which is ensured by the existence of a graph morphism $t_{NTG} : NTG \rightarrow DiSy$, called a *network type graph* in the following.

Definition 1 (network graph). *A* network type graph *is a graph morphism* $t_{NTG} : NTG \rightarrow DiSy$. *Each graph morphism* $t_G : G \rightarrow NTG$ *in* NTG *is a* network graph *w.r.t.* NTG. *We often write simply* NTG *for the network type graph* t_{NTG} *and* G *for a network graph w.r.t.* NTG *if no confusion is possible.*

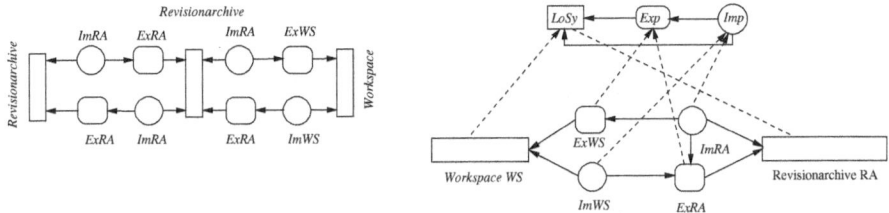

Fig. 2. A network graph modeling two revision archives with a workspace connected to one of the revision archives (left) and its network type graph (right).

Example 1. The network type graph for the distributed system of the case study as introduced in Section 2 is shown in Figure 2 on the right-hand side. It contains two kinds of local systems, namely *Workspace* and *Revisionarchive*, two different kinds of export interfaces, namely *ExWS* for workspaces and *ExRA* for revision archives, and two different types of import interfaces *ImWS* and *ImRA* for workspaces and revision archives, respectively.[2] Via the interfaces the replication between different archives and the checking in and out between archive and workspace can be realized. In Figure 2 on the left-hand side a network graph w.r.t. the network type graph on the right is shown. It represents a snapshot of a small project with two revision archives and one workspace, together with their interfaces.

The internal state of local systems and interfaces is described by a labeled graph. To simplify the presentation of the paper, we assume a common label set $\mathcal{L}$ for all local systems and interfaces.

Example 2. The internal states of workspaces and revision archives contain the revisions of documents (text files, code files, etc.) the project partners work on. Documents belonging together are packed into configurations that can contain other configurations. Each time a document or a whole configuration is changed the old version is kept and the new one is stored as a new revision of the old one. Documents or configurations cannot be changed in the revision archive. To change a document, it has to be checked out into a workspace. However, only one document can be checked out into the workspace and configurations have to be checked out completely. An example of a revision archive's local graph and its label alphabet is given in Figure 3. It consists of two configurations, where one configuration is a revision of the other. Each configuration contains two documents with one existing in both configurations.

Given a network graph G, each node in G_V is refined to an $\mathcal{L}$-labeled graph. If the node is a local system, this labeled graph represents the local state of the local system. If the node is an interface, it is the local state of the interface. An edge in G_E indicates a connection between local systems and their interfaces

---

[2] It may be also possible to use just one export interface of a revision archive for workspaces and other revision archives.

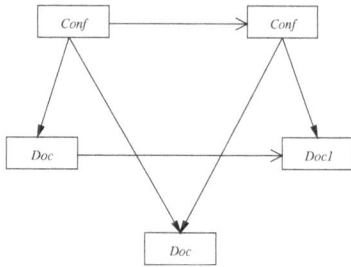

Fig. 3. A local graph consisting of one configuration with two documents and its revisions.

or between interfaces. This connection is refined to a label-preserving graph morphism that shows the relation between the local state of the source node of the network edge and its target node.

The refinement of a network graph $t_G : G \rightarrow NTG$ w.r.t. a network type graph NTG is formally defined by a functor from the small category **G** induced by the graph G into the category **Graph**$(\mathcal{L})$ of $\mathcal{L}$-labeled graphs and label preserving graph morphisms. A *distributed graph* integrates topological and local state aspects.

Definition 2 (distributed graph). *Given a network type graph* $t_{NTG} : NTG \rightarrow DiSy$ *and a set* $\mathcal{L}$ *of labels. A* distributed graph *over* G *is a pair* $\hat{G} = \langle t_G, \mathcal{G} \rangle$ *where* $t_G : G \rightarrow NTG$ *is a network graph w.r.t.* NTG *and* $\mathcal{G} : G \rightarrow G(\mathbf{Graph}(\mathcal{L}))$ *is a graph morphism from* G *to the underlying graph* $G(\mathbf{Graph}(\mathcal{L}))$ *of the category* $\mathbf{Graph}(\mathcal{L})$.

For a distributed graph $\hat{G}$, the $\mathcal{L}$-graph $\mathcal{G}(v)$ for each v in G_V is called the *local state* of v. For each edge e in G_E, the graph morphism $\mathcal{G}(e)$ is called the *local graph morphism* for e. A morphism between distributed graphs $\hat{G}$ and $\hat{H}$ over G resp. H relates the network graphs by means of a graph morphism between G and H, and relates the local state of each node in G to a local state in H by means of a label-preserving graph morphism. It can be seen as a natural transformation from $\mathcal{G}$ to $\mathcal{H} \circ f$.

Definition 3 (distributed morphism). *A distributed morphism* between distributed graphs $\hat{G}$ over G and $\hat{H}$ over H is a pair $\hat{f} = \langle f, \tau \rangle$ where $f : G \rightarrow H$ is a graph morphism and $\tau : \mathcal{G} \rightarrow \mathcal{H}$ is a family of arrows $\{\tau(a) | a \in G_V\}$ such that

- *for each node a in G, $\tau(a) : \mathcal{G}(a) \rightarrow \mathcal{H}(f(a))$ is a label preserving graph morphism and*
- *for each edge $e : i \rightarrow j$ in G, $\tau(j) \circ \mathcal{G}(e) = \mathcal{H}(f(e)) \circ \tau(i)$.*

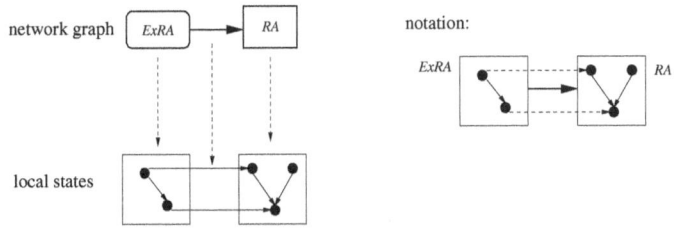

Fig. 4. A distributed graph and its notation.

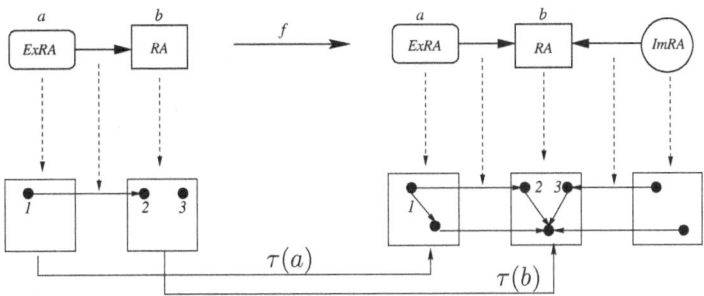

Fig. 5. A distributed morphism.

Example 3. An example of a distributed graph is shown in Figure 4. The level refinement is given in the left-hand side of the figure. Additionally, the notation used in the following is shown. It shows the local states of network nodes and the local graph morphisms for network edges, but omits the explicit representation of the network graph itself and its refinement. However, a name at the local state indicates the network node is also found in a local state. The local system of type RA and the interface of type $ExRA$ are refined to labeled graphs that represent their local state (for the sake of readability we omit labels in the figures). The network edge is refined to a graph morphism. In the example of a distributed morphism $\hat{f} : \hat{G} \to \hat{H}$ in Figure 5, the graph morphism f and the local morphisms $\tau(a)$ and $\tau(b)$ map elements to elements with the same name.

For a given network type graph NTG and a label set $\mathcal{L}$, distributed graphs and distributed morphisms form a category $\mathbf{Distr}(NTG, \mathcal{L})$. Composition of distributed morphisms is defined componentwise for graph morphisms between network graphs and local graph morphisms. The identity for each distributed graph $\hat{G}$ is given by $\hat{id}_{\hat{G}} = \langle id_G, \tau \rangle$ where $\tau(a) = id_{\mathcal{G}(a)}$ for each a in G_V.

The pushout of two distributed morphisms $\hat{f} = \langle f, \tau^f \rangle : \hat{A} \to \hat{B}$ and $\hat{g} = \langle g, \tau^g \rangle : \hat{A} \to \hat{C}$ is constructed by constructing first the pushout of f and g in $\mathbf{Graph}$. Then the pushout of $\tau^f(a)$ and $\tau^g(a)$ for each node a in A_V in category $\mathbf{Graph}(\mathcal{L})$ is constructed. The pushout object and the pushout morphisms for $\hat{f}$ and $\hat{g}$ in $\mathbf{Distr}(NTG, \mathcal{L})$ are then made up of these pushout components and the resulting unique pushout morphisms. However, this construction does not yield

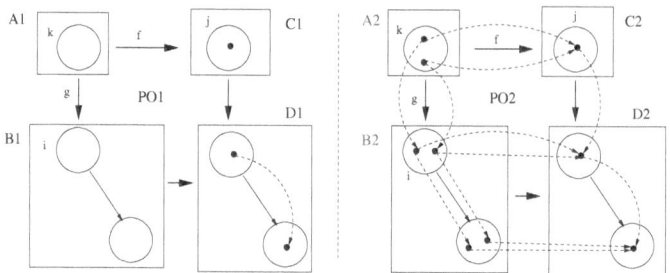

Fig. 6. Examples of a non componentwise pushout construction.

a pushout for all distributed morphisms as the counter examples in Figure 6 show.

In the example on the left, a node is added to a local source graph. In order to make the local graph morphism total, a node is inserted in the target graph as well. The second example shows that the gluing of two nodes in a local source graph is propagated to the target graph in order to maintain the well-definedness of the local graph morphism.

We provide *locality conditions* ensuring that the pushout over two distributed morphisms can be constructed componentwise for the network and all pairs of local morphisms. (For the proof consider the corresponding proof in [2] for category $\mathbf{Mod}_C$ with C being category $\mathbf{Graph}(\mathcal{L})$.)

Definition 4 (locality conditions). *Two distributed morphisms*

$$\hat{f} = \langle f, \tau^f \rangle : \hat{A} \to \hat{B} \ \text{and} \ \hat{g} = \langle g, \tau^g \rangle : \hat{A} \to \hat{C}$$

satisfy the locality conditions *if and only if*

- $f : A \to B$ *and* $g : A \to C$ *are injective,*
- *for each edge* $e \in B_E - f(A_E)$ *and for each node* $y \in A_V$, $f(y) = s(e)$ *implies* $\tau^g(y)$ *is bijective and*
- *for each edge* $e \in C_E - g(A_E)$ *and for each node* $y \in A_V$, $g(y) = s(e)$ *implies* $\tau^f(y)$ *is bijective.*

In the next section it is described how the pushout construction on distributed graphs can be used to model transformation rules along the lines of [8].

4   Distributed Graph Transformation

Distributed rules are given by a span of injective distributed morphisms. A transformation includes the transformation of the network graph as well as the local states. A local transformation is performed in each network node, which is preserved by the network transformation. The network as well as the local transformations are formulated as double-pushouts on graph morphisms. The local graph morphisms between transformed local graphs are induced as universal pushout morphisms. Furthermore, each network node and edge deleted, sees its local graphs and graph morphisms deleted as well. Creating a network node or edge is combined with the creation of a corresponding local graph or graph morphism. The result of a distributed graph transformation is again a distributed graph, since a distributed graph transformation can be characterized by a double-pushout ([8]).

Definition 5 (distributed rule). *A distributed rule* p *consists of a span*

$$(\hat{L} \xleftarrow{\hat{l}} \hat{I} \xrightarrow{\hat{r}} \hat{R})$$

of injective distributed morphisms.

A production p can be applied to a distributed graph $\hat{G}$, if there is an occurrence $\hat{m} = \langle m, \tau^m \rangle : \hat{L} \to \hat{G}$ of the left-hand side of the production in the distributed graph. The derivation of a distributed graph via a distributed rule is given by two pushouts in category $\mathbf{Distr}(NTG, \mathcal{L})$. In order to guarantee the existence and uniqueness of the pushout complement as well as the component-wise construction of the pushouts, the morphism $\hat{m}$ has to satisfy the so-called *distributed gluing condition*. This condition is satisfied by $\hat{m}$ if m is injective and satisfies the gluing condition for $(L \xleftarrow{l} I \xrightarrow{r} R)$ (defined e.g. in [1]), $\tau^m(l(x))$ satisfies the gluing condition for $(\mathcal{L}(l(x)) \xleftarrow{\tau^l(x)} \mathcal{I}(x) \xrightarrow{\tau^r(x)} \mathcal{R}(x))$ for all nodes $x \in I_V$ and $\hat{m}$ satisfies the *connection* and *network condition*. The connection condition is satisfied if whenever p deletes objects in some source local graph $\mathcal{G}(s(e))$ resp. in some target graph $\mathcal{G}(t(e))$, the local mapping $\mathcal{G}(e)$ must be changed correspondingly. If p adds new objects to a local graph $\mathcal{G}(s(e))$, the local mapping $\mathcal{G}(e)$ must be extended, too. The network condition is satisfied if a network node is only deleted together with its entire local graph and a network edge e can be only deleted if the local graph $\mathcal{G}(s(e))$ is completely mentioned in the rule, so that the local mapping can also be deleted. A new edge e is only inserted at an existing node v becoming the source node of e if $\mathcal{G}(v)$ is completely mentioned in the rule, i.e. the local morphism $\mathcal{G}(e)$ is completely specified.

Definition 6 (direct derivation). *Given a distributed rule* p *and a distributed morphism* $\hat{m} : \hat{L} \to \hat{G}$ *that satisfies the distributed gluing condition, then a direct derivation* $\hat{G} \xRightarrow{p, \hat{m}} \hat{H}$ *via* p *is given by two pushouts (1) and (2) in category* $\mathbf{Distr}(NTG, \mathcal{L})$.

$$
\begin{array}{ccccc}
\hat{L} & \xleftarrow{\hat{l}} & \hat{I} & \xrightarrow{\hat{r}} & \hat{R} \\
\downarrow{\hat{m}} & (1) & \downarrow & (2) & \downarrow \\
\hat{G} & \longleftarrow & \hat{C} & \longrightarrow & \hat{H}
\end{array}
$$

Example 4. In Figure 7, a rule is given modeling the check in of a document from the workspace into the revision archive as a new revision. This document was

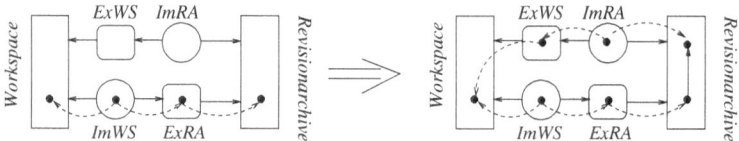

Fig. 7. Inserting a revision into the archive.

exported from the revision archive and imported from the workspace where it was changed. Then it is exported by the workspace and imported by the revision archive where the changed document becomes a successive revision of the version originally used for export.

We omit the intermediate graph in the notation of rules. Only the left-hand and the right-hand side of a rule are shown. The span can be achieved from this notation due to the position of nodes and edges in the graphs.

5   Local Views on a Distributed System

Distributed rules as defined in Def. 5 allow one to specify actions within a distributed system affecting several local systems in one rule application. An example is the rule in Figure 7, where an action affects the workspace and the revision archive. Thus, distributed actions are specified in a global view. In the early stage of system development, a global view on the entire distributed system is desirable to get an overview. Once the network structure and its reconfiguration possibilities as well as the interfaces are fixed, the system developers are supposed to take a local view on network nodes and local system parts running on them. Therefore, we are going to restrict distributed rules to so-called *local view rules*. A *local view graph* for a local system in the distributed system specifies the visible parts of the local system. It contains the local states of the local system itself and of export and import interfaces to which the local system has connections. While it is natural that a local system knows the exports from where it imports, it is not as clear that it knows the local states of imports that are connected to its exports. This information can be advantageously used to inform the imports immediately when the export changes which is not possible otherwise.

Definition 7 (local view). *Let $t_{NTG} : NTG \to DiSy$ be a network type graph as defined in Def. 1. Let $\hat{G} = \langle t_G, \mathcal{G} \rangle$ over G be a distributed graph and v a node in G_V such that $t_{NTG}(t_G(v)) = LoSy$. An interface node $w \in G_V$, i.e. $t_{NTG}(t_G(w)) = Imp$ or $t_{NTG}(t_G(w)) = Exp$, is an interface for v if w is directly connected to v, i.e. there is an edge $e \in G_E$ such that $s^G(e) = w$ and $t^G(w) = v$. Otherwise, the interface w is a remote interface w.r.t. v.*

The distributed graph $\hat{G}$ is called local view graph *w.r.t. v if G is connected and there does not exist a v' in G_V such that $v' \neq v$ and $t_{NTG}(t_G(v')) = LoSy$.*

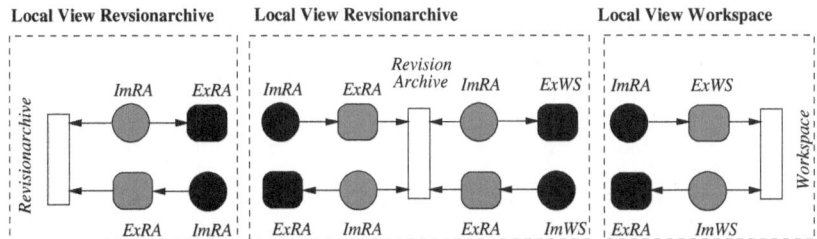

Fig. 8. Three local view graphs for Figure 2.

To include remote import interfaces into the view of a local system is also formally motivated. If remote import interfaces would not be visible for a local system, the local system cannot delete any object of its own export interfaces as long as other systems import these objects, what is formally forced by the distributed gluing condition.

Example 5. Taking Figure 2, three local systems are shown which yields three local view graphs (Figure 8). In these graphs the local systems are white, interfaces are grey and remote interfaces are filled black.

Given a local system, we now introduce distributed rules for the local system. The knowledge of the distributed system's state restricted to the local view of the local system is sufficient to apply those rules. We distinguish rules for transforming the local states of a local system, rules for creating new interfaces or new local systems and rules for the deletion of the local system or parts of it.

Definition 8 (local transformation rule). *A* local transformation rule *is a distributed rule* $p = (\hat{L} \xleftarrow{\hat{l}} \hat{I} \xrightarrow{\hat{r}} \hat{R})$ *where*

- *the network graph morphisms* l *and* r *are the identity on* I, *i.e.* $l = r = id_I$ *and*
- $\hat{L}$, $\hat{I}$ *and* $\hat{R}$ *are local view graphs w.r.t.* $v \in I_V$.

The definition of a local transformation rule ensures that the network graph remains unchanged and that only local states are transformed. The transformation of local states includes deletion, preservation and creation of objects in the state. With local transformation rules different kinds of actions can be described. If remote interfaces are not involved, local actions are described. Moreover, asynchronous or synchronous actions can be modeled. In these cases the remote interfaces are just read or, in the synchronous case, are allowed to be changed. Here, the same actions have to be performed by those local systems which own these interfaces.

In order to additionally transform the network graph, *local deletion* and *local creation rules* are introduced. A local system is allowed to delete only itself and its interfaces, but not remote interfaces. By means of the *local creation rule* new local systems with their interfaces can be created.

174   Ingrid Fischer et al.

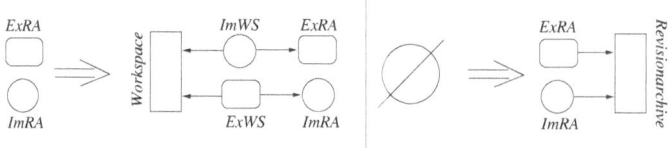

Fig. 9. Local creation rules for a workspace (left) and for an archive (right).

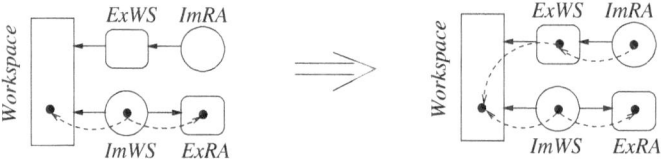

Fig. 10. The workspace exports a previously imported document again.

Definition 9 (local creation rule and local deletion rule). *Rule p is a local creation rule if*

- $\hat{L}$ *does not contain any local system node,*
- $\hat{R}$ *is a local view graph w.r.t. v in R_V,*
- $\hat{l}$ *is the identity on $\hat{I}$ and $\hat{I} = \hat{L}$ are subgraphs of $\hat{R}$,*
- *all remote interface nodes for v in L are preserved, i.e. if $w \in L_V$ is an remote interface for v then $w \in l_V(I)$, and*
- *no new remote interface node is created, i.e. there is not any interface node v which is not in the codomain of r.*

Rule p is a local deletion rule *if it the inverse rule of a local creation rule.*

Definition 10 (local view rule and local view grammar). *Given a network type graph $t_{NTG} : NTG \to DiSy$ and a local system node x in NTG, i.e. $t_{NTG}(x) = LoSy$, a local view rule for x is each local transformation, local deletion and local creation rule such that each local system node v occuring in the rule is of type x, i.e. $t_L(v) = x$ or $t_R(v) = x$. A local view grammar $LGG(x) = P(x)$ for x is given by a set of local view rules for x.*

Example 6. Two different kinds of local view grammars are necessary for the running example: one handles rules concerning workspaces, the second one is for revision archives. Local view rules for the workspace must include rules for checking in from the workspace into a revision archive as shown by the local transformation rule in Fig. 10. A document previously imported by the workspace, is exported again.[3] In Figure 9, a local creation rule for a workspace is given on the left. On the right the local creation rule for revision archives is shown.

In the local view grammar of the revision archive a rule corresponding to Fig. 10 can be found. It is shown in Figure 11 and used to import a revision exported by the workspace and adding it as a successor in the revision archive.

---

[3] The case where a new document is created is omitted here.

Fig. 11. The revision archive requests a document from the workspace to import it.

The application of the local rules in Fig. 11 and Fig. 10 on the same document should lead to the same result as the application of the global rule in Fig. 7, i.e. it must be ensured somehow that a changed document in the workspace is inserted into the revision archive as a direct successor after the revision from which it was originally checked out. To achieve this, the local rules have to be synchronized.

6 Synchronizing Local Rules for Communication between Local Systems

This section is concerned with communication between local views. Whereas asynchronous communication is modeled by sequential application of local view rules, synchronous communication is expressed by amalgamating local view rules over a common subrule in category **Distr**$(NTG, \mathcal{L})$ [5]. While the asynchronous communication describes the usage of a service already available in some interface, the synchronization models a kind of service request for objects not already available in the interface (cf. the rules in Figures 10 and 11).

For amalgamation so called *interface subrules* are necessary. An interface subrule p_s of a local view rule p for a local system v is a distributed rule, where all graphs of p_s are subgraphs of the corresponding graphs in p such that they contain only interfaces of v, but not v itself. Since we intend to describe synchronization via interface subrules and communication takes place over export and import interfaces, we require *handles* of export and import interfaces in the network graphs of the subrule. More exactly, a handle consists of an export interface and an import interface connected by an edge. This requirement prohibits export interfaces without an import interface connected, and vice versa. Considering local deletion and creation rules which have to be synchronized, handles can be found only on the left or on the right-hand side of an interface rule.

Definition 11 (interface subrule). Let $p_s = (\hat{L}_s \xleftarrow{\hat{l}_s} \hat{I}_s \xrightarrow{\hat{r}_s} \hat{R}_s)$ be a distributed rule such that

- L_s, I_s and R_s contain interface nodes only,
- there is an $X \in \{L, I, R\}$ such that for each export node $v \in X_s$, i.e. $t_{NTG}(t_X(v)) = Exp$, there is an import node $v' \in X_s$, i.e. $t_{NTG}(t_X(v')) = Imp$, and an edge $e : v' \to v$ and

- *there is an $X \in \{L, I, R\}$ such that for each import node $v' \in X_s$, i.e. $t_{NTG}(t_X(v')) = Imp$, there is an export node $v \in X_s$, i.e. $t_{NTG}(t_X(v)) = Exp$, and an edge $e : v' \to v$.*

Then, p_s is an interface subrule *of $p = (\hat{L} \xleftarrow{\hat{l}} \hat{I} \xrightarrow{\hat{r}} \hat{R})$ if there are injective distributed morphisms $\hat{in}_L : \hat{L}_s \to \hat{L}$, $\hat{in}_I : \hat{I}_s \to \hat{I}$ and $\hat{in}_R : \hat{R}_s \to \hat{R}$, called* subrule embeddings *such that $\hat{in}_L \circ \hat{l}_s = \hat{l} \circ \hat{in}_I$ and $\hat{in}_R \circ \hat{r}_s = \hat{r} \circ \hat{in}_I$.*

A distributed rule has to be synchronized with others, if it contains one or more remote interfaces changed by the rule. This part of the rule just reflects what has to be done by remote systems within their interfaces (cf. the rules in Figures 10 and 11). To synchronize two rules an interface subrule is needed which contains at most the intersection of the two rules, but is also allowed to be smaller. If the distributed rule resulting from a synchronization step still contains remote interfaces, it can and has to be further synchronized. A distributed rule not fully synchronized is not applicable because of the distributed gluing condition (namely the connection condition).

Definition 12 (synchronization of rules). *Two distributed rules p_1 and p_2 are* synchronized *w.r.t. s if s is an interface subrule embedding of p_1 and p_2 consisting of interface subrule p_s and the subrule embeddings $\hat{in}_{X_1}$, $\hat{in}_{X_2}$ satisfy the locality conditions in Def. 4 for $X \in \{L, I, R\}$. Moreover, for each interface w in $\hat{X}_s$, either w is a remote interface for all local systems in $\hat{X}_1$ or w is a remote interface for all local systems in $\hat{X}_2$.*

The synchronized rule *of p_1 and p_2 via s is given by the amalgamated rule $p_1 \oplus_s p_2 = (\hat{L} \xleftarrow{\hat{l}} \hat{I} \xrightarrow{\hat{r}} \hat{R})$. The distributed graphs $\hat{L}$, $\hat{I}$ and $\hat{R}$ are the pushout objects given by the pair $\hat{in}_{L_1}$ and $\hat{in}_{L_2}$, the pair $\hat{in}_{I_1}$ and $\hat{in}_{I_2}$ and the pair $\hat{in}_{R_1}$ and $\hat{in}_{R_2}$ in* $\mathbf{Distr}(NTG, \mathcal{L})$*. The rule morphisms $\hat{l}$ and $\hat{r}$ are obtained as the universal pushout morphisms.*

Example 7. Regarding the check-in of a revision from the workspace into the revision archive, its intended semantics is described as a revision which becomes a successor of that revision in the archive from which it was checked out. In order to achieve this, we have to synchronize the local view rule for a workspace in Figure 10 and the local view rule for a revision archive in Figure 11. These rules are not applicable separately because of the distributed gluing condition. E.g. the revision archive rule in Figure 11 is not applicable without synchronization, because of the insertion of the revision in the export interface of the workspace. Because of the local creation rule for a workspace there is always an edge from an export of a workspace to the workspace itself. Therefore, the revision inserted has to be assigned to a revision in the workspace as well, which, however, is not specified by the rule.

The synchronized rule of the rules in Figure 11 and Figure 10 using the interface subrule in Figure 12 looks like that in Fig. 7.

Next, we construct all possible rules for a given set of local view rules used to model the operational semantics of the distributed system. Here, the rules are

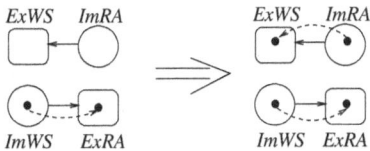

Fig. 12. Interface subrule for Fig. 10 and 11 to construct the rule in Fig. 7

synchronized as much as possible, i.e. the largest interface subrule embeddings are chosen for synchronization. An interface subrule embedding is one of the largest if there no other interface subrule which contains the rule of the first embedding as subrule. Since an interface subrule may be the empty rule, the largest interface subrule always exists.

Definition 13 (set of synchronized rules). *Let LV Rules be a set of local view rules, then, $Syn(LV Rules)$ is the smallest set of rules such that*

- *$p \in Syn(LV Rules)$ for all $p \in LV Rules$, and*
- *$p_1 \oplus_s p_2 \in Syn(LV Rules)$ for all $p_1, p_2 \in Syn(LV Rules)$ where s is one of the largest interface subrule embedding of p_1 and p_2, i.e. there is no interface subrule embedding s' of p_1 and p_2 such that p_s is a subrule of $p_{s'}$.*

A *distributed system grammar* w.r.t. a network type graph $t_{NTG} : NTG \to DiSy$ consists of a global start graph and a local view grammar for each local system type in NTG, i.e. for each v in NTG_V such that $t_{NTG}(t_G((v)) = LoSy$. For the operational semantics the set of all rules that can be constructed by the local view rules according to Definition 13 are considered.

Definition 14 (distributed system grammar). *Let $t_{NTG} : NTG \to DiSy$ be a network type graph and $LS = \{v \in NTG_V | t_{NTG}(v) = LoSy\}$ be the set of all local system types in NTG. A distributed system grammar w.r.t. NTG is a pair $DSG(NTG) = \langle \hat{G}_0, (LGG(v))_{v \in LS} \rangle$ where $\hat{G}_0$ is a distributed graph, $LGG(v) = P(v)$ is a local view grammar for each $v \in LS$.*

The operational semantics for a distributed system grammar $DSG(NTG)$ is given by the set of all distributed derivations starting at $\hat{G}_0$ using rules of $Syn(\bigcup_{v \in LS} P(v))$.

Example 8. The set of synchronized rules *SynRules* of our case study contains for example the synchronized rule in Figure 7 created by the synchronization of the local view rules in Figures 11 and 10. It may also contain the corresponding local view rules, but these are never applicable due to the distributed gluing condition. Also the rules in Figure 9 should be in *SynRules*.

7 Conclusion

In this paper, we introduced local views on distributed graph transformation. A local view is concerned with one local system, its import and export interfaces, and remote import and export interfaces to which the local system may have connections. Local systems can communicate asynchronously by simply applying local view rules sequentially, or synchronously by constructing amalgamated distributed rules from local view rules. The concepts of local views are presented on graphs without attributes. We expect that they can be directly lifted to attributed graphs. Moreover, application conditions for rules should be integrated to support a convenient specification of distributed systems by distributed graph transformation. The integration – formally as well as informally – of graph transformation with attributes and application conditions has been done in [7]. We use the amalgamated rule construction to describe the synchronization between local view activities. In general, the resulting rules are not local view rules anymore, but distributed rules in a global setting. For the operational semantics definition of a distributed system all possible synchronizations of rules are computed.

References

1. A. Corradini, U. Montanari, F. Rossi, H. Ehrig, R. Heckel, and M. Löwe. *Handbook of Graph Grammars and Computing by Graph Transformations. Vol. I: Foundations*, chapter Algebraic Approaches to Graph Transformation Part I: Basic Concepts and Double Pushout Approach. World Scientific, 1997. 171
2. M. Koch. *Integration of Graph Transformation and Temporal Logic for the Specification of Distributed Systems*. PhD thesis, Technische Universität Berlin, FB 13, 1999. to defend. 170
3. Manuel Koch. Bedingte verteilte Graphtransformation und ihre Anwendung auf verteilte Transaktionen. Technical Report 97-11, TU Berlin, 1997. 164
4. Manuel Koch and Gabriele Taentzer. Distributing Attributed Graph Transformations. In *Proc. Workshop on "General Theory of Graph Transformation Systems"*, *Bordeaux*, 1997. 164
5. G. Taentzer. Parallel high-level replacement systems. *Theoretical Computer Science*, (186), 1997. 165, 175
6. G. Taentzer, I. Fischer, M. Koch, and V. Volle. *Handbook of Graph Grammars and Computing by Graph Transformations*, volume III, chapter Distributed Graph Transformation with Application to Visual Design of Distributed Systems. World Scientific, 1998. to appear. 164
7. G. Taentzer, I. Fischer, M Koch, and V. Volle. Visual design of distributed systems by graph transformation. In G. Rozenberg, U. Montanari, H. Ehrig, and H.-J. Kreowski, editors, *Handbook of Graph Grammars and Computing by Graph Transformation, Volume 3: Concurrency and Distribution*. World Scientific, 1999. to appear. 178
8. Gabriele Taentzer. *Parallel and Distributed Graph Transformation: Formal Description and Application to Communication-Based Systems*. PhD thesis, TU Berlin, 1996. Shaker Verlag. 164, 170, 171
9. Karsten Victor Volle. Verteilte Konfigurationsverwaltung: *COMAND*. Technical report, Basys GmbH, Am Weichselgarten 4, 91058 Erlangen, 1997. 165

Dynamic Change Management by Distributed Graph Transformation: Towards Configurable Distributed Systems

Gabriele Taentzer[1], Michael Goedicke[2], and Torsten Meyer[2]

[1] Computer Science Department, Technical University of Berlin
D–10587 Berlin, Germany
gabi@cs.tu-berlin.de
[2] Dept. of Mathematics and Computer Science, University of Essen
D–45117 Essen, Germany
{goedicke,tmeyer}@informatik.uni-essen.de

Abstract. In this contribution we consider the application of distributed graph transformation to the problem of specifying dynamic change in distributed systems. Change in distributed systems is related to at least two levels. One is the management of change in a local node of the distributed system and how such a local change is then propagated to those nodes which need to know about the change. The other aspect is changing the structure of the distributed system itself. This implies e.g. to add and/or remove a local node or an entire subsystem to/from the distributed system. In some important application areas such operations must be done during runtime without disturbing the unmodified rest of the distributed computing system. We first give an overview of our model of change and how exactly the two aspects of change interact. We describe distributed graph transformation as a technique to realize our change model. An example - a ring database - then shows how our approach can be applied to a small but nontrivial distributed system. This example shows nicely how the two aspects of change can be described uniformly using graph transformation rules and how the interaction of the two change aspects can be defined in an adequate way. Since this is ongoing work we conclude with an assessment of our approach and a brief discussion of further work.

1 Introduction and Related Work

Graphical representations are an obvious means to describe various aspects of software systems. In the case of distributed systems, graphs are often used to describe models of the system's network structure. Graph transformations which here means the rule-based manipulation of graphical structures, can be conveniently used to model dynamic changes of the network architecture.

However, in the case of distributed systems, graph transformation can be employed twice: (1) to describe the dynamic reconfiguration of distributed systems and (2) to model evolving data and object structures in the local systems.

H. Ehrig et al. (Eds.): Graph Transformation, LNCS 1764, pp. 179–193, 2000.
© Springer-Verlag Berlin Heidelberg 2000

The usage of graph transformation at the local level can include the description of local actions as well as remote interaction, such as object migration, replication, communication and synchronization. All these activities are described in a rule-based manner by defining parts of the graph structure before and after a transformation. Thus the application of one or more graph rules, i.e. performing distributed graph transformations, results in a model of distributed computations.

Our approach of distributed system design by distributed graph transformation allows to export parts of the internal structures as well as dynamic behaviour which means to export also parts of the semantics. This information is extremely useful to find, for example, adequate services on remote systems, since syntactical information like names or signatures are often not expressible enough to find the right service. Another advantage is that interfaces may change dynamically which takes care of the fact that not all services can be offered all the time. This property can be described in a natural way using our approach since interfaces are again defined by graphs and graph rules. A service is available if a certain interface rule is applicable to its interface graph.

The concepts as well as the formal definition of distributed graph transformation based on the algebraic approach [3] are introduced in [16]. Distributed graph transformation is extended by an attribution concept for graphs and the concept of application conditions for graph rules in [17]. Using graph transformation as underlying formal framework, basic consistency properties of the resulting distributed system structures are ensured.

An obvious application area for distributed graph transformation is distributed systems design where the two levels mentioned above exist in a natural way. The design of such distributed systems involves to layout the architecture of the system which means to find the overall structure and the elements of the structure. Since the system runs in a distributed way, care must be taken to achieve a certain notion of consistent state, although not all components of the distributed system's state can be accessed all the time. In the case of dynamic (re)-configuration of such systems additional measures must be taken into account that the exchange, introduction and removal of system components is done in a smooth way without disturbing the rest of the system too much. In general such an architectural change causes the set of the involved components into a kind of silent state such that the architectural change actions can be performed and the new parts may be activated in turn.

The research area which is primarily involved in system design of this kind is the area of configurable distributed systems. In [11] J. Kramer and J. Magee propose a basic model for dynamic change management which separates structural issues from application specific ones, thus distinguishing change rules at the architecture configuration level and change actions at the component level. In [12] self-configuring architectures are investigated and in [6,13] a graphical user interface and a notation for component configurations based on the architecture description language REGIS/DARWIN [14] are presented. An approach to realize configurable distributed systems based on negotiation of component semantics with the help of graph transformation techniques is presented in [7]. The ring database example modeled in this paper was first introduced in the world of dynamic change using DARWIN in [10]. In [1] J. Purtilo et al. present a framework for dynamic reconfiguration of application software, including a notation to express reconfiguration plans and a prototype system.

Another work to describe self organizing software architectures is given in [20]. With the Chemical Abstract Machine [2] an architecture specification and programming model is presented and for covering dynamic issues graph rewrite rules [15] are introduced. However, this approach includes no distribution or modularization aspects.

In the following we will give a brief account of dynamic change management in distributed systems in Chapter 2. In chapter 3, we will present distributed graph transformation as the technique to realize these goals. We will show the applicability of our approach by modeling a well-known example in the world of dynamic change in chapter 4 and 5.

2   Dynamic Change Management and System Reconfiguration

Change in distributed systems has at least two aspects. On one side there is the problem to maintain a certain - application dependent - notion of consistency without having access to the entire state of the distributed system all the time. On the other side there is the general need to evolve a distributed system by adding, removing or substituting software components during the runtime of the system. We follow the model developed in [10,11]. This model implies a strict separation of the two change aspects mentioned above.

Separation of concerns applied to the design of distributed systems implies the separation of designing the various components from designing the structure of the system. This is naturally transferred to the management of change in such systems. The change of the distributed system structure has to be done at the level of the structure and the node interfaces only. This implies that the interaction between the change actions and the nodes of system is minimal. As a consequence the evolution of the system can be performed in a structured and systematic way.

The idea outlined above requires a certain view on the system structure and its nodes. The nodes offer an interface which provides their publicly available services which are linked to each other according to the system structure via communication channels.

The actions at the structural level - which we will refer to as the network level later on - are *addition* and / or *deletion* of a node or subsystem and *linking* and / or *unlinking* of nodes. This is the original model of change actions according to [11]. It minimizes the interaction between the change actions at the network level and the related actions necessary within the involved nodes. I.e. involved nodes have to be in a quiescent state in order to allow the execution of the change action at network level. A node is said to be in a *quiescent* state if the application's state is consistent and there is no communication in progress between nodes affected by a change nor between them and their environment. This means that a distributed node

1. is not currently engaged in a communication connection that it initiated,
2. will not initiate new communication connections,
3. is not currently engaged in servicing a communication connection request, and
4. no communication connections have been or will be initiated by other nodes which require service from this node.

If only the first requirement is satisfied a node is said to be in a *passive* state. Further, the nodes must remain quiescent while the entire change is executed.

Thus the only interaction between the structural change actions and the nodes are services which activate and / or passivate a node.

This view, however, has a certain drawback. It does not deal with the situation that a node is split into more than one node. Also in the case of adding a new node there is the problem of informing the new node about the state it has to know in order to cooperate smoothly with the existing nodes of the system. Thus we would like to extend the model by means to describe the interaction between nodes more in detail. The interaction has still to be done through interfaces which hide internal details from the (mis)use by other nodes. Our approach provides the possibility for multiple interfaces which allow to deal with different aspects using specialized interfaces.

Below we now introduce the technique used to realize dynamic change management: distributed graph transformation.

3   Distributed Graph Transformation

In the following, we introduce a recent specification technique, called *distributed graph transformation*, which is structured graph transformation at two abstraction levels: the *network* and the *object* (or local) level. The network level contains the description of a system's network topology by a network graph and its dynamic reconfiguration during runtime by network rules. At the object level, graph transformation is used to manipulate local data and object structures. To describe distributed objects, replication and migration of objects to remote systems, synchronization and communication between distributed systems, as well as dynamic system configuration the combination of network graph transformation and local transformations is needed. Distributed graph transformation has exactly this purpose. The concepts as well as the formal definition of distributed graph transformation are based on the *double-pushout approach* to graph transformation [3] where basic concepts from *category theory* are applied. Distributed graph transformation is introduced in [16] and further developed in [17].

The topological structure of a distributed system's network is presented graphically by a *network graph*. Nodes of the network graph represent any kind

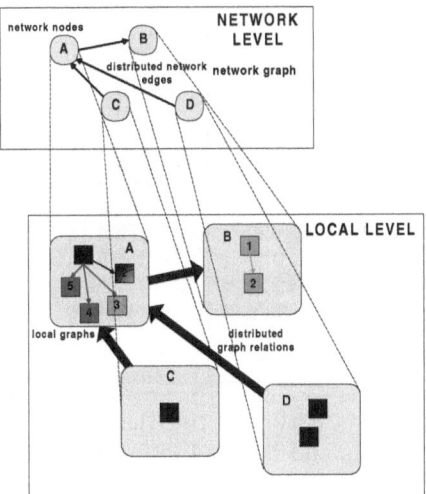

Fig. 1. This example of a distributed graph shows a local view of the system A. In this and in the following distributed graphs body elements of local graphs are depicted in medium gray, exported graph elements are depicted in light gray, imported graph elements are depicted in **black**, and common parameters graph elements (i.e. graph elements which are imported and exported) are depicted in mixed light gray / **black**. Relations between local graph elements are not explicitly depicted, but we assume a mapping between graph elements with identical labels.

of processing unit with some memory, edges indicate communication paths between nodes. The combination of the network graph structure and the local object structures is specified by distributed graphs in order to describe distributed object structures. A *distributed graph* consists of one *network graph* where each network node is equipped with a *local graph* containing data and object structures. Each network edge is equipped with a relation on local graphs defining replicated data and object structures in that way.

In the early stages of system development, a global view on the entire distributed system is desirable to get an overview. Once the network structure is fixed, the system developers are supposed to take a local view on network nodes and local system parts running on them to facilitate concurrent development. The *local view* [4] of a local system contains this local system together with remote import and export interfaces to which it may have connections. Communication between local systems takes place via export and import interfaces. In *export interfaces*, local systems present data and objects accessible for other local systems and *import interfaces* contain data and objects from remote export interfaces. Network nodes being the source of a network edge may be interpreted as interfaces, while network nodes being the target of network edges only may be considered as local system nodes. Figure 1 shows an example of a distributed

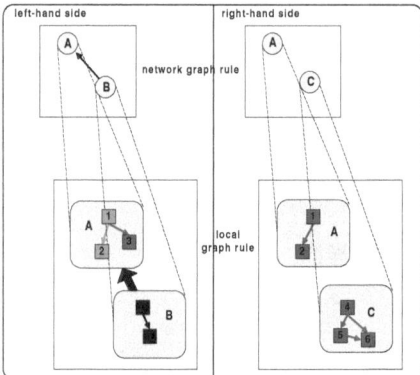

Fig. 2. In this example of a distributed rule the network node B, its local graph as well as the network edge and its local graph relation are deleted. The local graph determines the state in which such a network node is allowed to be deleted. The network node A is preserved. Its local graph has to be manipulated by the local rule given. On the right-hand side a new network node C has to be created and its initial state is given by its local graph.

graph where import and export graphs are distinguished by different gray shades of local graph elements.

Each node and edge in a distributed graph may be typed and attributed by further data types. Usually, each local system contains different sets of node, edge and attribute types. The interfaces may contain any subsets of types for communication (compare [17,5] for more details wrt. attribution of distributed graphs).

Distributed graph rules are useful to describe dynamic network reconfiguration by a network rule as well as dynamic local object structures in each component by a set of local rules. Furthermore, remote object interactions can be described by distributed rules. In this case, all network nodes where parts of the interaction should take place, are incorporated into the network graphs of the rule. A *distributed graph rule* consists of a *network rule* and *local rules* for all network nodes which are preserved by the network rule. All newly created network nodes as well as those which should be deleted are equipped with local graphs. See figure 2 for a schematic representation of a distributed rule. In the following, rules are presented by their graphs, the relation between left and right-hand sides are installed unambiguously obeying compatibility of source, target, type and attribute mappings.

Moreover, it is possible to split and glue network nodes. The underlying local graphs have to be split and glued, too. The information how to split has to and can be given by distributed rules, too. The local graph of a network node to be split has to be mapped to graph splits. If a node is split, the rule must also contain the specification where to put the adjacent edges, i.e. the context of

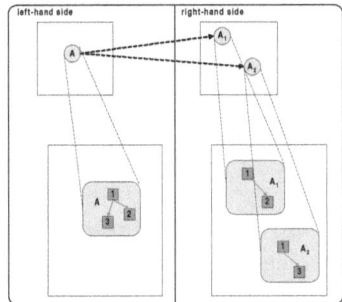

Fig. 3. Example of a splitting and copying distributed rule. The local node 1 in the left-hand side of the rule is copied into both local graphs on the right-hand side whereas the local nodes 2 and 3 as well as the local edges are split into the two local graphs on the right-hand side.

nodes to be split has to be completely specified by the rule. Local nodes and edges

- which are not mapped at all are deleted as usual,
- which are mapped to one local graph item, are assigned to that item,
- which are mapped several times to local graph items, are copied as often as needed.

Consider figure 3 for an example of a splitting and copying rule.

Rules may contain variables for attribute values which are globally defined within a rule. Some of these variables are considered as rule parameters which are set from outside while all the others are instantiated by a rule match. In chapter 4, variables for node attributes are given by sets NV and parameters are listed in a rule's head after its name.

To apply a distributed rule, several *application conditions* have to be checked to ensure that the resulting graph structure is again a distributed graph, i.e. no dangling edges occur and no replication relation becomes partial after application. Moreover, the application of distributed rules should not produce implicit side effects on the graph structure. The set of application conditions consists of the well-known *gluing condition* of the double-pushout approach [3] for network and local rules, a *connection condition*, a *network condition*, a *splitting condition*, and a set of *negative application conditions* (NACs).

The connection condition states that local nodes and edges are not allowed to be deleted, if they are used in an interface, i.e. if they have correspondents in an adjacent interface. Moreover, interface nodes and edges are not allowed to be created if there are not any correspondents in their adjacent local system nodes.

To ensure the network condition the corresponding local state graph of a network node to be deleted (created) has to be completely determined within the distributed rule. Moreover, to install (remove) a network edge, the source graph of its local morphism has to be completely determined, too, such that all mappings of the morphism are handled by the rule.

The splitting condition states that the edge context of the network nodes to be split have to be fully determined by the distributed rule.

A detailed description of these application conditions can be found in [3,16,17]. Compare figures 7, 10, and 11 for examples of NACs stated on the right of a rule.

In the next section we will model a sample system well-known in the world of dynamic change management by distributed graph transformation.

4   A Sample System: Distributed Ring Database

We show the applicability of distributed graph transformation as a natural formalism for realizing dynamic change management by modeling the ring database example introduced by J. Kramer and J. Magee [10]. Let us consider a ring of connected replicated databases where each database can autonomously initiate an update of its containing data. This update is then propagated around the ring. Copy inconsistencies while different update requests are circulating in the ring simultaneously are tolerated. However, to ensure long term consistent updates in the presence of node autonomy and concurrency we impose as a requirement to system integrity identity of all databases' states when no updates are circulating and no database is locally initiating an update. Thus all database copies are consistent when the entire ring is quiescent.

Let us consider a sample distributed ring database comprising three interconnected databases storing the data of a distributed software development environment. Figure 4 sketches the network level of a distributed graph representing the ring: it comprises three local systems, the distributed databases DB_A, DB_B, and DB_C. Each local system is represented by a network node the local graph of which contains graph elements describing the local state of the database (body) as well as graph elements describing its interfaces (export and import). distributed network edges between them. Interconnections between local systems are modeled by network edges. In the following we will assume that node label strings (like those depicted in figure 4) are comprised of a name variables (e.g., DB_A).

The body graph elements in a network node's local graph represent the distributed copies of the database and changes to a database's state resulting from updates are propagated over its interfaces.

Now let us consider a simplified sample update propagation assuming Entity-Relationship (ER) diagrams as data structure. Within this example, ER diagrams serve as a data representation which intuitively demonstrates the possibilities of our approach. Of course, if it is not appropriate to model the entire distributed system with graphical means, it is possible to abstract from local states as far as needed. For example, the data involved in the communication processes may have a graph representation only. On the other hand, graphs may be labeled by complex attributes like databases again, or pictures, etc. [17,7]

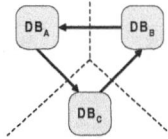

Fig. 4. Ring database network graph. The distributed network edges indicate the direction of update propagations, i.e. DB_B imports an update which is exported by DB_A, etc.

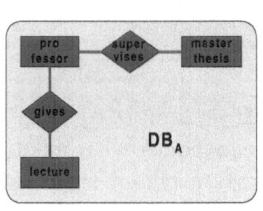
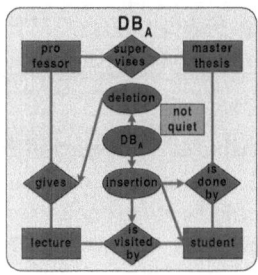

Fig. 5. The left part of the figure depicts the initial database state of the update propagation, the right part shows an update introduced in the originator (local view of DB_A).

but this is beyond the scope of this paper. The initial state of all replicated databases comprises the entities *professor*, *lecture*, and *master thesis*, as well as the relations *gives* and *supervises*. The left part of figure 5 shows the local view of the database DB_A in a quiescent state, there exist no export or import graph elements as well as no distributed graph relations to other databases.

For controlling an update propagation we introduce the nodes *originator*, *insertion*, and *deletion* representing the originator of an update and information about ER structures to be deleted and created. This representation allows us to handle many update requests simultaneously via one communication connection. Now within database DB_A a local update is initiated: three update propagation control nodes are inserted, an additional entity *student*, the relations *is visited by* and *is done by* connecting *student* with the entities *lecture* and *master thesis* are added, and the relation *gives* is deleted (cf. the right part of figure 5). The rules by which the update is introduced are shown in [19] in detail. An export node representing a 'not quiet' flag which is used by the application / change management interface is also introduced in DB_A (cf. chapter 5).

The next step is to publish in DB_A the update propagation nodes as well as the ER structure connected with them (including possible interface entities of copied relations, cf. the left part of figure 6). A sample rule for exporting entities is depicted in figure 7, the complete set of rules for exporting entities, relations, and update propagation control nodes are presented in [19].

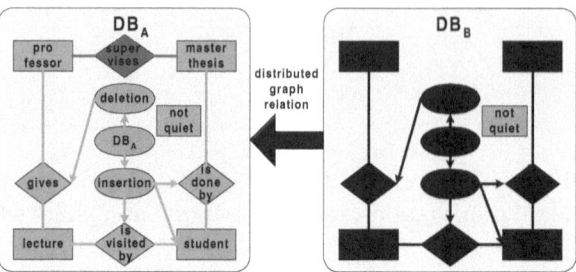

Fig. 6. The update is exported by the originator and imported by the successor database (local view of DB_A). Wrt. the distributed graph relation a mapping between graph elements with identical labels is assumed.

Now a distributed use relation from DB_B's import to DB_A's export has to be established (assuming a mapping between nodes with identical names). As we have presumed synchronous communication channels (cf. above and [10]), the application of the update to DB_B's body graph has to be done at the same time as the update is imported (cf. figure 6). Again, a sample rule for importing entities is shown in figure 7 and further import rules can be found in [19].

Then DB_B's import graph elements and DB_A's export graph elements are deleted (i.e. interfaces are emptied). Finally, the update is performed in DB_A's body graph and the control nodes as well as the 'not quiet' flag are deleted (cf. figure 8). The rules by which this is done are depicted in [19], sample rules for deleting export / import entities, as well as sample update perform rules are sketched in figure 7.

The update propagation is continued by exporting the update again from DB_B in the same way as sketched above for DB_A (cf. figure 8). This cycle is terminated when a database recognizes itself as the update originator. Then, its predecessor's export structure is directly deleted without importing the update. Figure 9 shows the termination rule which has to be applied by all databases before importing an update.

5   Dynamic Change within the Ring Database: Automated Database Inclusion and Removal at Runtime

While in the previous chapter we have sketched how the ring database example can be described by distributed graph transformation, we now discuss how dynamic accommodation of evolutionary changes to the ring's structure can be modeled, i.e. automated inclusion and removal of databases at runtime. Thus distributed graph transformation can be regarded as a natural underlying formalism to realize configurable distributed systems where modifications not envisaged at design time can be handled at runtime while those system parts not affected by the change are not disturbed in their operation.

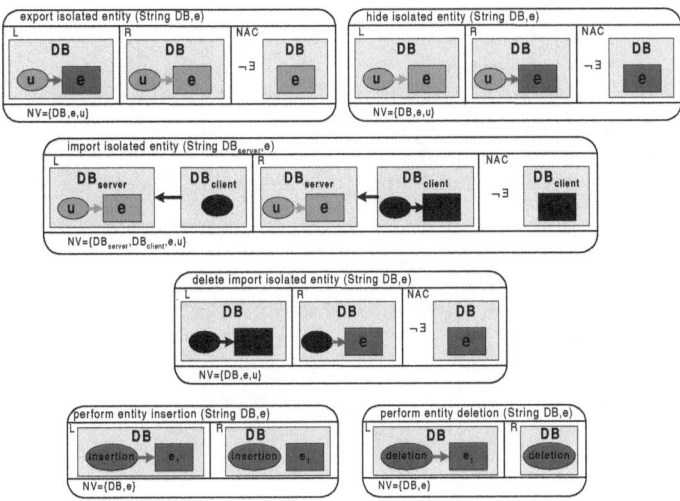

Fig. 7. Rules for exporting and hiding entities, for importing entities and deleting import relations, and for performing entity insertions and deletions (local view of *DB*).

Following the change model presented in chapter 2 changes to a node - i.e. the deletion or creation of a replicated database - are only permitted when the node is quiescent. Assuming that channels accept/transfer update messages from/to the nodes synchronously and do not loose any messages, node quiescence is given when the node is passive: the database is not updating its data. In this passive state change management can delete a node, or if two nodes are in a passive state, a new node can be inserted between them.

We define *node passivity* as the state in which a database is not updating its data - it is not locally *initiating an update* nor it is *servicing an update request* originating from another database of the ring. Thus a node is in a passive state

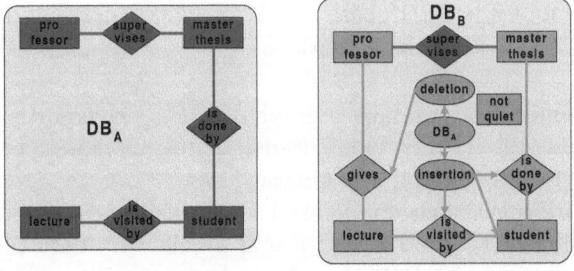

Fig. 8. The update is finally performed in database DB_A (local view of DB_A) and then exported from DB_B (local view of DB_B).

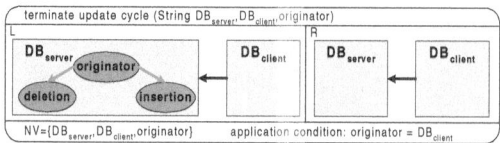

Fig. 9. Rule for terminating a update propagation (local view of DB_{client}).

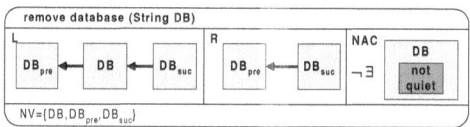

Fig. 10. Change management rule for removing databases (local view of database DB).

if *only* the graph rewrite rules for adding control nodes or importing control nodes [19] are applicable, i.e. the only allowed operation is initiating or importing an update. This means that neither a database imports update propagation control nodes (the database is servicing an update request) nor its body contains control nodes (the database is initiating an update). Quiescence management rules [19] set a *not quiet flag*, if body or import *originator* nodes exist in a database and delete the flag if such nodes do no longer exist. The *not quiet* flag is modeled as an export node so that the change management rules (cf. figures 10 and 11) do not need to inspect a database's body graph to determine its quiescence status.

Until now we have sketched the application's graph rewrite rules. Now we investigate how graph rewrite rules for change management look like. Figure 10 sketches a network rule for deleting a database DB. The application condition presupposes that a *not quiet* flag does not exist for DB, i.e. the database is not within a data update transaction. Assuming synchronous transfer of update messages, there is no need for connected components to also be passive [10] (please note that if an update initiated by a database just deleted is still circulating within the ring, the originator label has to be set to the successor database).

The creation of a new database DB between two existing databases DB_{pre} and DB_{suc} is done within three steps:

1. First the entire body database graph of DB_{pre} is exported by the help of export rules [19]. This is done because change management rules must not access the local body graph of a database.
2. The network rule sketched in figure 11 is applied to the ring's network graph presupposing DB_{pre} and DB_{suc} are in a quiescent state (i.e., DB_{pre} must not be exporting and DB_{suc} must not be importing an update [11]). The dotted arrows in figure 11 denote the copying of network nodes as introduced in chapter 3 and figure 3.

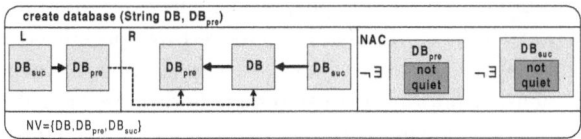

Fig. 11. Change management rule for creating databases (local view of database DB).

3. Finally, DB_{pre}'s export and DB's import graph elements are deleted (i.e. set transferred back to body graph elements) using rules for deleting export / import relations [19].

6 Conclusions and Further Work

In this contribution, we have introduced our approach to the problem of dynamic change management in distributed systems applying distributed graph transformation. Distributed graph transformation nicely satisfies all the requirements postulated in [11]:

- *changes are specified in terms of the system structure,*
- *change specifications are declarative,*
- *change specifications are independent of algorithms, protocols, and states of the application,*
- *changes leave the system in a consistent state,*
- *changes minimize the disruption to the application system.*

Moreover, by using graph transformation coherently at both levels - the network level and the local node level - more detailed and specialized change actions can be described. This is done while the encapsulation provided by the nodes is preserved using only the node's interfaces. The chosen example of a ring database storing ER-diagrams shows the natural way to express the change actions at the two levels. Of course our approach can also be applied to other diagrammatic notations, e.g., in [8] we apply distributed graph transformation to SDL diagrams.

Our approach provides several benefits compared to architecture description languages used for dynamic change management (such as DARWIN as presented in [10]). In addition to general advantages of a graph-based approach in this application area such as combining intuitive usability with a formal basis and the ability to express structural information distributed graph transformation introduces concepts for visually managing modularization and distribution.

This is ongoing work and we will improve our notions through implementation and more examples. One aspect - the interaction of change actions at the network level and the node level - leaves still room for improvement. Instead of providing related interfaces for this purpose the necessary mapping of local node state to the network level could also be achieved by specialized morphisms.

As stated in chapter 2, consistency is a very important aspect in dynamic change management. A notion of consistency conditions, logical formulas on the existence and non-existence of graph structures, and a constructive approach to ensure these conditions are presented in [9]. It is due to future work to extend this work to distributed graph transformation.

Currently we are working on an implementation of our ideas which will be based on the current AGG - machine [18]. Of course, the implementation will also make use of a distributed computing system. In a first approach, we plan to support the definition of import and export interfaces for graphs and rules realized as substructures of those. The main attention lays on the synchronization of local transformations related by subtransformations. The communication will be realized by remote method invocations (RMI) supported by Java.

References

1. Agnew, B., Hofmeister, C.R., and Purtilo, J.M.: Planning for Change: A Reconfiguration Language for Distributed Systems, Proc. 2nd Int. Conf. on Distributed Computing Systems, IEEE Computer Society Press, Pittsburgh, U.S.A., 1994. 181
2. Berry, G. and Boudol, G.: The chemical abstract machine, Theoretical Computer Science, (96) p. 217-248, 1992. 181
3. Corradini, A., Montanari, U., Rossi, F., Ehrig, H., Heckel, R., Löwe, M.: Algebraic Approaches to Graph Transformation, in Rozenberg, G. (ed.), Handbook of Graph Grammars and Computing by Graph Transformation, Vol. 1 Foundations, pp. 163-245, World Scientific, Singapore, 1997. 180, 182, 185, 186
4. Fischer, I., Koch, M., Taentzer, G.: Local Views on Distributed Systems and their Communication, TAGT 98, tr-ri-98-201 Dept. of Computer Science, University of Paderborn, Germany. 183
5. Fischer, I., Koch, M., and Volle, V.: Modeling Distributed Configuration Management by Distributed Attributed Graph Transformation, to appear as technical report at University of Erlangen, Germany, 1998. 184
6. Fossa, H. and Sloman, M.: Implementing Interactive Configuration Management for Distributed Systems, Proc. of the 3rd ICCDS, IEEE Computer Society Press, Annapolis, U.S.A, pp. 44-51, 1996. 180
7. Goedicke, M. and Meyer, T.: Dynamic Semantics Negotiation in Distributed and Evolving CORBA Systems: Towards Semantics-Directed System Configuration, Proc. 4th ICCDS, Annapolis, USA, 1998. 181, 186
8. Goedicke, M., Meyer, T., and Taentzer, G.: ViewPoint-oriented Software Development by Distributed Graph Transformation: Towards a Basis for Living with Inconsistencies, Proc. 4th IEEE Intl Symp. on Requirements Engineering, Limerick, Ireland, 1999. 191
9. Heckel, R. and Wagner, A.: Ensuring Consistency of Conditional Graph Grammars – A constructive Approach, Proc. of SEGRAGR'95 'Graph Rewriting and Computation', Electronic Notes of TCS, Vol. 2, http://www.elsevier.nl/locate/entcs/volume2.html, 1995. 192
10. Kramer, J. and Magee, J.: Analysing Dynamic Change in Software Architectures: A Case Study, Proc. 4th ICCDS, Annapolis, USA, 1998. 181, 186, 188, 190, 191

11. Kramer, J. and Magee, J.: The Evolving Philosophers Problem: Dynamic Change Management, IEEE TSE, SE-16, 11, pp. 1293-1306, 1990. 180, 181, 190, 191
12. Magee, J. and Kramer, J.: Self Organizing Software Architectures, Proc. SIGSOFT Workshops, ACM Press, pp.35-38,1996. 180
13. Magee, J. and Kramer, J.:Dynamic Structure in Software Architectures, Proc. 4th ACM SIGSOFT Symp. Foundations of Software Engineering, ACM Press, San Francisco, U.S.A., 1996. 180
14. Magee, J., Dulay, N., and Kramer, J.: Regis: A Constructive Development Environment for Distributed Programs, IEE/IOP/ BCS Distributed Systems Engineering Journal 1(5),304-312,1994. 180
15. Métayer, D.L.: Software architecture styles as graph grammars, Proc. 4th ACM SIGSOFT Symp. Foundations of Software Engineering, p. 15-23, ACM Press, 1996. 181
16. Taentzer, G.: Parallel and Distributed Graph Transformation: Formal Description and Application to Communication-Based Systems, PhD thesis, TU Berlin, Shaker Verlag, 1996. 180, 182, 186
17. Taentzer, G., Fischer, I., Koch, M., and Volle, V.: Distributed Graph Transformation with Application to Visual Design of Distributed Systems, to appear in Graph Grammar Handbook 3 Concurrency and Distribution, 1998. 180, 182, 184, 186
18. Taentzer, G., Ermel, C., and Rudolf, C.: AGG-Approach: Language and Tool Environment, in Graph Grammar handbook 2: Specification and Programming, 1998. 192
19. Taentzer, G., Goedicke, M., and Meyer, T.: Dynamic Change Management by Distributed Graph Transformation: Towards Configurable Distributed Systems, Tech. Rep., Dept of Math and Computing, Univ. of Essen, 1998. 187, 188, 190, 191
20. Wermelinger, M.: Towards a Chemical Model for Software Architecture Reconfiguration, Proc. 4th ICCDS, Annapolis, USA, 1998. 181

A Framework for NLC and ESM: Local Action Systems[*]

Nico Verlinden and Dirk Janssens

Department of Mathematics and Computer Science, U.I.A.
Universiteitsplein 1, B–2610 Antwerp, Belgium
{dmjans,verlind}@uia.ua.ac.be

Abstract. NLC and ESM are two types of graph rewriting based on node replacement and embedding. Their embedding mechanisms are quite different, and no unifying framework for them was known so far. It is demonstrated that Local Action Systems provide such a framework. Since rewriting in an NLC grammar is not confluent, this leads to the notion of an ordered LAS processes: a process equipped with a complete order on the local actions. It is shown that NLC graph rewriting can be described in terms of these ordered processes, and that the order can be omitted if the sets of available local actions satisfy a simple algebraic property. ESM graph rewriting fits into this unordered version of the theory.

1 Introduction

NLC (Node Label Controlled) graph grammars ([6], [2]) are a well known type of graph rewriting systems, mainly intended as a generalization of the familiar string grammars from the language theory. ESM (Extended Structured Morphisms) systems were introduced in [4] as a model of concurrent computation, and a process theory for them has been developed in [5]. Local Action Systems (LAS) originated as a generalization of ESM systems ([3]), and have with the latter in common that their theory is entirely based on the notion of a process. The semantics of a system is given by characterizing its set of processes, and the derivation sequences can be derived from that if desired. Hence the semantics considered here is at least as detailed as the more traditional semantics based on derivation sequences.

LAS processes describe rewriting histories. In their definition one takes the point of view that rewriting has two aspects: replacement (removing and creating) and transformation (change). Perhaps the most basic idea of the LAS approach is that the transformation is the result of *local actions*: actions that transform information about just one node x into information about just one other node y, which is then causally dependent on x. A LAS process consists

---

[*] Partially supported by the EC TMR Network GETGRATS (General Theory of Graph Transformation Systems) and Esprit Working Group APPLIGRAPH through Universitaire Instelling Antwerpen.

H. Ehrig et al. (Eds.): Graph Transformation, LNCS 1764, pp. 194–215, 2000.

of all nodes that occur in the rewriting, the local actions (the dynamic structure) and the information that is either created or initially present (the static structure). The notion of a LAS process is defined for given alphabets of local actions, node labels and edge labels, but is not dependent on the choice of a particular system (set of productions). Each system determines a subclass of the class of LAS processes over the alphabets involved, and this subclass is regarded as its process semantics. In this respect LAS processes differ from other process notions, e.g. the one introduced in [1] for graph grammars based on the double pushout approach: there each graph grammar determines a class of processes, but there is no general notion of a process, common to all graph grammars over a given alphabet. A detailed comparison of these process notions is however beyond the scope of this paper.

In Section 2 and 3 some basic notions and terminology is recalled, and in Section 4 and 5 the basic theory of LAS, based on ordered processes, is developed. In Section 6 it is shown that NLC graph rewriting can be encoded into the framework, and in Section 7 it is investigated under which conditions the temporal order of processes can be omitted. The encoding of ESM systems satisfies these conditions.

2   Preliminaries

1. For a set V, the powerset of V is denoted by $\mathcal{P}(\mathcal{V})$ and the set $V \times V$ is denoted by V^2. For a relation $R \subseteq V^2$, R^* denotes the transitive and reflexive closure of R. For a set $A \subseteq V$, $R(A)$ denotes the set $\{v \in V \mid$ there exists an $a \in A$ such that $(a, v) \in R\}$, and for a set $B \subseteq V^2$, $R^{(2)}(B)$ denotes the set $\{(v, w) \in V^2 \mid$ there exists a pair $(a, b) \in B$ such that (a, v), $(b, w) \in R\}$.

2. For a partial function $f : X \to Y$, the domain of f is denoted by $dom(f)$. Let $f^{(2)} : X \times X \to Y \times Y$ be defined by $f^{(2)}((x_1, x_2)) = (f(x_1), f(x_2))$, for each x_1, $x_2 \in dom(f)$. The free commutative monoid on a set X is denoted by $FCM(X)$. Its elements are functions from a finite subset of X into the set of natural numbers, and the neutral element is the empty function.

3. Let $f_1 : X_1 \to Y$ and $f_2 : X_2 \to Y$ be partial functions with Y a commutative monoid with monoid operation $+$. Then $f_1 \oplus f_2$ is the partial function $f : X_1 \cup X_2 \to Y$ such that $dom(f) = dom(f_1) \cup dom(f_2)$ and

$$f(x) = \begin{cases} f_1(x), & \text{if } x \in dom(f_1) \setminus dom(f_2) \\ f_2(x), & \text{if } x \in dom(f_2) \setminus dom(f_1) \\ f_1(x) + f_2(x), & \text{if } x \in dom(f_1) \cap dom(f_2) \end{cases}$$

4. Let V be a set and let $R \subseteq V \times V$. A *path in* R is a sequence $p = x_0, x_1, \ldots, x_n$ of elements of V such that $n \geq 0$ and, for each $0 \leq i \leq n-1$, $(x_i, x_{i+1}) \in R$. n is the *length* of the path p and p is a path *from* x_0 *into* x_n. The fact that p is a path from x into y is denoted by $p : x \to y$. $Pairs(p)$ denotes the set $\{(x_i, x_{i+1}) \mid 0 \leq i \leq n - 1\}$. The concatenation of paths is denoted by $\odot$: if $p = x_0, x_1, \ldots, x_n$ and $q = x_n, x_{n+1}, \ldots, x_{n+k}$, then $p \odot q$ is the

path $x_0, x_1, \ldots, x_{n+k}$. The relation R is *acyclic* if there are no nontrivial cycles; i.e. if $p = x_0, x_1, \ldots, x_n$ is a path in R such that $x_0 = x_n$, then $n = 0$.

5. Let V be a set and let R be a partial order on V. A *cut of R* is a maximal subset K of V such that, for each $x, y \in K$, $(x, y) \notin R$ and $(y, x) \notin R$.

3   NLC and ESM Graph Rewriting

In this section the formal definitions of NLC grammars and ESM systems are recalled.

An NLC graph grammar consists of an alphabet Σ of node labels, a set of productions, and a common initial graph for all the derivations considered. An NLC graph over Σ is a 3–tuple $g = (V_g, Ed_g, lab_g)$, where V_g is a finite set (of nodes), Ed_g is a set of 2–element subsets $\{x, y\}$ of V (where $x \neq y$), and lab_g is a function from V into Σ. The set of node labels is divided into terminal and nonterminal labels, and only nodes with a nonterminal label are replaced. An NLC production is a pair (A, r) such that A is a nonterminal label and r is an NLC graph. The production (A, r) can only be applied to nodes with label A. The graph h resulting from applying a production (A, r) to an A–labeled node x of a graph g is constructed as follows:

- x is removed, together with all edges incident with x.
- an isomorphic copy $\bar{r}$ of the graph r is created; it is assumed that $\bar{r}$ has no nodes in common with the graph g.
- edges between nodes of $\bar{r}$ and nodes of the remaining part of g are established according to the embedding mechanism of the grammar.

The embedding mechanism of an NLC grammar is specified by a relation $Conn \subseteq (\Sigma)^2$, which is fixed for the whole grammar. Formally one has the following.

Definition 1. *Let g be an NLC graph, let (A, r) be an NLC production, let x be an A–labeled node in g and let $f : r \to \bar{r}$ be an isomorphism such that $V_{\bar{r}} \cap V_g = \emptyset$. Then the result graph of the application of (A, r) to x in g via f is the NLC graph h such that*

$$V_h = (V_g \setminus \{x\}) \cup V_{\bar{r}},$$

$$lab_h(z) = \begin{cases} lab_g(z), & \text{if } z \in V_g \setminus \{x\} \\ lab_{\bar{r}}(z), & \text{if } z \in V_{\bar{r}} \end{cases},$$

and $e = \{v, w\} \in Ed_h$ if and only if either (i) $e \in Ed_g$ and $v \neq x$, $w \neq x$, or (ii) $e \in Ed_{\bar{r}}$, or (iii) $v \in V_{\bar{r}}$, $w \in V_g \setminus \{x\}$, $\{x, w\} \in Ed_g$ and $(lab_g(w), lab_{\bar{r}}(v)) \in Conn$.

Example 1. Fig. 1 depicts the application of the production (A, r) to x, where $\Sigma = \{a, A, b, c\}$, A is a nonterminal label, $r = (\{w, y\}, \{\{w, y\}\}, lab_r)$, $lab_r(w) = b$, $lab_r(y) = a$, and $Conn = \{(a, b), (c, a)\}$ (to simplify the situation it is assumed that $r = \bar{r}$).

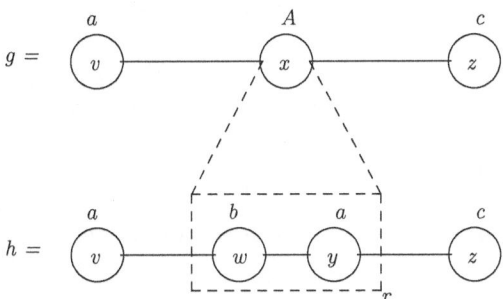

Fig. 1. The application of an NLC production

An ESM system is defined as follows. An ESM system consists of an alphabet Σ_1 of node labels, an alphabet Σ_2 of edge labels, and a set of productions. An ESM graph over (Σ_1, Σ_2) is a triple (V_g, Ed_g, lab_g), where V_g is a set of nodes, $Ed_g \subseteq (V_g \times \Sigma_2 \times V_g)$ is the set of (directed, labeled) edges, and lab_g is a function from V_g into Σ_1.

An ESM production is a 4–tuple (l, r, in, out), where l and r are graphs, and where in and out are relations specifying the embedding mechanism: $in \subseteq (V_l \times V_r)$ and $out \subseteq (V_l \times \Sigma_2) \times (V_r \times \Sigma_2)$. The effect of applying such a production to a graph g is that an occurrence $\bar{l}$ of l in g is removed, replaced by an isomorphic copy $\bar{r}$ of r, and that the edges between $\bar{r}$ and the remaining part of g are established according to the embedding mechanism: $(x, y) \in in$ means that the node corresponding to y in the result graph inherits the incoming edges of the node corresponding to x, and $((x, \alpha), (y, \beta)) \in out$ means that y in the inherits the outgoing edges of the node x that are labeled by α, and relabels these edges by β. Thus the out relation specifies a combination of edge inheritance and edge relabeling. Formally, one has the following.

Definition 2. *Let g be an ESM graph, let (l, r, in, out) be an ESM production, let $f_l : l \to \bar{l}$ and $f_r : r \to \bar{r}$ be isomorphisms such that $\bar{l}$ is a subgraph of g and $V_{\bar{r}} \cap V_g = \emptyset$. Then the result graph of the application of (l, r, in, out) to $\bar{l}$ in g via f_l and f_r is the ESM graph h such that*

$$V_h = (V_g \setminus V_{\bar{l}}) \cup V_{\bar{r}},$$

$$lab_h(x) = \begin{cases} lab_g(x), & \text{if } x \in V_g \setminus V_{\bar{l}} \\ lab_{\bar{r}}(x), & \text{if } x \in V_{\bar{r}} \end{cases}$$

$$Ed_h = \bigcup_{e \in Ed_g} E(e) \cup Ed_{\bar{r}},$$

where for each $e = (x, \alpha, y) \in Ed_g$ the set $E(e)$ is defined by: $(u, \beta, w) \in E(e)$ if and only if (i) either $(x, \alpha) = (u, \beta)$ or $((f_l^{-1}(x), \alpha), (f_r^{-1}(u), \beta)) \in out$, and (ii) either $y = w$ or $(f_l^{-1}(y), f_r^{-1}(w)) \in in$.

4   Local Actions and Ordered Processes

Local Action Systems use structured alphabets: throughout the paper (Δ^v, Δ^e) denotes a pair of commutative monoids (both monoid operations are denoted by $+$). Δ^v is used as the set of node labels and Δ^e is used as the set of edge labels. Moreover, it is assumed that Δ^v and Δ^e are equipped with partial orders $\leq_v$ and $\leq_e$, respectively.

T^v and T^e denote the sets of automorphisms on Δ^v or Δ^e, respectively; local actions are chosen from these sets. The composition of elements of T^v or T^e is denoted by $\langle\cdot\rangle$: for each t_1, $t_2 \in T^v$ and $a \in \Delta^v$, or t_1, $t_2 \in T^e$ and $a \in \Delta^e$, $\langle t_1.t_2\rangle(a) = t_2(t_1(a))$. The sum of elements in T^v or T^e is defined by $(t_1+t_2)(a) = t_1(a)+t_2(a)$. The identity and the constant zero mapping (mapping everything into the neutral element) are automorphisms, they are denoted by 1 and 0, respectively. As the LAS approach is based on the manipulation of labels, it is convenient to think of graphs as sets of nodes equipped with two partial labeling functions.

Definition 3. *A graph on a set of nodes V is a pair $g = (g^v, g^e)$, where $g^v : V \to \Delta^v$ and $g^e : V^2 \to \Delta^e$ are partial functions.*

The set $dom(g^e)$ may be viewed as the set of edges of g. For a subset X of V, the subgraph of g induced by X is the graph (h^v, h^e) on X, where h^v and h^e are the restrictions of g^v and g^e to X and X^2, respectively.

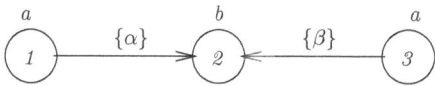

Fig. 2. A graph

Example 2. Let $\Sigma_1 = \{a, b, c\}$, $\Sigma_2 = \{\alpha, \beta\}$, $\Delta^v = FCM(\Sigma_1)$ and $\Delta^e = \mathcal{P}(\Sigma_2)$. Let $\leq_v$ be the equality on Δ^v and let $\leq_e$ be set inclusion. Then Fig. 2 depicts a graph g, where $dom(g^v) = \{1, 2, 3\}$ and $dom(g^e) = \{(1, 2), (3, 2)\}$.

A central part of the description of a rewriting history is the description of the local actions that take place between causally dependent nodes. For a pair (x, y) of nodes, the local action d associated with it consists of four components: $d = (d^v, d^s, d^t, d^{st})$, where d^v acts on the node label of x, d^s acts on the label of outgoing edges of x, d^t acts on the label of incoming edges of x, and d^{st} acts on the label of loops in x. The result of the action d is taken into account in the node label of y and all the labels of the edges incident with y. This idea is captured by the notion of a dynamic structure.

Definition 4. *Let V be a set. A dynamic structure on V is a partial function $D : V \times V \to T^v \times (T^e)^3$.*

The empty function is a dynamic structure on V. Slightly abusing the notation, it is denoted by $\emptyset$. A system run or rewriting history of a LAS is described by an *ordered process*. An ordered process specifies all the nodes occurring in the course of a run, the structure that is either initially present or created, the causality relation between the nodes, and the dynamic structure. Additionally, one also equips a process with a (total) order on the local actions. This order is called the *temporal order* of the process; it specifies the order in which the local actions occur, and evidently it should be compatible with the causality relation. It will be shown in Section 7 that the temporal order is not needed if the local actions satisfy certain algebraic properties. This special case is the one considered in [3].

Definition 5. *An* ordered process *is a 5–tuple* $(V, C, D, \prec, S)$, *where* V *is a finite set,* $C \subseteq V \times V$ *is an acyclic relation,* D *is a dynamic structure on* V *such that* $C = dom(D)$, $\prec$ *is a total order on* C *such that, for each* $(x, y), (y, z) \in C$, $(x, y) \prec (y, z)$, *and* S *is a graph on* V.

The partial order corresponding to C is called the *causal order* of *proc*, and it is denoted by $<_C$. Thus $x <_C y$ if there is a path $p : x \to y$ of length at least one in C. $Min(proc)$ and $Max(proc)$ denote the sets of minimal and maximal nodes of V with respect to $<_C$, respectively.

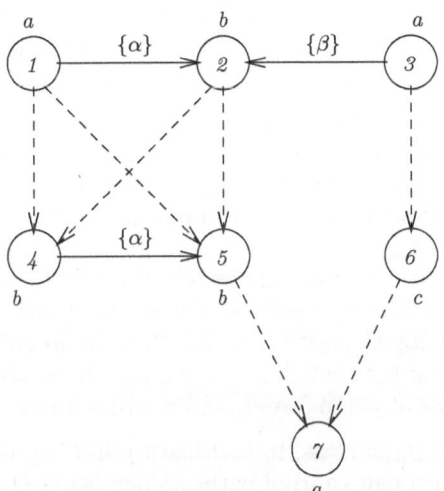

Fig. 3. An ordered process

Example 3. Let Δ^v, Δ^e be the monoids of Example 2 and let, for each $\gamma, \delta \in \Sigma_2$, $t_{\gamma\delta}$ be the automorphisms on Δ^e determined by

$$t_{\gamma\delta}(\{x\}) = \begin{cases} \{\delta\}, & \text{if } x = \gamma \\ \emptyset, & \text{otherwise.} \end{cases}$$

Let $V = \{1, 2, 3, 4, 5, 6, 7\}$, $C = \{(1,4), (1,5), (2,4), (2,5), (3,6), (5,7), (6,7)\}$, and $\prec$ is the lexicographic order on C. Then $proc = (V, C, D, \prec, S)$ is an ordered process, where S is the graph on V depicted in Fig. 3 (the dashed arrows represent C) and where all components of the dynamic structure D are assumed to be constant zero, except the following ones: $D^s(1,4) = t_{\alpha\beta}$, $D^s(3,6) = t_{\beta\beta}$, $D^s(6,7) = t_{\beta\alpha}$, $D^t(1,4) = D^t(2,5) = D^t(5,7) = 1$, $D^{st}(1,4) = t_{\alpha\beta}$.

Remark 1. The graph (g^v, g^e) on a set V may be identified with the ordered process $(V, \emptyset, \emptyset, \emptyset, (g^v, g^e))$. Thus operations defined on ordered processes (such as the sequential composition, Definition 8) can be applied to graphs.

4.1   The Initial Graph and the Result Graph of an Ordered Process

For an ordered process $(V, C, D, \prec, S)$ describing a rewriting history, C describes the direct causality relation between nodes, and thus the minimal and maximal nodes with respect to the causal order $<_C$ are the nodes of the initial graph and of the final graph of the rewriting history, respectively. Since the local actions have no effect on the initial nodes, the initial graph of the process is the graph obtained by restricting the underlying graph S to the input nodes. The result graph, on the other hand, consists of the structure on the output nodes that is generated by applying the local actions specified by D to the underlying graph S. In the LAS approach, two basic ideas are used to define this structure.

- The result of applying the local actions to S is the sum of the results of applying them to the individual node and edge labels of S: indeed S is the union of graphs consisting of one labeled node (for S^v) or two unlabeled nodes and one labeled edge (for S^e), and the results obtained by applying local actions to these elementary components of S are combined using the sum on node and edge labels.
- For each node x, its effect on a node u of the result graph is the sum of the contributions obtained by applying the local actions on each path from x into u. Similarly, for an edge (x, y), its effect on an edge (u, v) of the result graph is the sum of the contributions obtained by applying the local actions on each pair (p, q) of paths from x and y into u and v, respectively.

In order to formalize these ideas, to each path p in C an operation D_p of T^v is associated, and to each pair (p, q) of paths an operation D_{pq} of T^e is associated. In the definition of the latter, the temporal order $\prec$ is used to determine the relative order in which the actions from p and q occur. One also takes into account that the actions of p are associated with the source of an edge, and that the actions of q are associated with the target of an edge. If a pair (x, y) occurs on both p and q, i.e. p and q join in x, then (x, y) is associated with an action on both the source and target of an edge. Formally one has the following.

Definition 6. *Let $(V, C, D, \prec, S)$ be an ordered process. For paths p and q in C, D_p and D_{pq} are defined as follows.*

$$D_p = \begin{cases} 1_{T^v}, & \text{if } length(p) = 0 \\ \langle D_{p_1} \cdot D^v(d) \rangle, & \text{if } p = p_1 \odot d \end{cases}$$

$$D_{pq} = \begin{cases} 1_{T^e}, & \text{if } length(p) = length(q) = 0 \\ \langle D_{p_1 q} \cdot D^s(d) \rangle, & \text{if } p = p_1 \odot d \text{ and } e \prec d, \text{ for each } e \in Pairs(q) \\ \langle D_{p q_1} \cdot D^t(d) \rangle, & \text{if } q = q_1 \odot d \text{ and } e \prec d, \text{ for each } e \in Pairs(p) \\ \langle D_{p_1 q_1} \cdot D^{st}(d) \rangle, & \text{if } p = p_1 \odot d \text{ and } q = q_1 \odot d \end{cases}$$

The initial graph and the result graph of an ordered process are defined as follows.

Definition 7. *Let $proc = (V, C, D, \prec, S)$ be an ordered process.*

1. *The* initial graph *of proc, denoted by Init(proc), is the subgraph of S induced by Min(proc).*
2. *The* result graph *of proc, denoted by Res(proc), is the graph (g^v, g^e) on Max(proc) such that g^v and g^e are defined by*

$$g^v(x) = \sum_{z \in dom(S^v)} \sum_{p: z \to x} D_p(S^v(z))$$

$$g^e(x, y) = \sum_{(u,v) \in dom(S^e)} \sum_{\substack{p: u \to x, \\ q: v \to y}} D_{pq}(S^e(u, v)).$$

where p and q range over paths in C.

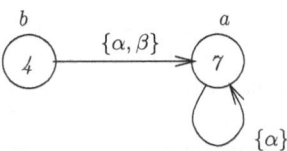

Fig. 4. The result graph

Example 4. Let *proc* be the ordered process from Example 3. *Init(proc)* is the graph from Example 2. If p and q are the paths $3, 6, 7$ and $2, 5, 7$, then $D_{pq} = t_{\beta\alpha}$. *Res(proc)* is the graph depicted in Fig. 4.

Two ordered processes $proc_1$ and $proc_2$ can be sequentially composed if they overlap only in nodes that are maximal in $proc_1$ and minimal in $proc_2$, i.e, if no node of $proc_1$ is causally dependent on a node of $proc_2$. The temporal order $\prec$ of the composed process expresses the fact that the local actions of $proc_1$ precede those of $proc_2$.

Definition 8. *Let* $proc_1 = (V_1, C_1, D_1, \prec_1, S_1)$ *and* $proc_2 = (V_2, C_2, D_2, \prec_2, S_2)$ *be ordered processes such that* $V_1 \cap V_2 \subseteq Max(proc_1) \cap Min(proc_2)$. *The* sequential composition *of* $proc_1$ *and* $proc_2$, *denoted by* $proc_1; proc_2$, *is the ordered process* $(V, C, D, \prec, S)$ *where* $V = V_1 \cup V_2$, $C = C_1 \cup C_2$, $D = D_1 \oplus D_2$, $\prec$ *is the total order on* C *such that, for each* $e_1, e_2 \in C$, $e_1 \prec e_2$ *if either* $e_1, e_2 \in C_1$ *and* $e_1 \prec_1 e_2$, *or* $e_1, e_2 \in C_2$ *and* $e_1 \prec_2 e_2$, *or* $e_1 \in C_1$ *and* $e_2 \in C_2$, $S^v = S_1^v \oplus S_2^v$, *and* $S^e = S_1^e \oplus S_2^e$.

4.2   Configurations

In order to characterize the set of ordered processes generated by a given set of productions (in Section 5), it is needed to take into account what is usually called the left–hand side of a production: a graph that should occur in the configuration that is reached at the time the production is applied; i.e. a production may only be applied to occurrences of its left–hand side. Thus one first needs a characterization of the configurations. A configuration is the result of an initial part of a rewriting history; this initial part evidently contains all minimal nodes. Thus one may define the configurations of an ordered process in the following way.

Definition 9. *Let* $proc$, $proc_1$ *and* $proc_2$ *be ordered processes such that* $V_1 \cap V_2$ $= Min(proc_2) \subseteq Max(proc_1)$ *and* $proc = proc_1; proc_2$. *The* configuration *of* $proc$ *corresponding to* $proc_1$ *is the graph* $Res(proc_1)$.

The requirement $Min(proc_2) \subseteq Max(proc_1)$ implies that $Min(proc_1; proc_2) \subseteq$ $Min(proc_1)$. An important special case of the sequential composition $proc_1; proc_2$ is that where $proc_1$ is a graph; i.e. $proc_1$ is of the form $(V, \emptyset, \emptyset, \emptyset, S)$. Then $proc_2$ can be viewed as an operation transforming $proc_1$ into the graph $Res(proc_1; proc_2)$: informally, this operation consists in executing the rewriting history described by $proc_2$ starting from the graph $proc_1$. Thus, for a graph g and an ordered process $proc$ such that $V_g \cap V_{proc} = Min(proc)$, the process $g; proc$ is called the *application of proc to* g.

In general, it is not necessarily true that $g = Init(g; proc)$, due to the fact that the underlying graph of $proc$ may be nontrivial on the minimal nodes. However, it seems natural to consider the situation where this is not the case, i.e. to consider the case where the underlying graph of $proc$ is only nontrivial on nodes that are causally dependent on the minimal nodes. Informally, this comes down to assuming that the initial graph of a rewriting history is not changed before any rewriting takes place. This is the view taken when defining productions and systems (Section 5): productions have no structure on their minimal nodes, and the same is true for ordered processes that are built from production occurrences. Formally, one has the following.

Proposition 1. *Let* g *be a graph on* V_g *and let* $proc$ *be an ordered process such that* $V_g \cap V_{proc} = Min(proc)$, $dom(S_{proc}^v) \cap Min(proc) = \emptyset$ *and* $dom(S_{proc}^e) \cap (Min(proc))^2 = \emptyset$. *Then* $Init(g; proc) = g$.

The notions of sequential composition, result graph, configuration and the application of a graph make sense only if, for a process of the form $proc_1; proc_2$ where $Min(proc_2) \subseteq Max(proc_1)$, the application of $proc_2$ to the result graph of $proc_1$ yields the result graph of $proc_1; proc_2$. Next it is demonstrated that this is the case; first a technical lemma is shown.

Lemma 1. *Let $proc_1 = (V_1, C_1, D_1, \prec_1, S_1)$ and $proc_2 = (V_2, C_2, D_2, \prec_2, S_2)$ be ordered processes such that $V_1 \cap V_2 = Min(proc_2) \subseteq Max(proc_1)$. Let $p = x_1, \ldots, x_n$ be a path in $C_1 \cup C_2$ such that $x_1 \in V_1$ and $x_n \in V_2$. Then there is exactly one i, $1 \leq i \leq n$, such that $x_i \in Max(proc_1)$.*

Proof. Let i be minimal such that $x_i \in V_2$ (i exists because $x_n \in V_2$). Then $x_i \in V_1 \cap V_2$ because either $i = 1$ or $(x_{i-1}, x_i) \in C_1$. Thus $x_i \in Max(proc_1)$, and either $i = n$ or $(x_i, x_{i+1}) \in C_2$. In the latter case, $x_{i+1} \notin Min(proc_2)$ and thus $x_{i+1} \notin V_1$. Since there are obviously no paths from $V_2 \setminus V_1$ into V_1 in $C_1 \cup C_2$, the result follows.

Theorem 1. *Let $proc_1 = (V_1, C_1, D_1, \prec_1, S_1)$ and $proc_2 = (V_2, C_2, D_2, \prec_2, S_2)$ be ordered processes such that $V_1 \cap V_2 = Min(proc_2) \subseteq Max(proc_1)$. Then*

$$Res(Res(proc_1); proc_2) = Res(proc_1; proc_2).$$

Proof. Let $(V, C, D, \prec, S) = (proc_1; proc_2)$, $k = Res(proc_1)$, $h = Res(k; proc_2)$ and $g = Res(proc_1; proc_2)$. It is obvious that $V_h = V_g = Max(proc_1; proc_2)$. Next it is shown that $h^e = g^e$; the proof of $h^v = g^v$ is similar but simpler.

Let $(u, v) \in dom(k^e)$ and $(w, z) \in Max(proc_1; proc_2)$. Consider the effect $E(u, v, w, z)$ of $k^e(u, v)$ on $h^e(w, z)$:

$$E(u, v, w, z) = \sum_{(p_2, q_2)} D_{p_2 q_2}(k^e(u, v)),$$

where (p_2, q_2) ranges over all pairs of paths $p_2 : u \to w$ and $q_2 : v \to z$ in C_2. On the other hand,

$$k^e(u, v) = \sum_{(x,y) \in dom(S_1^e)} \left(\sum_{(p_1, q_1)} D_{p_1 q_1}(S_1^e(x, y)) \right),$$

where (p_1, q_1) ranges over all pairs of paths $p_1 : x \to u$ and $q_1 : y \to v$ in C_1. It follows from Lemma 1 that, for $x \in V_1$ and $w \in V_2$, there is a one–to–one correspondence between the set of paths $p : x \to w$ in C and the set of 3–tuples (p_1, u, p_2) where $p_1 : x \to u$ is a path in C_1, $u \in Max(proc_1)$ and $p_2 : u \to w$ is a path in C_2. Moreover, if $p = p_1 \odot p_2$ and $q = q_1 \odot q_2$, for paths p_1, q_1 in C_1 and p_2, q_2 in C_2, then $D_{pq} = \langle D_{p_1 q_1} \cdot D_{p_2 q_2} \rangle$. It follows that

$$E(u, v, w, z) = \sum_{(x,y) \in dom(S_1^e)} \sum_{\substack{(p_1, q_1) \\ (p_2, q_2)}} \langle D_{p_1 q_1} \cdot D_{p_2 q_2} \rangle (S_1^e(x, y))$$

$$= \sum_{(x,y) \in dom(S_1^e)} \sum_{(p_u, q_v)} D_{p_u q_v}(S_1^e(x, y))$$

where (p_u, q_v) ranges over all pairs of paths $p_u : x \to w$ and $q_v : y \to z$ in C such that u is the unique node of $Max(proc_1)$ on p_u and v is the unique node of $Max(proc_1)$ on q_v. It follows that

$$\sum_{(u,v)\in dom(k^e)} E(u,v,w,z) = \sum_{(u,v)\in dom(k^e)} \left(\sum_{(x,y)\in dom(S_1^e)} \sum_{(p_u,q_v)} D_{p_u q_v}(S_1^e(x,y)) \right)$$

$$= \sum_{(x,y)\in dom(S_1^e)} \sum_{(p,q)} D_{pq}(S_1^e(x,y))$$

where (p,q) ranges over all pairs of paths $p : x \to w$ and $q : y \to z$ in C. Finally,

$$h^e(w,z) = \sum_{(x,y)\in dom(k^e \oplus S_2^e)} \sum_{(p,q)} D_{pq}((k^e \oplus S_2^e)(x,y))$$

where (p,q) ranges over all pairs of paths $p : x \to w$ and $q : y \to z$ in C_2. However, for $(x,y) \in dom(S_2^e)$, each path $p : x \to w$ in C is a path in C_2, and similarly for paths $q : y \to z$. Thus one has

$$h^e(w,z) = \sum_{(u,v)\in dom(k^e)} E(u,v,w,z) + \sum_{(x,y)\in dom(S_2^e)} \sum_{(p,q)} D_{pq}(S_2^e(x,y))$$

where (p,q) ranges over all pairs of paths $p : x \to w$ and $q : y \to z$ in C instead of C_2. Thus

$$h^e(w,z) = \sum_{(x,y)\in dom(S_1^e)} \sum_{(p,q)} D_{pq}(S_1^e(x,y)) + \sum_{(x,y)\in dom(S_2^e)} \sum_{(p,q)} D_{pq}(S_2^e(x,y))$$

$$= \sum_{(x,y)\in dom(S^e)} \sum_{(p,q)} D_{pq}(S^e(x,y)) = g^e(w,z)$$

5   Local Action Systems and Process Semantics

The aim of this section is to introduce the notion of a system, and the ordered process semantics of a system. A production is a pair consisting of an ordered process of a somewhat restricted form, and a graph on the minimal nodes of it. This graph serves as left–hand side of the production: an occurrence of the production is only allowed if the left–hand side occurs in the graph rewritten by that occurrence. A system is simply a set of productions.

Definition 10. A production is a pair $(proc, L)$ such that $proc = (V, C, D, \prec, S)$ is an ordered process, $(Min(proc), Max(proc))$ is a partition of V, $C = Min(proc) \times Max(proc)$, $dom(S^v) \subseteq Max(proc)$, $dom(S^e) \subseteq (Max(proc))^2$, and L is a graph on $Min(proc)$. A system is a set of productions.

For a production π, its components are denoted by $proc_\pi$ and L_π, respectively. The components of $proc_\pi$, in turn, are denoted by $V_\pi, C_\pi, D_\pi, \prec_\pi$ and S_π.

Example 5. Three productions π_1, π_2 and π_3 are depicted in Fig. 5. The dashed arrows represent the causality relation and the temporal orders are assumed to coincide with the lexicographic order. All components of the dynamic structures are assumed to be constant zero, except the following ones: $D^s_{\pi_1}(1,3) = t_{\alpha\beta}$, $D^s_{\pi_2}(1,2) = t_{\beta\beta}$, $D^s_{\pi_3}(2,3) = t_{\beta\alpha}$, $D^t_{\pi_1}(1,3) = D^t_{\pi_1}(2,4) = D^t_{\pi_3}(1,3) = 1$, $D^{st}_{\pi_1}(1,3) = t_{\alpha\beta}$.

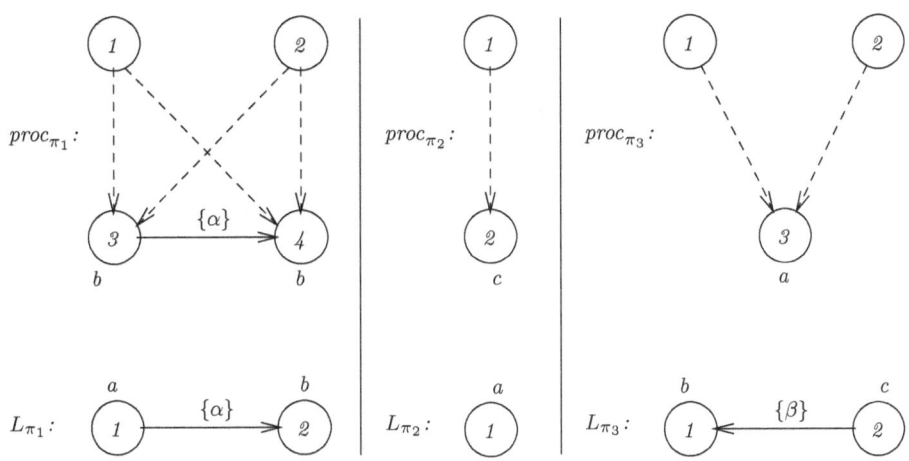

Fig. 5. The productions π_1, π_2 and π_3

The ordered processes corresponding to a given system are built by composing occurrences (isomorphic copies) of the processes $proc_\pi$ using sequential composition, and applying them to an initial graph. Each output node of a production occurrence is directly causally dependent on each input node, and hence each cut of a process built from production occurrences, using sequential composition, divides the production occurrences of that process into occurrences that causally precede the cut, and production occurrences that follow it. In a rewriting step new structure is only created on the newly introduced nodes, not on the ones that are rewritten. Note also that the notion of a system allows infinite sets to be viewed as systems. Writing down, or implementing, such an infinite system would of course require some form of parametrization, but that aspect is not treated in this paper. Formally the notion of a production occurrence is defined as follows.

Definition 11.

1. *Let* $proc_1 = (V_1, C_1, D_1, \prec_1, S_1)$ *and* $proc_2 = (V_2, C_2, D_2, \prec_2, S_2)$ *be ordered processes. An* isomorphism *from* $proc_1$ *into* $proc_2$ *is a bijective function* f :

$V_1 \rightarrow V_2$ such that $C_2 = f^{(2)}(C_1)$, $D_1 = D_2 \circ f^{(2)}$, $\prec_2 = f^{(2)}(\prec_1)$, $S_1^v = S_2^v \circ f$ and $S_1^e = S_2^e \circ f^{(2)}$.

2. Let π and ν be productions. An isomorphism from π into ν is an isomorphism f from $proc_\pi$ into $proc_\nu$ such that the restriction of f to $Min(proc_\pi)$ is an isomorphism from L_π into L_ν.

3. An occurrence of a production π is a production ν such that there exists an isomorphism from π into ν.

Observe that, in 2 of the above definition, one uses the fact that the graphs L_π and L_ν can be viewed as processes.

Next the semantics of a system is defined. It consists of ordered processes of the form $g; proc_1; \ldots; proc_n$, where the $proc_i$ are the first components of production occurrences. Thus the process $g; proc_1; \ldots; proc_n$ describes the application of a process that is built from production occurrences to an initial graph g. However, not each ordered process of this form can be considered as valid in the sense that it describes a possible rewriting history: the left–hand sides of the productions should also be taken into account. So consider a production occurrence ν_i in a process $proc$. Informally, the meaning of the second component L_{ν_i} is that the occurrence ν_i is only allowed if the graph L_{ν_i} occurs in the configuration that is reached just before the rewriting step described by ν_i. To formalize the notion that a graph "occurs in" another graph, the partial orders $\leq_v$ and $\leq_e$ on Δ^v and Δ^e are used to define a partial order on graphs in the following way.

Definition 12. Let g and h be graphs on V_g and V_h, respectively. Then $g \leq h$ if $V_g \subseteq V_h$, $dom(g^v) \subseteq dom(h^v)$, $dom(g^e) \subseteq dom(h^e)$ and, moreover, for each $x \in dom(g^v)$ and each $(x,y) \in dom(g^e)$, $g^v(x) \leq_v h^v(x)$ and $g^e(x,y) \leq_e h^e(x,y)$.

An ordered set of productions that can be composed sequentially is called an ordered covering; the notion is formally defined as follows.

Definition 13. Let Sys be a system. An ordered Sys–covering is an ordered set $\{\nu_1, \ldots, \nu_n\}$ of occurrences of productions from Sys such that, for each $1 \leq i < j \leq n$, $V_{\nu_i} \cap V_{\nu_j} \subseteq Max(proc_{\nu_i}) \cap Min(proc_{\nu_j})$.

Obviously, for an ordered Sys–covering $\{\nu_1, \ldots, \nu_n\}$ and $0 \leq i \leq n$, the ordered process $proc_{\nu_1}; \ldots; proc_{\nu_i}$ is defined (for the case $i = 0$ one gets the empty ordered process).

A rewriting history in which a sequence of productions is applied to a graph is described by an ordered process of the form $g; proc$, where g is the graph and $proc$ describes the production occurrences. It is assumed that $V_g \cap V_{proc} = Min(proc)$, i.e. g is the entire initial configuration of the rewriting history. $proc$ corresponds to an ordered covering $\{\nu_1, \ldots, \nu_n\}$. Moreover, in the rewriting history described by a process of the form $g; proc_{\nu_1}; \ldots; proc_{\nu_n}$, where $V_g = Min(g; proc_{\nu_1}; \ldots; proc_{\nu_n})$, the configuration reached by the time $proc_{\nu_i}$ is applied is the configuration that results from $g; proc_{\nu_1}; \ldots; proc_{\nu_{i-1}}$, and hence one has the following definition for the ordered process semantics of a system.

Definition 14. *Let Sys be a system. The set of ordered processes of Sys is the set of ordered processes proc such that there exists a graph g and an ordered Sys–covering $\{\nu_1, \ldots, \nu_n\}$ for which $V_g \cap V_{(proc_{\nu_1}; \ldots; proc_{\nu_n})} = Min(proc_{\nu_1}; \ldots; proc_{\nu_n})$, proc $= g; proc_{\nu_1}; \ldots; proc_{\nu_n}$, and, for each $1 \leq i \leq n$, $L_{\nu_i} \leq Res(g; proc_{\nu_1}; \ldots; proc_{\nu_{i-1}})$.*

For a production π, a π–step describes a rewriting history consisting of just one production application.

Definition 15. *For a production $\pi \in Sys$, a π–step is an ordered process of Sys such that the corresponding ordered covering is a singleton $\{\nu\}$, where ν is an occurrence of π.*

As a consequence of Theorem 1, for each ordered Sys–process proc $= g; proc_{\nu_1}; \ldots; proc_{\nu_n}$, one has $Res(proc) = Res(\ldots Res(Res(g; proc_{\nu_1}); proc_{\nu_2}); \ldots; proc_{\nu_n})$. Thus the result graph of proc can be obtained by a sequence of production applications.

Example 6. Let π_1, π_2, π_3 be the productions from Example 5, let $Sys = \{ \pi_1, \pi_2, \pi_3 \}$ and let g be the graph from Example 2. Then the ordered process proc from Example 3 is an ordered Sys–process: there exist occurrences ν_1, ν_2, ν_3 of π_1, π_2, π_3 (with node sets $\{1, 2, 4, 5\}$, $\{3, 6\}$ and $\{5, 6, 7\}$, respectively) such that proc $= g; proc_{\nu_1}; proc_{\nu_2}; proc_{\nu_3}$.

In the next section *unordered* processes are considered; i.e, processes without a temporal order. It is shown that in that case the coverings need not be ordered. For that reason it is useful to give an alternative formulation of the last condition of Definition 14. This characterization is based on the observation that one can, for an ordered process, construct the graph in which the effects of all the local actions are combined. The configurations $Res(g; proc_{\nu_1}; \ldots; proc_{\nu_{i-1}})$ are induced subgraphs of that graph. Formally one has the following.

Definition 16. *Let proc $= (V, C, D, \prec, S)$ be an ordered process. The computed structure of proc, denoted by $CS(proc)$, is the graph (V, F^v, F^e), where $dom(F^v) = C^*(dom(S^v))$, $dom(F^e) = (C^*)^{(2)}(dom(S^e))$ and, for each $x \in dom(F^v)$ and $(x, y) \in dom(F^e)$,*

$$F^v(x) = \sum_{z \in dom(S^v)} \sum_{p:z \to x} D_p(S^v(z))$$

$$F^e(x, y) = \sum_{(u,v) \in dom(S^e)} \sum_{\substack{p:u \to x, \\ q:v \to y}} D_{pq}(S^e(u, v)).$$

Let proc $= g; proc_{\nu_1}; \ldots; proc_{\nu_n}$ be an ordered process as in Definition 14, let $0 \leq i \leq n$ and let $proc_1 = g; proc_{\nu_1}; \ldots; proc_{\nu_i}$ ($proc_1 = g$ if $i = 0$). Since, for $j > i$, there are no nontrivial paths in C_{proc} from nodes of $proc_{\nu_j}$ into nodes of $Max(proc_1)$, and since the $proc_{\nu_j}$ have no structure on their minimal nodes, it is easily verified that the graph $Res(g; proc_{\nu_1}; \ldots; proc_{\nu_i})$ is the subgraph of $CS(proc)$ induced by $Max(proc_1)$. Thus the following is equivalent to the last condition of Definition 14: for each $1 \leq i \leq n$, $L_{\nu_i} \leq CS(proc)$.

6   NLC Grammars and Local Action Systems

In this section it is demonstrated that NLC grammars can be encoded into the LAS framework. The encoding is relatively simple, in spite of the fact that NLC graph rewriting is rather different from LAS graph rewriting: not only does the NLC approach deal with undirected graphs, but moreover their embedding mechanism is based on node labels, whereas in LAS graph rewriting edges and nodes are treated separately.

Consider a fixed NLC grammar G, where Σ is the alphabet of node labels and $Conn \subseteq (\Sigma)^2$ is the connection relation. In order to show that G can be encoded into the LAS framework, one has to provide the following.

- The monoids Δ^v, Δ^e, and the partial orders $\leq_v$ and $\leq_e$.
- An encoding cod, mapping each NLC graph g into a LAS graph $cod(g)$.
- An encoding of productions mapping each NLC production (A, r) into a LAS production $cod(A, r)$.

Then it has to be shown that these encodings are such that the steps of an NLC grammar correspond exactly to those of its encoding.

The node labels of the LAS system corresponding to G are essentially the node labels of G, i.e. the elements of Σ. One cannot simply have $\Delta^v = \Sigma$, because Σ is not a commutative monoid. However, one can consider Σ as part of the monoid $FCM(\Sigma)$, and define $\Delta^v = FCM(\Sigma)$. Since a production (A, r) can only be applied to an A–labeled node (i.e. the node labels have to match exactly), the partial order $\leq_v$ is chosen to be the equality.

The edge labels Δ^e are used to encode information about the nodes incident to an edge: each (directed) edge in the encoding of an NLC graph is labeled by a pair (a, b), where a is the label of its source and b is the label of its target. Thus one needs edge labels in $(\Sigma)^2$. To get the required monoid structure, one embeds $(\Sigma)^2$ into $FCM((\Sigma)^2)$. The partial order $\leq_e$ is again the equality.

An NLC graph $g = (V_g, E_g, lab_g)$ over Σ is encoded by the LAS graph $cod(g) = (g^v, g^e)$ on V_g defined by: $g^v = lab_g$ and, for each $(x, y) \in (V_g)^2$,

$$g^e(x, y) = \begin{cases} 0, & \text{if } \{x, y\} \notin Ed_g \\ (lab_g(x), lab_g(y)), & \text{if } \{x, y\} \in Ed_g. \end{cases}$$

For the encoding of NLC productions, the following notation is used. Let $a, b \in \Sigma$ and $H \subseteq \Sigma$. Then $[a/b, H]$ and $[H, a/b]$ denote the automorphisms on Δ^e determined by

$$[a/b, H](x, y) = \begin{cases} (a, y), & \text{if } x = b \text{ and } y \in H \\ 0, & \text{otherwise.} \end{cases}$$

$$[H, a/b](x, y) = \begin{cases} (x, a), & \text{if } x \in H \text{ and } y = b \\ 0, & \text{otherwise.} \end{cases}$$

For an NLC production $\pi = (A, r)$, $cod(\pi) = (proc_\pi, L_\pi)$, where V_π, L_π, C_π, D_π, $\prec_\pi$ and S_π are defined as follows. $V_\pi = V_r \cup \{x\}$, where $x \notin V_r$, $dom(L_\pi^v) = \{x\}$,

$dom(L_\pi^e) = \emptyset$, $L_\pi^v(x) = A$, $C_\pi = \{x\} \times V_r$, $\prec_\pi$ is an arbitrary order on C_π, S_π is such that $dom(S_\pi^v) = V_r$, $dom(S_\pi^e) = (V_r)^2$ and the subgraph of S_π induced by V_r is $cod(r)$, and, for each $(x, w) \in C_\pi$, $D_\pi^v(x, w) = D_\pi^{st}(x, w) = 0$,

$$D_\pi^s(x, w) = [lab_r(w)/A, H], \text{ and}$$
$$D_\pi^t(x, w) = [H, lab_r(w)/A],$$

where $H = \{a \in \Sigma \mid (a, lab_r(w)) \in Conn\}$.

Next it is shown that each rewriting step in an NLC grammar corresponds exactly to a step in its encoding, and vice versa.

Proposition 2. *Let g and h be NLC graphs, let (A, r) be an NLC production, let $f : r \to \overline{r}$ be an isomorphism such that $V_{\overline{r}} \cap V_g = \emptyset$, and let $x \in V_g$ be such that $lab_g(x) = A$. Let h be the result of the application of (A, r) to x in g via f. Let ν be the occurrence of $cod(A, r)$ such that $V_\nu = V_{\overline{r}} \cup \{x\}$, and such that for the corresponding isomorphism $\overline{f} : cod(A, r) \to \nu$, f is the restriction of $\overline{f}$ to V_r. Then $Res(cod(g); proc_\nu) = cod(h)$.*

Proof. Let $(V, C, D, \prec, S) = (cod(g); proc_\nu)$ and let $k = Res(cod(g); proc_\nu)$. One has to show that $k = cod(h)$. Since $C_\nu = \{x\} \times V_{\overline{r}}$, one has $V_k = Max(cod(g); proc_\nu) = (V_g \setminus \{x\}) \cup V_{\overline{r}} = V_h$, as desired. It also is easy to prove that $k^v = lab_h$: since $D^v(w, y) = 0$ for each $(w, y) \in C$, one has, for each $y \in V_h$,

$$k^v(y) = \sum_{w \in dom(S^v)} \sum_{p:w \to y} D_p(S^v(w)) = S^v(y)$$

$$= \begin{cases} lab_g(y), \text{ if } y \in V_g \setminus \{x\} \\ lab_{\overline{r}}(y), \text{ if } y \in V_{\overline{r}} \end{cases} = lab_h(y).$$

For the edges, consider 4 cases. Let $(y, w) \in (V_h)^2$.

1. If $y, w \in V_g \setminus \{x\}$, then there are only trivial paths to y and w in C. Hence

$$k^e(y, w) = \sum_{(u,v) \in dom(S^e)} \sum_{\substack{p:u \to y, \\ q:v \to w}} D_{pq}(S^e(u, v))$$

$$= S^e(y, w) = g^e(y, w) = h^e(y, w).$$

2. If $y, w \in V_{\overline{r}}$, then since $S^e(x, x) = 0$,

$$k^e(y, w) = \sum_{\substack{p:y \to y, \\ q:w \to w}} D_{pq}(S^e(y, w)) = S^e(y, w) = \overline{r}^e(y, w) = h^e(y, w).$$

3. If $y \in V_g \setminus \{x\}$ and $w \in V_{\overline{r}}$, then since $(y, w) \notin dom(S^e)$,

$$k^e(y, w) = \sum_{\substack{p:y \to y, \\ q:x \to w}} D_{pq}(S^e(y, x)) = D^t(x, w)(g^e(y, x)).$$

Hence, if $g^e(y,x) = 0$, then $k^e(y,w) = h^e(y,w) = 0$. On the other hand, if $g^e(y,x) \neq 0$, then $g^e(y,x) = (a,A)$, where $a = lab_g(y)$, and hence

$$k^e(y,w) = \begin{cases} (a, lab_{\overline{r}}(w)), & \text{if } (a, lab_{\overline{r}}(w)) \in Conn \\ 0, & \text{otherwise} \end{cases} = h^e(y,w).$$

4. The case where $y \in V_{\overline{r}}$ and $w \in V_g \setminus \{x\}$ is analogous to case 3.

Theorem 2. *Let g and h be NLC graphs. Let (A,r) be an NLC production. Then the application of (A,r) to a node x in g via an isomorphism $f : r \to \overline{r}$ yields h if and only if there exists a $cod(A,r)$-step proc such that $Init(proc) = cod(g)$ and $Res(proc) = cod(h)$.*

Proof. Let h be the result of the application of (A,r) to a node x in g via an isomorphism $f : r \to \overline{r}$. Let ν be the occurrence of $cod(A,r)$ such that $V_\nu = V_{\overline{r}} \cup \{x\}$, and such that for the corresponding isomorphism $\overline{f} : cod(A,r) \to \nu$, f is the restriction of $\overline{f}$ to V_r. Obviously $lab_g(x) = A$ and $(cod(g); proc_\nu)$ is a $cod(A,r)$-step. One has $Init(cod(g); proc_\nu) = cod(g)$ and, by Proposition 2, $Res(cod(g); proc_\nu) = cod(h)$.

Conversely, let proc be a $cod(A,r)$-step such that $Init(proc) = cod(g)$ and $Res(proc) = cod(h)$. Thus $proc = g'; proc_\nu$ for some graph g' and some occurrence ν of $cod(A,r)$. Since $Init(proc) = g'$, one has $g' = cod(g)$. Moreover, since proc is a $cod(A,r)$-step, $lab_g(x) = g^\nu(x) = A$, where $x = V_g \cap V_\nu$. Let $\overline{f} : cod(A,r) \to \nu$ be the isomorphism corresponding to the production occurrence ν, and let $f : r \to \overline{r}$ be the restriction of $\overline{f}$ to V_r. Then it follows from Proposition 2 that $Res(proc) = cod(h)$, where h is the result of the application of (A,r) to x in g via f.

The encoded version of the NLC rewriting step of Fig. 1 is depicted in Fig. 6, where an edge label (a,b) is denoted by ab.

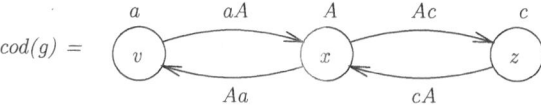

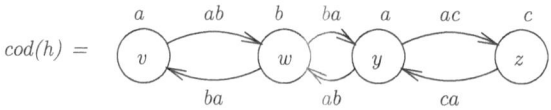

Fig. 6. The encoding of the NLC graphs g and h

7   The Unordered Case and ESM Systems

The notion of an ordered process, investigated in the previous sections, has the major disadvantage that it does not allow an obvious way to define the concurrent (as opposed to sequential) composition of processes. The difficulty is caused by the temporal order $\prec$: it is not clear how the temporal orders of two ordered processes can be combined in such a way that the result can be viewed as the concurrent composition of those processes. To put things differently, the combined effect of the local actions of a process should be determined by the causal order alone. In this section it is investigated how the framework developed in Section 4 and 5 can be adapted so that the temporal order is not needed. As an example of the unordered case, ESM systems are considered.

7.1   Unordered Processes

In this subsection it is shown that the temporal order can be omitted from the ordered processes under certain conditions and it is demonstrated what is the effect of this change on the framework developed in Section 4 and 5. Unordered processes are defined as follows.

Definition 17. *An* unordered process *is a 4–tuple* (V, C, D, S)*, where* V *is a finite set,* $C \subseteq V^2$ *is an acyclic relation,* D *is a dynamic structure on* V *such that* $C = dom(D)$*, and* S *is a graph on* V*.*

The composition of unordered processes is a commutative operation; it is not required that one process precedes the other.

Definition 18. *Let* $proc_1 = (V_1, C_1, D_1, S_1)$ *and* $proc_2 = (V_2, C_2, D_2, S_2)$ *be unordered processes such that* $V_1 \cap V_2 \subseteq (Max(proc_1) \cap Min(proc_2)) \cup (Min(proc_1) \cap Max(proc_2))$*, and* $C_1 \cup C_2$ *is acyclic. The* composition *of* $proc_1$ *and* $proc_2$*, denoted by* $proc_1 \oplus proc_2$*, is the process* (V, C, D, S) *where* $V = V_1 \cup V_2$*,* $C = C_1 \cup C_2$*,* $D = D_1 \oplus D_2$*,* $S^v = S_1^v \oplus S_2^v$*, and* $S^e = S_1^e \oplus S_2^e$*.*

The sequential composition now appears as a special case of $\oplus$: the case where $V_1 \cap V_2 \subseteq (Max(proc_1) \cap Min(proc_2))$.

An apparent problem arises in defining the result graph of a process; in the definition of D_{pq} (Definition 6), the temporal order $\prec$ is used to determine the relative order of the local actions on the paths p and q. There is, however, a situation where the relative order does not matter: assume that the actions of the path p (the source side) belong to a subset T^s of T^e, that those of the path q (the target side) belong to a subset T^t of T^e, and that the elements of T^s and T^t commute; i.e, for each $a \in T^s$, $b \in T^t$, $\langle a \cdot b \rangle = \langle b \cdot a \rangle$. Moreover, assume that one uses only dynamic structures D such that $D^{st}(e) = \langle D^s(e) \cdot D^t(e) \rangle$ for each $e \in dom(D)$. Then the second part of Definition 6 can be replaced by:

$$D_p^s(d) = \begin{cases} 1_{T^s}, & \text{if } length(p) = 0 \\ \langle D_{p_1}^s \cdot D^s(d) \rangle, & \text{if } p = p_1 \odot d \end{cases},$$

$$D_p^t(d) = \begin{cases} 1_{T^t}, & \text{if } length(p) = 0 \\ \langle D_{p_1}^s \cdot D^t(d) \rangle, & \text{if } p = p_1 \odot d \end{cases},$$

$$D_{pq} = \langle D_p^s \cdot D_q^t \rangle.$$

In this restricted case D is a partial function into $T^v \times T^s \times T^t$, instead of $T^v \times (T^e)^3$, because there is no need to specify D^{st}. The definitions of productions and systems need not to be changed (except, of course, that one uses unordered processes). The process semantics can be defined using unordered coverings.

Definition 19. *Let Sys be a system. A Sys–covering is a set cov of occurrences of productions from Sys such that, for each μ, $\nu \in cov$, $\mu \neq \nu$ implies $V_\mu \cap V_\nu \subseteq (Max(proc_\mu) \cap Min(proc_\nu)) \cup (Min(proc_\mu) \cap Max(proc_\nu))$, and $\bigcup_{\nu \in cov} C_\nu$ is acyclic.*

Definition 20. *Let Sys be a system. The set of unordered processes of Sys is the set of unordered processes proc such that there exists a graph g and a Sys–covering cov for which $Min(proc) = V_g$, $proc = g \oplus (\bigoplus_{\nu \in cov} proc_\nu)$, and for each $\nu \in cov$, $L_\nu \leq CS(proc)$.*

7.2   ESM Systems

Throughout this subsection, let Σ_1 and Σ_2 be fixed but arbitrary alphabets. The elements of Σ_1 and Σ_2 are used as node labels and edge labels, respectively.

ESM graphs and systems are encoded as follows. The node alphabet Σ_1 has no particular algebraic structure, and is taken over in the LAS framework in the same way as is the case for NLC grammars: Σ_1 is identified with its image in $\Delta^v = FCM(\Sigma_1)$ and $\leq_v$ is the equality in Δ^v. The edge alphabet of the LAS encoding is the powerset of Σ_2; this is a commutative monoid if the set union serves as $+$, and the ESM notion of "g occurs in h" is captured by using the set inclusion $\subseteq$ for $\leq_e$.

An ESM graph $g = (V_g, Ed_g, lab_g)$ is encoded by the LAS graph $cod(g) = (g^v, g^e)$ on V_g, where $g^v = lab_g$ and g^e is defined by $g^e(x, y) = \{\alpha \mid (x, \alpha, y) \in Ed_g\}$, for each $(x, y) \in (V_g)^2$.

For an ESM production $\pi = (l, r, in, out)$, $cod(\pi) = (proc_\pi, L_\pi)$, where $L_\pi = cod(l)$, $V_\pi = V_l \cup V_r$ (assume $V_l \cap V_r = \emptyset$), $C_\pi = V_l \times V_r$, and S_π is such that $dom(S_\pi^v) = V_r$, $dom(S_\pi^e) = (V_r)^2$ and the subgraph of S_π induced by V_r is $cod(r)$. D_π is defined by $D_\pi^v(x, w) = 0$,

$$D_\pi^s(x, w) = \sum_{((x,\alpha),(w,\beta)) \in out} t_{\alpha\beta}, \text{ and}$$

$$D_\pi^t(x, w) = \begin{cases} 1, \text{ if } (x, w) \in in \\ 0, \text{ otherwise} \end{cases},$$

where $t_{\alpha\beta}$, for each α, $\beta \in \Sigma_2$ is defined as in Example 3.

The production π_1 of Example 5 is the encoding of the ESM production (l, r, in, out), where l and r are the graphs depicted in Fig. 7, $in = \{(1, 3), (2, 4)\}$, and $out = \{((1, \alpha), (3, \beta))\}$. Since the only automorphisms that occur in the definition of $D_{\pi_1}^t$ are 1 and 0, one may choose $T^t = \{0, 1\}$ and clearly T^t commutes with any other set of automorphisms on Δ^e.

Fig. 7. The left–hand side and the right–hand side of an ESM production

Proposition 3. *Let g and h be ESM graphs and let $\pi = (l, r, in, out)$ be an ESM production, let $f_l : l \to \bar{l}$ and $f_r : r \to \bar{r}$ be isomorphisms such that $\bar{l}$ is a subgraph of g and $V_{\bar{r}} \cap V_g = \emptyset$. Let h be the result of the application of π to $\bar{l}$ in g via f_l and f_r. Let ν be the occurrence of π such that $V_\nu = V_{\bar{l}} \cup V_{\bar{r}}$, and such that for the corresponding isomorphism $\bar{f} : \pi \to \nu$, f_l is the restriction of $\bar{f}$ to V_l, and f_r is the restriction of $\bar{f}$ to V_r. Then $Res(cod(g); proc_\nu) = cod(h)$.*

Proof. Let $(V, C, D, \prec, S) = (cod(g); proc_\nu)$ and let $k = Res(cod(g); proc_\nu)$. One has to show that $k = cod(h)$.

Since $C_\nu = V_{\bar{l}} \times V_{\bar{r}}$, one has $V_k = Max(cod(g); proc_\nu) = (V_g \setminus V_{\bar{l}}) \cup V_{\bar{r}} = V_h$, as desired. Since the D^v are zero, it is easily seen that $k^v = lab_h$ (see also the proof of Proposition 2). This implies that there exists an ESM graph k' such that $k = cod(k')$: let $V_{k'} = V_k$, $lab_{k'} = k^v$ and for each $(y, w) \in (V_h)^2$, $(y, \beta, w) \in Ed_{k'}$ if and only if $\beta \in k^e(y, w)$. It remains to show that $Ed_{k'} = Ed_h$.

Let $(y, w) \in (V_h)^2$ and $\beta \in \Sigma_2$. Then $(y, \beta, w) \in Ed_{k'}$ if and only if there exists a pair $(u, v) \in dom(S^e)$ and a label $\alpha \in S^e(u, v)$ such that $\beta \in \bigcup_{\substack{p:u \to y, \\ q:v \to w}} D_{pq}(\alpha)$.

Since $S^e = g^e \oplus \bar{r}^e$ and $V_{\bar{r}} \subseteq Max(cod(g); proc_\nu)$, $(y, \beta, w) \in Ed_{k'}$ if and only if either (i) $(y, \beta, w) \in Ed_{\bar{r}}$, or (ii) there exists a pair $(u, v) \in dom(g^e)$ and a label $\alpha \in g^e(u, v)$ such that $\beta \in \bigcup_{\substack{p:u \to y, \\ q:v \to w}} D_{pq}(\alpha)$. Since there is at most one path from u to y and from v to w, respectively, (ii) is equivalent to the following: there exists an edge $(u, \alpha, v) \in Ed_g$ such that $\beta \in D_{(u,y)(v,w)}(\alpha)$. However it follows from the definition of $D_{cod(\pi)}$ that $\beta \in D_{(u,y)(v,w)}(\alpha)$ if and only if $(y, \beta, w) \in E(u, \alpha, v)$ (the notation $E(u, \alpha, v)$ is introduced in Definition 2). Thus one concludes that $(y, \beta, w) \in Ed_{k'}$ if and only if either (i) $(y, \beta, w) \in Ed_{\bar{r}}$, or (ii) $(y, \beta, w) \in \bigcup_{(u,\alpha,v) \in Ed_g} E(u, \alpha, v)$, i.e. $(y, \beta, w) \in Ed_{k'}$ if and only if $(y, \beta, w) \in Ed_h$. This concludes the proof.

Theorem 3. *Let g and h be ESM graphs. Let $\pi = (l, r, in, out)$ be an ESM production. Then the application of π to a subgraph $\bar{l}$ in g via isomorphisms $f_l : l \to \bar{l}$ and $f_r : r \to \bar{r}$ yields h if and only if there exists a $cod(\pi)$–step proc such that $Init(proc) = cod(g)$ and $Res(proc) = cod(h)$.*

Proof. Let h be the result of the application of π to a subgraph $\bar{l}$ in g via isomorphisms $f_l : l \to \bar{l}$ and $f_r : r \to \bar{r}$. Let ν be the occurrence of $cod(\pi)$ such that $V_\nu = V_{\bar{l}} \cup V_{\bar{r}}$, and such that for the corresponding isomorphism $\bar{f} :$

$cod(A, r) \rightarrow \nu$, f_l is the restriction of $\overline{f}$ to V_l and f_r is the restriction of $\overline{f}$ to V_r. Then $cod(\overline{l}) \leq cod(g) \leq CS(cod(g); proc_\nu)$ and thus $(cod(g); proc_\nu)$ is a $cod(\pi)$–step. One has $Init(cod(g); proc_\nu) = cod(g)$ and, by Proposition 3, $Res(cod(g); proc_\nu) = cod(h)$.

Conversely, let $proc$ be a $cod(\pi)$–step such that $Init(proc) = cod(g)$ and $Res(proc) = cod(h)$. Thus $proc = g'; proc_\nu$ for some graph g' and some occurrence ν of $cod(\pi)$. Since $Init(proc) = g'$, one has $g' = cod(g)$. Moreover, since $proc$ is a $cod(\pi)$–step, $cod(\overline{l}) \leq CS(proc)$, where $V_{\overline{l}} = V_g \cap V_\nu$. Since all nodes in $cod(\overline{l})$ are minimal in $proc$, $cod(\overline{l}) \leq cod(g)$ or $\overline{l}$ is an ESM subgraph of g. Let $\overline{f} : cod(\pi) \rightarrow \nu$ be the isomorphism corresponding to the production occurrence ν, let $f_l : l \rightarrow \overline{l}$ be the restriction of $\overline{f}$ to V_l, and let $f_r : r \rightarrow \overline{r}$ be the restriction of $\overline{f}$ to V_r. Then it follows from Proposition 3 that $Res(proc) = cod(h)$, where h is the result of the application of π to $\overline{l}$ in g via f_l and f_r.

Conclusion

It has been shown that LAS rewriting provides a unifying framework for NLC and ESM graph rewriting. The possibility to make other choices for the monoids of labels and the automorphisms suggests that many concrete models (e.g. Petri nets) fit into it. The process notion used is fairly simple, and the characterization of the unordered case by a commutativity property of the sets of automorphisms may help to clarify whether a given type of graph rewriting is suitable for describing concurrency or not. It may also be useful to compare the approach with other approaches to true concurrency, such as Pomsets [7] or Event Structures [8], where one usually represents an action by a point, rather than by a (labeled) edge of the causality relation, as is the case in LAS processes.

References

1. A. Corradini, U.Montanari and F.Rossi, Graph Processes, *Fundamenta Informaticae*, **26** (1996), 241-265. 195
2. J. Engelfriet and G. Rozenberg, Node Replacement Graph Grammars, in *Handbook of Graph Grammars and Computing by Graph Transformation*, **Vol. 1**, World Scientific (1997), 1-94. 194
3. D. Janssens, Actor Grammars and Local Actions, in *Handbook of Graph Grammars and Computing by Graph Transformation*, **Vol. 3**, World Scientific (1999). 194, 199
4. D. Janssens, ESM Systems and the Composition of Their Computations, in *Graph Transformations in Computer Science*, Lecture Notes in Computer Science, **Vol. 776**, Springer-Verlag, Berlin, 1994, 203-217. 194
5. D. Janssens and T. Mens, Abstract semantics for ESM systems, *Fundamenta Informaticae*, **26** (1996), 315-339. 194
6. D. Janssens and G. Rozenberg, On the Structure of Node Label Controlled Graph Languages, Information Sciences **20** (1980), 191-216. 194
7. V.R. Pratt, Modelling Concurrency with Partial Orders, *J. of Parallel Programming*, **15** (1987), 33-71. 214

8. G. Winskel, Events in Computation, Ph. D. Thesis, Comp. Sci. Dept., University of Edinburgh (1980).   214

Redundancy and Subsumption in High-Level Replacement Systems[*]

Hans-Jörg Kreowski[1] and Gabriel Valiente[2]

[1] Mathematics and Computer Science Department, University of Bremen, Germany
`kreo@informatik.uni-bremen.de`
[2] Department of Software, Technical University of Catalonia, Spain
`valiente@lsi.upc.es`

Abstract. System verification in the broadest sense deals with those semantic properties that can be decided or deduced by analyzing a syntactical description of the system. Hence, one may consider the notions of redundancy and subsumption in this context as they are known from the area of rule-based systems. A rule is redundant if it can be removed without affecting the semantics of the system; it is subsumed by another rule if each application of the former one can be replaced by an application of the latter one with the same effect. In this paper, redundancy and subsumption are carried over from rule-based systems to high-level replacement systems, which in turn generalize graph and hypergraph grammars. The main results presented in this paper are a characterization of subsumption and a sufficient condition for redundancy, which involves composite productions.

1 Introduction

High-level replacement systems [6] generalize the algebraic approach to graph transformation, both the double-pushout approach [3] and the single-pushout approach [5], to other classes of replacement systems. They provide a common categorical framework for different classes of replacement systems, such as grammars on graphs, relational structures, and algebraic specifications, based on categories and pushouts.

This paper deals with aspects of verification of both double-pushout (DPO) and single-pushout (SPO) high-level replacement systems. System verification in the broadest sense is concerned with those semantic properties that can be decided or deduced by analyzing a syntactical description of the system. The properties studied in this paper are redundancy and subsumption, as they are known from the area of rule-based systems (see, for instance, [2], [11], [12]).

[*] Partially supported by the EC TMR Network GETGRATS (General Theory of Graph Transformation Systems) and by the ESPRIT Working Group APPLIGRAPH (Applications of Graph Transformation) through the University of Bremen and by the Spanish DGES project PB96-0191-C02-02 and CICYT project TIC98-0949-C02-01 HEMOSS.

H. Ehrig et al. (Eds.): Graph Transformation, LNCS 1764, pp. 215–227, 2000.

Consider a high-level replacement system, that is, a set of productions, an initial object, and a class of terminal objects. A production is subsumed by another if any application of the former is mimicked by the latter. As a first result, we present a sufficient condition for subsumption. If a production p' is covered by a production p, that is, p' is directly derived by p, then p' is also subsumed by p. This is interesting because covering is easier to check than subsumption, since it is defined by a single direct derivation. It turns out that covering is not only sufficient, but also necessary for subsumption in the case of single-pushout high-level replacement systems, whereas this is not true in the double-pushout case. Moreover, we consider subsumption of a production by composite productions. Even in this case, one can show that the subsumed production is redundant, that is, the semantics of the given system does not change if the production is removed. Altogether, one obtains a procedure that removes some redundancy from a given system: Enumerate composite productions, check them for covering, and remove every covered production.

2   High-Level Replacement Systems

In this section, we recall the basic notions and notations of high-level replacement systems the rewriting of which is based on both double-pushout (DPO) and single-pushout (SPO) constructions.

Starting with the DPO case, let C be a category, whose objects will be regarded as high-level structures and whose morphisms will be regarded as structure-preserving mappings between these objects. The morphisms in a distinguished class $\mathcal{M}$ will be used in productions, while general morphisms in C will be used to define application of productions to high-level structures.

Definition 1 (DPO HLR system). *Let C be a category with a distinguished class $\mathcal{M}$ of morphisms.*

1. *A DPO production $p = (L \leftarrow K \rightarrow R)$ in C consists of a pair of objects (L, R), called* left-hand side object *and* right-hand side object *respectively, an object K, called* interface object, *and two morphisms $K \rightarrow L$ and $K \rightarrow R$ belonging to $\mathcal{M}$.*

2. *An object G can be directly derived into an object H using a DPO production $p = (L \leftarrow K \rightarrow R)$, denoted by $G \Rightarrow H$ via p if there are pushout squares*

$$
\begin{array}{ccccc}
L & \longleftarrow & K & \longrightarrow & R \\
\downarrow & & \downarrow & & \downarrow \\
G & \longleftarrow & D & \longrightarrow & H
\end{array}
$$

in C.

3. *Let $\boldsymbol{P}$ be a set of DPO productions. A derivation $G \Rightarrow^* H$ from G to H in $\boldsymbol{P}$ is a sequence of $n \geqslant 0$ direct derivations $G = G_0 \Rightarrow G_1 \Rightarrow \cdots \Rightarrow G_n = H$ via $(p_1, \ldots, p_n)$ provided that $p_1, \ldots, p_n \in \boldsymbol{P}$.*

4. A high-level replacement system $H = (S, \boldsymbol{P}, \boldsymbol{T})$ in $\mathcal{C}$ consists of a start object S in $\mathcal{C}$, a set $\boldsymbol{P}$ of DPO productions, and a class $\boldsymbol{T}$ of terminal objects in $\mathcal{C}$.

5. The language $L(H)$ of a high-level replacement system $H = (S, \boldsymbol{P}, \boldsymbol{T})$, also denoted by $L(S, \boldsymbol{P}, \boldsymbol{T})$, is given by the set of all terminal objects in $\mathcal{C}$ derivable from S by $\boldsymbol{P}$, that is, $L(H) = \{G \in \boldsymbol{T} \mid S \Rightarrow^* G\}$. □

In a part of the literature, less general high-level replacement systems are considered where all objects are accepted as terminal objects and the generated languages are the exhaustive ones. Our following considerations work for the more general case in the same way as for the special case.

The following example presents high-level replacement system (which is actually a graph grammar) that allows the generation and recognition of all Eulerian graphs, based on [13, Sect. 3.2]. Recall that a graph is Eulerian if it has an Euler circuit, that is, a circuit that contains every edge of the graph exactly once. Eulerian graphs are characterized by being connected and having only nodes of even degree [1].

Example 2. Let $\mathcal{C}$ be the category of undirected graphs and $\mathcal{M}$ be the class of all injective graph morphisms. Consider the following double-pushout high-level replacement system $(S, \boldsymbol{P}, \boldsymbol{T})$ for generating and recognizing all Eulerian graphs, where S is a graph with a single node and no edge, $\boldsymbol{P} = \{p_1, \ldots, p_9\}$, and $\boldsymbol{T}$ is the class of all graphs.

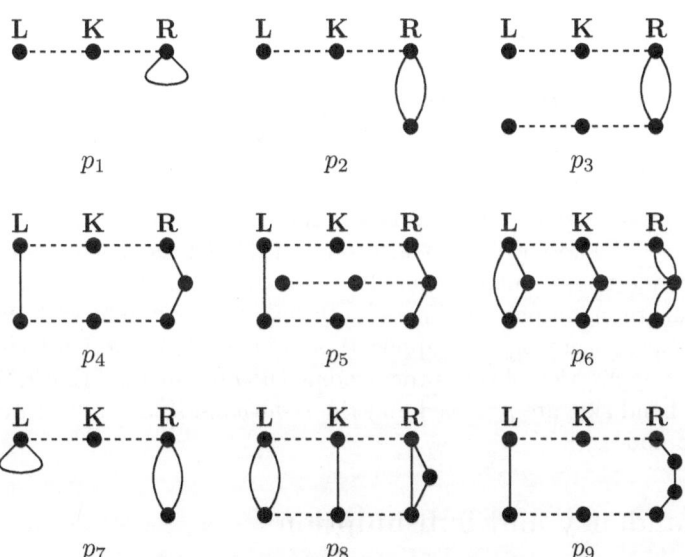

The dotted lines indicate the morphisms from the interface object to the left-hand side object and to the right-hand side object, respectively. □

The SPO case differs from the DPO case in two aspects: A production consists of a single morphism and a direct derivation of a single pushout. In typical examples like graphs, the morphism of a production may be a partial mapping (where nodes and edges outside the domain of definition specify the items to be removed) while the occurrence of the left-hand side in the host graph should be a total mapping. To formalize such situations, a category and a subcategory are assumed.

Definition 3 (SPO HLR system). *Let C be a category and let O be a subcategory of C such that O coincides with C on objects.*

1. *An* SPO production $p = (L \to R)$ *in C consists of a pair of objects (L, R), called* left-hand side object *and* right-hand side object *respectively, and a morphism $L \to R$ in C.*
2. *An object G can be* directly derived *into an object H using an SPO production $p = (L \to R)$, denoted by $G \Rightarrow H$ via p if there is a pushout square*

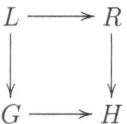

   in C such that $L \to G$ is in O.
3. *Let $\boldsymbol{P}$ be a set of SPO productions. A* derivation $G \Rightarrow^* H$ *from G to H in $\boldsymbol{P}$ is a sequence of $n \geqslant 0$ direct derivations $G = G_0 \Rightarrow G_1 \Rightarrow \cdots \Rightarrow G_n = H$ via $(p_1, \ldots, p_n)$ provided that $p_1, \ldots, p_n \in \boldsymbol{P}$.*
4. *A* high-level replacement system $H = (S, \boldsymbol{P}, \boldsymbol{T})$ *in C consists of a* start object S *in C, a set $\boldsymbol{P}$ of SPO productions, and a class $\boldsymbol{T}$ of* terminal objects *in C.*
5. *The* language $L(H)$ *of a high-level replacement system $H = (S, \boldsymbol{P}, \boldsymbol{T})$, also denoted by $L(S, \boldsymbol{P}, \boldsymbol{T})$, is given by the set of all terminal objects in C derivable from S by $\boldsymbol{P}$, that is, $L(H) = \{G \in \boldsymbol{T} \mid S \Rightarrow^* G\}$.* □

Example 4. If one ignores the interface graphs in the productions of Example 2 and interprets the dotted lines as inclusions of the left-hand side nodes into the right-hand side nodes, one obtains SPO productions in the category of graphs with partial graph morphisms. Choosing the category of graphs with total graph morphisms as subcategory, the application of an SPO production has the same effect as the application of the corresponding DPO production, that is, the edges of the left-hand side are removed and the right-hand side is added by merging the related nodes. □

3   Redundancy and Subsumption

Verification of high-level replacement systems is concerned with formal properties of the systems. Some of the formal properties to be verified arise from the rule-based paradigm itself; cf. [13]. In particular, redundancy and subsumption are generalized in this section from rule-based systems to high-level replacement systems.

A high-level replacement system is redundant if it contains a production that can be removed without affecting the semantics of the system. In particular, such productions may be subsumed by other productions. A production subsumes (or is more general than) another production if the subsuming production can be applied and it yields the same result whenever the subsumed production can be applied.

Definition 5 (Redundancy and Subsumption). *Let P be a set of productions (either of the DPO or SPO type).*

1. *A production $q \in P$ is* redundant *if there is a derivation $G \Rightarrow^* H$ in $P - \{q\}$ whenever there is a derivation $G \Rightarrow^* H$ in P.*
2. *A production $p \in P$* subsumes *a production $q \in P$, denoted by $p \leqslant q$, if there is a direct derivation $G \Rightarrow H$ via p whenever there is a direct derivation $G \Rightarrow H$ via q.* □

Obviously, a production q is redundant if it is subsumed by a production p. Moreover, a high-level replacement system $H = (S, \mathbf{P}, \mathbf{T})$ for some start object S and class of terminal objects $\mathbf{T}$ generates the same language as $H - q = (S, \mathbf{P} - \{q\}, \mathbf{T})$ if q is redundant or, in particular, if q is subsumed by some $p \in \mathbf{P} - \{q\}$. The other way round, a production $q \in \mathbf{P}$ is redundant if $L(S, \mathbf{P}, \mathbf{T}) = L(S, \mathbf{P} - \{q\}, \mathbf{T})$ for any start object S and any class $\mathbf{T}$ of terminal objects.

As the generated language is preserved by the removal of redundant productions, the same would be true for the generated binary relation between initial and terminal objects if we would replace the start object by a class of initial objects in the definition of high-level replacement systems. But semantic invariance could not be expected if we would consider the derivation process or coresponding transition and transformation systems as semantics (cf. [7]), because they decrease whenever applicable productions are removed.

Example 6. Productions p_6 to p_9 in Example 2 are redundant. Production p_6 is subsumed by production p_5, and productions p_7 and p_8 are subsumed by production p_4. Production p_9 is discussed in Example 16. □

Redundancy and subsumption may be undesirable for several reasons. First, the presence of redundant and subsumed productions may degrade execution efficiency. Second, and most important, it may make system validation more difficult.

4   A Characterization of Subsumption

Subsumption can be regarded as a *local* form of redundancy, since it only involves two productions: the subsuming production and the subsumed production. Nevertheless, subsumption is difficult to check because the definition involves arbitrary objects to be derived by the involved productions. To avoid this obstacle, we introduce the notion of a covering, which relates two productions by means of a single direct derivation and is, therefore, much easier to check. Coverings are also called *strict production morphisms* in [9]. Using the sequential composition of pushouts, covering implies obviously subsumption. Moreover, it turns out that covering and subsumption are equivalent notions in the SPO case, but not in the DPO case.

Definition 7 (Covering in DPO HLR systems). *Let $p = (L \leftarrow K \rightarrow R)$ and $p' = (L' \leftarrow K' \rightarrow R')$ be two productions. Then p covers p' if there exist morphisms $L \rightarrow L'$, $K \rightarrow K'$ and $R \rightarrow R'$ such that $L \rightarrow L' \leftarrow K'$ is a pushout of $L \leftarrow K \rightarrow K'$ and $K' \rightarrow R' \leftarrow R$ is a pushout of $K' \leftarrow K \rightarrow R$.*

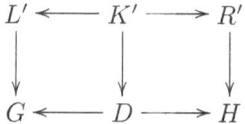

$\square$

Observation 8. *Let $p = (L \leftarrow K \rightarrow R)$ and $p' = (L' \leftarrow K' \rightarrow R')$ be two productions. Then $p \leqslant p'$ if p covers p'.* $\square$

Proof. Consider a direct derivation $G \Rightarrow H$ via p' of the form

$$L' \longleftarrow K' \longrightarrow R'$$
$$\downarrow \qquad \downarrow \qquad \downarrow$$
$$G \longleftarrow D \longrightarrow H$$

By hypothesis, there exist pushout squares (1) and (2), and by (vertical) composition of pushout squares, there also exist pushout squares (3) and (4), that is, there exists a direct derivation $G \Rightarrow H$ via p.

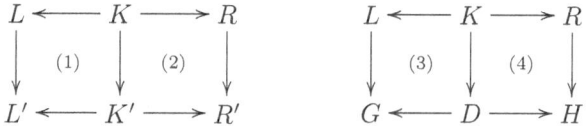

$\square$

Observation 9. *There are DPO productions p and p' such that $p \leqslant p'$, but p does not cover p'.* $\square$

Proof. Consider the following counter-example in the category **Gra** of graphs and graph morphisms. Let $p = (\bullet \leftarrow \bullet \rightarrow \bullet)$ and $p' = (\bullet \leftarrow \emptyset \rightarrow \bullet)$ be two productions given by identities and by empty morphisms, respectively.

There exists a derivation $G \Rightarrow H$ via p' if and only if the left-hand side node is mapped to an isolated node of G. Moreover, in this case, we have $H = G$ because the isolated node is removed and added again. Using the same occurrence, we obtain a derivation $G \Rightarrow G$ via p. In other words, $p \leqslant p'$. However, p does not cover p' because there is no morphism $\bullet \to \emptyset$ in **Gra**, and, therefore, no double pushout of the form

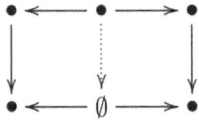

□

In the case of single-pushout high-level replacement systems, however, subsumption is characterized by covering.

Definition 10 (Covering in SPO HLR systems). *Production $p = (L \to R)$ covers production $p' = (L' \to R')$ if there exist morphisms $L \to L'$ in $\mathcal{O}$ and $R \to R'$ such that $L' \to R' \leftarrow R$ is a pushout of $L' \leftarrow L \to R$.*

$$
\begin{array}{ccc}
L & \longrightarrow & R \\
\vdots & & \vdots \\
 & (PO) & \\
\downarrow & & \downarrow \\
L' & \longrightarrow & R'
\end{array}
$$

□

Theorem 11. *Let $p = (L \to R)$ and $p' = (L' \to R')$ be two productions. Then $p \leqslant p'$ if and only if p covers p'.* □

Proof. (If part) The proof is the same as for Observation 8 for single pushout squares if one takes into account that the morphisms of $\mathcal{O}$ are closed under composition.

(Only-if part) Since the hypothesis holds for any direct derivation through production p', in particular it holds for the direct derivation $L' \Rightarrow R'$ via p', that is, for the direct derivation given by pushout square (1) where the vertical morphisms are the identities. Then, there also exists a direct derivation $L' \Rightarrow R'$ via p, that is, there exist morphisms $L \to L'$ and $R \to R'$ such that square (2) is a pushout square. Then p covers p'.

$$
\begin{array}{ccccccc}
L' & \longrightarrow & R' & \qquad & L & \longrightarrow & R \\
\downarrow & (1) & \downarrow & & \downarrow & (2) & \downarrow \\
L' & \longrightarrow & R' & & L' & \longrightarrow & R'
\end{array}
$$

□

Contrary to the case of (linear) rule-based systems, however, mutual subsumption does not mean isomorphic productions.

Observation 12. $p \leqslant p'$ and $p' \leqslant p$ does not imply $p = p'$. □

Proof. Consider the following counter-example in the category **Set** of sets and functions. Let $A = \{a\}$ and $B = \{b, c\}$ be two sets, and let $p = (A, A, A)$ and $p' = (B, B, B)$ be two double-pushout productions given by identities.

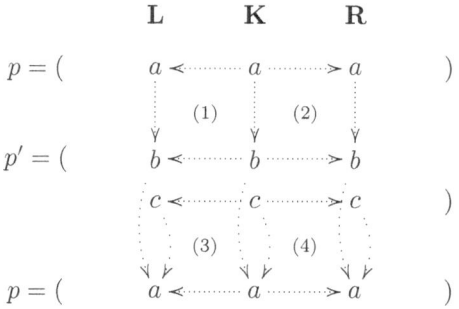

There exist morphisms $L \to L'$, $K \to K'$ and $R \to R'$ given by $a \mapsto b$ such that (1) and (2) become pushout squares, and there exist morphisms $L' \to L$, $K' \to K$ and $R' \to R$ given by $b \mapsto a, c \mapsto a$ such that (3) and (4) also become pushout squares. However, p and p' are not isomorphic productions.

A similar counter-example applies to single-pushout productions.   □

5   A Sufficient Condition for Redundancy

While subsumption between productions of a high-level replacement system can be regarded as a local form of redundancy, one obtains more global forms of redundancy by the combination and composition of several production to subsume other productions.

Actually, the most general notion of composition is given by the construction of concurrent productions, which consists of the composition of two productions over a dependency relation D between the right-hand side object of the first production and the left-hand side object of the second production. The resulting composite production is called D-concurrent production. We recall the notion for double-pushout high-level replacement systems as given in [4]. To guarantee that all necessary constructions exist, we assume so-called HLR2 categories, which are defined in the Appendix (according to [6]).

Definition 13 (Concurrent DPO production).

1. *Let $p = (L \leftarrow K \to R)$ and $p' = (L' \leftarrow K' \to R')$ be two productions and let D be an object together with two morphisms $D \to R$ and $D \to L'$. The pair $(D \to R, D \to L')$, or short D, is called a* dependency relation *for (p, p') if the pushout object H^* of $D \to R$ and $D \to L'$ exists and if there are unique pushout complements of $K \to R \to H^*$ and $K' \to L' \to H^*$ up to isomorphism.*

2. *Given a dependency relation $(D \rightarrow R, D \rightarrow L')$ for (p, p'), the D-concurrent production $p *_D p' = (L^* \leftarrow K^* \rightarrow R^*)$ of p and p' is given by the construction in the following diagram, where:*

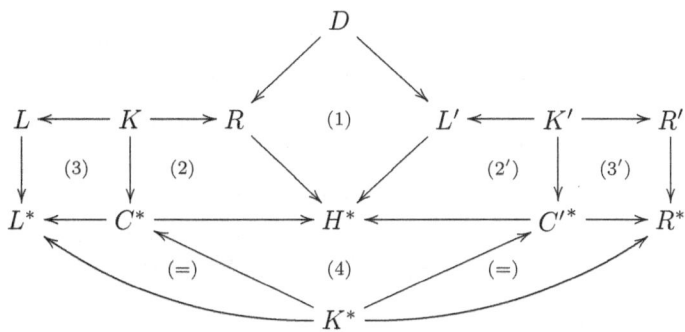

(a) *H^* is the pushout object in diagram (1);*
(b) *C^* and C'^* are the pushout complements in diagrams (2) and (2'), respectively;*
(c) *L^* and R^* are the pushout objects in diagrams (3) and (3'), respectively; and*
(d) *K^* is the pullback object in diagram (4) with $K^* \rightarrow L^* = K^* \rightarrow C^* \rightarrow L^*$ and $K^* \rightarrow R^* = K^* \rightarrow C'^* \rightarrow R^*$.*

3. *Two productions $p = (L \leftarrow K \rightarrow R)$ and $p' = (L' \leftarrow K' \rightarrow R')$ are composable if there exists a dependency relation D for (p, p'), and in such a case their composite production is given by the D-concurrent production $p *_D p' = (L^* \leftarrow K^* \rightarrow R^*)$. If D is not needed explicitly, the composite production is denoted by $p * p'$.* □

Notice that two productions are always composable if $\mathcal{C}$ has an initial object, which is the case of a HLR2 category; see the Appendix. The D-concurrent production $p *_D p'$ becomes the parallel composition $p + p'$ when D is the initial object in $\mathcal{C}$.

A special case of composition, namely via a dependency relation $D = L'$, is particularly interesting from a verification point of view, since the matching algorithm of the high-level replacement system can then be used to test for redundant productions, namely by finding a match of L' in R.

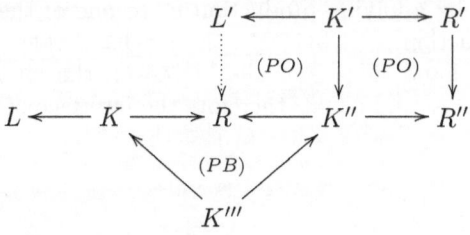

In other words, this kind of composition is obtained by applying the second production to the right-hand side of the first one.

Using the following fact, which is proved for double-pushout high-level replacement systems in [4] as analysis step of the so-called Concurrency Theorem, we can show that a production q is redundant if it is subsumed by a composite production.

Fact 14. *Given a direct derivation $G \Rightarrow H$ via a composite production $p * q$ there is a derivation sequence $G \Rightarrow X \Rightarrow H$ via (p, q).* □

Theorem 15. *A production q is* redundant *if there is a composite production $p_0 * \cdots * p_n$ in $\mathbf{P} - \{q\}$ ($n \geqslant 0$) such that $p_0 * \cdots * p_n \leqslant q$.* □

Proof. Let $p_0 * \cdots * p_n$ be a composite production in $\mathbf{P} - \{q\}$ ($n \geqslant 0$) such that $p_0 * \cdots * p_n \leqslant q$, and assume $G \Rightarrow^* H$ in $\mathbf{P}$. If q does not belong to this derivation, then $G \Rightarrow^* H$ in $\mathbf{P} - \{q\}$ as well. Otherwise, let $G \Rightarrow^* H$ in $\mathbf{P}$ be given by $G \Rightarrow \cdots \Rightarrow G_i \Rightarrow G_j \Rightarrow \cdots \Rightarrow H$ with $G_i \Rightarrow G_j$ via q. By Definition 5, there is also a direct derivation $G_i \Rightarrow G_j$ via $p_0 * \cdots * p_n$, which by Fact 14 can be decomposed into a derivation $G_i \Rightarrow^* G_j$ via $(p_0, \ldots, p_n)$. Therefore, there is a derivation $G \Rightarrow^* H$ in $\mathbf{P} - \{q\}$. □

Example 16. Production p_9 of Example 2 is redundant because it is subsumed by the composite production

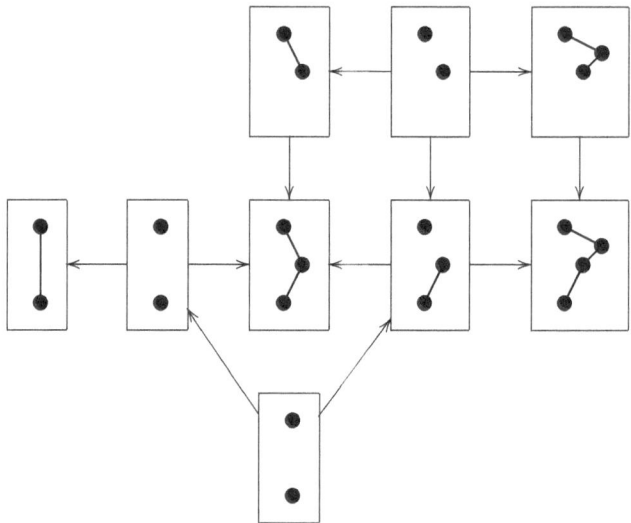

which is obtained by applying production p_4 to one of the edges of the right-hand side of production p_4. Obviously, the left-hand sides of $p_4 * p_4$ and p_9 as well as their right-hand sides coincide. Moreover, the interface of p_4 becomes the interface of $p_4 * p_4$ and equals therefore the interface of p_9, and then $p_4 * p_4$ and p_9 are equal. □

6   Conclusion

The notions of redundancy and subsumption are generalized in this paper from rule-based systems to high-level replacement systems. In particular, a characterization of subsumption in single-pushout high-level replacement systems and a sufficient condition for redundancy in double-pushout high-level replacement systems are presented in this paper.

The sufficient condition for redundancy can be understood as a first step towards the minimization of high-level replacement systems, aiming at minimal high-level replacement systems free of redundant productions. It remains an open problem to find a more general form of redundancy than direct subsumption and subsumption by a composite production.

The notions of redundancy and subsumption can be extended by taking into account not only the set of productions but also the start object and the class of terminal objects of the high-level replacement systems.

Finally, it also remains open to find an efficient verification procedure based on the sufficient condition for redundancy presented in this paper.

It should be noted that similar techniques are used in [8] to study temporal refinements of graph transformation systems. It may be interesting to investigate the relationship between both approaches.

Acknowledgment

We thank W. Bartol, A. Corradini, M. Große-Rhode, M. Llabrés, and F. Parisi-Presicce for their helpful comments on an earlier version of this paper.

References

1. B. Bollobás. *Graph Theory: An Introductory Course.* Springer-Verlag, 1979. 217
2. W. Buntine. Generalized subsumption and its applications to induction and redundancy. *Artificial Intelligence*, 36(2):149–176, 1988. 215
3. A. Corradini, U. Montanari, F. Rossi, H. Ehrig, R. Heckel, and M. Löwe. *Algebraic Approaches to Graph Transformation. Part I: Basic Concepts and Double Pushout Approach*, chapter 3, pages 163–245. Volume 1: Foundations of Rozenberg [10], 1997. 215
4. H. Ehrig, A. Habel, H.-J. Kreowski, and F. Parisi-Presicce. Parallelism and concurrency in high-level replacement systems. *Mathematical Structures in Computer Science*, 1:361–404, 1991. 222, 224
5. H. Ehrig, R. Heckel, M. Korff, M. Löwe, L. Ribeiro, A. Wagner, and A. Corradini. *Algebraic Approaches to Graph Transformation. Part II: Single Pushout Approach and Comparison with Double Pushout Approach*, chapter 4, pages 247–312. Volume 1: Foundations of Rozenberg [10], 1997. 215
6. H. Ehrig and M. Löwe. Categorical principles, techniques and results for high-level replacement systems in computer science. *Applied Categorical Structures*, 1:21–50, 1993. 215, 222
7. M. Grosse-Rhode. Algebra transformation systems and their composition. *Lecture Notes in Computer Science*, 1382:107–122, 1998. 219
8. M. Große-Rhode, F. Parisi-Presicce, and M. Simeoni. Spatial and temporal refinement of typed graph transformation systems. In *Proc. 23rd Int. Symposium on Mathematical Foundations of Computer Science*, volume 1450 of *Lecture Notes in Computer Science*, pages 553–561. Springer-Verlag, 1998. 225

9. F. Parisi-Presicce. Transformations of graph grammars. In J. E. Cuny, H. Ehrig, G. Engels, and G. Rozenberg, editors, *Proc. 5th Int. Workshop on Graph Grammars and their Application to Computer Science*, volume 1073 of *Lecture Notes in Computer Science*, pages 428–442. Springer-Verlag, 1996. 220

10. G. Rozenberg, editor. *Handbook of Graph Grammars and Computing by Graph Transformation*, volume 1: Foundations. World Scientific, 1997. 225

11. J. Tepandi. Comparison of expert system verification criteria: Redundancy. In M. Ayel and J.-P. Laurent, editors, *Validation, Verification and Test of Knowledge-Based Systems*, chapter 4, pages 49–62. John Wiley & Sons, 1991. 215

12. G. Valiente. Verification of knowledge base redundancy and subsumption using graph transformations. *Int. Journal of Expert Systems*, 6(3):341–355, 1993. 215

13. G. Valiente. *Knowledge Base Verification using Algebraic Graph Transformations*. PhD thesis, University of the Balearic Islands, 1994. 217, 218

Appendix: HLR2 Categories

Let $\mathcal{C}$ be a category and let $\mathcal{M}$ be a class of morphisms of $\mathcal{C}$. The pair $(\mathcal{C}, \mathcal{M})$ is said to satisfy condition *HLR2* when it satisfies the following properties:

- *Existence of semi-$\mathcal{M}$-pushouts.* There exists a pushout of any pair of co-initial morphisms when at least one of them is in $\mathcal{M}$.
- *Inheritance of $\mathcal{M}$-morphisms under pushouts.* For any pushout square

$$\begin{array}{ccc} A & \xrightarrow{f} & B \\ {\scriptstyle g}\downarrow & & \downarrow{\scriptstyle g'} \\ C & \xrightarrow{f'} & D \end{array}$$

if $f \in \mathcal{M}$ then $f' \in \mathcal{M}$.
- *Existence of binary coproducts.* $\mathcal{C}$ has all binary coproducts, and if $f : A \to A'$ and $g : B \to B'$ are two homomorphisms in $\mathcal{M}$ then the coproduct morphism $f + g : A + B \to A' + B'$ is also in $\mathcal{M}$.
- *Existence of $\mathcal{M}$-pullbacks.* There exists a pullback of any pair of co-final $\mathcal{M}$-morphisms.
- *Inheritance of $\mathcal{M}$-morphisms under pullbacks.* For any pullback square

$$\begin{array}{ccc} A & \xrightarrow{f} & B \\ {\scriptstyle g}\downarrow & & \downarrow{\scriptstyle g'} \\ C & \xrightarrow{f'} & D \end{array}$$

if $f', g' \in \mathcal{M}$ then $f, g \in \mathcal{M}$.

– *M-pushout-pullback decomposition.* In each diagram of the form

$$
\begin{array}{ccccc}
A & \xrightarrow{\ f\ } & B & \xrightarrow{\ h\ } & E \\
\downarrow{\scriptstyle g} & (1) & \downarrow{\scriptstyle g'} & (2) & \downarrow{\scriptstyle g''} \\
C & \xrightarrow[\ f'\ ]{} & D & \xrightarrow[\ h'\ ]{} & F
\end{array}
$$

if $(1)+(2)$ is a pushout square, (2) is a pullback square and $g, g', g'', h, h' \in \mathcal{M}$ then (1) is a pushout square.
– *Closure of $\mathcal{M}$.* The class $\mathcal{M}$ is closed under composition.
– *Cube-pushout-pullback property.* Given any commutative cube such that all morphisms in the top and bottom squares are in $\mathcal{M}$, the top diagram is a pullback square and the front and right-hand side diagrams are pushout squares, then the bottom diagram is a pullback square if and only if the back and left-hand side diagrams are pushout squares.

Knowledge Representation and
Graph Transformation

Stefan Schuster

University of Bremen – TZI
Universitätsallee 21-23, 28359 Bremen
+49-421-218-2731
sts@tzi.org

Abstract. In this paper, the knowledge representation language ALC is modeled by means of graph transformation. This yields a formally defined graphical or visual version of ALC where ALC-concepts and ALC-sentences are represented as graphs and their syntactically correct generation is specified by graph grammar rules. The semantics of ALC-sentences, which is expressed by satisfiability, is compatible with the graphical representation. Moreover, the tableau calculus, which provides an algorithm for checking the satisfiability of constraints, can be carried over to the level of graph transformation, thus obtaining a visual verification procedure.

1 Introduction

In this paper, the knowledge representation language ALC is modeled by means of graph transformation. This knowledge representation language ALC (see e.g. [2,3,7]) builds a common basis of many knowledge representation systems and provides a set of language constructs which keep all calculations (in ALC) decidable. Although visualizations exist, the language itself and all related knowledge representation systems are textual. The modeling by means of graph transformation (see, e.g., [10]) yields a formally defined graphical or visual version of ALC where ALC-concepts and -sentences are represented as graphs and their syntactically correct generation is specified by graph grammar rules. The semantics of ALC-sentences, which is expressed by satisfiability, is compatible with the graphical representation. Moreover, the tableau calculus, which provides an algorithm for checking the satisfiability of constraints, can be carried over to the level of graph transformation, thus obtaining a visual verification procedure. With our approach it is possible to use graphical input (e.g. for knowledge representation systems), have a direct interpretation of the graphs, and direct execution of calculations including all representations. There is no explicit transformation from a textual into a graphical notation and vice versa necessary in order to represent textual constructs graphically and to provide calculations on graphical representations. This new graphical interface allows both the editing of the concepts as well as a visual verification without the need of the former textual notation. In order to reach this goal, the following is needed.

H. Ehrig et al. (Eds.): Graph Transformation, LNCS 1764, pp. 228–237, 2000.

1. Definition of graphical representations of the text-based constructs of ALC,
2. specification of graph grammar rules, capable of generating and manipulating the graphical constructs of ALC including inference,
3. semantic interpretation of the resulting graphs compatible with the interpretation of corresponding textual constructs.

This paper is organized as follows: The next section introduces the knowledge representation language ALC and the tableau calculus. Section 3 illustrates how ALC-constructs can be modeled by graph grammars. The paper ends with the conclusion.

2   The Knowledge Representation Language ALC

ALC is a knowledge representation language which allows the specification of terminologies as systems of related concepts by means of logical expressions and sentences of a certain form (see, e.g., [2,3,7]). Starting from identifiers of individual objects, atomic concepts, the basic bricks or primitives, representing sets of objects, and roles representing relations between objects and using the usual logical connectives for combining concepts, one can formulate terminological and assertional sentences in ALC. While a terminological sentence expresses the subsumption of a concept by another one or the equivalence of two concepts, an assertional sentence states is-element relations between objects and concepts and is-in-relation relations between objects. The meaning of ALC-expressions is given in the usual way of mathematical logic by an interpretation function for combined concepts and a satisfaction relation for sentences. Interesting enough, the satisfiability of assertional sentences of the is-element type can be checked algorithmically by means of the tableau calculus [11].

2.1   Syntax and Semantics of ALC

Let $\mathcal{I}$ be a finite set of identifiers (for *individual objects*), $\mathcal{S}$ be a finite set of *atomic concepts*, and $\mathcal{R}$ be a finite set of *roles*. Then the set $\mathcal{C}$ of *concept expressions*, $\mathcal{T}$ of *terminological sentences*, and $\mathcal{A}$ of *assertional sentences* are defined recursively as follows:

    (i) $\bot, \top \in \mathcal{C}$ and $\mathcal{S} \subseteq \mathcal{C}$,
    (ii) $(C \wedge D), (C \vee D), (\neg C), (\forall R.C), (\exists R.C) \in \mathcal{C}$ for all $C, D \in \mathcal{C}$ and $R \in \mathcal{R}$,
    (iii) $(C \leq D) \in \mathcal{T}$ for all $C \in \mathcal{S}$ and $D \in \mathcal{C}$,
    (iv) $(C \equiv D) \in \mathcal{T}$ for all $C, D \in \mathcal{C}$,
    (v) $(a\ is\text{-}a\ C), (a\ rel\ b) \in \mathcal{A}$ for all $a, b \in \mathcal{I}$, $C \in \mathcal{C}$, and $R \in \mathcal{R}$.

Let $\mathcal{D}$ be some domain of individual objects of any kind. Let $I(a) \in \mathcal{D}$ for all $a \in \mathcal{I}$, $I(C) \subseteq \mathcal{D}$ for all $C \in \mathcal{S}$, and $I(R) \subseteq \mathcal{D}x\mathcal{D}$ for all $R \in \mathcal{R}$ be given or chosen meanings of the elements of $\mathcal{I}$, $\mathcal{S}$, and $\mathcal{R}$. Then I can be extended to an *interpretation function* on $\mathcal{C}$ in the following straightforward way.

(i) $I(\bot) = \emptyset$ and $I(\top) = \mathcal{D}$,

(ii) $I(C \wedge D) = I(C) \cap I(D)$,
$I(C \vee D) = I(C) \cup I(D)$,
$I(\neg C) = \mathcal{D} - I(C)$,
$I(\forall R.C) = \{x \mid [(\forall y) : (x, y) \in I(R)] \Rightarrow y \in I(C)\}$,
$I(\exists R.C) = \{x \mid (\exists y) : [(x, y) \in I(R) \text{ and } y \in I(C)]\}$ for $C, D \in \mathcal{C}, R \in \mathcal{R}$.

An interpretation function I *satisfies* a terminological or assertional sentence α, denoted by $\models_I \alpha$, in the following cases:

(iii) $\models_I C \leq D$ if $I(C) \subseteq I(D)$,

(iv) $\models_I C \equiv D$ if $I(C) = I(B)$,

(v) $\models_I a$ *is-a* C if $I(a) \in I(C)$ and $\models_I a$ *rel* b if $(I(a), I(b)) \in I(R)$.

A set T of terminological sentences is called a *terminology* or a *TBox*, a set A of assertional sentences is called a *constraint system* or an *ABox*. I satisfies T (respectively A) if I satisfies each sentence of T (respectively A). Accordingly, a constraint system A is called *satisfiable* if there is an interpretation function I satisfying A.

2.2   Tableau Calculus

There is an algorithm checking the satisfiability of constraints of the form x *is-a* C, i.e. is-element constraints, if C is *simple* in the sense that all negations occuring in C are negations of atomic concepts. This algorithm is based on the tableau calculus which consists of the following four rules and allows to derive constraint systems from constraint systems(see, e.g. [4,5,11]).

1. $S \rightarrow_\wedge \{x$ *is-a* C_1, x *is-a* $C_2\} \cup S$ if x *is-a* $C_1 \wedge C_2$ is in S and x *is-a* C_1 and x *is-a* C_2 are not both in S.

2. $S \rightarrow_\vee \{x$ *is-a* $D\} \cup S$ if x *is-a* $C_1 \vee C_2$ is in S, but neither x *is-a* C_1 nor x *is-a* C_2 are in S and $D = C_1$ or $D = C_2$.

3. $S \rightarrow_\exists \{xRy, y$ *is-a* $C\} \cup S$ if x *is-a* $\exists R.C$ is in S, there is no z so that xRz holds in S, z *is-a* C is in S and y is a new variable.

4. $S \rightarrow_\forall \{y$ *is-a* $C\} \cup S$ if x *is-a* $\forall R.C$ is in S, xRy holds in S and y *is-a* C is not in S.

The rules preserve satisfiability. Starting from a finite constraint system, there is never an infinite sequence of rule applications. Hence, each sequence can be terminated resulting in a *complete constraint system* which is reduced with respect to the four rules. Such a complete system is satisfiable if and only if it does not contain *clashes*, i.e. constraints of the form x *is-a* $\bot$ or x *is-a* C as well as x *is-a* $\neg C$. Altogether the satisfiability of the constraint x *is-a* C with a simple concept C can be checked by applying the rules as long as possible and looking for clashes.

3   Graph-Transformational Version of ALC

A graphical variant of the syntax, semantics, and verification of ALC is modeled by a structured graph transformation system of the form depicted in Figure 1.

$G_{Concepts}$ is a graph grammar generating so-called ALC-trees which are tree representations of concept expressions. The other five graph-transformation systems are graph grammar schemes each with a fixed set of rules, but variable start graphs. The choices of start graphs are indicated by the edges in Figure 1. An edge $G_x \rightarrow G_y$ means that each graph generated by G_x can be used as start graph by G_y. If Z is such a graph, then $G_y(Z)$ denotes the graph grammar consisting of the rules of G_y and the start graph Z. Its generated language $L(G_y(Z))$ contains all graphs which are derivable from Z through rules of G_y and which are reduced with respect to these rules.

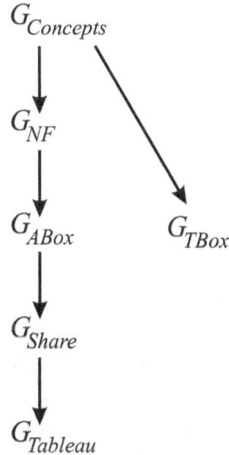

Fig. 1. Structure of the graph-transformational version of ALC

ALC-expressions and ALC-sentences are represented by directed labeled graphs. The labels include all the ALC-symbols introduced in Subsection 3.1. The rules used in the following have the form $N \supseteq L \rightarrow R$ where the left-hand side L is a subgraph of N and some nodes of L and the right-hand side R may be numbered. If L is a proper subgraph of N, the latter is used as negative context condition. For each numbered node of L there is a corresponding node in R with the same number. Such a rule is applied to a graph in the sense of the so-called double-pushout approach with negative context conditions (see, e.g., [6,8]) where the numbered nodes form the gluing graph.

In the following subsections, the components $G_{Concepts}$ and $G_{Tableau}$ are discussed in detail while the other five components are only sketched. The full description of the graph-transformational version of ALC can be found in [14].

3.1   Generation of ALC-Trees by $G_{Concepts}$

The graph grammar $G_{Concepts}$ comprises the rules

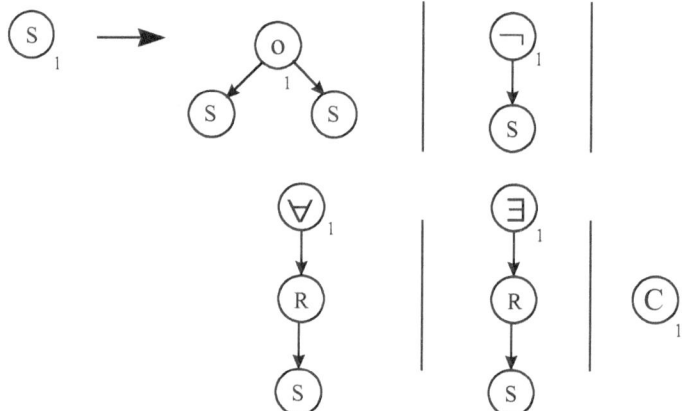

Fig. 2. The Rules of $G_{Concepts}$

where S is an extra nontermional symbol and $o \in \{\wedge, \vee\}$, $R \in \mathcal{R}$, and $C \in \mathcal{S}$. The start graph is the single-node graph labeled with S. The generated language is denoted by $L(G_{Concepts})$ and contains obviously only trees, called ALC-*trees*. See Figure 3 for an example of a derivation in $G_{Concepts}$, where the applied rules are numbered according to the order above.

Without going into the technical details, there is a well-known correspondence between functional and relational expressions and trees as graphical representations. This principle applies to ALC-trees in such a way that each ALC-tree t specifies an expression $xpr(t)$. Using this notion it is not difficult to prove by induction on the recursive structure of concept expressions on one hand and length of derivations on the other hand that the set of concept expressions $\mathcal{C}$ defined in Subsection 2.1 and the set of expressions underlying ALC-trees coincide.

Theorem 1. $xpr(L(G_{Concepts})) = \mathcal{C}$.

3.2   Generation of Normal Forms, Assertional and Terminological Sentences, and Shared Graphs

In this subsection G_{NF}, G_{ABox}, G_{TBox}, and G_{Share} are described with respect to their generative effect without giving the rules explicitly.

1. The component G_{NF} takes ALC-trees as start graphs. Each ALC-tree is transformed into a normal form by pushing down all negations as far as

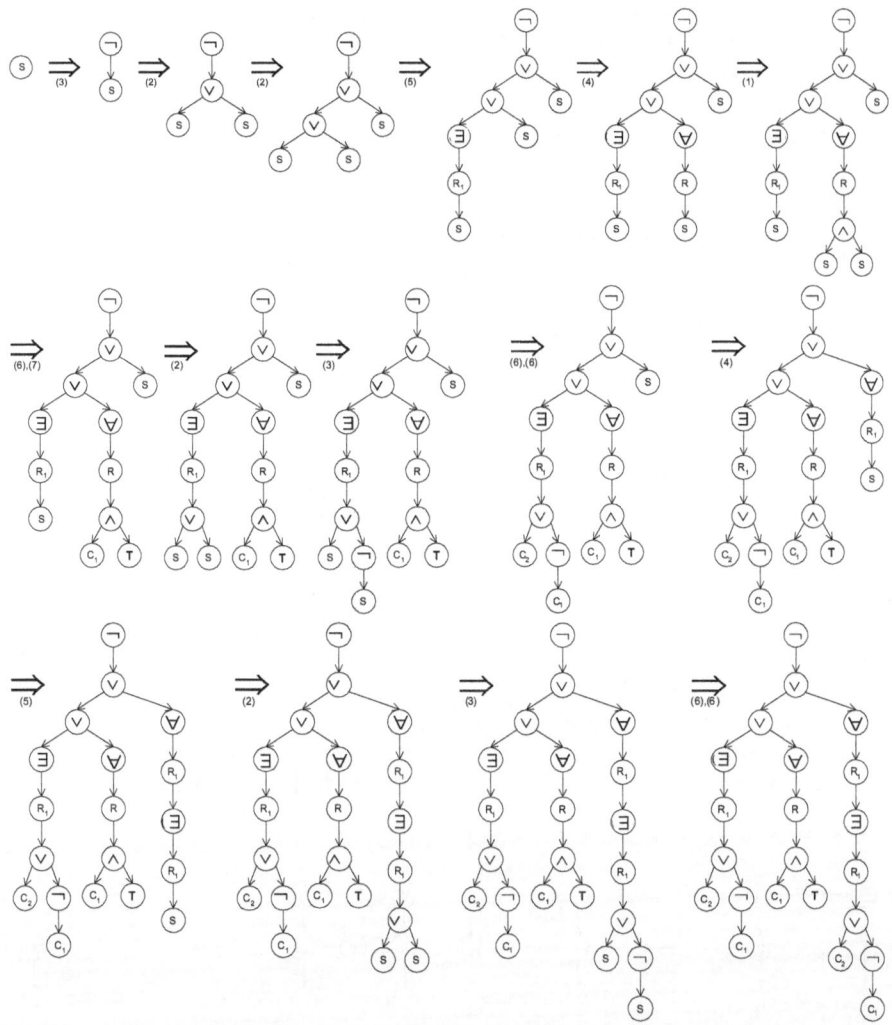

Fig. 3. Example of a derivation with rules out of $G_{Concepts}$

possible, i.e. the daughter node of each negation node in the resulting ALC-tree is labeled with an atomic concept. In other words, the expression of the resulting tree is a simple concept. The rules of G_{NF} reflect the usual logical laws which describe the interaction of negation with other logical connectives. Moreover, the transformation in G_{NF} is invariant with respect to the interpretation function. This means, in particular, that the constraint x is-a $xpr(t)$ is satisfiable if and only if x is-a $xpr(t')$ is satisfiable provided that t' is derived from t in in G_{NF}.

2. G_{ABox} takes ALC-trees in normal form and transforms them into assertional sentences by adding identifier nodes and is-element edges. The underlying ALC-expression is not changed and, therefore, satisfiability is preserved by derivation.

3. Similarly, G_{TBox} takes ALC-trees and transforms them into terminological sentences by adding subsumption and equivalence nodes and relating edges.

4. Finally, G_{Share} takes results of G_{ABox} and collapses identical subgraphs as much as possible, but in such a way that the underlying ALC-expression is kept invariant. In particular, the derivations in G_{Share} preserve satisfiability.

3.3   The Verification Component $G_{Tableau}$

The component $G_{Tableau}$ takes the results of G_{Share} as start graphs. Its rules are given in Figure 4. The dotted edges and nodes denote the subgraph N of the left-hand side L of a rule, that means the negative context condition.

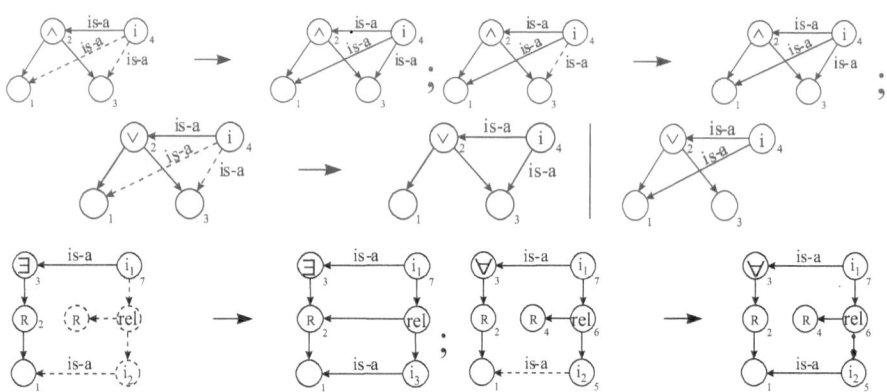

Fig. 4. The Rules of $G_{Tableau}$

See Figure 5 for an example of a derivation in $G_{Tableau}$, where again the applied rules are numbered according to the order above.

By construction, a start graph Z of $G_{Tableau}$ represents a constraint of the form x is-a C where C is a simple concept expression. In other words, Z corre-

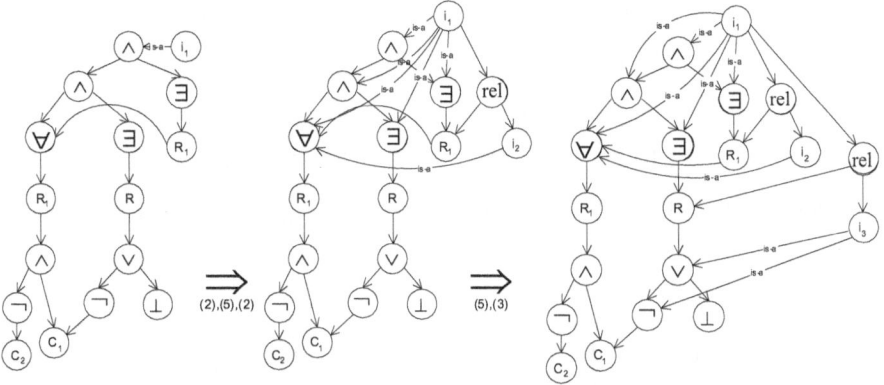

Fig. 5. Example for a derivation with rules out of $G_{Tableau}$

sponds to an input of the tableau calculus. A derivation in $G_{Tableau}$ adds edges of the is-element type and is-in-relation items according to the structure of concept expressions in the same way as tableau calculus rules add new constraints. This can be proved by induction on the length of derivations in $G_{Tableau}$ on one hand and in the tableau calculus on the other hand. As a consequence, one gets the following result stating the correctness of $G_{Tableau}$ with respect to the tableau calculus.

Theorem 2. *Let Z be a start graph of $G_{Tableau}$ and x is-a C the represented constraint. Then x is-a C is satisfiable iff there is a derivation in $G_{Tableau}$ so that G is reduced with respect to the $G_{Tableau}$-rules and G does not contain any subgraph of the form depicted in Figure 6.*

Fig. 6. A graphical represented clash

For a better understanding the relation between ALC and the tableau calculus on one hand and between the graphical version of ALC and the graph grammar $G_{Tableau}$ on the other hand may be represented by the diagram in Figure 7.

4   Conclusion

In this paper, we have modeled the knowledge representation language ALC using graph transformation. We have presented two graph grammars explicitly, the

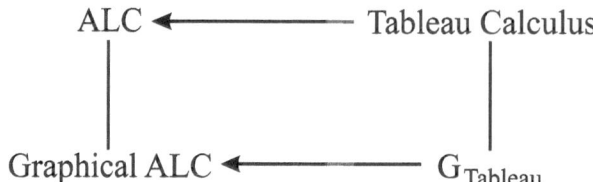

Fig. 7. The relation between the textual and graphical versions of ALC

first of which generates the set of ALC-concepts while the second implements the tableau calculus of ALC. Three further graph grammars for modeling ALC-constructs, like terminological sentences, are given in [14]. The presented graph grammars have been implemented with the PROGRES-system (see e.g. [12,13]) as a prototype. With some more effort a graphical user interface can be built on top of it (to be used within a knowledge representation system). In conclusion we can say, that there exists a semantically based definition of a graphical version of ALC including terminological sentences, assertions, a normalized representation and the rules of the (graphical) tableau calculus. In order to strengthen the link between knowledge representation and graph transformation, further investigations will be necessary including the following points.

1. The knowledge representation language may be extended by type definitions.
2. In subsection 3.1 the structuring of the graph-transformational version of ALC is given ad hoc. Instead one may try to apply the structuring concepts of the graph-transformation-based specification language GRACE in a more systematic way (see, e.g., [1,9] .
3. Optimization aspects, like the reduction of storage requirement may be considered.
4. Special inference algorithms (like the classifier) may be realized and their efficiency may be analyzed.
5. A graphical user interface to be used in a knowledge representation system may be realized.
6. Some case studies may help to illustrate that terminologies can be better understood and analyzed by means of the graphical ALC.

Acknowledgment

I would like to thank Hans-Jörg Kreowski, Sabine Kuske and the three anonymous referees for their helpful and detailed comments on the previous version of the paper which allowed me to provide a thoroughly improved version.

References

1. Marc Andries, Gregor Engels, Annegret Habel, Berthold Hoffmann, Hans-Jörg Kreowski, Sabine Kuske, Detlef Plump, Andy Schürr, and Gabriele Taentzer. Graph transformation for specification and programming. *Science of Computer Programming*, 34(1):1–54, 1999. 236
2. F. Baader. Logic-based knowledge representation. In M.J. Wooldridge and M. Veloso, editors, *Artificial Intelligence Today – Recent Trends and Developments*, volume 1600 of *Lecture Notes in Artificial Intelligence*, pages 13–41. Springer Verlag, 1999. 228, 229
3. F. Baader, H.-J. Bürckert, B. Hollunder, W. Nutt, and J. Siekmann. Concept logic. In *Proceedings of the Symposium on Computational Logic*, pages 177–201, Brussels (Belgien), 1990. 228, 229
4. F. Baader and B. Hollunder. A terminological knowledge representation system with complete inference algorithms. In *Proceedings of the First International Workshop on Processing Declarative Knowledge*, volume 572 of *Lecture Notes in Computer Science*, pages 67–85, Kaiserslautern (Germany), 1991. Springer–Verlag. 230
5. A. Borgida and P.F. Patel-Schneider. A semantics and complete algorithm for subsumption in the classic description logic. *JAIR*, 1:277–308, 1994. 230
6. Andrea Corradini, Ugo Montanari, Francesca Rossi, Hartmut Ehrig, Reiko Heckel, and Michael Löwe. Algebraic approaches to graph transformation - part I: Basic concepts and double pushout approach. In G. Rozenberg, editor, *Handbook of Graph Grammars and Computing by Graph Transformation. Vol. I: Foundations*, chapter 3, pages 163–246. World Scientific, Singapur, 1997. 231
7. F. Donini, M. Lenzerini, D. Nardi, and W. Nutt. The complexity of concept languages. In *Proceedings of the 1991 International Conference on Knowledge Representation (KR'91)*, Boston (USA), 1991. 228, 229
8. Annegret Habel, Reiko Heckel, and Gabriele Taentzer. Graph grammars with negative application conditions. *Fundamenta Informaticae*, 26 (3-4):287–313, 1996. 231
9. Hans-Jörg Kreowski and Sabine Kuske. On the interleaving semantics of transformation units—A step into GRACE. In J. Cuny, H. Ehrig, G. Engels, and G. Rozenberg, editors, *Proc. Fifth Intl. Workshop on Graph Grammars and Their Application to Comp. Sci.*, volume 1073 of *Lecture Notes in Computer Science*, pages 89–106. Springer, 1996. 236
10. G. Rozenberg, editor. *Handbook of Graph Grammars and Computing by Graph Transformation. Vol. I: Foundations*. World Scientific, Singapur, 1997. 228
11. M. Schmidt-Schauß and G. Smolka. Attributive concept descriptions with complements. *AI*, 48(1):1–26, 1991. 229, 230
12. A. Schürr. Introduction to progres, an attribute graph grammar based specification language. In M. Nagl, editor, *Proc. WG'89 Workshop on Graph-Theoretic Concepts in Computer Science*, volume 411 of *Lecture Notes in Computer Science*, pages 151–165. Springer Verlag, 1989. 236
13. A. Schürr, A. Winter, and A. Zündorf. Progres: Language and environment. In G. Rozenberg, editor, *Handbook on Graph Grammars Vol. II: Applications*. World Scientific, Singapur, 1999. 236
14. Stefan Schuster. Graphtransformation in der Wissensrepräsentation. Master's thesis, University of Bremen, 1997. 231, 236

Utilizing Constraint Satisfaction Techniques for Efficient Graph Pattern Matching

Michael Rudolf

TU Berlin,
Franklinstr. 28/29, D-10587 Berlin
`mich@cs.tu-berlin.de`

Abstract. This paper presents a way to represent and solve the problem of graph matching – also known as the *subgraph homomorphism problem* – as a *constraint satisfaction problem (CSP)*, opening up direct access to the large variety of research findings on optimized solution algorithms for CSPs. By decoupling the solution algorithm from the concrete graph model, this approach allows for variations of the model without affecting the algorithm. Furthermore, complementing the standard CSP definition, a query concept is introduced to allow for abstract representation of concrete implementation properties for optimization purposes.

1 Introduction

In the common rule-based approaches to graph transformation, we use a graph to describe the state of a modelled system at a given time, and we have graph rules to describe operations on the system's state in a "before/after" manner. To apply a graph rule to a state graph, first we have to find an embedding, also called a *match*, of the rule's left side into the state graph. In the field of graph theory, this problem is known as the *subgraph homomorphism problem*[1], and it is known to be NP complete [Meh84]. As a consequence in practice, finding a match for a graph rule has a worst-case complexity growing exponentially with the size of the rule's left-hand side[2]. Since the actual rewriting operation, where the embedding of the left side of a rule is replaced by its right side, can typically be performed in about linear time, it is obvious that matching performance will be decisive to the overall performance of any implemented graph transformation system.

Of course, an algorithm with exponential time complexity is inacceptable for serious system implementation. While we won't be able to decrease the worst-case complexity of the matching problem unless we prove $P = NP$, it is fortunately possible to considerably reduce its average-case complexity at least. Implemented systems like PROGRES [Zün95,Sch97] have shown that it is possible to cut down the cost of graph matching to a level well suited for application development. In general, to achieve acceptable performance it is inevitable

---

[1] Or *subgraph isomorphism problem*, if we want to restrict ourselves to injective embeddings.

[2] The size of a graph is given by the sum of its nodes and arcs.

H. Ehrig et al. (Eds.): Graph Transformation, LNCS 1764, pp. 238–252, 2000.

to consequently exploit the special properties of both the concrete graph and graph transformation model, and even of their implementation. Since it is almost impossible to reuse an algorithm developed like that, it is common practice today to re-develop a new algorithm from scratch whenever the need for matching functionality arises in the context of a new software project. Still there is a striking commonality in all those efforts: they are all based on the backtracking paradigm. Knowing that the overwhelming majority of theoretical as well as empirical studies on the optimization of backtracking algorithms is done in the context of constraint satisfaction problems (CSPs), it is near at hand to open this knowledge base for graph matching algorithms by reformulating the subgraph homomorphism problem as a CSP.

Apart from gaining direct access to the rich findings of the research work done in the field of CSPs, solving the graph homomorphism problem as a CSP has another significant advantage: by introducing a new level of abstraction, it allows for variation of the graph model without affecting the actual matching algorithm. To avoid performance drawbacks which often come with abstraction, in Sect. 5 we propose a *query concept* to make crucial properties of the graph model and its implementation available in the abstract CSP representation.

Note that all the ideas and concepts presented in the following sections have been successfully applied in the implementation of the matching subsystem of AGG, a graph transformation system developed at TU Berlin [AGG98,Rud97] which is implemented in Java.

2   Constraint Satisfaction Problems

A lot of problems in computer science, most notably in artifical intelligence, can be interpreted as special cases of the *constraint satisfaction problem (CSP)*. Research work in the CSP area is focussed on the theoretical as well as the empirical investigation of methods to solve such problems. While research activity is lively in the area, results are manifold already. Hence, representing and solving a given problem as a CSP pays off for complex problems where no efficient analytical solutions are known, as we will then be able to directly apply the latest findings in CSP technology.

Our definition of a CSP is based on [DvB97] (called *constraint network* there).

Definition 1 (CSP). *A constraint satisfaction problem (CSP) consists of*

- *a finite set of variables $X = \{x_1, \ldots, x_n\}$,*
- *a finite and discrete domain D_i of possible values for every variable $x_i \in X$, and*
- *a finite set R of constraints on the variables of X.*

Definition 2 (Constraint). *A constraint C_S on a tuple of variables $S = (x_1, \ldots, x_r)$ is a relation on these variables' domains: $C_S \subseteq D_1 \times \cdots \times D_r$.*

The number r of variables a constraint is defined upon is called arity *of the constraint.*

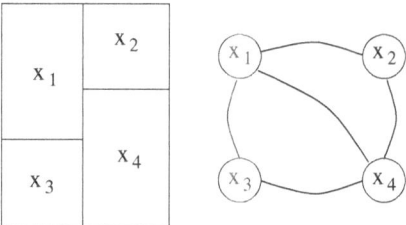

Fig. 1. A sample map coloring problem and its representation as a constraint graph

A constraint comprises the values a variable is allowed to take with respect to other variables. Of special interest are the constraints with arity two, the so-called *binary constraints*, as any arbitrary n-ary CSP can be transformed into an equivalent CSP with only unary[3] and binary constraints [RDP90]. Hence, most literature only cares for *binary CSPs*, and so do we. A nice property of binary CSPs is that they can easily be depicted as simple graphs, where the variables are denoted as nodes and the constraints as directed arcs between them; the direction of an arc gives the order of the variables in the corresponding constraint's tuple. The graph representation of a CSP is called *constraint graph*.

To explain how to solve a CSP, we need some more definitions:

Definition 3 (Instantiation of Variables). *Let $X = \{x_1, \ldots, x_n\}$ be a set of variables with their respective domains $D_i, i \in \{1, \ldots, n\}$. Then any n-tuple $\Gamma = (a_1, \ldots, a_n)$, $a_i \in D_i$ denotes an* instantiation *of each variable x_i with the corresponding value a_i. We also write $\Gamma(x_i) = a_i$ for the value of x_i under an instantiation Γ.*

Definition 4 (Satisfied Constraint). *A constraint C_S on a tuple of variables $S = (x_1, \ldots, x_r)$ is* satisfied *by an instantiation Γ if $(\Gamma(x_1), \ldots, \Gamma(x_r)) \in C_S$.*

Definition 5 (Solution of a CSP). *An instantiation Γ is a* solution *of a CSP if it satisfies all the constraints of the problem.*

The following example illustrates the definitions.

Example 1 (Map Coloring). In the map coloring problem, we want to color the regions of a map in a way that no two adjacent regions have the same color. The actual problem is that only a certain limited number of colors is available. To cast a CSP out of this problem, we take the regions to be colored as the variables, and the set of colors as the variables' domains. Let's say we have four regions as shown in Fig. 1, and only three colors available. We get a set of variables $X = \{x_1, x_2, x_3, x_4\}$ with domains, e.g., $D_1 = D_2 = D_3 = D_4 = \{\text{red}, \text{green}, \text{blue}\}$. Now we just have to find a set of constraints to express our

---

[3] Note that a unary constraint can be represented by a binary one containing the same variable twice.

demand for adjacent regions having different colors. To ensure that x_1 gets another color than x_2, we may define $C_{(x_1,x_2)} = \{(u,v) \in D_1 \times D_2 \mid u \neq v\}$, and just the like for $C_{(x_1,x_3)}, C_{(x_1,x_4)}, C_{(x_3,x_4)}$ and $C_{(x_2,x_4)}$. A solution for the CSP would then be $\Gamma = (\text{red}, \text{green}, \text{green}, \text{blue})$, for instance.

Note that since our constraints are symmetric, the direction of the constraint arcs is insignificant in this example. $\triangle$

3   Graph Matching as a CSP

In the graph matching problem, we're trying to find a mapping between the object sets of two graphs. Several restrictions apply to make this mapping a valid graph morphism. The exact definition of these restrictions depends on the concrete graph model, but usually at least a morphism has to preserve the graph structure in some way, and also the typing information of the graph objects if available. The analogy to the constraint satisfaction problem is quite obvious: Here we are also looking for a mapping between two sets, namely between a set of variables and a set of values, where some restrictions apply which are called constraints. So basically, what we have to do to obtain an equivalent CSP for a given graph matching problem is:

- take the objects of the graph to be matched as the CSP's set of variables,
- take the objects of the graph to be matched into as the variables' domain, and
- find a proper translation of the restrictions that apply to a graph morphism into a set of constraints.

This way, we'll get a CSP whose solutions are actually graph morphisms. To go into some more detail on how to construct an equivalent CSP, especially on how to find the proper constraints, first we have to fix the concrete graph model to consider. For the purpose of demonstration, we choose a simple model with typed graph objects:

Definition 6 (Simple Graph). *A simple graph G is a tuple $G = (G_V, G_E, L, s, t, \ell)$ consisting of*

- *a finite set of nodes (or "vertices") G_V and a finite set of arcs (or "edges") G_E with $G_V \cap G_E = \emptyset$,*
- *two total mappings $s, t : G_E \to G_V$ ("source" and "target"),*
- *a set of types (or "labels") L and*
- *a total mapping $\ell : G_V \cup G_E \to L$.*

The nodes and arcs of a graph are also collectively called the "objects" of the graph (or "graph objects").

A match of one graph into another is given by a graph morphism, which is a mapping of one graph's object sets into the other's, with some restrictions to preserve the graph's structure and its typing information:

Definition 7 (Graph Morphism). *A graph morphism* $m : L \to G$ *between two simple graphs* $L = (L_V, L_E, L_L, s_L, t_L, \ell_L)$ *and* $G = (G_V, G_E, L_G, s_G, t_G, \ell_G)$ *is a pair of total mappings* $m = (m_V : L_V \to G_V, m_E : L_E \to G_E)$, *where the following restrictions apply:*

1. $\forall v \in L_V : \ell_L(v) = \ell_G(m_V(v))$
2. $\forall e \in L_E : \ell_L(e) = \ell_G(m_E(e))$
3. $\forall e \in L_E :$
   - $m_V(s_L(e)) = s_G(m_E(e))$
   - $m_V(t_L(e)) = t_G(m_E(e))$

The image of a graph object o *with respect to a morphism* m *we simply denote by* $m(o)$, *since the index can always be derived from the context.*

Now that we have defined a concrete graph model and its morphisms, let us consider an example on how to construct a CSP out of a given graph matching problem:

Example 2 (CSP for graph matching). Figure 2 shows a very simple graph matching problem concerning the two graphs L and G; objects' type information is omitted for simplicity. Now we are seeking for all the possible morphisms $m : L \to G$. To construct a CSP out of this problem, we advance as outlined before:

- The set of variables X is given by $X = L_V \cup L_E = \{x_1, x_2, x_3\}$.
- As the variables' domain, we could simply use $D = G_V \cup G_E$. But since we don't want a node to be accidentally mapped to an arc or vice versa, we rather use different domains for the nodes and arcs of L: $D_1 = D_3 = G_V = \{d_1, d_3, d_4, d_6\}$ and $D_2 = G_E = \{d_2, d_5\}$.

Still missing in our CSP is the proper set of constraints to enforce morphism properties on the solutions. For example, it must not be allowed to assign x_2 to d_2 when x_1 is already mapped to d_3. Hence, we need a constraint to ensure source compatibility:

$$C^{\text{src}}_{(x_2, x_1)} = \{(d_E, d_V) \in D_2 \times D_1 \mid s(d_E) = d_V\}$$
$$= \{(d_2, d_1), (d_5, d_6)\}$$

Analogously, we need a constraint to ensure target compatibility for the mapping of x_2 w.r.t. x_3:

$$C^{\text{tar}}_{(x_2, x_3)} = \{(d_E, d_V) \in D_2 \times D_3 \mid t(d_E) = d_V\}$$
$$= \{(d_2, d_3), (d_5, d_6)\}$$

And as a matter of fact, for our tiny example these two constraints already do the trick; now our CSP yields only the solutions $\Gamma_1 = (d_1, d_2, d_3)$ and $\Gamma_2 = (d_6, d_5, d_6)$ which are valid morphisms. Furthermore, it is easy to see that these are actually the only two morphisms that can exist between the graphs L and G.

So far we deliberately ignored the typing of graph objects which is provided by our graph model. Did we care for type compatibility, we'd simply have to

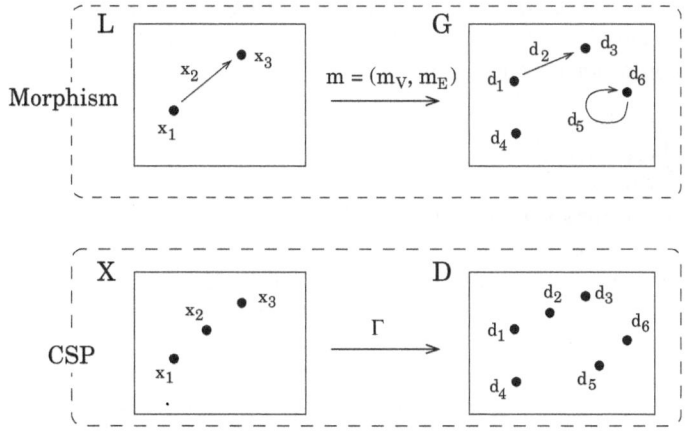

Fig. 2. Translation of a graph matching problem into CSP terminology

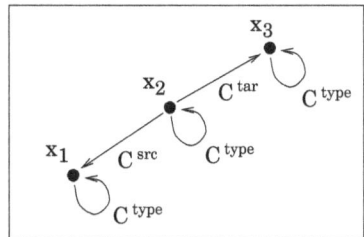

Fig. 3. The constraint graph of the CSP in Example 2

add another constraint $C^{\text{type}}_{(x_i)} = \{d \in D_i \mid \ell(x_i) = \ell(d)\}$ for every object $x_i \in X$. These are unary constraints which can of course be denoted by binary constraints $C^{\text{type}}_{(x_i,x_i)}$ as well. Figure 3 shows the complete constraint graph of the resulting CSP. $\triangle$

Instead of adding unary type constraints to our CSP, we could also simply remove those values from any domain D_i that will not satisfy the constraint $C^{\text{type}}_{(x_i)}$. This is the preferred method to deal with unary constraints, as it avoids dynamically checking the constraints over and over again in the backtracking solution process. It is known as establishing *node consistency*[4] [Mac77,Kum92] on a CSP. However in Sect. 5 we will introduce the concept of *queries* as a much more flexible approach to domain reduction.

We want to generalize the construction to be able to automatically generate the proper CSP for any two given graphs:

---

[4] this refers to the nodes of the constraint graph, i.e. the variables of a CSP

Construction 1 (CSP for graph matching). Given two graphs $L = (L_V, L_E, L_L, s_L, t_L, \ell_L)$ and $G = (G_V, G_E, L_G, s_G, t_G, \ell_G)$ according to Def. 6. Then we construct a CSP as follows:

- $X = L_V \cup L_E = \{x_1, \ldots, x_n\}$, $n = |X|$.
- $D_i = \begin{cases} G_V, & \text{when } x_i \in L_V \\ G_E, & \text{otherwise} \end{cases}$, $i \in \{1, \ldots, n\}$.
- The constraint set R is built according to Table 1: whenever a condition listed in the left column of the table holds for a given pair of variables (x_i, x_k), the corresponding constraint is to be included in R.

Table 1. Conditions for when to establish a constraint between two variables x_i, x_k

Condition	$\longleftrightarrow$	Constraint $\in R$
$x_i = x_k$		$C^{\text{type}}_{(x_i)} = \{d \in D_i \mid \ell_L(x_i) = \ell_G(d)\}$
$x_i \in L_E$, $x_k \in L_V$, $s_L(x_i) = x_k$		$C^{\text{src}}_{(x_i, x_k)} = \{(d_i, d_k) \in D_i \times D_k \mid s_G(d_i) = d_k\}$
$x_i \in L_E$, $x_k \in L_V$, $t_L(x_i) = x_k$		$C^{\text{tar}}_{(x_i, x_k)} = \{(d_i, d_k) \in D_i \times D_k \mid t_G(d_i) = d_k\}$

It can be easily shown that the resulting CSP is equivalent to the original graph matching problem. For a straight forward proof of this result, cf. [Rud97].

4   Variable Ordering and Intelligent Backtracking

In this section we give a brief sketch of some basic issues arising when it comes to solve a CSP algorithmically. These are issues well-known to the CSP community, which are discussed in more detail in the literature on the subject. We recommend [Kum92] as a good starting point for further reading.

The standard approach to solve a CSP is using backtracking. Given a CSP like the one in Fig. 4 with certain variable and value orders, backtracking essentially performs a depth-first search on the depicted search space of potential solutions.

The good thing about backtracking is that whenever a constraint check fails for a variable x_i, it eliminates an exponential subproblem. Let's elaborate on that. In a CSP with a number of $|X|$ variables and an average domain size of $|D_\emptyset|$, the total number of variable instantiations to be checked amounts to about $|D_\emptyset|^{|X|}$. Now if instantiation fails for a variable x_i with some value, we have $|D_\emptyset|^{|X|-i}$ instantiation variants less left to consider. Interesting enough, the optimization effect grows exponentially with decreasing i. This leads us directly to the most prominent issue in variable ordering: the *first fail principle* suggests that variables with the highest probability of instantiation failure should always

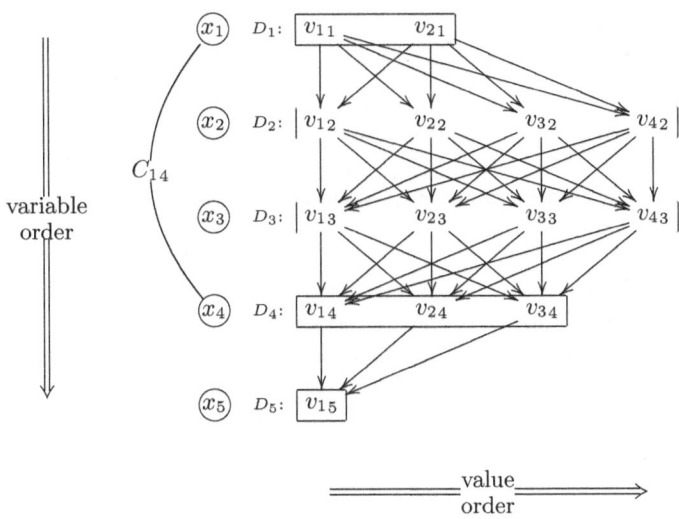

Fig. 4. The search space of a sample CSP

come first. A good starting point for how to determine a variable's "failure probability" is the number of constraints it is involved in, but in the end a good variable order is a matter of heuristics highly dependent on the particular application environment. Even the first fail principle, as evident as it may be, is not appropriate in all cases; if you are not interested in all the solutions, but want to find only one solution as fast as possible, you are likely to be better off using a *succeed first* strategy instead.

Since the impact of the applied variable ordering heuristic is so drastic, it is always worth to investigate and experiment. Of course there's a great variety of studies already available on the topic, e.g. [GMP+96,Smi97].

The bad thing about backtracking is, however, that it is dumb in that it often keeps looking for consistent instantiations in some subspace of the search tree where for some reason it is already evident that the search must fail. This kind of systematic misbehaviour is commonly known as *thrashing*, and it is mostly responsible for the poor performance of simple backtracking in practice. Diverse efforts have been made to overcome these deficiencies; they are subsumed under the collective term *intelligent backtracking*. As an example, we'll now have a look at a typical thrashing situation and at the idea how to resolve it.

Example 3 (Thrashing). Given the CSP constellation in Fig. 4, let's say we have successfully instantiated the variables x_1 through x_3 with values v_{11}, v_{12}, and v_{13}, respectively, using the standard backtracking procedure. Now suppose there's no consistent instantiation of x_4 w.r.t. C_{14} and the current value of x_1: $\forall v \in D_4 : (v_{11}, v) \notin C_{14}$. As a consequence, the subsequent attempt to instantiate x_4

will fail. A back step is due, and the next possible value v_{23} will be assigned to x_3. However, since the constraint C_{14} having caused the failure in the first place is independent of x_3, reattempting to instantiate x_4 will inescapably fail again. Obviously, the only chance to break out of the "dead end" at x_4 is to change the value of x_1. But the simple backtracking scheme will not revisit x_1 unless it has finished "thrashing" at all the possible value combinations of the variables x_2 and x_3. Evidently, this kind of thrashing situation imposes an overhead growing exponentially with the number of variables enclosed by the constraint which caused the failure.   △

This is where *backjumping* comes in. Backjumping is a simple intelligent backtracking algorithm that tries to be smart about just the kind of situation discussed above. Its basic idea is just as simple as it is appealing: "Whenever a dead-end occurs during variable instantiation, determine the culprit of failure and jump directly back to revise the variable concerned, without revisiting all the variables in between." But the backjumping algorithm has some drawbacks, too, and it only addresses one of the many causes for thrashing behaviour. Therefore, numerous approaches are out there trying to be much more intelligent about eliminating virtually any kind of thrashing. For further reading on intelligent backtracking, you may refer e.g. to [Dec97,Bru81,Kon94,Pro93].

But be aware that, as [Kum92] puts it, "a simple intelligent backtracking scheme may turn out to have less overall complexity than a more complicated intelligent backtracking"; this is due to a tendency towards an increasing expense on the maintenance of auxiliary data structures in the more sophisticated approaches.

5   Domain Reduction by Queries

By now, we have focussed our attention on the optimization of only one dimension of the search space: both intelligent backtracking and variable ordering address optimization *in depth*. These approaches are most promising as they directly affect the exponential factor of the worst-case complexity term $|D_\emptyset|^{|X|}$, so they are likely to have substantial impact on the actual runtime performance.

Although it may seem less profitable to consider optimization *in breadth* since it is only concerned with the base factor $|D_\emptyset|$, it is still worthwhile to have a closer look. What optimization in breadth actually means is trying to cut down on the overall number of constraint checks by reducing the domains in some way. In particular, the query approach to dynamic domain reduction presented below will often be able to cut down a variable's domain by orders of magnitude, having considerable implications on runtime performance.

Of course, we may only remove a value from a domain if it is certainly not part of any of the CSP's solutions. Obviously, for a value $v \in D_{x_i}$ to be part of a solution, it is a necessary condition that v satisfies every constraint $C_{(x_i,x_k)}$ when x_k has been instantiated previously. We call these conditions *dynamic* as they depend on the state of other variables which have been instantiated previously in the solution process. The observation is that sometimes, the value domain is structured in a way providing us with operations to obtain exactly the set of all values satisfying such a condition; this set can then be used as a properly reduced domain. Let us consider an example:

Example 4 (Dynamic domain reduction). Suppose that in the situation depicted in Fig. 2, x_2 is instantiated by d_2 while no other variables have any values assigned. As a necessary condition for a consistent value $\Gamma(x_3)$ for x_3, we have $(d_2, \Gamma(x_3)) \in C^{\text{tar}}_{(x_2, x_3)}$ (cf. Fig. 3). With the definition of the target constraint, we get $\Gamma(x_3) = t(d_2)$. The term $t(d_2)$, however, can be evaluated in constant time yielding the only consistent value d_3, whereas the standard procedure would have taken us linear time to find the same result in the domain D_3 by trial and error.

Things look different, though, when we start the other way round, looking for a consistent instantiation of x_2 while d_3 is already assigned to x_3. The necessary condition imposed by $C^{\text{tar}}_{(x_2, x_3)}$ leads to $d_3 = t(\Gamma(x_2))$. By "looking close" we can easily see that only d_2 matches this condition; but since there is no inverse counterpart to the target operation available in our graph model, we cannot resolve the equation to $\Gamma(x_2)$. At this time, we have to descend from the abstract model down to the implementation level. Considering e.g. AGG, its graph model implementation does in fact provide direct access to the outgoing and incoming arcs of a node, representing exactly the concept of inverse relations to the source and target operations we were missing in the abstract model. Thus, we can still get constant time access to the reduced domain we were looking for.   △

The example teaches us that in the end, to do this kind of dynamic domain reduction efficiently we need to know about the properties of the domain's implementation. This gives rise to a problem seeming contradictory: On one hand, we want to exploit concrete knowledge about the properties of domain implementation for the sake of optimization, on the other hand we want to settle the solution algorithms which are to utilize that knowledge on the abstract level of a general CSP, just to decouple them from the issues of the concrete graph model. Fortunately, this kind of problem has been solved before: hiding their concrete definition behind the abstract notion of satisfaction, the constraints themselves are the perfect example for the concept of abstract representation of concrete model properties. Following this example, we introduce the notion of a *query*[5] to the abstract framework of the CSP:

Definition 8 (Query). *Given a CSP Z with variable set $X = \{x_1, \ldots, x_n\}$ and corresponding domains D_i, $i \in \{1, \ldots, n\}$. Also given a subset $Y \subset X$, a consistent instantiation Γ of Y and a variable $x_t \in X \setminus Y$. A Query $Q_{Y, x_t}(\Gamma)$ is then a subset of D_t so that the following condition holds:*

$$\forall d \in D_t \setminus Q_{Y, x_t}(\Gamma):$$
$$\not\exists \Gamma_X \quad \text{such that} \quad \forall x \in Y : \Gamma_X(x) = \Gamma(x),$$
$$\Gamma_X(x_t) = d \quad \text{and}$$

---

[5] Name and idea have been inspired by the use of queries in [Zün96].

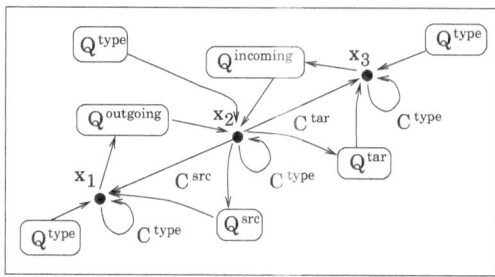

Fig. 5. The constraint graph of Fig. 3 with some typical queries

Γ_X *is a solution of* Z.

The elements of Y *are called the* sources *of* $Q_{Y,x_t}(\Gamma)$, *while* x_t *is known as the* target *of the query.*

A query $Q_{Y,x_t}(\Gamma)$ *is called* constant *if and only if* $Y = \emptyset$. *We may also write* Q_{x_t} *in that case.*

Queries may be used for the purpose of dynamic as well as static domain reduction, where the latter is the job of constant queries. Note that static domain reduction corresponds to the notion of *node consistency* mentioned in Sect. 3.

Furthermore, note that the use of queries is not limited to situations where the concrete domain a priori provides the data structures for efficient querying. It may well be worth the effort to do some linear time preprocessing on the domain to build up the structures for additional queries to operate upon.

In a constraint graph, a query Q_{Y,x_t} is represented by a hyperedge originating in the variables of Y and terminating in x_t. Figure 5 shows the constraint graph from Fig. 3, complemented by some simple queries. Since none of these queries has more than one source variable, there are no actual hyperedges displayed. As an example on how the definition of a query may look like, we consider $Q^{\mathrm{tar}}_{\{x_2\},x_3}$:
$Q^{\mathrm{tar}}_{\{x_2\},x_3}(\Gamma) = \{t\,(\Gamma(x_2))\}$. Note that the queries for incoming and outgoing arcs cannot be defined in terms of the definition of our graph model.

6   CSP Construction Revisited

The way we constructed a CSP out of a graph matching problem in Sect. 3 was pretty much straight forward in that we had one variable in the CSP for each object of the graph to be matched. Of course, this is not the only possible way to do the translation. One interesting alternative is motivated by the observation that the matching of any arc uniquely determines the matching of its source and target nodes. The idea is then that we only need to match the arcs of a graph, since the nodes are matched implicitly – except for isolated nodes. As a consequence, we can choose the set of arcs of the graph to be matched as the set of variables of the CSP, augmented by one variable for each isolated node.

Then for any two connected arcs we need a constraint to enforce structural compatibility on the solutions. As type compatibility is concerned, we need two different kinds of constraints to achieve this. For the isolated nodes we have the same type constraints as known from Sect. 3. The type constraints for the arcs, however, are different in that they have to include type compatibility checks for their source and target nodes as well, since these are not explicitly represented in the CSP.

Using the approach as outlined above, we may get a CSP representation of a graph matching problem which has typically about half the number of variables compared to the straight forward representation. At first, this looks like a tremendous improvement, for it reduces the worst-case complexity term by orders of magnitude. On a closer look, however, the approaches appear to be just two implementations of the same idea. The difference in the size of the variable sets is made up solely by the source and target nodes of arcs which are not explicitly represented in the second approach. But in the first approach, we were able to find consistent instantiations for these variables without any searching too, by using the dedicated source and target queries once the corresponding arc was instantiated. This means that these variables cannot be taken into account for the exponential factor of the worst-case complexity term either.

To conclude, both approaches are addressing the fact that that matching an arc uniquely determines the matching of its source and target nodes, but they do so in different ways. The first approach implements this explicitly by using queries, while the second one proposes an implicit solution by not representing the source and target nodes at all. However, this doesn't mean that the second approach renders the idea of queries obsolete: it will benefit from additional type and structure queries just as well. In particular, we'd want a query similar to the former incoming/outgoing queries, yielding the candidates for adjacent arcs.

Note that this alternative CSP representation has not yet been tested within the AGG system. Due to the above-mentioned reasons, no significant performance improvement is to be expected.

7   Conclusion

In this paper, we proposed a way to represent and solve the graph matching problem as a CSP. The main benefit of this approach is that we gain direct access to the rich research findings in the CSP area; instead of inventing new algorithms for graph matching from scratch, we can now apply well-elaborated CSP solution algorithms right "out of the box".

Another important advantage is that the actual solution algorithm becomes independent of the concrete graph model, allowing us to change the model without affecting the algorithm; we only have to adapt the translation step from the graph model into the CSP representation. In spite of this additional level of abstraction, solution algorithms can still be highly optimized by exploiting crucial properties of the graph model implementation. To accomplish this, we introduced the query concept to represent properties of the concrete model on the more abstract CSP level. As an additional means to provide model knowledge for optimized heuristics, we may assign weighting information to constraints, queries, and variables [BMR97], reflecting e.g. the "tightness" of a constraint or the estimated size of a query result.

As an effect of the decoupling of the graph model from the solution process, it becomes feasible now to develop and reuse software libraries of algorithms for CSP solution and variable ordering. To some extent, the CSP framework implemented for the AGG system [Rud97] can be seen as a prototype of such a library.

As for the drawbacks of the CSP approach to graph matching, we have to accept a linear computation overhead for the translation steps between the graph model and the CSP representation. In general, this overhead is neglectable, since the problem itself dominates the complexity. But even then, since the constraint graph of the CSP can be generated from the graph rule's left-hand side alone, most of the overhead can be moved away from matching time to editing time.

We could only give a very limited insight into the vast field of optimization issues. For example we did not consider *relational consistency algorithms* at all [Mac77] (also known as *path consistency* or *constraint propagation*), which are frequently used to reduce the amount of backtracking needed in the solution process. Based on a minimal graph model, McGregor [McG79] did some experimental comparison of several relational consistency algorithms when applied to the subgraph isomorphism problem.[6] He found that only the simplest consistency algorithms were cost-effective. However, recent studies [FD96] suggest that the larger and harder the problem instances are, the more constraint propagation may pay off. Note that the query concept introduced here may be seen as a very limited, yet extremely cost-effective, form of constraint propagation as well.

A lot of future work remains to be done, most notably systematical empirical testing is due in order to compare the performance of various solution algorithms and ordering heuristics on typical instances of the graph matching problem.

Acknowledgements

Reiko Heckel pointed me to the CSP field right on time. Mark Minas and Gabriel Valiente contributed ideas and comments.

References

AGG98.   AGG Homepage, 1998. http://tfs.cs.tu-berlin.de/agg. 239
BMR97.   S. BISTARELLI, U. MONTANARI, and F. ROSSI. Semiring-based Constraint Solving and Optimization. *Journal of the ACM*, 44(2):201–236, March 1997. 250

---

[6] Actually, he used a simple CSP representation of the problem, without starting any systematical investigation on that matter.

Bru81.   M. BRUYNOOGHE. Solving Combinatorial Search Problems by Intelligent
            Backtracking. *Information Processing Letters*, 12(1):36–39, 1981. 246
Dec97.   R. DECHTER. *Backtracking algorithms for constraint satisfaction problems –
            a survey.* Technical Report, University of California, Irvine, January 1997.
            ftp://ftp.ics.uci.edu/pub/CSP-repository/papers/backtracking.ps. 246
DvB97.   R. DECHTER and P. VAN BEEK. Local and Global Rela-
            tional Consistency. *Theoretical Computer Science*, 173:283–308, 1997.
            ftp://ftp.cs.ualberta.ca/pub/vanbeek/papers/tcs97.ps. 239
FD96.   D. FROST and R. DECHTER. Looking at full look-ahead. In *Proc. 2nd
            Intl. Conf. on Constraint Programming (CP-96)*, 1996. 250
GMP⁺96.   I. P. GENT, E. MACINTYRE, P. PROSSER, B. M. SMITH, and T. WALSH. An
            Empirical Study of Dynamic Variable Ordering Heuristics for the Constraint
            Satisfaction Problem. In E. C. FREUDER (ed.), *Proc. 2nd Intl. Conf. on
            Principles and Practice of Constraint Programming*, volume 1118 of *Lec-
            ture Notes in Computer Science*, pages 179–193. Springer Verlag, 1996.
            http://www.cs.strath.ac.uk/ apes/papers/CRCcp96gmpsw.ps.gz. 245
Kon94.   G. KONDRAK. *A Theoretical Evaluation of Selected Backtracking Al-
            gorithms.* Technical Report TR94-10, University of Alberta, 1994.
            ftp://ftp.cs.ualberta.ca/pub/TechReports/TR94-10.ps.Z. 246
Kum92.   V. KUMAR. Algorithms for Constraint Satisfaction Problems: A Survey. *AI
            Magazine*, 13(1):32–44, 1992.
            http://www.cirl.uoregon.edu/constraints/archive/kumar.ps. 243, 244, 246
Mac77.   A. K. MACKWORTH. Consistency in Networks of Relations. *Artificial In-
            telligence*, 8(1):99–118, 1977. 243, 250
McG79.   J. J. MCGREGOR. Relational Consistency Algorithms and Their Applica-
            tion in Finding Subgraph and Graph Isomorphisms. *Information Sciences*,
            19:229–250, 1979. 250
Meh84.   K. MEHLHORN. *Graph Algorithms and NP-Completeness.* Springer Verlag,
            1984. 238
Pro93.   P. PROSSER. Hybrid Algorithms for the Constraint Satisfaction Problem.
            Computational Intelligence, 9(3):268–299, 1993. 246
RDP90.   F. ROSSI, V. DAHR, and C. PETRIE. On the equivalence of constraint satis-
            faction problems. In *Proc. Europ. Conf. on Artificial Intelligence (ECAI90)*,
            August 1990. Also MCC Technical Report ACT-AI-222-89. 240
Rud97.   M. RUDOLF. *Konzeption und Implementierung eines Interpreters
            für attributierte Graphtransformation.* Master's thesis, Technical Uni-
            versity of Berlin, Dep. of Comp. Sci., 1997. http://tfs.cs.tu-
            berlin.de/publikationen/public1997.html. 239, 244, 250
Sch97.   A. SCHÜRR. Programmed Graph Replacement Systems. In G. ROZENBERG
            (ed.), *Handbook of Graph Grammars and Computing by Graph Transforma-
            tions, Volume 1: Foundations.* World Scientific, 1997. 238
Smi97.   B. M. SMITH. Succeed-first or Fail-first: A Case Study in Variable and
            Value Ordering Heuristics. In G. SMOLKA (ed.), *Proc. 3rd Intl. Conf. on
            Principles and Practice of Constraint Programming*, volume 1330 of *Lec-
            ture Notes in Computer Science*, pages 321–330. Springer Verlag, 1997.
            file://www.scs.leeds.ac.uk/scs/doc/reports/1996/96_26.ps.Z. 245
Zün95.   A. ZÜNDORF. *Eine Entwicklungsumgebung für PROgrammierte GRaphEr-
            setzungsSysteme.* PhD thesis, RWTH Aachen, 1995. 238
Zün96.   A. ZÜNDORF. Graph Pattern Matching in PROGRES. In *Proc. 5th
            Intl. Workshop on Graph Grammars and their Application to Computer*

Science, volume 1073 of *Lecture Notes in Computer Science*, pages 454–468. Springer Verlag, 1996.   247

Conceptual Model of the Graphical Editor GenGEd for the Visual Definition of Visual Languages

Roswitha Bardohl and Hartmut Ehrig

Department of Computer Science
Technical University Berlin, Germany
{rosi,ehrig}@cs.tu-berlin.de

Abstract This contribution presents a conceptual model of GenGEd, an editor supporting the visual definition of visual languages (VLs). A VL is defined by an alphabet and a grammar. These constituents are the input of a syntax-directed graphical editor allowing the manipulation of visual sentences over VL.

The conceptual framework of GenGEd is based on algebraic graph transformation and algebraic specification techniques. Starting with a type signature METAVISUAL for visual alphabets, the user of GenGEd can define a specific VL-alphabet including graphical constraints. The VL-grammar definable on top of the VL-alphabet consists of a start sentence and a set of VL-rules. Each VL-rule is defined by a graph grammar rule in the sense of algebraic graph transformation.

Keywords: visual definition of visual languages, generation of syntax-directed graphical editors, typed algebraic graph grammars

1   Introduction

Visual languages (VLs) are used within various application areas: teaching children and adults, programming for non-programmers, adaption of standard software to individual requirements, development of graphical user interfaces, etc. VLs are also used for software development, especially for the analysis and design of software systems. Well known examples of visual modelling and specification languages are UML, the Unified Modelling Language ([Cor98]), automata, Petrinets and so on. All these languages can be supported by graphical tools. Common to all graphical tools is the fact, that they offer a VL instead of a textual one allowing the manipulation of visual sentences. But most of these tools do not allow the visual definition of a visual language. Of course, most of the graphical editors do not allow the definition of a VL; they are not intended to allow this, as they are tuned to a specific VL. Problems arise when graphical means of a VL are changed. Then, fixed implemented graphical editors must be redesigned and a reimplemented, leading to a time and cost intensive affair. Such problems are avoided by the GenGEd environment.

H. Ehrig et al. (Eds.): Graph Transformation, LNCS 1764, pp. 252–266, 2000.
© Springer-Verlag Berlin Heidelberg 2000

The GENGED **environment** supports the visual definition of VLs on the one hand and the generation of syntax-directed graphical editors for a defined VL on the other hand. Similar to textual languages a VL consists of an alphabet and a grammar, called VL-alphabet and VL-grammar, respectively. In contrast to textual languages the alphabet of a VL is rather complex consisting of textual and graphical components. The VL-alphabet can be defined using an alphabet editor, whereas a grammar editor allows the definition of a VL-grammar based on the defined alphabet. A VL-grammar consists of one VL-sentence (the start graph) and a set of VL-rules (graph grammar rules). These VL-rules are the edit commands of a generated syntax-directed graphical editor allowing the correct manipulation of sentences wrt. the defined VL. The main components of GENGED are illustrated by Figure 1 (a).

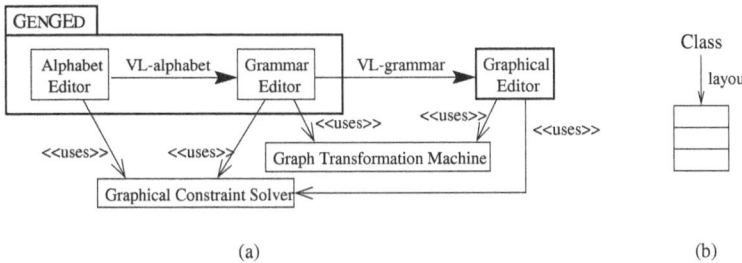

(a)   (b)

Figure1. GENGED-components (a) and one graphical symbol (b).

A VL-alphabet comprises graphical symbols and links between them. A graphical symbol consists of a name (type) and a graphic, both are connected by a layout operation. One graphical symbol for a class occuring in class diagrams (cf. [Cor98]) is presented in Figure 1 (b). Links between graphical symbols are defined by unary operations and graphical constraints. These constraints define equations between the layout of symbols. The alphabet definition results in an algebraic specification, called alphabet specification. Visual sentences occuring within a VL-grammar, that is the start graph as well as the constituents of a VL-rule, e.g., its left– and right-hand-side, are algebras wrt. the alphabet specification. The structure of a visual sentence is twofold: it consists of a logical level (the abstract syntax) and a visual level (the concrete syntax). Both levels are connected by layout operations. The manipulation of visual sentences is done by a graph transformation machine responsible for the logical level together with a graphical constraint solver which is responsible for the graphical layout.

The first ideas towards GENGED are presented in [BT97] where algebraic specification techniques and graph grammars are used to describe VLs. Additionally it is briefly discussed which kind of editors may support the visual definition of VLs. Based on these ideas the first implementation of GENGED started in

October 1997 within a student's project where we gathered a lot of experiences[1]. These experiences are not only concerned with implementation issues but additionally with aspects of how to define a VL using GENGED. These topics are discussed in [Bar98]. Both articles, however, do not provide a conceptual model of GENGED based on typed algebraic signatures and meta-modelling.

In this contribution we present such a conceptual model of GENGED which uses concepts of algebraic graph transformation [EHK+97], algebraic specifications [EM85] and typed algebraic graph signatures in analogy to typed graphs [CEL+96].

The paper is organized as follows: In Section 2 we present our conceptual algebraic framework for GENGED which is illustrated by an example of a visual alphabet. In Section 3 the alphabet editor is discussed wrt. to this example, i.e. it is shown which part of the alphabet is built up by the components of the alphabet editor. The visual grammar will be discussed in Section 4 together with the grammar editor and the generated graphical editor. Related work is presented in Section 5 and the conclusion in Section 6.

2   Conceptual Algebraic Framework

The conceptual framework of the generic graphical editor GENGED is based on algebraic graph transformation (cf. [EHK+97]) and algebraic specifications (cf. [EM85]). In this section we review the notion of algebraic graph signatures corresponding to graphs. In analogy to typed graphs (cf. [CEL+96]) they are extended to typed algebraic graph signatures (similar to [Löw97]) which are suitable for meta-modelling. This allows to formulate a meta-modelling concept on GENGED which is additionally explained in this section and which will be used in the subsequent sections.

Typed Algebraic Graph Signatures. An **algebraic graph signature** SIG = (S, OP) with sort symbols S and operation symbols OP is a signature in the sense of algebraic specifications (cf. [EM85]) where all operation symbols are unary. This implies that an algebraic graph signature can be considered as a graph, where the vertices are the sorts and the edges are the operation symbols.

Given an algebraic graph signature TSIG, called **type signature**, a typed algebraic graph signature (SIG, t: SIG → TSIG) is an algebraic graph signature SIG together with a signature morphism t: SIG → TSIG, called **type morphism**. In this case SIG is called to be **typed over** TSIG. Since signature morphisms are closed under composition we immediately have

Fact 21 (*Transitivity of Typing*)
Given signatures SIG_1, SIG_2 and SIG_3. If SIG_3 is typed over SIG_2 and SIG_2 is typed over SIG_1 then also SIG_3 is typed over SIG_1.   △

---

[1] The reimplementation of GENGED starting now, is partially supported by the German Research Council (DFG), the TMR network GETGRATS, and the ESPRIT Basic Research Working Group APPLIGRAPH.

Another important observation is the fact that each SIG-Algebra A for an algebraic graph signature SIG can be transformed into a signature $T(A)$ typed over SIG:

Fact 22 (*Transformation of SIG-Algebras into Algebraic Graph Signatures*)
Given an algebraic graph signature SIG and a SIG-algebra A we obtain a new algebraic graph signature $T(A)$ typed over SIG as follows:

If A consists of domains A_s ($s \in S$) and operations op_A ($op \in OP$) for SIG $=$ (S, OP) then $T(A)$ consists of

- sort symbols $S_{T(A)}$ defined by the disjoint union of all domains A_s ($s \in S$),
- operation symbols $OP_{T(A)}$ defined for $a_i \in A_{s_i}$, ($i = 1, 2$) by
  $OP_{T(A)}\ a_1, a_2 = \{(op, a_1, a_2) \mid op \in OP_{s_1, s_2}$ and $op_A(a_1) = a_2\}$,
- typing $t(a) = s$ for $a \in A_s$, $s \in S$ and $t(op, a_1, a_2) = op$.

△

Meta-Modelling Concepts on GENGED. The notions defined above form the basis for the meta-modelling concepts on GENGED (cf. Figure 2): Given META$_{\mathrm{VISUAL}}$, a signature for VL-alphabets, the alphabet editor (cf. Figure 1) allows the user to define an algebra wrt. META$_{\mathrm{VISUAL}}$ and further constraints. The algebra will be transformed into a SIG-algebra, called alphabet signature, which is typed over META$_{\mathrm{VISUAL}}$. Constraints which are used to link graphical symbols on the visual level are interpreted as equations, which yields an alphabet specification, called VL-alphabet. Visual sentences defined over the VL-alphabet are first of all algebras wrt. the VL-alphabet satisfying the equations. Again these algebras can be transformed into SIG-algebras, called sentence signature, which are typed over the alphabet signature. Because typing is transitiv visual sentences are also typed over META$_{\mathrm{VISUAL}}$. Note, visual sentences occur in a VL-grammar as well as in VL-sentences which are derived by applying VL-grammar rules.

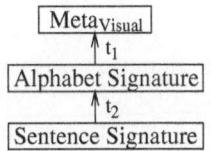

Figure2. Meta-modelling concepts on GENGED.

Definition 23 (META$_{\mathrm{VISUAL}}$)
Let us regard the algebraic graph signature META$_{\mathrm{VISUAL}}$ which is implemented by GENGED:

$\underline{\text{META}_{\text{VISUAL}}} =$

Graph Part	**sorts** V, E, AC$_\text{V}$, AC$_\text{E}$
	opns source, target: E $\rightarrow$ V;
	op$_\text{V}$: AC$_\text{V}$ $\rightarrow$ V;
	op$_\text{E}$: AC$_\text{E}$ $\rightarrow$ E; $\tilde{[}$.1in]
Data Part	**sorts** Data
Attribution Part	**opns** attr$_\text{V}$: AC$_\text{V}$ $\rightarrow$ Data;
	attr$_\text{E}$: AC$_\text{E}$ $\rightarrow$ Data;
Graphic Part	**sorts** Graphic
Layout Part	**opns** layout$_\text{V}$: V $\rightarrow$ Graphic
	layout$_\text{E}$: E $\rightarrow$ Graphic
	layout$_{\text{AC}_\text{V}}$: AC$_\text{V}$ $\rightarrow$ Graphic
	layout$_{\text{AC}_\text{E}}$: AC$_\text{E}$ $\rightarrow$ Graphic
	layout$_\text{Data}$: Data $\rightarrow$ Graphic

$\triangle$

Example 24 (*VL-Alphabet for Class Diagrams*)
An alphabet for a visual language VL in GENGED is defined to be an algebra of signature META$_{\text{VISUAL}}$. As specific example we consider a simple version of class diagrams (cf. [Cor98]) defining a visual language *CD*. The alphabet for *CD* is given by a META$_{\text{VISUAL}}$-algebra A_{CD}. The algebra restricted to the graph part, for example, is given by

$$A_{CD/GraphPart} = (A_{CD,V}, A_{CD,E}, A_{AC_V}, A_{AC_E}, source^A, target^A, op_V^A, op_E^A)$$

with the carrier sets

$$A_{CD,V} = \{CD, Class\}, \quad A_{CD,E} = \{Incl\_In, Assoc\},$$
$$A_{AC_V} = \{CN, AL, ML\}, \quad A_{AC_E} = \{AN\},$$

and operations defined on the carrier sets:

$$source^A(Incl\_In) = Class; \ target^A(Incl\_In) = CD;$$
$$source^A(Assoc) = Class; \quad target^A(Assoc) = Class;$$
$$op_V^A(AN) = Assoc;$$
$$op_E^A(CN) = op_E^A(AL) = op_E^A(ML) = Class;$$

According to Fact 22 the algebra A_{CD} is transformed into an algebraic graph signature T(A_{CD}). This signature with its sort and operation symbols is illustrated in the upper part of Figure 3. Note, the signature is typed over META$_{\text{VISUAL}}$; the typing is not illustrated in this figure. Vertex and attribute carrier sorts are depicted by rectangles, edge sorts by ovals, and the operations are depicted by arrows. The sort symbol CD is used for a *Class Diagram* whose graphic should include all classes and associations. A class (Class) is defined together with its attributes, i.e. the attribute carriers together with the corresponding data types String and ListOfString (short LOS). These attributes are used for a class name (CN), an attribute list (AL), and a method list (ML). An association (Assoc) is attributed by an association name (AN). Associations connect classes; this is expressed by the source and target operations. Each graphical object shown in the lower part of Figure 3 is connected (dashed arrows) with an item of the logical level, i.e. an item of the graph part, data part or the attribute part wrt. META$_{\text{VISUAL}}$. These dashed lines respresent the layout operations. On top of the alphabet signature graphical constraints can be used to express the links between graphical objects according to the operations op$_V$ and op$_E$ wrt. META$_{\text{VISUAL}}$. The constraints, not shown in Figure 3, are equations and extend the alphabet signature VISUALSIG to an alphabet specification VISUALSPEC, called VL-alphabet for the language *CD*, in the sense of Definition 25.   △

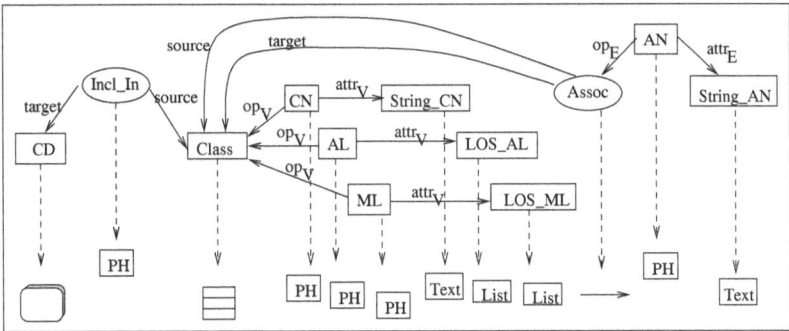

Figure3. Visualization of an alphabet signature for the visual language of class diagrams.

Definition 25 (*Alphabet Signature and VL-Alphabet*)
An alphabet signature for GENGED is an algebraic graph signature which is typed over META$_{\text{VISUAL}}$ and denoted as VISUALSIG. A visual specification VISUALSPEC, called VL-alphabet, is an extension of VISUALSIG by equations expressing graphical constraints.   △

Visual sentences are algebras wrt. VISUALSPEC. They occur in VL-grammars as well as they are VL-sentences which are visual sentences derived by applying grammar productions. A visual sentence, for example, is an algebra consisting only of two class symbols; the other carrier sets are empty. Then, all operations are given by empty mappings. VL-grammars and the application of grammar productions are discussed in Section 4. However, according to Fact 22 visual sentences are transformed into sentence signatures, which are typed over the alphabet signature VISUALSIG.

Definition 26 (*Sentence Signature*)
A sentence signature for GENGED is an algebraic graph signature typed over VISUALSIG. VISUALSIG is again typed over META$_{\text{VISUAL}}$ and because of transitivity of typing also VISUALSIG is typed over META$_{\text{VISUAL}}$.   △

The algebraic framework discussed above allows to formulate the following meta-modelling concept for GENGED:

Definition 27 (*Meta-Modelling Concept*)

Step 1:   Define a META$_{\text{VISUAL}}$-algebra A. By Fact 22 above we obtain an algebraic graph signature T(A), called VISUALSIG, which is typed over META$_{\text{VISUAL}}$. The alphabet signature VISUALSIG for GENGED is fixed for each visual language VL.

Step 2:   Each VISUALSIG-algebra A_{VL} can be represented by Fact 22 as an algebraic graph signature T(A_{VL}) typed over VISUALSIG. By Fact 21 T(A_{VL}) is also typed over META$_{\text{VISUAL}}$.

$\triangle$

Note, each VISUALSPEC algebra is additionally a VISUALSIG algebra (cf. [EM85]).

The conceptual model of GENGED in the following sections is based on this meta-modelling concept. Furthermore, in Section 3 we will extend the signature VISUALSIG to a specification VISUALSPEC using graphical constraints as discussed above, such that all visual sentences of a visual language VL become VISUALSPEC-algebras.

3   Visual Alphabet and Alphabet Editor

The alphabet editor allows to define VL-alphabets. More precisely, we first consider an algebra A wrt. META$_{\text{VISUAL}}$. This algebra is transformed into a signature VISUALSIG, typed over META$_{\text{VISUAL}}$. Furthermore, as we will show in this section, the alphabet editor allows to define graphical constraints. These constraints are transformed into equations, such that VISUALSIG is extended to a visual specification VISUALSPEC which is the VL-alphabet established by using the alphabet editor. The alphabet editor consists of several components illustrated in Figure 4 and discussed below.

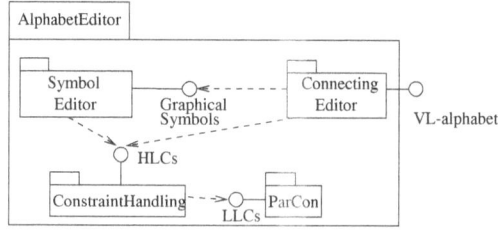

Figure4. Components of the alphabet editor.

Symbol Editor. The symbol editor (SE) is an implementation of a fixed graphic algebra G, i.e. SE allows the drawing of graphical objects as given in the lower part of Figure 3 for establishing graphical symbols.

A graphical object consists either of one primitive object or of several primitive objects which are connected by graphical constraints. Graphical primitives are given by rectangles, circles, and in addition by place holders as "PH" in Figure 3. Such place holders can be used to denote a graph sort which should not be visible for further progress. Graphical objects are described by terms and defining equations. They are generic because of the position and the size.

Not only the drawing of graphical objects is possible using the SE but additionally the definition of graph sorts typed over the graph part and data sorts typed over the data part of META$_{\mathrm{VISUAL}}$. All these sorts have to be connected with a generic graphic constituting the layout definitions of VISUALSIG. Based on built-in data types for strings and lists of strings a specific data type for each VL is constructed leading to the data sorts of VISUALSIG. This yields a data signature DATASIG and a corresponding algebra D. Corresponding data objects are represented graphically by "Text" resp. "List".

Altogether, the symbol editor provides a fixed graphic algebra G which is the implementation of a graphic signature GRAPHICSIG. It allows for each VL the establishment of specific graphical symbols as e.g. the symbol called CD in Figure 3 together with the layout operation and the graphic.

Connecting Editor. The connecting editor (CE) allows to define the operations typed over the operations in the graph part and the attribution part of META$_{\mathrm{VISUAL}}$. This definition is based on graphical symbols which are established using the symbol editor. In the connecting editor all source and target operations, as well as attribution operations have to be defined visually (by selection) together with some layout constraints. These layout constraints are interpreted as equations over a VISUALSIG which yields an algebraic specification VISUAL-SPEC.

We have to mention that usually we do not differ between common graphical symbols and data type symbols in the CE. Both are interpreted as graphical symbols, but a data type symbol can only be connected with a graphical symbol which represents a vertex or an edge. I.e. two data type symbols cannot be connected.

Constraint Solver. The integrated graphical constraint solver PARCON [Gri96] works as a server. I.e. each editing area is a client of this server and gets its own solver incarnation. PARCON is implemented in Objective C, whereas all other components discussed in this paper are implemented in Java. The purpose of PARCON within GENGED is to solve the layout constraints for the construction of graphical objects.

Constraint Handling. The constraint handling component offers a high-level constraint language. Each high-level constraint (HLC) will be translated into corresponding low-level constraints (LLC) the constraint solver PARCON works with. The predefined HLCs are collected within a library which can be seen as a super set of the constraint language used by PARCON[2]. A HLC is e.g. given by "Rect1 Above Rect2". Such a HLC is transformed into several LLCs PARCON works with; i.e., it is breaked down into several LLCs under consideration of the x,y coordinates and the height of the involved rectangles.

Summing up, a VL-alphabet given by a visual specification VISUALSPEC is established when using the alphabet editor. This visual specification consists of

- a visual signature VISUALSIG defined in Step 1 of the meta-modelling concept for GENGED,
- a data signature DATASIG built over the data part of VISUALSIG with fixed DATASIG-algebra D for each visual language VL,
- a graphic signature GRAPHICSIG built over the graphic part of VISUALSIG with a GRAPHICSIG-algebra G fixed within GENGED. This GRAPHICSIG-algebra G includes generic graphical objects $g_1, \ldots, g_n$ of the graphic sort *Graphic*. These objects are specific for each visual language VL,
- layout constraints for each operation of VISUALSIG's graph and attribute part which yields equations extending VISUALSIG to VISUALSPEC.

A VISUALSPEC-algebra A is then an algebra with a fixed interpretation D for the data part and G for the graphic part of VISUALSIG. All rules in the following section are built over VISUALSPEC-algebras and all visual sentences of each visual language VL become VISUALSPEC-algebras.

4   Visual Grammar, Grammar and Graphical Editor

The grammar editor supports the definition of a VL-grammar, i.e. one visual sentence (the start graph) and a set of VL-rules. The edit commands of the grammar editor are alphabet rules automatically generated from the specification VISUALSPEC. The user defined VL-grammar will be the input of the generated graphical editor (cf. Figure 1) allowing the manipulation of VL-sentences by applying VL-rules.

4.1   Automatically Generated Alphabet Rules

The grammar editor uses a concrete VISUALSPEC (the output of the alphabet editor) to generate specific rules, called alphabet rules. Alphabet rules are generated allowing the insertion and deletion of graphical symbols. The algorithm behind the generation of alphabet rules is based on VISUALSPEC, or more concretely, on the graph part of VISUALSPEC. Once established the logical level of a rule, the graph items are extended by the layout descriptions. According to

---

[2] Surely, these predefined HLCs may not be sufficient to support all necessities for the definition of graphical symbols and their connections. Therefore, a user can extend the library by own constraints.

insertion rules for each sort symbol of the graph part a rule is generated with an empty lhs. The rhs comprises the graphical symbol, i.e. the graph item which is equipped with its layout. In a second step the operations of the graph part are considered. Whenever a sort symbol occurs in the domain of an operation, both sides of the rule are extended by the codomain of this operation. I.e., the corresponding graphical symbols (logical item together with its layout) are put into both sides. Additionally, the corresponding operation is defined in the rhs, and the connection constraints are solved, such that both sides of the alphabet rule are algebras wrt. VISUALSPEC. Deletion rules are built up inverse to the insertion rules; i.e., both sides of the alphabet rules are exchanged, but in addition we need negative application conditions (cf. Section 4.2.

Example 41 (*Alphabet Rule*)
Given the alphabet signature of Example 24. Figure 5 illustrates a generated alphabet rule allowing the insertion of an association which was defined to be an edge. In this figure L (resp. R) denotes the lhs (resp. rhs) of the rule. Furthermore, the upper parts of the figure denote the logical level (the graph part of VISUALSPEC) which is connected by layout operations (dashed arrows) with the layout of each element. Because each side of the rule is an algebra wrt. VISUAL-SPEC it satisfies the equations. I.e., for each operation (source, target) at least one equation was established from the user defined constraint. We assume, that according to our example for the source (target) operation an `equalPosition` constraint is used to express that the association arrow begins (ends) at the east corner (west corner) of the class layout. These equations are satisfied in the rhs of the alphabet rule. Note, although some of the carrier sets are empty, both sides of the rule are algebras; the corresponding operations are then given by empty mappings.

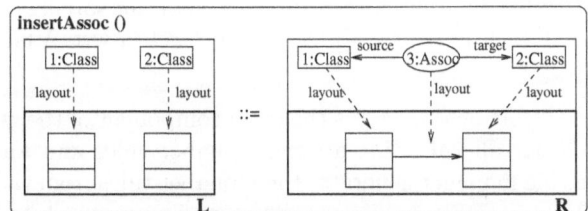

Figure5. Alphabet rule describing the insertion of an association.

△

Note, some of the automatically generated alphabet rules may express already intended VL-rules as one may think in the case of the alphabet rule presented in Figure 5. But sometimes it is useful to have more powerful VL-rules for editing VL-sentences in a generated graphical editor. The definition of VL-rules is supported by the grammar editor.

4.2   Grammar Editor

The grammar editor supports the definition of a VL-grammar by application of the automatically generated alphabet rules in a syntax-directed way. How a VL-grammar can be defined is explained along the components of the grammar editor illustrated in Figure 6.

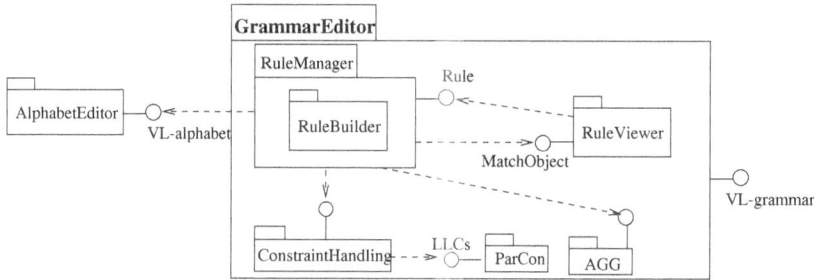

Figure6. Components of the Grammar Editor.

Rule Viewer. The RuleViewer supports the visualization of one rule. This rule can be one of the generated alphabet rules or one already constructed VL-rule. Furthermore, the rule viewer exports the objects which are selected by the user for a match.

Rule Builder. The RuleBuilder allows the definition of VL-rules consisting of a lhs, a rhs and a morphism between items of both sides. I.e., the RuleBuilder consists of two working graphs, one for the lhs and one for the rhs of a VL-rule. It supports the application of one rule visualized by the RuleViewer in both sides. The application of rules is based on graph transformation implemented by the AGG-system ([Rud97,TER99]) and graphical constraint solving (cf. Section 4.3). This implies that each part of a VL-rule is an algebra wrt. VISUALSPEC in all actual working states.

Rule Manager. The RuleManager is the main component of the grammar editor. It is responsible for all rules, i.e. for the alphabet rules and for the VL-rules. Especially, the rule manager generates the alphabet rules from the VL-alphabet.

AGG. The AGG-system (cf. [Rud97,TER99]) supports the graph transformation on the logical level of each algebra. The transformation is done using the single-pushout (SPO) approach as described in [EHK+97].

Constraint Handling and Constraint Solver. These components are already discussed in Section 3.

Using the grammar editor for defining VL-rules means to allow the definition of rule names and rule parameters for certain data type values which is supported by the RuleManager. According to our alphabet rule of Figure 5 the VL-rule describing the insertion of an association would be named like `insertAssoc(an: String)` where **an** denotes the variable and `String` the data type. The visual level of this rule is illustrated in Figure 7.

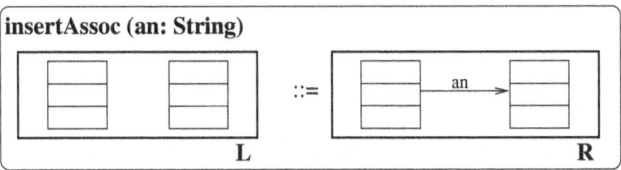

Figure7. Visual level of VL-rule describing the insertion of an association.

In the grammar editor we also have the possibility to define so-called negative application conditions (NAC) as proposed in [HHT96]. These NACs are bound by morphisms with the lhs of a rule. According to VLs such conditions allow to express that some situations must not occur when applying a VL-rule. This topic is mainly concerned with deletion rules, one is illustrated in Figure 8. It is required that the class symbol which should be deleted when applying the rule must not be connected with an association (arrow).

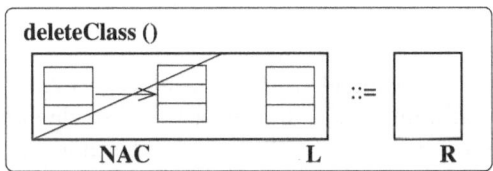

Figure8. VL-rule describing the deletion of a class.

4.3   Graphical Editor

The graphical editor allows the manipulation of visual sentences over a specific visual language VL defined by a VL-alphabet and a VL-grammar. According to Figure 1 it results from GENGED for a given VL. As for the grammar editor this manipulation is based on algebraic graph transformation and graphical constraint solving. The edit commands are the user-defined VL-rules which are correct in the sense of our framework, i.e. all sentences are algebras wrt. VISU-ALSPEC.

The transformation of VISUALSPEC-algebras in GENGED is based on algebraic graph transformation [EHK+97], but there is no formal description of attribute and layout handling in GENGED up to now. This is due to the fact that we use implemented components, e.g. the AGG-attribute component [Mel99] and the graphical constraint solver PARCON [Gri96]. This means that the transformation of VISUALSPEC-algebras is done by the following steps:

Step 1:   The graph part of VISUALSPEC-algebras A is transformed according to the SPO-approach which is implemented by the AGG-system.

Step 2:   Attribute handling is done via Java in AGG.

Step 3:   Layout handling is done via constraint solving supported by PARCON.

5   Related Work

Many different formalisms have been proposed for the definition of VLs [MM96]. In contrast to existing approaches using either algebraic techniques [Üsk94] or graph grammar approaches as presented by Göttler [Göt87], Andries, Engels, Rekers, Schürr [Rek94,AER96] and Minas, Viehstaedt [Min93,MV95] our approach uses both for defining an alphabet and a language grammar. We use algebraic specification techniques to define graphical symbols, links and layout constraints in an axiomatic way. This seems to meet the definition of the very basic issues of a visual language in a natural way. The proper language, i.e. its grammatical structure, is described by graph transformation which is again a very natural formalism for this purpose.

Many different tools just as formalisms have been proposed supporting the definition of VLs. In contrast to other approaches GENGED allows the explicit definition of alphabets and grammars for VLs. G.Costagliola et.all introduced the vlcc-environment [CDLOT97] supporting the visual definition of VLs, too. A symbol editor can be used to define terminal and non-terminal symbols. The defined symbols are then available within a production editor allowing the definition of context-free grammar rules. In contrast, we use algebraic graph grammars which are not restricted to be context-free. Furthermore, to be still in one formalism, we present automatically generated alphabet rules, which can be used for editing the VL-rules. In some cases these alphabet rules already define VL-rules, which are correct wrt. VL-alphabet. Nevertheless, vlcc offers not only a nice possibility to define VLs visually. Moreover, it generates free-hand editors for defined VLs.

6   Summary and Conclusions

In this paper we have introduced a conceptual model of the graphical editor GENGED supporting the visual definition of VLs described by an algebraic framework. In this framework VLs are based on algebraic graph grammars. The manipulation of sentences is done by the graph transformation system AGG ([Rud97,TER99]) in connection with the graphical constraint solver PARCON ([Gri96]).

GENGED can be used to define a great variety of VLs because it allows any kind of symbols and connections. Furthermore, GENGED offers a hybrid language where the visual sentences consist of graphical symbols as well as string sentences. These are the data attributes of nodes or edges.

Moreover, the theory of algebraic graph transformation offers analysis techniques, for example for consistency checks. Consistency conditions concerning the existence and non-existence of graph parts can be proven for all graphs within a graph grammar produced language ([HW95,Wag97]). Using consistency conditions it is possible to ensure, e.g. that two classes do not have the same name.

A first prototype of GENGED is realized within a student's project which is the basis for the current (re)implemention of GENGED. The following extensions of the GENGED environment are planned:

- Specification of application conditions.
- Specification of user interactions.
- Specification of the connection of several generated editors.
- Animation component which is useful for visual sentences modelling dynamical aspects such as firing in Petri-nets or automata.

References

AER96.   M. Andries, G. Engels, and J. Rekers. How to represent a Visual Program? In *[TVL96]*, 1996. 264

Bar98.   R. Bardohl. GENGED - A Generic Graphical Editor for Visual Languages based on Algebraic Graph Grammars. In *[VL'98]*, pages 48–55, 1998. 254

BT97.   R. Bardohl and G. Taentzer. Defining Visual Languages by Algebraic Specification Techniques and Graph Grammars. In *[TVL97]*, pages 27–42, 1997. 253

CDLOT97.   G. Costagliola, A. De Lucia, S. Orefice, and G. Tortora. A Framework of Syntactic Models for the Implementation of Visual Languages. In *[VL'97]*, pages 58–65, 1997. 264

CEL+96.   A. Corradini, H. Ehrig, M. Löwe, U. Montanari, and J. Padberg. The Category of Typed Graph Grammars and its Adjunctions with Categories of Derivations. In *LNCS 1073 , Proc. Williamsburg, U.S.A.*, pages 56–74, Springer, 1996. 254

Cor98.   Rational Software Corporation. UML – Unified Modeling Language. Technical report, Rational Software Corporation, http://www.rational.com, 1998. 252, 253, 256

EHK+97.   H. Ehrig, R. Heckel, M. Korff, M. Löwe, L. Ribeiro, A. Wagner, and A. Corradini. Algebraic Approaches to Graph Transformation II: Single Pushout Approach and Comparison with Double Pushout Approach. In G. Rozenberg, editor, *The Handbook of Graph Grammars, Volume 1: Foundations.* World Scientific, 1997. 254, 262, 263

EM85.   H. Ehrig and B. Mahr. *Fundamentals of Algebraic Specifications 1: Equations and Initial Semantics*, volume 6 of *EACTS Monographs on Theoretical Computer Science.* Springer, Berlin, 1985. 254, 258

Göt87.   H. Göttler. Graph Grammars and Diagram Editing. In H. Ehrig, M. Nagl, G. Rozenberg, and A. Rosenfeld, editors, *Graph Grammars and Their Application to Computer Science*, volume 291 of *Lecture Notes in Computer Science*, pages 216–231, 1987. 264

Gri96.   P. Griebel. *ParCon - Paralleles Lösen von grafischen Constraints.* PhD thesis, Paderborn University, February 1996. 259, 263, 264

HHT96.   A. Habel, R. Heckel, and G. Taentzer. Graph Grammars with Negative Application Conditions. *Special issue of Fundamenta Informaticae*, 26(3,4), 1996. 263

HW95.   R. Heckel and A. Wagner.   Ensuring Consistency of Conditional Graph Grammars – A constructive Approach. *Proc. of SEGRAGRA'95 "Graph Rewriting and Computation", Electronic Notes of TCS*, 2, 1995. http://www.elsevier.nl/locate/entcs/volume2.html. 265

Löw97.   M. Löwe. Evolution Patterns. Habilitation, TU-Berlin, 1997. also Technical Report No. 98–4. 254

Mel99.   B. Melamed. Grundkonzeption und -implementierung einer Attributkomponente für ein Graphtransformationssystem. Diplomarbeit, Informatik, TU Berlin, 1999. 263

Min93.   M. Minas. Spezifikation von Diagrammeditoren mit automatischer Layoutanpassung. In H.Reichel, editor, *Proc. 23. GI-Jahrestagung, Dresden, Reihe 'Informatik aktuell'*, pages 334–339. GI, September 1993. 264

MM96.   K. Marriott and B. Meyer. Towards a Hierarchy of Visual Languages. In *[TVL96]*, 1996. 264

MV95.   M. Minas and G. Viehstaedt. DiaGen: A Generator for Diagram Editors Providing Direct Manipulation and Execution of Diagrams. In *[VL'95]*, 1995. 264

Rek94.   J. Rekers. On the use of Graph Grammars for defining the Syntax of Graphical Languages. Technical Report tr94-11, Leiden University, Dep. of Computer Science, 1994. 264

Roz99.   G. Rozenberg, editor. *Handbook of Graph Grammars and Computing by Graph Transformaitons, Volume 2: Applications*. World Scientific Publishing, 1999. to appear. 266

Rud97.   M. Rudolf. Konzeption und Implementierung eines Interpreters für attributierte Graphtransformation. Diplomarbeit, Informatik, TU Berlin, 1997. 262, 264

TER99.   G. Taentzer, C. Ermel, and M. Rudolf. The AGG Approach: Language and Tool Environment. In *[Roz99]*. 262, 264

TVL96.   *Proc. of the AVI'96 Workshop Theory of Visual Languages*, Gubbio, Italy, May 30. 1996. 265, 266

TVL97.   *Proc. Workshop on Theory of Visual Languages*, Capri, Italy, 27 September 1997. 265

Üsk94.   S.M. Üsküdarlı. Generating Visual Editors for Formally Specified Languages. In *[VL'94]*, pages 278–285, 1994. 264

VL'94.   *Proc. Symp. on Visual Languages*, St. Louis, Missouri, October, 4-7 1994. IEEE Computer Society Press. 266

VL'95.   *Proc. Symp. on Visual Languages*, Darmstadt, Germany, September, 5-9 1995. IEEE Computer Society Press. 266

VL'97.   *Proc. Symp. on Visual Languages*, Capri, Italy, September 1997. IEEE Computer Society Press. 265

VL'98.   *Proc. Symp. on Visual Languages*, Halifax, Canada, September 1998. IEEE Computer Society Press. 265

Wag97.   A. Wagner. *A Formal Object Specification Technique Using Rule-Based Transformation of Partial Algebras*. PhD thesis, TU Berlin, 1997. 265

From Formulae to Rewriting Systems*

Paolo Bottoni, Francesco Parisi-Presicce, and Marta Simeoni

Department of Computer Science, University of Rome *La Sapienza*
Via Salaria 113, 00198 Roma, Italy
{bottoni,parisi,simeoni}@dsi.uniroma1.it

Abstract. We present an algorithmic method to obtain pure rewriting systems from logical descriptions of languages. The method, presented in an SPO framework, is applicable to visual language specification.

1 Introduction

Pure rewriting systems are a way to specify languages by defining the transformations that sentences of a language can go through, rather than by defining their deep structure, as in classical Chomsky-like grammars. A typical example is that of Lindenmayer systems, originally intended to model growth processes in simple organisms and subsequently studied as language-defining devices on their own. Stripped of the parallelism inherent to L-systems, the pure rewriting mechanism gives pure grammars [12], where the evolution of a string is restricted to take place in a single location at a time.

Pure systems may be used to define incremental processes, where transformations of a sentence occur by modifying existing elements in it or by creating new ones, without ever destroying existing elements. Such a model is of interest when defining visual languages for human-computer interactive systems [1].

On the other hand, the absence of non-terminals makes it more difficult to constrain the rewriting process to happen in specific locations, as well as to build deterministic parsers, or in general to state the defined language. This problem is obviated by defining some condition on the generated sentences.

We propose a method to derive a pure rewriting system and a characterising condition from a formula defining a language. The discussion is restricted to formulae which describe a limited but significant set of local and global properties of hypergraphs. We represent hypergraphs as terms and we allow for subterms whose value denotes other sets of terms, so that relations among terms can be directly represented.

The rules of the derived attributed rewriting system are of an injective type and are set in a Single Pushout (SPO) framework, which allows specification of both positive and negative application conditions through graph morphisms (see [7]). The resulting rules can be the basis for the specification of interactive visual systems from user requirements, in the case where the languages to be

---

* Partially supported by the EC under TMR Network GETGRATS and Esprit Working Group APPLIGRAPH.

H. Ehrig et al. (Eds.): Graph Transformation, LNCS 1764, pp. 267–280, 2000.

defined are visual languages (see [2]). For graph rewriting rules, a straightforward implementation is given by the Algebraic Graph Grammar (AGG) system (see [14]). The formal construction guarantees that the resulting system meets user requirements. Moreover, the characterising condition obtained by the proposed construction is verifiable by inspection of attribute values, usually reducing the complexity of checking the satisfaction of the whole original formula.

2   Related Work

Constraint-based definition of languages in the field of visual languages has led to so-called "unification-based" grammars (e.g. [15], [8]), as well as to methods to translate static constraints on the sentences into dynamic constraints on their interactive construction [13]. Constraint Multiset Grammars (CMG) [8], [11] augment a context-sensitive way of rewriting with conditions and attribute updating. They have full computational power, using non-terminal symbols and allowing unrestricted attribute evaluation. A stricter version, with only attribute copying, still has sufficient power to simulate several other formalisms for visual language definition. Visual unification grammars are based on linear strings and context-free-like rules [15]. Chok and Marriott provide an algorithm to create a control system from a CMG, so that each sentence satisfies the whole set of constraints [4]; however, rules are not local and determine global re-evaluation of constraints. Serrano analyses by hand the whole set of constraints defining a visual language to identify allowed violations during sentence derivation, but does not provide an algorithm [13]. In graph rewriting, negative application conditions prevent the creation of sentences which do not represent valid states of a distributed system [7]. We capitalise on the representation of visual languages via hypergraphs, as proposed by Minas and exploited in the generation of diagram editors in [10], according to which spatial relations among different entities can be expressed by suitable hyperedges among the nodes representing the entities.

3   An Example

We consider the graphical language for representation of Petri Nets without marking (the left side of Figure 1 constitutes a sentence), informally specified by requesting that at least a place and a transition exist, and that every arrow goes from a place (circle) to a transition (box) or vice versa. Between a place and a transition only one connection can be placed and no place or transition can remain isolated: the right hand side of Fig.1 represents an incorrect net.

Formally, this can be specified by the following formula, where Arr_{pt} and Arr_{tp} indicate the sets of arrows, Plc the set of places and Trn the set of transitions.

$(Plc \neq \emptyset) \wedge (Trn \neq \emptyset) \wedge (\forall a \in Arr_{pt} \ \exists!p \in Plc, \exists!t \in Trn, pt(a,p,t)) \wedge$
$(\forall a \in Arr_{tp} \ \exists!p \in Plc, \exists!t \in Trn, tp(a,t,p)) \wedge$
$(\forall(p,t) \in Plc \times Trn, pt(a,p,t) \wedge pt(b,p,t) \implies a = b) \wedge$
$(\forall(p,t) \in Plc \times Trn, pt(a,p,t) \implies \neg(\exists b \ tp(b,t,p))) \wedge$

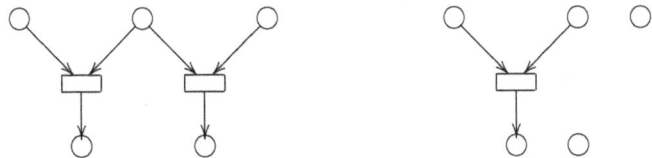

Fig. 1. Left: a Petri net. Right: an incorrect net.

$(\forall(p,t) \in Plc \times Trn, \; tp(a,t,p) \wedge tp(b,t,p) \implies a = b) \; \wedge$
$(\forall(p,t) \in Plc \times Trn, \; tp(a,p,t) \implies \neg(\exists b \; pt(b,p,t))) \; \wedge$
$(\forall p \in Plc \; \exists a \in Arr_{tp} \cup Arr_{pt} \; \exists t \in Trn, \; pt(a,p,t) \vee tp(a,t,p)) \; \wedge$
$(\forall t \in Trn \; \exists a \in Arr_{tp} \cup Arr_{pt} \; \exists p \in Plc, \; pt(a,p,t) \vee tp(a,t,p)).$

The first two components of the formula do not place any constraint on the characteristics of the individual places or transitions, which can be freely created with the following two rules (the null antecedent indicates that no precondition is needed for creation of places and transitions).

$Rule_1: \quad \longrightarrow \quad \bigcirc$ $\qquad Rule_2: \quad \longrightarrow \quad \square$

If we require that no dangling arrow can be created, the third and fourth components require that arrows be created only to connect places and transitions. The predicate pt indicates that the arrow goes from the place to the transition, the predicate tp the opposite direction. This gives rise to the two rules:

The next four components produce application conditions on rules 3 and 4. In order to be able to check the application conditions, we associate places and transitions with two attributes I (for *input*) and O (for *output*), indicating the places or transitions connected by incoming or outgoing arrows, respectively. The application conditions are then expressed as negative formulae on the attributes I and O of elements in the antecedent. Two more attributes, called TP and PT denote the arrows connected to these elements. Moreover, an attribute M (for *members*) is associated with each arrow, to indicate the elements connected by the arrow. Collectively, all these attributes allow global information on net composition to be distributed among its components. The updating of these attributes is associated with the rules in a consistent way. Hence, rules are augmented with semantic actions, producing the following four rules:

$Rule_1: \quad \longrightarrow \quad \bigcirc \quad \{place(Id, \{\}, \{\}, \{\}, \{\})\} \qquad Id = newIdPlace()$

$Rule_2: \quad \longrightarrow \quad \square \quad \{trans(Id, \{\}, \{\}, \{\}, \{\})\} \qquad Id = newIdTrans()$

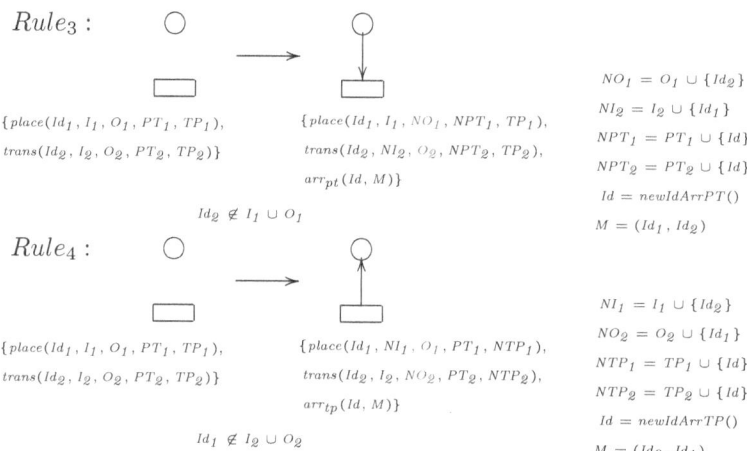

$Rule_3$:

$\{place(Id_1, I_1, O_1, PT_1, TP_1),$
$trans(Id_2, I_2, O_2, PT_2, TP_2)\}$

$\{place(Id_1, I_1, NO_1, NPT_1, TP_1),$
$trans(Id_2, NI_2, O_2, NPT_2, TP_2),$
$arr_{pt}(Id, M)\}$

$Id_2 \notin I_1 \cup O_1$

$NO_1 = O_1 \cup \{Id_2\}$
$NI_2 = I_2 \cup \{Id_1\}$
$NPT_1 = PT_1 \cup \{Id\}$
$NPT_2 = PT_2 \cup \{Id\}$
$Id = newIdArrPT()$
$M = (Id_1, Id_2)$

$Rule_4$:

$\{place(Id_1, I_1, O_1, PT_1, TP_1),$
$trans(Id_2, I_2, O_2, PT_2, TP_2)\}$

$\{place(Id_1, NI_1, O_1, PT_1, NPT_1),$
$trans(Id_2, I_2, NO_2, PT_2, NPT_2),$
$arr_{tp}(Id, M)\}$

$Id_1 \notin I_2 \cup O_2$

$NI_1 = I_1 \cup \{Id_2\}$
$NO_2 = O_2 \cup \{Id_1\}$
$NTP_1 = TP_1 \cup \{Id\}$
$NTP_2 = TP_2 \cup \{Id\}$
$Id = newIdArrTP()$
$M = (Id_2, Id_1)$

where $trans$, $place$, arr_{pt}, arr_{tp} are *term constructors*. The association of a term with a graphical element yields a *graphical term*. Rewriting occurs in a consistent way both at the graphical level (where nodes and arrows are created) and at the textual one (where attributes are updated and conditions checked) [1].

The last two components of the formula require that places (resp. transitions) be connected to transitions (resp. places) via arrows of the proper type. They give rise to four rules which simultaneously create an element and an arrow. We show only the two rules creating a place–transition arrow. The other two differ for the orientation of the created arrow and the corresponding updating of attributes.

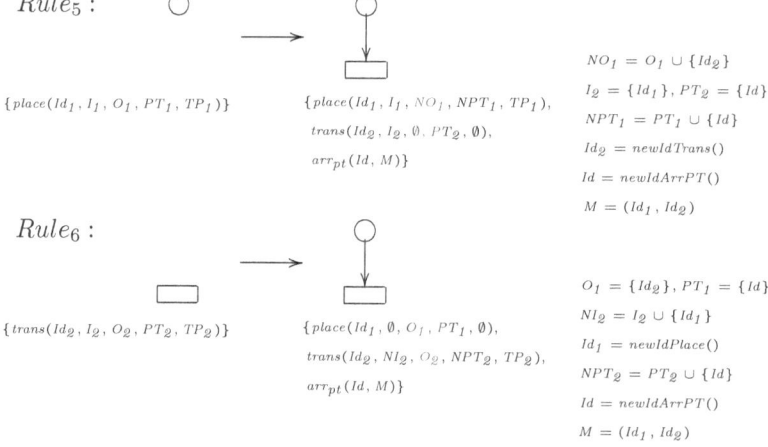

$Rule_5$:

$\{place(Id_1, I_1, O_1, PT_1, TP_1)\}$

$\{place(Id_1, I_1, NO_1, NPT_1, TP_1),$
$trans(Id_2, I_2, \emptyset, PT_2, \emptyset),$
$arr_{pt}(Id, M)\}$

$NO_1 = O_1 \cup \{Id_2\}$
$I_2 = \{Id_1\}, PT_2 = \{Id\}$
$NPT_1 = PT_1 \cup \{Id\}$
$Id_2 = newIdTrans()$
$Id = newIdArrPT()$
$M = (Id_1, Id_2)$

$Rule_6$:

$\{trans(Id_2, I_2, O_2, PT_2, TP_2)\}$

$\{place(Id_1, \emptyset, O_1, PT_1, \emptyset),$
$trans(Id_2, NI_2, O_2, NPT_2, TP_2),$
$arr_{pt}(Id, M)\}$

$O_1 = \{Id_2\}, PT_1 = \{Id\}$
$NI_2 = I_2 \cup \{Id_1\}$
$Id_1 = newIdPlace()$
$NPT_2 = PT_2 \cup \{Id\}$
$Id = newIdArrPT()$
$M = (Id_1, Id_2)$

The application of rules from the generated rewriting system does not ensure that no isolated element exists, which requires a final check. Verifying directly this condition would require a check linear in the product of the number of terms for each type (for each place inspect all transitions and see whether a connection exists and vice versa); by using the attributes introduced above, this can be solved in time linear in the sum of the number of terms for each type: simply inspect the value of the attributes TP and PT for each place and transition, to check that their union is not empty.

To recap, from a language specification, expressed as a logical formula, we have obtained: (1) a set of rewriting rules augmented with conditions and semantic actions which guarantee that only legal sentences are generated; and (2) a predicate to be satisfied by individual elements in the sentence. By *legal* we mean that, even if the sentence does not belong to the language, it can be transformed into a language sentence without deleting any existing element [13].

In the next sections we provide a general methodology to obtain rewriting systems and validity checks from a formula describing the language.

4   An SPO Framework

The example of Section 3 is based on a representation of attributed (hyper)graphs through graphical terms, pairs consisting of a graphical representation of a node or (hyper)edge and a term describing its properties (values of the associated attributes). Terms are built up from an alphabet K of *term constructors* and a set V of attributes, defined over an attribute algebra D. A distinguished attribute is assumed to be present in each term, acting as an identifier. Attributes are *individual*, taking values in usual domains, or *relational*, taking values in the powerset of the set of identifiers.

Let T be the set of hypergraphs representable with sets of graphical terms over (K, V, D); let F be a formula expressing conditions on hypergraphs in T through corresponding conditions on graphical terms over (K, V, D). We call $F(K, D) \subset T$ the language of hypergraphs specified by F, i.e. the language of the hypergraphs in T satisfying F. The formula F may specify: a) existence of elements of a given type, possibly with some individual properties; b) the types of participants in the relations; c) existence and uniqueness conditions on relations and entities; d) mutual exclusion of relations. It cannot force participation in a relation to depend on a chain of relations of arbitrary length (e.g. transitive closure) .

This construction will be used in Section 5 to set the informal procedure of Section 3 in the framework of the Single Pushout (SPO) approach.

An SPO production is a partial graph morphism $p : L \to R$ where L and R are the left and right hand–side graphs, respectively. Objects (nodes and edges) where p is undefined are deleted, while objects preserved by p form the application context. The objects in R which are not in the preimage of p are the newly created objects.

In the SPO approach, both the rules and the application conditions resulting from a given formula can be expressed with graphs and graph morphisms. In fact, *positive* and *negative* application conditions (i.e conditions on the existence or non existence of some object in the given graph in order to apply the rule) can be specified through morphisms. In particular, negative application conditions are specified through morphisms $c_i : L \to \overline{L}_i$ called *constraints*, where the elements in $\overline{T}_{i} - c_{i}(L)$ represent the forbidden structure. For example, the SPO version of $Rule_3$ of the Petri net example is:

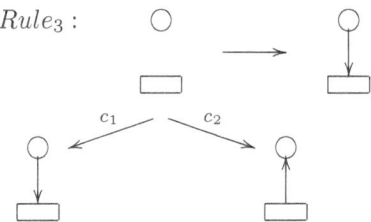

$Rule_3$:

where constraint c_1 prevents the application of the rule if an arrow from the place to the transition already exists, while c_2 prevents the application if an arrow in the opposite direction already exists.

SPO was extended to attributed (hyper)graphs [9], where attributes are specified using algebraic specifications (in the sense of [5]), i.e attributes are labels of graphical objects taken from an attribute algebra. An attributed graph morphism consists of a partial graph morphism (to relate the graphical parts) and a total algebra homomorphism (to relate the algebraic parts) such that the graphical and algebraic part are compatible. Since the application conditions are expressed in categorical terms, they can be easily adapted for attributed graphs and hypergraphs (see [7]).

For example, suppose that two nodes can be connected by multiple arcs. To be able to visualize all of them, we associate a *curving* attribute to each arc. The following SPO rule (together with the signature for the attribute algebra written on the right) can be used for adding an arc:

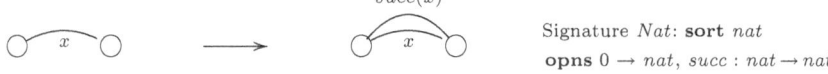

Signature Nat: **sort** nat
opns $0 \to nat$, $succ : nat \to nat$

The construction presented in this paper produces injective morphisms on attributed hypergraphs in which negative conditions mention nodes only if already present in L and positive conditions are expressed on attribute values.

5   The Construction

In this section we formalise and generalise the procedure informally described in Section 3. For simplicity, we present the procedure using graphical terms and set rewriting rules (instead of attributed hypergraphs and SPO rules).

Let T be the set of hypergraphs constructed using graphical terms over (K, V, D). Let F be a formula on graphical terms over (K, V, D) of the type discussed above. We construct a rewriting system $S_F = ((K, V, D), P, \Longrightarrow)$ where P is a set of rules of the form $p : L \to R$, **IF** ϕ, with L and R attributed graphs on T and ϕ an application condition on the antecedent. The rewriting relation $\Longrightarrow$ stipulates that rules are applied sequentially, by replacing the antecedent with the consequent, provided that ϕ is satisfied. Given a hypergraph $ax \in T$ and S_F, the language $L(S_F, ax) = \{x \mid ax \Longrightarrow^* x\}$ is defined. Given a set $Ax \subset T$, $L(S_F, Ax) = \bigcup_{ax \in Ax} L(S_F, ax)$.

For $p : L \rightarrow R$, **IF** ϕ, let T_L and T_R be the sets of graphical terms in L and R, respectively. Let $constructor : T_L \cup T_R \rightarrow K$ be a function mapping a graphical term $k(t_1, \ldots, t_m)$ to its constructor k (its type). p is n-increasing if

1. $\exists \mu : T_L \longrightarrow T_R, \mu$ injective function;
2. $\forall t \in T_L, constructor(t) = constructor(\mu(t))$;
3. $\mid T_R \mid \leq \mid T_L \mid + n, n \geq 0$.

The graphical terms in the antecedent are also present, up to the transformation μ which cannot change their type, in the consequent, which contains no more than n new graphical terms; the rules in Section 3 are 2-increasing. While n-increasing rules can be derived more directly from the requirements, 1-increasing rules can be directly translated into the control automaton governing the interaction in a syntax-directed interactive system [2]. This gap can be bridged by a systematic transformation of n-increasing rules into structured collections of 1-increasing ones, adapting to graphs the construction in [3].

Let (K, V, D) be as before; we reformulate the objective of this study as follows.

Problem: given a formula F defining a language $F(K, D)$, find an n-increasing rewriting system S_F, a set of axioms Ax and a predicate $VALID$ such that $L(S_F, Ax) \cap L(VALID) = F(K, D)$, where $L(VALID)$ indicates the language of sentences which satisfy $VALID$.

We develop the construction using a normalisation of F so that F is considered to be composed only of subformulae having the following types:

I $\exists y \in Y \quad p(y)$, with p some predicate on individual attributes of y;
II $\forall x \in X \; \exists! y_1, \ldots, y_n \in Y_1 \times \ldots \times Y_n \quad rel_X(x, y_1, \ldots, y_n)$;
III $\forall y \in Y \; \exists x \in X \exists! y_1 \ldots, y_n \in Y_1 \times \ldots \times Y_n \quad rel_X(x, y, y_1, \ldots, y_n)$;
IV $\exists y \in Y \; \forall y_1 \ldots, y_n \in Y_1 \times \ldots \times Y_n \; \exists! x \in X \quad rel_X(x, y, y_1, \ldots, y_n)$;
V $\forall y_1 \ldots, y_k \in Y_1 \times \ldots \times Y_k \; (\exists x_1 \in X_1 \; \exists y_{k+1} \ldots, y_n \in Y_{k+1} \times \ldots \times Y_n$
$rel_{X_1}(x_1, y_1, \ldots, y_n) \implies \Phi)$ where Φ may be of kind:
  - $\nexists x_2 \in X_1, \; x_2 \neq x_1, \quad rel_{X_1}(x_2, y_1, \ldots, y_n)$
  - $\nexists rel_{X_2} : \exists x_2 \in X_2 \quad rel_{X_2}(x_2, y_1, \ldots, y_n)$
  - $\exists x_2 \in X_1, \; x_2 \neq x_1 \quad rel_{X_2}(x_2, y_1, \ldots, y_n) \implies p(x_1, x_2)$ with p predicate

and analogous formulae of type I, III, or IV asserting uniqueness of an instance, i.e. with $\exists!$ instead of $\exists$. X is a set of instances denoting relations and Y_i are sets of instances denoting entities to be put in relation. In formulae of type II, III, IV, for every relation x, suitable relational attributes have to assume consistent values throughout the derivation.

The mentioned normalisation consists of considering a subformula of type $Y \neq \emptyset$ as a special case of a subformula of type I, and of producing a different subformula for each term of a disjunction or of a conjunction. Hence, a subformula with $\forall x \in X \; \exists y \in Y_1 \cup Y_2$ gives origin to two subformulae with $\forall x \in X$ $\exists y_1 \in Y_1$ and $\forall x \in X \; \exists y_2 \in Y_2$. Note that this does not change the language generated since these two subfomulae will give rise to two different rules. Hence, the non-determinism given by the disjunction is subsumed in the non-determinism typical of the rewriting mechanism. Analogously, a subformula with $\forall y \in Y$ $\exists y_1 \in Y_1 \wedge \forall y \in Y \exists y_2 \in Y_2$ gives origin to the two subformulae $\forall y \in Y$ $\exists y_1 \in Y_1$ and $\forall y \in Y \exists y_2 \in Y_2$. The formulae derived from this normalisation (and the rules obtained from these formulae) must be indexed so that the new subformulae and rules can be traced back to the original complex subformulae.

In this case, the procedure for obtaining rules and conditions is as follows:

procedure ConstructRules(*formula*) {
    Normalise(*formula*)
    EnrichSymbolsWithAttributes
    ConstructRulesForFreeGeneration
    ConstructRulesForContextualGeneration
    ConstructConditionsOnRules
    AmalgamateRules
    ConstructAxiom }

We assume that *formula* is a suitable coding of F and that **Normalise** performs normalisation and indexing as required above, producing a new coding of F to which the other procedures called within **ConstructRules** have access.

procedure EnrichSymbolsWithAttributes {
    foreach type of relation x with members in $Y_1 \times \ldots \times Y_n$
      augment symbol x with attribute *members*
//$D(members)$={n-tuples of identifiers of members of x}
      foreach type Y_i in $Y_1, \ldots, Y_n$
        augment symbol y_i with attribute in_X
//$D(in\_X)$={sequence of identifiers of instances of x}
        augment symbol y_i with attribute $partners\_in\_X$
//$D(partners\_in\_X)$={sequence of tuples of instance identifiers in $Y_1 \times \ldots \times Y_{i-1} \times Y_{i+1} \ldots \times Y_n$}
      endfor
    endfor }

This procedure creates attributes allowing distribution of the global graph information to individual elements. The attributes I, O, M, PT and TP of the example in Section 3 are generated in this way. For instance, the attribute M of arrows corresponds to *members*, while the attribute O for places corresponds to $partners\_in\_PT$.

procedure ConstructRulesForFreeGeneration {
    foreach formula of type $\exists y \in Y \quad p(y)$
      construct a rule $\emptyset \rightarrow \{y\}$
      add $\exists y \in Y \quad p(y)$ to *ValidityCheck*
    endfor }

The rules created by this procedure allow the free creation of elements whose presence is required, while the check is meant to verify that these rules have been applied: $Rule_1$ and $Rule_2$ in Section 3 have been created in this way.

procedure ConstructRulesForContextualGeneration {
   foreach formula of type
   $\forall x \in X \ \exists! y_1, \dots, y_n \in Y_1 \times \dots \times Y_n \quad rel_X(x, y_1, \dots, y_n)$
     construct a rule $\{y_1, \dots, y_n\} \rightarrow \{y_1, \dots, y_n, x\}$ and
     $x.members = (y_1.id, \dots, y_n.id)$; $y_i.in\_X = y_i.in\_X \circ x.id$;
     $y_i.partners\_in\_X =$
                $y_i.partners\_in\_X \circ (y_1.id, \dots y_{i-1}.id, y_{i+1}.id, \dots, y_n.id)$
   endfor
   foreach formula of type
   $\forall y \in Y \ \exists x \in X \ \exists! y_1, \dots, y_n \in Y_1 \times \dots \times Y_n \quad rel_X(x, y, y_1, \dots, y_n)$
     construct a rule $\{y_1, \dots, y_n\} \rightarrow \{y_1, \dots, y_n, y, x\}$ and
     $x.members = (y.id, y_1.id, \dots, y_n.id)$; $y.in\_X = x.id$;
     $y.partners\_in\_X = (y_1.id, \dots, y_n.id)$ and
     functions for each y_i analogous to those above
     if the existential on x asserted uniqueness **then**
        add negative condition
        $(y_1.id, \dots, y_{i-1}.id, y_{i+1}.id, \dots, y_n.id) \notin y_i.partners\_in\_X$
        for $i = 1, \dots, n$ to rule $\{y_1, \dots, y_n, y\} \rightarrow \{y_1, \dots, y_n, y, x\}$
     endif
   endfor
   foreach formula of type
   $\exists y \in Y \ \forall y_1 \dots, y_n \in Y_1 \times \dots \times Y_n \ \exists! x \quad rel_x(x, y, y_1, \dots, y_n)$
     construct a rule $\{y\} \rightarrow \{y, y_1, \dots, y_n, x\}$ and
     $x.members = (y.id, y_1.id, \dots, y_n.id)$; $y.in\_X = y.in\_X \circ x.id$;
     $y.partners\_in\_X = y.partners\_in\_x \circ (y_1.id, \dots, y_n.id)$
     and functions for each y_i analogous to those above
     if existential on y asserted uniqueness **then**
        generate boolean attribute x_source;
        place y in axiom with $y.x\_source = true$;
        add condition $y.x\_source = true$ to rule $\{y\} \rightarrow \{y, y_1, \dots, y_n, x\}$;
        add semantic action $y.source = false$ to all other rules for creation of y
     endif
   endfor }

This procedure generates rules which comply with the requirements of sub-formulae of types II, III and IV. In particular, the first cycle creates $Rule_3$ and $Rule_4$ of the example of Section 3, the second cycle creates $Rule_5$ and $Rule_6$.

procedure ConstructConditionsOnRules {
   foreach formula of type
   $\forall y_1 \dots, y_k \ (\exists x_1 \in X_1 \ \exists y_{k+1} \dots, y_n \ rel_{X_1}(x_1, y_1, \dots, y_n) \Longrightarrow \Phi)$
     case TypeOfPhi **of**
        $\nexists x_2 \in X_1, \ x_2 \neq x_1, \ rel_{X_1}(x_2, y_1, \dots, y_n)$:
        add negative condition
        $(y_1.id, \dots, y_{i-1}.id, y_{i+1}.id, \dots, y_n.id) \notin y_i.partners\_in\_X_1$
        for $i = 1, \dots, n$ to rule $\{y_1, \dots, y_n\} \rightarrow \{y_1, \dots, y_n, x_1\}$
        $\nexists rel_{X_2}$ s.t. $\exists x_2 \in X_2 \ rel_{X_2}(x_2, y_1, \dots, y_n)$:

add negative conditions
$(y_1.id, \dots, y_{i-1}.id, y_{i+1}.id \dots, y_n.id) \notin y_i.partners\_in\_X_1$
for $i = 1, \dots, n$ to rule $\{y_1, \dots, y_n\} \to \{y_1, \dots, y_n, x_2\}$ and
$(y_1.id, \dots, y_{i-1}.id, y_{i+1}.id, \dots, y_n.id) \notin y_i.partners\_in\_X_2$
for $i = 1, \dots, n$ to rule $\{y_1, \dots, y_n\} \to \{y_1, \dots, y_n, x_1\}$
$\exists x_2 \in X_1, x_2 \neq x_1 \ rel_{X_2}(x_2, y_1, \dots, y_n) \implies p(x_1, x_2)$:
generate rule $\{y_1, \dots, y_n, x_1\} \to \{y_1, \dots, y_n, x_1, x_2\}$
with condition $x_1.members = (y_1, \dots, y_n)$;
define attribute att_p; add semantic actions $x_2.att_p = f_{p_2}(x_1.att_p)$ and
$x_1.att_p = f_{p_1}(x_1.att_p)$, where f_{p_2} and f_{p_1} derive from the form of p;
add negative condition
$(y_1.id, \dots, y_{i-1}.id, y_{i+1}.id, \dots, y_n.id) \notin y_i.partners\_in\_X_1$
for $i = 1, \dots, n$ to rule $\{y_1, \dots, y_n\} \to \{y_1, \dots, y_n, x_1\}$
and semantic action $x_1.att_p = f_p(y_1, \dots, y_n)$
    default : add formula to *ValidityCheck*
  endcase
endfor }

Note that the construction of negative conditions of the first type corresponds, in terms of the SPO approach, to placing a copy of the consequent in $\overline{L}$. The first two cycles create the negative conditions on $Rule_3$ and $Rule_4$ in the example of Section 3.

procedure AmalgamateRules {
  foreach rule $\emptyset \to \{y\}$
    if a rule $\{y_1, \dots, y_n\} \to \{y_1, \dots, y_n, x, y\}$ **AND** NonCircular(y) exists **then**
      remove rule $\emptyset \to \{y\}$
    endif
  endfor
  foreach set of rules producing elements of type y and derived from an original
  conjunction of formulae
    replace the set by a rule having the union of antecedents as antecedent
    and the union of consequents as consequent
  endfor
  foreach set of rules which induce a circularity and such that no element
  can be generated otherwise
    replace the set by a rule having as antecedent the union of all the
    non circular elements of the antecedents, and as consequent the union
    of the consequents
  endfor
  foreach set of rules derived from formulae of type III and IV, all creating
  an element of type Y
    replace the set with a rule having as antecedent the difference between
    the union of antecedents and the set of new elements appearing
    in the consequents, and as consequent the union of the consequents
  endfor }

Rule amalgamation generally allows a reduction in size of the difference $L(S_F, Ax) \setminus F(K, D)$. The third cycle of the procedure is instead necessary for maintaining the correspondence between the generated rules and the given formula.

The algorithm NonCircular(*type*) verifies that there are no chains of rules in which the presence of an element of a given type is precondition for the existence of instance of other types which are ultimately preconditions for its own existence, such as in $\{y_1\} \to \{y_1, y_2, x_1\}$, $\{y_2\} \to \{y_2, y_3, x_2\}$ and $\{y_3\} \to \{y_3, y_1, x_3\}$. Non circularity is checked to avoid that the result of the amalgamation prevents the creation of elements of that type. For example, in the case of Petri nets, $Rule_1$ for free creation of a place could be amalgamated with $Rule_6$ for simultaneous creation of a place and an arrow provided that a transition exists, and similarly for $Rule_2$ and $Rule_5$. This amalgamation, however, would induce a circularity on rules creating transitions and places.

The algorithm **ConstructAxiom** prescribes the presence in the axiom of all elements for which unique existence is required and removes all rules which possibly generate such elements. If no such element exists, the set of axioms is composed of only the empty sentence: $Ax = \Lambda$.

6   Correctness of the Construction

Let S_F be the rewriting system, Ax the axiom, and *VALID* the predicate constructed by the procedure **ConstructRules** for a consistent formula F. We have to show that $L(S_F, Ax) \cap L(VALID) = F(K, D)$, i.e that each model of the formula can be generated by the rules of the obtained rewriting system.

Submitting the formula to the various steps of the procedure (up to the amalgamation) means creating rules and application conditions according to the subformulae. Each rule adds elements (nodes and hyperedges) only if the proper context for them exists, i.e. they are created consistently w.r.t. the formula. The proper context is also expressed by negative conditions associated to the rules to prevent the creation of contextual elements violating the constraints required by formulae of tye V. The procedure **AmalgamateRules** may delete and/or replace some of the generated rules, so that we are left with showing that the correspondence with the given formula is maintained, i.e that it does not modify the generated language. We examine in detail its four steps:

- the first step eliminates the rules for free creation of elements, but only if they are redundant, i.e if there are other rules creating the same elements;
- the second step glues together rules creating an element of type y deriving from an original conjunction of formulae. For instance the formula

  ($\forall box \; \exists circle \; \exists x_1 : \; rel_{X_1}(x_1, box, circle)) \wedge$
  ($\forall box \; \exists triangle \; \exists x_2 : \; rel_{X_2}(x_2, box, triangle))$

  expressing the fact that each box has to be related both with a circle (via x_1) and with a triangle (via x_2), gives rise to the two rules:

and their equivalent amalgamated rule is:

- the third step substitutes each set of rules inducing a circularity and such that no element can be generated otherwise with a single equivalent rule. Such a rule creates all the elements in the consequents of the circular rules from the non circular elements in their antecedents. This step is necessary for creating sentences satisfying the formula but that cannot be created due to the circularity of rules. For example the rules corresponding to the formula

$(\forall box \ \exists x \ \exists triangle, circle : \ rel_X(x, box, circle, triangle)) \land$
$(\forall circle \ \exists x' \ \exists triangle, box : \ rel_{X'}(x', box, circle, triangle))$

are the following:

Note that the rules are circular because the circle is a precondition for creating the box and the box is a precondition for creating the circle. Their amalgamation yields the rule:

- the last step amalgamates rules that create the same elements but starting from formulae of type III and IV. In this way a chain of dependent rules collapses into a single rule having exactly the same effect of the original ones. For example, the properties of a rooted tree (i.e a tree where each leaf is connected directly with the root) can be expressed by:

$(\exists root \ \forall leaf \ \exists! x_1 : \ rel_{X_1}(x_1, root, leaf)) \land$
$(\forall leaf \ \exists node \ \exists! x_2 : \ rel_{X_2}(x_2, leaf, node))$

It gives rise to the rules:

and their amalgamation yields the following equivalent rule:

Let now h be a sentence obtained starting from an axiom in Ax and applying rules in S_F, and suppose that $h \in L(S_F, Ax) \setminus F(K, D)$, i.e. h does not satisfy the *VALID* predicate. We have to show that starting from h and applying rules in S_F it is possible to obtain another sentence $h' \in F(K, D)$. Note that, since the rules do not create elements outside a proper context, the only reason why h may not satisfy the formula is the lack of some nodes and/or hyperedges.

If an hyperedge is missing, it can be created using rules resulting from formulae of type II, III or IV. If a node is missing, it can be created using rules resulting from formulae of type I, III or IV. In both cases (hyperedge or node missing), if none of the rules can be applied it means that either a) their context is not present (i.e some other nodes are missing), or b) some application conditions are not satisfied.

The key point to prove that one such h' can be obtained is that the chain of dependences between missing elements in h is finite. This follows from the consistency of F, the absence of any circular dependency between rules (eliminated by procedure **AmalgamateRules**), and the the fact that the negative application conditions associated with the rules cannot induce a circularity on the application of rules, since they may only refer to nodes in the antecedent. Hence, a layered structure can be defined on the set of types, such that layer 1 contains types present in the axiom or freely generable, while elements at layer i are generated using elements at layer $i - 1$.

In fact, the first two types of subformulae in formulae of type V may prevent the creation of an edge among some nodes only if an edge of a given type already exists among those nodes, so that the negative condition generated may not prevent the creation of a new node. Subformulae of the third type may instead 1) prescribe properties of a new edge among some already created elements, or 2) prevent the creation of edges among previously unrelated elements. The first case does not prevent the creation of any edge. The second case may instead require the creation of some new context in order to create the required new element at layer k. Assuming that the model for F is not empty and that no circular element exists, such a context can be generated by generating the required elements at layers lower than k.

7   Conclusions and Discussion

We have presented an algorithm to translate a logical specification of a language into a specification expressed through injective rewriting systems admitting negative and positive application conditions, a set of axioms and a characterising condition, as is typical in formal language theory. The obtained rewriting system can be used as a basis for the specification of an interactive translator for the construction of sentences in the desired language, as studied for visual languages [2]. The setting in the SPO approach indicates possible extensions to higher level forms of rewriting.

The construction has been presented for the case where syntactic restrictions exist on the formulae. The resulting language of formulae is sufficiently rich to express several visual languages of practical interest. Extending the construction to richer languages should lead to the definition of additional attributes.

n-increasing rules can be refined to 1-increasing rules in such a way that the resulting 1-increasing system can be seen as an implementation of the n-increasing one (in the sense of [6]). Each rule of the n-increasing system is equal to the sequential composition (i.e temporal refinement) of its associated 1-increasing rules, and the new attributes introduced in these rules define the refinement instructions needed to compose them in the correct way.

References

1. P. Bottoni, M. F. Costabile, S. Levialdi, P. Mussio. "Visual Conditional Attributed Rewriting Systems in Visual Language Specification", *Proc. IEEE Symp. on Visual Languages '96*, 156 – 163. 267, 270

2. P. Bottoni, M.F. Costabile, S. Levialdi, P. Mussio, "Specifying dialog control in Visual Interactive Systems", *Journal of Visual Languages and Computing*, 9, 535-564, 1998. 268, 273, 279

3. P.Bottoni, M.F.Costabile, P.Mussio, "Specification and Dialogue Control of Visual Interaction through Visual Rewriting Systems", to appear in *ACM TOPLAS.* 273

4. S.S. Chok, K. Marriott. "Automatic Construction of User Interfaces from Constraint Multiset Grammars". *Proc. IEEE VL'95*, 242 – 249. 268

5. H. Ehrig, B. Mahr. *Fundamentals of Algebraic Specifications 1: Equations and Initial Semantics*, Springer Verlag, Berlin, 1985. 272

6. M. Große–Rhode, F. Parisi–Presicce, M. Simeoni. "Spatial and Temporal Refinement of Typed Graph Transformation Systems", *Proc. MFCS'98*, LNCS 1450. 279

7. A. Habel, R. Heckel, G. Taentzer. "Graph grammars with negative application conditions", *Fundamenta Informaticae*, 26, 287-313, 1996. 267, 268, 272

8. R. Helm, K. Marriott. "A Declarative Specification and Semantics for Visual Languages",*Journal of Visual Languages and Computing*, 2, 311–332, 1991. 268

9. M. Löwe, M. Korff, A. Wagner. "An algebraic framework for the transformation of attributed graphs". In *Term Graph Rewriting: Theory and Practice*, 185–199, Wiley, 1993. 272

10. M. Minas, G. Viehstaedt, "DiaGen: A generator for diagram editors providing direct manipulation and execution of diagrams", *Proc. IEEE Symp. on Visual Languages '95*, 203 – 210. 268

11. K. Marriott, B. Meyer. "Classification of visual languages", *Journal of Visual Languages and Computing*, 8, 375-402, 1997. 268

12. A. Salomaa, *Formal languages*, Academic Press, 1973. 267

13. J.A. Serrano. "The Use of Semantic Constraints on Diagram Editors". *Proc. IEEE VL'95*, 211-216. 268, 271

14. G. Taentzer, M. Beyer. "Amalgamated Graph Transformations and Their Use for Specifying AGG - an Algebraic Graph Grammar System". In *Graph Transf. in CS*, LNCS 776, 1994. 268

15. K. Wittenburg, L. Weitzmann. "Visual grammars and incremental parsing for interface languages". *Proc. IEEE VL'90*, 111–118. 268

Hypergraphs as a Uniform Diagram Representation Model

Mark Minas

Lehrstuhl für Programmiersprachen, Universität Erlangen-Nürnberg
Martensstr. 3, 91058 Erlangen, Germany
minas@informatik.uni-erlangen.de

Abstract. When working with diagrams in visual environments like graphical diagram editors, diagrams have to be represented by an internal model. Graphs and hypergraphs are well-known concepts for such internal models. This paper shows how hypergraphs can be uniformly used for a wide range of different diagram types where hyperedges are used to represent diagram components as well as spatial relationships between components. This paper also proposes a procedure for translating diagrams into their hypergraph model, i.e., a graphical scanner, and a procedure to check the hypergraph against a hypergraph grammar defining the diagrams' syntax, i.e., a parsing procedure. Such procedures are necessary to make use of such a hypergraph model in visual environments that support free-hand editing where the user can modify diagrams arbitrarily.

1 Introduction

Diagrams play an important role whenever complex situations have to be represented. In computer science, which is only one field of diagram applications, Nassi-Shneiderman Diagrams [17], Message Sequence Charts [10], and Entity-Relationship Diagrams are only some examples. When used on computers, e.g., for editing and interactions, diagrams have internally to be represented by a formal model which abstracts from diagrams' redundant visual information and which makes informations about the diagram readily available.

Several concepts have been used as internal models. Among others, multi-sets of tokens [12], attributed symbols [5], and different kind of graphs and hypergraphs. Typical graph models are graphs where nodes represent tokens and edges represent relationships between tokens [20], special graphs where nodes have distinct connection points which are then used by edges for representing connections [22], and hypergraphs where visual tokens (diagram components) are represented by hyperedges and connections between them by nodes [14,16]. Graphs and hypergraphs have the advantage that they are a formal and yet visual concept. Furthermore, there are powerful mechanisms like graph transformation theory and the existence of (hyper) graph parsers for syntactic analysis.

This paper extends the use of hypergraphs in the context of graphical diagram editors offering *free-hand editing*. Free-hand editing—in contrast to *syntax-*

H. Ehrig et al. (Eds.): Graph Transformation, LNCS 1764, pp. 281–295, 2000.

directed editing allows the user to arbitrarily arrange and modify diagram components on the computer screen. It is then the editor's task to distinguish syntactically correct diagrams from incorrect ones and to (re)construct the diagram's syntactic information. Previous work [14] has described how diagrams are internally represented by hypergraphs and how a diagram language is specified by a hypergraph grammar. However, this approach was limited due to two reasons:

1. Hypergraph grammars were restricted to context-free ones (with optional *embedding productions* which add single edges to a certain context) in order to allow for efficient parsing.
2. Hypergraphs could not be used to represent diagrams that make use of arbitrary spatial relationships like *inside* or *above*.

This paper describes how hypergraphs can now represent diagrams using arbitrary spatial relationships, too. Hypergraphs become thus a uniform representation model for a wide range of diagram languages. In order to use this internal model, this paper describes two specific tasks of a diagram editor, *scanning* and *parsing*:

1. The user somehow arranges a set of diagram components. The editor has to create resp. update a hypergraph which represents this arrangement of components ("Scanning step").
2. After creating resp. updating the hypergraph, the editor has to check whether the hypergraph is syntactically correct according to some hypergraph grammar ("Parsing step").

The rest of the paper is organized as follows: The next section describes two diagram languages which are used as demonstration examples throughout the paper. Section 3 then outlines how hypergraphs can be used to represent diagrams which use arbitrary spatial relationships, too. Sections 4 and 5 describe the scanning and parsing methods which create the internal hypergraph model and check its syntactic correctness. Related work is discussed in Sect. 6, Sect. 7 concludes.

2   Two Diagram Language Examples

Throughout this paper, we will use two kinds of diagrams: *Message Sequence Charts* (MSC) and *Visual Logic Diagrams* (VLD). This section briefly describes these two diagram languages.

MSC is a language for the description of interaction between entities [10]. A diagram in MSC describes which messages are interchanged between process instances, and what internal actions they perform. Figure 1a shows a sample diagram for MSC.

Visual logic diagrams (VLDs) [1] have been developed as an alternative visual notation for Horn clauses, a subset of first order logic. The following paragraph roughly sketches the idea of VLD syntax. For a complete description, see [1,18].

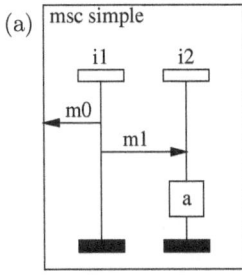

(a)

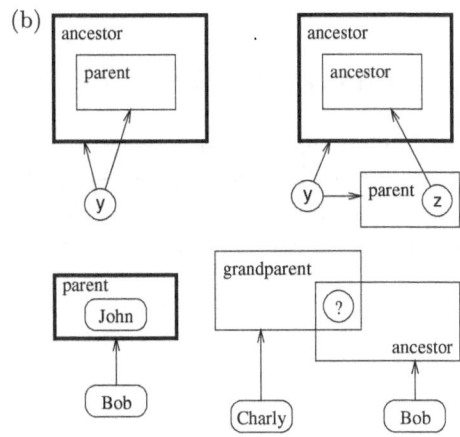

(b)

Fig. 1. A Message Sequence Chart (a) and four visual logic diagrams (b). The letters y and z in (b) are not part of the diagram; they are used for reference in the text.

Terms in a VLD are visually represented by directed acyclic graphs (DAGs) where circles stand for variables and ovals for constants and functions. *Predicates* are visually represented by labeled boxes, meaning the set comprehension over one of its arguments, i.e., each predicate $P(x_1, x_2, \ldots, x_n)$ is defined by a box whose area represents a set $\mathbf{P}(x_2, \ldots, x_n) = \{x_1 \mid P(x_1, x_2, \ldots, x_n)\}$. Arguments $x_2, \ldots, x_n$ are explicitly represented by DAGs which are connected by arrows with the box's border. A VLD is composed of various predicates and terms put together in a Venn-diagram-like fashion. Each diagram has a special predicate (the predicate which is defined by the diagram, i.e., the Horn clause) which is highlighted by a thick border. Figure 1b shows some examples: The top-left diagram represents $\mathbf{parent}(y) \subseteq \mathbf{ancestor}(y)$, i.e., parent$(x, y) \Rightarrow$ ancestor(x, y). The top-right diagram visually represents $z \in \mathbf{parent}(y) \Rightarrow \mathbf{ancestor}(z) \subseteq \mathbf{ancestor}(y)$, i.e., parent$(z, y) \wedge$ ancestor$(x, z) \Rightarrow$ ancestor(x, y). The other diagrams represent a fact that John is a parent of Bob and a query who (represented by the variable with the question mark) is a grandparent of Charly and an ancestor of Bob.

Each diagram generally consists of a set of atomic diagram components which are spatially related. The following shows that MSC and VLDs use different concepts which hypergraphs can represent in a uniform and straightforward way.

For MSC diagrams, components are surrounding boxes, start and end boxes, vertical lifelines, message arrows, action boxes, and labeling text. For VLDs, we have circles, ovals, arrows, rectangular boxes, and labeling text. Spatial relationships which are used for composing a diagram from its components are very different for MSCs and VLDs: MSC components are simply 2-dimensional shapes (boxes, text) which are related by connection objects (lines and arrows). MSC components are combined by attaching lines and arrows to boxes in a graph-like manner and by putting text near arrows and boxes. However, the ways of relating components in VLDs are more versatile: boxes can intersect or contain each other, circles and ovals may lie inside of boxes. Arrows can connect circles, ovals, and boxes.

3   Hypergraph Representation of Diagrams

Hypergraphs have proved to be an intuitive means for internally representing diagrams [14,15,16,21]. A hypergraph is a generalization of a graph, in which edges are *hyperedges* which can be connected to any (fixed) number of nodes [8]. Each hyperedge has a type and a number of connection points ("tentacles") that attach to nodes. We say the hyperedge *visits* these nodes. The familiar directed graph can be seen as a hypergraph in which all hyperedges visit exactly two nodes.

A hypergraph-based specification of a diagram language has to consist of a mapping between diagrams and their hypergraph representation. A hypergraph grammar (see Sect. 5) specifies the set of all hypergraphs that represent valid diagrams. Mappings between diagrams and hypergraphs are specified in two steps. First, it is specified how each atomic diagram component is represented, and second, how these diagram components resp. their hyperedge representations are linked together. Each atomic diagram component is modeled by a hyperedge; hyperedge tentacles represent the component's "attachment areas", i.e., the areas which can actually connect to other components' attachment areas. However, only compatible attachment areas may connect. Such connections generally are established by overlapping attachment areas. For certain diagram languages, it is sufficient to represent all connections by "common" nodes: two tentacles are connected to the same node if the corresponding attachment areas are connected. This idea has been used in previous work [14,15,16,21] and is applicable to MSC as described in the next paragraph. However, this representation is not sufficient for other diagram languages like VLDs which will be described afterwards.

MSC components are surrounding boxes, start and end boxes, vertical lifeline segments between events and actions, message arrows, action boxes, and labeling text. They are modeled by hyperedge types *msc, start, end, lifeline, message, action,* and *text*. A surrounding box's attachment areas are its borderline (*env*), its area for lifelines (*contains*), and its subarea for labeling text (*name*). A message's attachment areas are head and tail of the arrow (*to* resp. *from*) as well as an area for the labeling text (*name*). The other diagram components have similar attachment areas. E.g., a message's *to* attachment area is compatible to attachment areas representing end points of lifeline segments (*from* resp. *to*), i.e., corresponding hyperedges visit a common node if these attachment areas overlap. Figure 2 shows the resulting hypergraph which represents the MSC of Fig. 1a. Tentacles carry the names of their corresponding attachment areas.

This easy way of representing component connections is not sufficient for more general spatial relationships among diagram components as in VLDs. VLDs consist of ovals, circles, labeling texts, arrows, and rectangular boxes which are represented by corresponding hyperedges. *Text* hyperedges, which visit a single node (the text's area), represent text labels. Arrow hyperedges visit two nodes

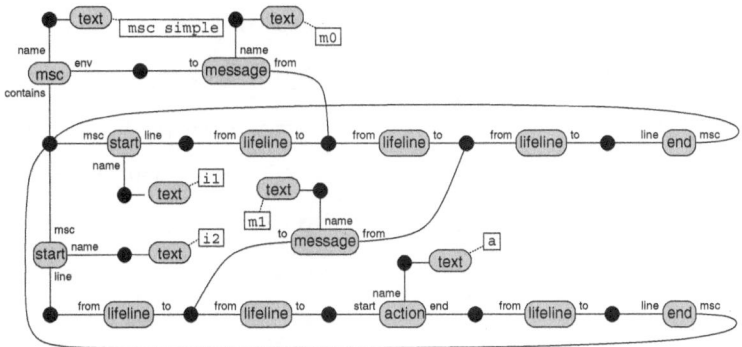

Fig. 2. Hypergraph model of the MSC diagram depicted in Figure 1a. Nodes are drawn as black dots, hyperedges as ovals, and tentacles as labeled lines connecting hyperedges with their visited nodes. m0, m1, ... are textual attributes containing labeling text represented by *text* hyperedges.

which represent the arrow's end points. In a VLD, circles, ovals, and boxes have two attachment areas: their borderlines and their area. Therefore, circles, ovals, and boxes are represented by (directed) edges which connect the two nodes that represent these attachment areas.

VLDs' main spatial relationships are inclusion and intersection: Two boxes may overlap, one may contain another one, circles and ovals may lie inside of boxes. The situation where one box contains another one cannot be described by simply visiting the same area node. It would not be clear which box is the inner one. In order not to loose information by representing a diagram by an internal hypergraph, additional hyperedge types are necessary which directly represent spatial relationships. For VLD, *overlap* and *inside* hyperedges are used. The first one denotes an undirected edge (which is internally represented by two directed edges with opposite directions) connecting the area nodes of overlapping boxes, the latter one denotes a directed edge from the area node of a box to the area node of another box which contains the first one. Figure 3 shows the according hypergraph representations for the top-right and bottom-right VLD of Fig. 1b.

As a result of this section, the hypergraph model of a diagram consists of hyperedges which represent diagram components and nodes which represent the components' attachment areas. Spatial relationships between a diagram's components are either modeled by visiting the same nodes or by inserting additional hyperedges which represent the spatial relationships explicitly.

4   Scanning

In diagram editors which offer free-hand editing, diagram components are directly arranged by the user, i.e., created, deleted, dragged around etc. The internal hypergraph model has to be kept up-to-date. This is a graphical scanner's

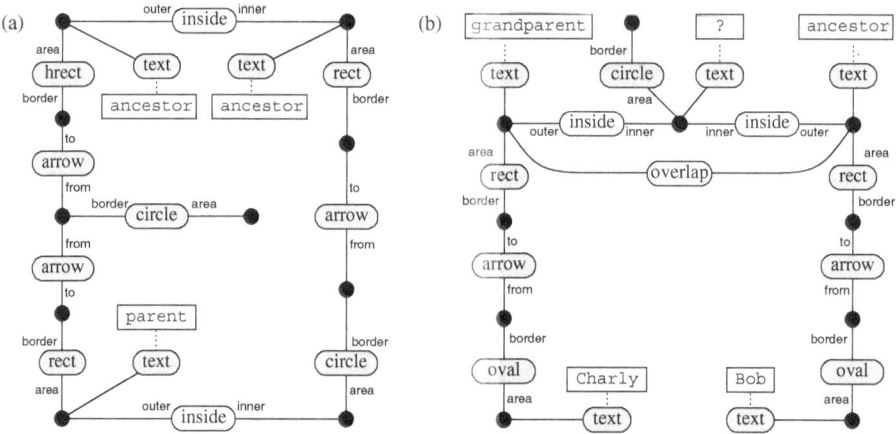

Fig. 3. Hypergraph representations of the top-right (a) and bottom-right (b) VLD of Fig. 1b. Hyperedges for diagram components have a gray label, spatial relationship edges have a white one. 'rect' specifies rectangular boxes, 'hrect' highlighted ones. `ancestor`, `parent` etc. are textual attributes containing labeling text represented by *text* hyperedges.

task. This section describes a scanning method and how it is adapted to specific diagram languages. As in the previous sections, we use MSC and visual logic diagrams as demonstration examples.

4.1   Intersecting Attachment Areas

The input of the scanning process is a set of atomic diagram components, each of them with a specific set of attachment areas. These are used for connections, i.e., spatial relationships between diagram components. For each component type (e.g., a box in VLD), there are different attachment area types with different shapes (e.g., the borderline and the inner area of boxes in VLDs). Attachment areas get connected if they intersect in a specific way. Given two attachment areas A and B, basically three different kinds of intersection are possible (the case where A and B do not intersect is omitted, i.e., $A \cap B \neq \emptyset$ holds for all three cases):

- $A \subset B$ or $B \subset A$ (Type "C" for *containing*).
- $A \not\subset B \wedge B \not\subset A$ and $A \cap B$ is not separated into disconnected pieces (Type "S" for *single intersection*).
- $A \not\subset B \wedge B \not\subset A$ and $A \cap B$ is separated into disconnected pieces (Type "M" for *multiple intersection*)

In VLDs, the differences between the intersection types are essential: If one box contains another one, we have type "C" intersection of the attachment areas of the boxes' areas. An arrow ending at a box's border causes an "S" or "C" intersection of the arrow's end point and the box's borderline (Requiring a "C" intersection exclusively would be too restrictive if the arrow does not exactly end at the borderline). Finally, two intersecting boxes can be detected by an "M" intersection of their borderlines.[1]

In [14], the following scanning procedure has been used to create the hypergraph model from the set of diagram components: For each diagram component together with its attachment areas, we have a hyperedge which visits a set of nodes, one node for each attachment area. Now check each intersection between any pair of attachment areas. Depending on the types of attachment areas and the type of intersection, do nothing or unify the corresponding nodes, i.e., the former two distinct nodes become a single one. This method works for the MSC example, but has to be extended for the extended hypergraph model of this paper: In some cases, additional hyperedges which represent general spatial relationships have to be inserted. However, additional context also has to be taken into account as the following example with VLD shows:

Consider the top-left diagram in Fig. 1b with the 'ancestor' box containing the 'parent' box. The 'parent' text belongs to the inner box which has to be represented by a single node that is visited by the text and the box hyperedge. The extended scanning procedure has to contain the rule to unify a box's area node with a text's node if the text's attachment area lies inside of the box's area. However, the outer box contains the text's attachment area, too. The same rule would require to unify the text's node, which is already unified with the inner box's area node, with the area node of the outer box, too. Obviously, this rule must not be applied in this context.

A similar problem arises in VLDs with many nested boxes. The method described above would create a large number of *inside* edges. Actually, the result would be the transitive closure of the set of actually required *inside* edges.

4.2   A Two-Step Scanning Method

An obvious solution to these problems would be to consider nesting level: if there is an intersection with the attachment areas of some box, all boxes containing this one are left out from further considerations. In the following, we present a more general solution which allows for specification of exactly this behavior.

For a correct scanning method, we do not create the hypergraph directly as in the previous method. Instead, we first create an intermediate graph which is then used for creating the hypergraph. The intermediate graph simply consists of all attachment areas as nodes and all detected intersections as explicit edges which are labeled with the intersection type. This *intersection graph* makes it possible to check the context for forbidden context graphs before unifying nodes or adding hyperedges to the hypergraph.

The scanning method now works as follows: Create the intersection graph by adding all attachment areas of the diagram as nodes. For each pair of intersect-

---

[1] This is equivalent to an "S" intersection of the boxes' areas.

ing attachment areas, draw an edge between the corresponding nodes. Use the intersection type as the edge's label. When this graph is completed, check each edge in the intersection graph. Depending on its label and its context, unify the corresponding hypergraph nodes, connect them with an appropriate additional hyperedge, or do nothing. This is specified by function

$$F : T \times I \times T \rightarrow Action \times 2^{\mathcal{G}}$$

where T is the set of attachment area types, $I = \{C, S, M\}$ the set of intersection types, $Action$ the set of actions for hypergraph modifications, and $\mathcal{G}$ is the set of all graphs. $2^{\mathcal{G}}$ denotes the powerset of $\mathcal{G}$. For each pair $t_1, t_2 \in T$ of attachment area types and for each intersection type $i \in I$, $(act, context) = F(t_1, i, t_2)$ consists of an action and a set of forbidden context graphs. Either $act = \bot$ which represents the null operation ("do nothing"), or $act = \lambda x.\lambda y.action(x, y)$ where $action$ is an action which refers to two hypergraph nodes x and y. Possible actions are "*Unify the nodes x and y*" or "*Draw an inside edge from node x to node y*", i.e., actions are actually transformation rules with two distinguished nodes. Finally, each element $G \in context$ is a (forbidden context) graph of attachment areas whose edges are labeled with intersection labels and with two of its nodes being labeled with α_1 and α_2, resp.

Function F describes how the scanning phase works after the intersection graph has been created: For each edge e of the intersection graph, determine the label i of edge e, source node c_1, and target node c_2 which are attachment areas that have some types t_1 and t_2, resp. Compute $(act, context) = F(t_1, i, t_2)$. If $act = \bot$, do nothing. Else try to match at least one graph $G \in context$ with the intersection graph such that G's nodes α_1 and α_2 match c_1 and c_2, resp. If such a match does exist, do nothing. Else determine the hypergraph nodes n_1 and n_2 which represent c_1 and c_2, resp., in the hypergraph model and use $act(n_1, n_2)$ in order to modify the hypergraph model.

For VLDs, attachment area types are $T = \{area, border, point, text\}$ which represent a box's, oval's, or circle's inner area, its borderline, an arrow's end points, and a labeling text's area. For brevity, we only show the specification of F for the case which has caused problems in the previous scanning method, i.e., a text is contained in a box:

$$F(text, C, area) = (\lambda x.\lambda y.unify(x, y), \{G\})$$

where $unify(x, y)$ means "Unify nodes x and y" and

Graph G detects the situation where the text lies in a nested box. Since the intersection graph contains the transitive closure of "C" edges, the text will get connected to the innermost box only.

4.3   Complexity Issues

In order to get an idea of the scanning method's speed, we will now estimate its complexity.

The first phase of the scanning method checks for each intersection of attachment areas. When doing preprocessing by searching for intersecting rectangles which serve as bounding boxes for the attachment areas, this phase has complexity $O(n \log n + k)$ where n is the number of attachment areas and k the number of (bounding box) intersections since a well know Plane-sweep-algorithm can be used [13]. The worst case of the second phase is to match graphs for each intersection edge. If e is the maximum number of edges in (forbidden context) graphs, the worst case of graph matching is $O(k^e)$. Therefore, the worst case complexity of the scanning procedure is $O(n \log n + k^{e+1})$. But the intersection graph is normally sparse which reduces complexity of graph matching a lot.

5   Parsing

Diagrams are represented by hypergraphs. In order to define a diagram language in terms of a hypergraph language, *hypergraph grammars* are an appropriate means. Hypergraph parsers are used to check syntactic correctness of diagrams in terms of their hypergraphs and to (re)construct their syntactic structure. Hypergraph grammars and restricted classes of hypergraph grammars which allow for efficient parsing are briefly introduced in the following. However, these restrictions turn out to be too rigid for many diagram languages, e.g., VLD. This section outlines a parsing procedure which is applicable to a wider range of hypergraph grammars.

Hypergraph grammars are special cases of algebraic graph grammars [7]: Each hypergraph grammar HG consists of two disjoint finite sets N, T of nonterminal resp. terminal hyperedge types, a finite set of hypergraph productions, and a starting hypergraph (*axiom*) S which contains only nonterminal hyperedges. Each production $p = (L \xleftarrow{p_1} G \xrightarrow{p_r} R)$ is an algebraic hypergraph transformation rule with hypergraphs L, G, R called left-hand side (LHS), interface, and right-hand side (RHS), resp., and hypergraph morphisms p_1, p_r where p_1 is injective, i.e., G is a sub-hypergraph of L. Production p is applied to a hypergraph H by finding L as a subgraph (*redex*) of H, removing the sub-hypergraph $L \setminus G$ from H and merging in $R \setminus G$ instead where G and p_r exactly describe how the RHS has to fit in. The resulting hypergraph H' is called *derived* from H, $H \rightarrow H'$, in one step. As usual, the hypergraph language $L(HG)$ is the set of all hypergraphs H which contain terminal hyperedges only and which can be derived from S in a finite number of steps, $S \xrightarrow{*} H$.

There exist efficient parsers for restricted classes of hypergraph grammars only. [14] has discussed efficient parsers for context-free hypergraph grammars with optional embedding productions: Similar to context-free string grammars, context-free hypergraph grammars are defined in terms of their productions: A context-free production consists of a single nonterminal hyperedge as LHS, a discrete interface containing each node of the LHS, and injective morphisms p_1, p_r [8].

290   Mark Minas

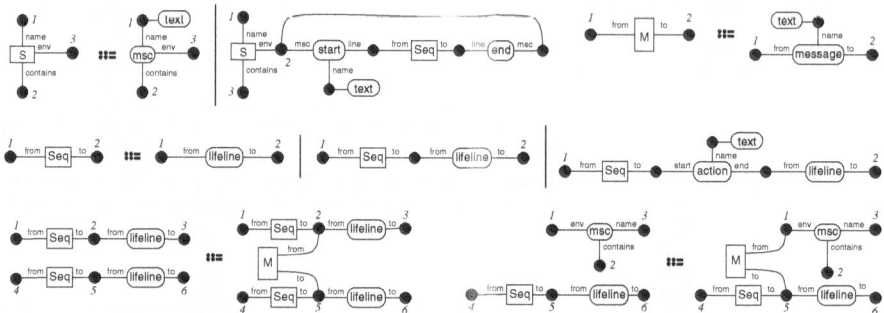

Fig. 4. Productions of a hypergraph grammar for MSCs. Ovals depict terminal
hyperedges, rectangles nonterminal ones. Interface hypergraphs are implicitly
defined by the labeled nodes and the hyperedges contained in the LHS and the
RHS at the same time. One production which is the same as the last depicted
one with a reversed message is omitted here.

Embedding productions add hyperedges to a certain context. Figure 4 shows the
productions of a context-free grammar with two embedding productions (the two
productions at the bottom) for MSC.

For VLD and the hypergraph model described in Sect. 3, we have not found
such a context-free hypergraph grammar with embeddings. However, a grammar according to the general hypergraph grammar definition is easily created.
Figure 5 shows a selection of eleven productions. The complete grammar consists of 32 productions. All shown productions except P_8 and P_{11} are actually
context-free ones. P_8's interface hypergraph consists of the nodes 1 and 2 and
of the 'Term' and 'Parameters' hyperedges as well. P_{11} is similar to P_8.

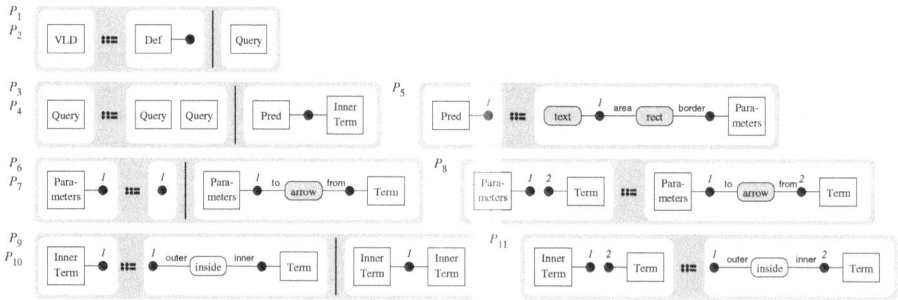

Fig. 5. Some selected hypergraph productions of the VLD grammar. The representation of productions is the same as in Fig. 4.

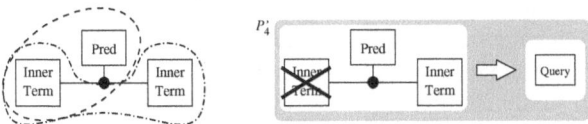

Fig. 6. Hypergraph with overlapping redexes (dashed borders) and appropriately extended reversed P_4 (see Fig. 5) with negative application condition (indicated by "X").

For a diagram editor with free-hand editing, a parsing algorithm has to try to reconstruct a derivation from the starting hypergraph to the internal hypergraph model of the current diagram. The diagram is valid if the parser succeeds, otherwise it is not. For context-free hypergraph grammars with optional embedding productions, this problem is more or less efficiently decidable [14]. For the general grammar, this problem is no longer decidable. For certain restricted graph grammars[2], e.g., *layered graph grammars* [19], there are special parsing algorithms, however they are quite inefficient. Recently, the class of *reserved graph grammars* (RGGs) [22] has been proposed which allows for a straight-forward way to parse hypergraphs.[3] The derivation for a hypergraph H is reconstructed by exchanging LHS and RHS of each production ("reversed productions"), by starting a derivation at H, and using reversed productions until the derivation stops. If the resulting hypergraph is the starting hypergraph, H is valid. For RGGs, parsing is always terminating since their is a well-founded ordering on hypergraphs which is decreasing for each derivation step during parsing [22]. Furthermore, the resulting system of reversed productions is confluent, i.e., for all derivations $H \xrightarrow{*} H_1$ and $H \xrightarrow{*} H_2$, there is always a hypergraph H' with $H_1 \xrightarrow{*} H'$ and $H_2 \xrightarrow{*} H'$. Therefore we can conclude that the simple RGG parser finds a derivation from any hypergraph H to the starting hypergraph if and only if there exists such a derivation. The parser cannot run into a dead end.

However, this confluence property is the crucial property which is frequently, e.g., for our VLD grammar, hard to fulfill. We propose a simple but effective extension of graph grammars: if the system of reversed productions is not confluent, we extend the productions by appropriate context and—if this is not yet sufficient—add appropriate *negative application conditions* (NACs) to affected productions. NACs have been motivated by Habel et al. [9]: a production with matching LHS is not applicable if one of its NACs is satisfied. A NAC is simply a hypergraph that is connected to the LHS of the reversed production. The NAC is satisfied for an embedding of the LHS into the host hypergraph if the NAC hypergraph can be embedded also.

---

[2] In the literature, mainly graphs as the simple form of hypergraphs are used. However, most results for graphs also apply to hypergraphs.

[3] *Reserved graph grammars* have been introduced by Zhang and Zhang for a special kind of graphs which are actually hypergraphs.

Of course, additional contexts and NACs modify and add further information to hypergraph grammars; the original grammar is no longer the only description of the hypergraph language's syntax. Tools may help here to find critical pairs and to assist in making the system of reversed productions confluent.

This paper does not present a fully fledged set of reversed productions for VLD due to its size (32 productions). Instead we give an example why we need NACs for VLDs. Figure 6 shows a hypergraph for which the reversed productions of P_4 as well as P_{10} are applicable. Reversed P_4 must not be applied since the inner terms have to be combined to a single inner term first. Otherwise, the second inner term could not be reduced. The correct behavior can be guaranteed by forbidding the existence of a second inner term if reversed P_4 is going to be applied. Figure 6 shows the appropriately extended reversed production P_4'.

6   Related Work

This paper is related to other work in the field of tools for creating diagram editors and in the fields of scanning algorithms and (hyper) graph parsing. The following list selects only some approaches.

In the field of frameworks for diagram editors there are several related approaches; the most closely related ones, which also allow to explicitly represent spatial relationships, are VLCC by Costagliola et al. [6], the proposed visual environment by Rekers et al. [2,20], and GenGEd by Bardohl and Taentzer [3]:

Costagliola et al. use an object-oriented hierarchy for representing diagrams according to their syntactic models instead of a uniform representation by hypergraphs as in this work. For connecting visual components, their VLCC system uses attachment points which can be connected by lines. This paper presents a similar, but more general approach which allows to represent arbitrary spatial relationships between components.

The approach by Rekers et al. actually uses two kinds of graphs as internal representations of diagrams: the *spatial relationship graph* (SRG) abstracts from the physical diagram layout and represents higher level spatial relationships. Additionally, an *abstract syntax graph* (ASG) represents the diagram's logical structure and is kept up-to-date with the SRG. Two different but connected context-sensitive graph grammars are used to define the syntax of SRGs and ASGs. Free-hand editing of diagrams is planned to modify the first graph, syntax-directed editing is going to modify the second. The actually other graph is modified accordingly. In this paper and in [14,15,16,21], we use hypergraphs instead of graphs which allows for a more "natural" diagram model. Furthermore, they restrict their discussions to graph-like diagrams. This work also considers diagrams with arbitrary spatial relationships.

GenGEd is an interactive tool for creating syntax-directed editors for graph-like diagram languages based on a powerful graph transformation system. However, free-hand editing based on some graph parser is not supported.

The ideas presented in Sect. 4 are also related to work of Blostein and Grbavec [4] on mathematics-recognition. They describe a scanning method which uses graph rewriting to create graphs as a representation of mathematical formulas. Application of graph transformation rules is controlled by spatial coordinates of recognized symbols. In contrast to our approach, they do not depend on overlapping attachment areas, but allow to take into account even distant components. Therefore, their approach is more general than ours at the expense of efficiency which is crucial in the context of interactive applications like diagram editors.

Furthermore, this work is related to other (hyper) graph parsing approaches. As an example, Lutz has presented a chart parser for flowgraphs [11] as a special kind of diagrams. Flowgraphs can be considered as a special case of hypergraphs used in this paper. The chart parser can be used top-down as well as bottom-up for context-free "*flow grammars*" only. Rekers and Schürr have proposed a graph parser for more general grammars, so called *layered graph grammars* [19]. Their parsing algorithm uses a bottom-up and then a top-down phase to (re)construct derivation sequences of the graph given. Finally, Zhang and Zhang have proposed an efficient parser even for context-sensitive *reserved graph grammars* based on Rekers and Schürr's one [22]. However, the properties required for these (hyper) graph grammars are quite restrictive. The hypergraph parsing procedure discussed in this paper actually uses the same parser, but relaxes these restrictions. However, additional information (*negative application conditions*) has to be provided.

7   Conclusions and Future Work

In this paper we have reconsidered modelling diagrams by hypergraphs as it is done by diagram editors in *DiaGen* [14,16]. In a diagram editor, the internal (hypergraph) model is then further processed for syntax checking, semantic evaluation, etc., but this was beyond the scope of this paper. The paper has discussed different ways of how to model diagrams and how to obtain diagrams by connecting and combining their components. When using such a model in a diagram editor which supports free-hand editing, a graphical scanner is needed which creates the hypergraph model of the diagram currently edited. Such a scanning method has been presented in the paper. The scanning method makes use of a specification of the diagram language which describes how relationships between diagram components have to be modeled by edges in the hypergraph depending on the diagram components' contexts. This approach makes hypergraphs a flexible modelling concept suitable for modelling a large number of different diagram languages.

The paper has also outlined an efficient method for syntax-checking. Diagrams' syntax is specified by a hypergraph grammar. Correctness of a diagram is checked by trying to reduce the diagram's hypergraph to the grammar's axiom by using reversed grammar productions as transformation rules. Additional negative application conditions are used to avoid dead ends. This simple yet efficient syntax-checking procedure is applicable to a wide range of hypergraph grammars not restricted to context-freeness or context-sensitiveness. However, negative application conditions have to be provided as further information.

So far, the scanner does not work incrementally, i.e., the whole hypergraph model is (re) created from scratch even if only small diagram parts are modified. Current work on this problem tries to obtain such an incremental scanner. Other work has to deal with tool support. As already pointed out, tool support for detecting critical pairs in reversed productions would simplify the task of creating a confluent system.

References

1. J. Agusti, J. Puigsegur, and D. Robertson. A visual syntax for logic and logic programming. *Journal of Visual Languages and Computing*, 9:399–427, 1998. 282
2. M. Andries, G. Engels, and J. Rekers. How to represent a visual program? In *Proc. 1996 Workshop on Theory of Visual Languages, Gubbio, Italy*, May 1996. 292
3. R. Bardohl and G. Taentzer. Defining visual languages by algebraic specification techniques and graph grammars. In *Proc. 1997 Workshop on Theory of Visual Languages, Capri, Italy*, Sept. 1997. 292
4. D. Blostein and A. Grbavec. Recognition of mathematical notation. In H. Bunke and P. Wang, editors, *Handbook of Character Recognition and Document Image Analysis*, chapter 21, pages 557–582. World Scientific, 1997. 292
5. P. Bottoni, M. Costabile, S. Levialdi, and P. Mussio. Formalising visual languages. In *[24]*, pages 45–52, 1995. 281
6. G. Costagliola, A. D. Lucia, S. Orefice, and G. Tortora. A framework of syntactic models for the implementation of visual languages. In *[26]*, pages 58–65, 1997. 292
7. H. Ehrig. Introduction to the algebraic theory of graph grammars. In V. Claus, H. Ehrig, and G. Rozenberg, editors, *Graph-Grammars and Their Application to Computer Science and Biology, LNCS* 73, pages 1–69, 1979. 289
8. A. Habel. *Hyperedge Replacement: Grammars and Languages, LNCS* 643, 1992. 284, 289
9. A. Habel, R. Heckel, and G. Taentzer. Graph grammars with negative application conditions. *Fundamenta Informaticae*, 26(3,4), 1996. 291
10. ITU-T, Geneva. *Recommendation Z.120: Message Sequence Chart (MSC)*. 281, 282
11. R. Lutz. Chart parsing of flowgraphs. In *Proc. 11th Int. Conf. on Artificial Intelligence (IJCAI'89), Detroit, Michigan*, pages 116–121, Aug. 1989. 293
12. K. Marriott. Constraint multiset grammars. In *[23]*, pages 118–125, 1994. 281
13. K. Mehlhorn. *Data Structures and Algorithms 3 – Multi-dimensional Searching and Computational Geometry*. Springer-Verlag, Berlin, 1984. 289
14. M. Minas. Diagram editing with hypergraph parser support. In *[26]*, pages 230–237, 1997. 281, 282, 284, 287, 289, 291, 292, 293
15. M. Minas and L. Shklar. A high-level visual language for generating web structures. In *[25]*, page 248f, 1996. 284, 292
16. M. Minas and G. Viehstaedt. DiaGen: A generator for diagram editors providing direct manipulation and execution of diagrams. In *[24]*, pages 203–210, 1995. 281, 284, 292, 293

17. I. Nassi and B. Shneiderman. Flowchart techniques for structured programming. *SIGPLAN Notices*, 8(8):12–26, Aug. 1973. 281

18. J. Puigsegur, W. M. Schorlemmer, and J. Agusti. From queries to answers in visual logic programming. In *[26]*, pages 102–109, 1997. 282

19. J. Rekers and A. Schürr. A graph grammar approach to graphical parsing. In *[24]*, pages 195–202, 1995. 291, 293

20. J. Rekers and A. Schürr. A graph based framework for the implementation of visual environments. In *[25]*, pages 148–155, 1996. 281, 292

21. G. Viehstaedt and M. Minas. Interaction in really graphical user interfaces. In *[23]*, pages 270–277, 1994. 284, 292

22. D.-Q. Zhang and K. Zhang. Reserved graph grammar: A specification tool for diagrammatic VPLs. In *[26]*, pages 288–295, 1997. 281, 291, 293

23. *1994 IEEE Symp. on Visual Languages, St. Louis, Missouri*, Oct. 1994. 294, 295

24. *1995 IEEE Symp. on Visual Languages, Darmstadt, Germany*, Sept. 1995. 294, 295

25. *1996 IEEE Symp. on Visual Languages, Boulder, Colorado*, Sept. 1996. 294, 295

26. *1997 IEEE Symp. on Visual Languages, Capri, Italy*, Sept. 1997. 294, 295

Story Diagrams: A new Graph Rewrite Language based on the Unified Modeling Language and Java

Thorsten Fischer, Jörg Niere, Lars Torunski, Albert Zündorf

AG-Softwaretechnik, Fachbereich 17, Universität Paderborn,
Warburger Str. 100, D-33098 Paderborn, Germany;
e-mail: [tfischer|nierej|torunski|zuendorf]@uni-paderborn.de

Abstract. Graph grammars and graph rewrite systems improved a lot towards practical usability during the last years. Nevertheless, there are still major problems to overcome in order to attract a broad number of software designers and developers to the usage of graph grammars and graph rewrite systems. Two of the main problems are, (1) that current graph grammar notations are too proprietary and (2) that there exists no seamless integration of graph rewrite systems with common (OO) design and implementation languages like UML and C++ or Java.

Story Diagrams are a new graph rewrite language that tries to overcome these deficiencies. Story Diagrams adopt main features from Progres, e.g. explicit graph schemes, programmed graph rewriting with parameterized rules, negative, optional and set-valued rule elements. Story diagrams extend common graph models by offering direct support for ordered, sorted, and qualified associations and aggregations as known from the object-oriented data model. Story Diagrams adopt UML class diagrams for the specification of graph schemes, UML activity diagrams for the (graphical) representation of control structures, and UML collaboration diagrams as notation for graph rewrite rules. Story Diagrams are translated to Java classes and methods allowing a seamless integration of object-oriented and graph rewrite specified system parts.

1 Introduction

At the last graph grammar conference in Williamsburg four years ago, Blostein stated a number of requirements for the industrial use of graph grammars and graph rewrite systems as a design and implementation means, cf. [BFG96]. They should be less difficult to learn. They should be expressive. It must be possible to use them for fractions of a software system (in order to get started). Even applied to larger fractions, they should work seamlessly with standard system parts. Their execution should be fast and environments are needed.

During the last four years theory, implementation and application of graph grammars and graph rewrite systems improved a lot. In theory, the expressive power of most ap-

H. Ehrig et al. (Eds.): Graph Transformation, LNCS 1764, pp. 296-309, 2000.

proaches was increased by attribute conditions, negative application conditions, general constraints, and control structures [Roz97]. Graph grammar and graph rewriting environments emerged and improved, meeting many of the requirements mentioned in [BFG96]. For example the AGG system was extended by a sophisticated graph pattern matching algorithm for the automatic execution of AGG rules, cf. [Rud97]. The Progres environment offers now means for the rapid prototyping of applications from their graph grammar specification [SWZ95a].

Despite these improvements, graph grammars and graph rewrite systems did not yet succeed to attract a broad number of software designers and developers. Two of the main problems are (1) that current graph grammar notations are too proprietary and (2) that there exists no seamless integration of graph rewrite systems and common (OO) design like UML (cf. [UML97]) and implementation languages like C++ or Java.

Story Diagrams are a new graph rewrite language that tries to overcome these deficiencies. Story Diagrams adopt main features from Progres [SWZ95a], e.g. directed, attributed, node and edge labeled graphs, explicit graph schemes, programmed graph rewriting with parameterized rules, negative, optional and set-valued rule elements. However, Story Diagrams extend the Progres graph model by direct support for ordered, sorted, and qualified associations and aggregations. Thus, the data model of Story Diagrams corresponds to the object-oriented data model.

Accordingly, Story Diagrams exploit UML class diagrams for the specification of graph schemes. Story Diagrams adopt UML activity diagrams for the (graphical) representation of control structures. The activities of a Story Diagram contain either program code (like in UML) or graph rewrite rules. The graph rewrite rules use an UML collaboration diagram like notation. Due to our experiences in several industrial projects, this notation looks quite familiar to software engineers.

Chapter 2 introduces the key features of our new graph grammar language. Chapter 3 gives a short introduction of the formal semantics of Story Diagrams. Chapter 4 outlines the translation of Story Diagrams to Java code. Chapter 5 summarizes our results and highlights some future work.

2 Story Diagrams, the language

Story Diagrams rely on an explicit graph scheme that defines static properties of the specified data structures and allows consistency checks of the dynamic specification. We use standard UML class diagrams for this purpose, since they are familiar to our target customers, i.e. software engineers. Figure 1 shows a screen shot of an UML class diagram for a lift simulator used as a running example within this paper. The screenshot is taken from the *Fujaba*[1] *environment* (cf. [FNT98]).

Note, the qualified association between houses and their levels. In standard graph models, multiple links of a certain type attached to a given node are indistinguishable and have no specific order, cf. [Roz97]. The object-oriented data model extends this concept by ordered and sorted and qualified associations. In case of an ordered associations, the

---

1. Fujaba is an acronym for "From UML to Java And Back Again"

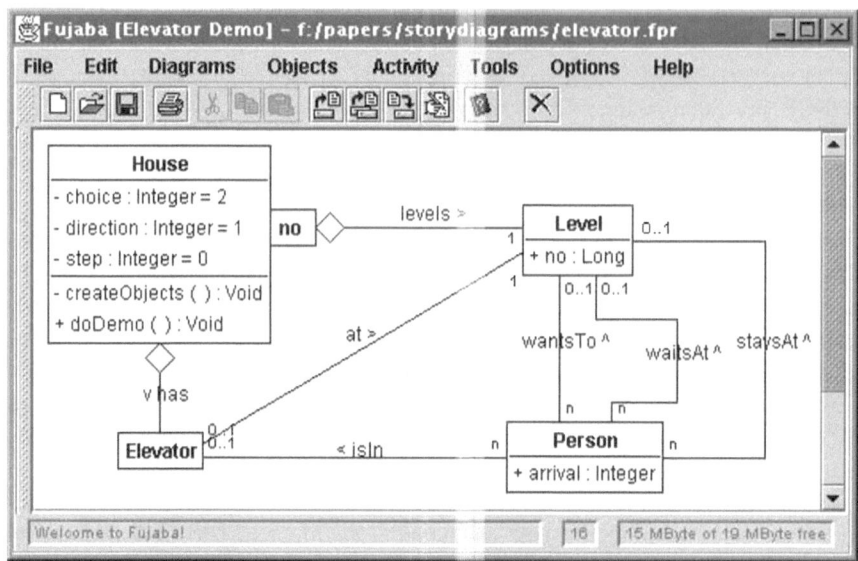

Figure 1 : Screen shot of the elevator class diagram specified with Fujaba

corresponding links attached to a given node build a list with a fixed order (defined at insertion time). Multiple links at a given node that belong to a sorted association are ordered according to a less-than operation on the reached neighbours. Qualified associations allow to distinguish between multiple links via a key value. We consider these extensions as very valuable and thus we incorporated them into Story Diagrams.

As one might have noticed, class House contains two methods createObjects and do-Demo. We use Story Diagrams to specify and implement such methods. Figure 2 shows the Story Diagram for method doDemo of class House. Story Diagrams may have formal parameters for passing attribute values and object references. Story Diagrams adapt UML activity diagrams to represent control flow graphically. Thus, the basic structure of a Story Diagram consists of a number of so-called activities shown by big rectangles with rounded left- and right-hand sides. For convenient referring, the activities are numbered at their upper right corner. Activities are connected by transitions, that specify the execution sequence. Execution starts at the unique start activity represented by a filled circle. Execution proceeds following the outgoing transition(s). Multiple outgoing transitions are guarded by mutual exclusive[1] boolean expressions shown in square brackets, e.g. [this.step<100]. Diamond shaped activities express branching. When the stop activity, represented by an "bulls eye", is reached, method execution terminates.

Despite Story Diagrams, there exists only one other notable graph rewrite language providing explicit control structures, i.e. the Progres environment. The Progres control structures preserve an atomic execution semantic for complex operations, i.e. either all

1. In general, it is not possible to check the mutual exclusiveness of different guards statically. One could check it at runtime. However, so far we generate code that just applies the different guards one after the other until the first evaluates to true.

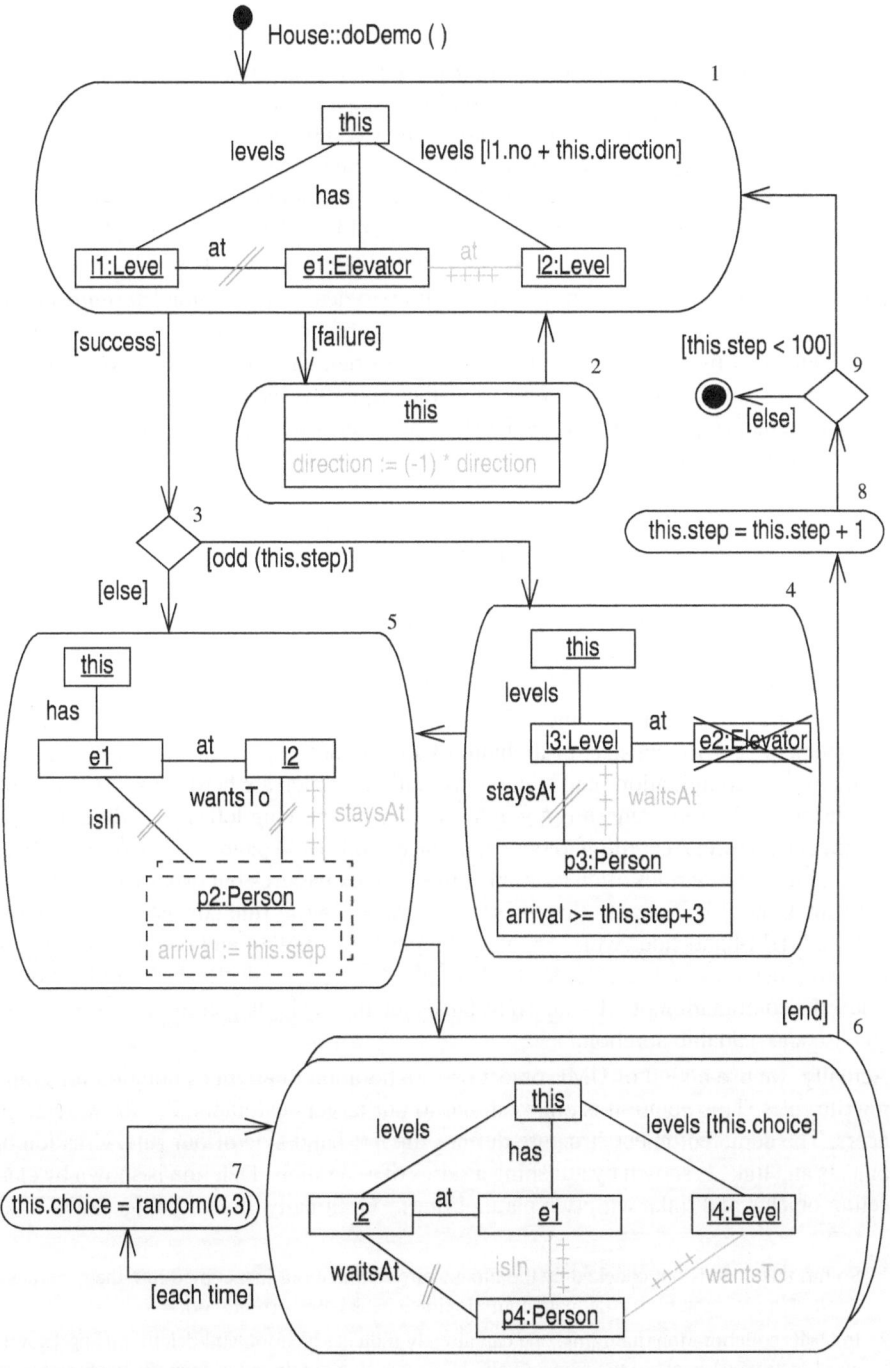

Figure 2 : The Story Diagram for the method doDemo of class House.

steps are executed successfully or none is executed at all. In addition, Progres control structures deal with the nondeterministic nature of graph rewrite rules. In general, a graph rewrite rule might have different matches within a given host graph. Usually, one of these matches is chosen randomly. Within a sequence of rewrite rules the random choice for the first rewrite rule might affect the executability of subsequent graph rewrite steps. The Progres semantics require that a sequence of graph rewrite steps must be successfully executed iff there exists choices for the application of each rule that allow the subsequent rules to be executed, too. To achieve this, Progres employs a complex backtracking mechanism for the execution of control structures.

Although these backtracking mechanisms allow to deal with the nondeterministic nature of graph rewrite rules, properly, they actually prevent the seamless integration of code generated by Progres with standard object-oriented program code. We consider this as a serious drawback with respect to the requirements stated in [BFG96]. Thus, Story Diagram control structures do NOT employ backtracking mechanisms. In case of multiple possible matches one is chosen randomly and this choice is not revisited. The user has to take care of this problem. However, this enables a direct and seamless translation of Story Diagrams into standard object-oriented code. In addition, there exists a reasonable number of big and medium sized Progres specifications of CASE tools, reengineering environments [JSZ96], and application programs originating from various authors summing up to more than 1000 pages. These real world specifications do not make use of the backtracking mechanisms of Progres. At least, there exist large application areas not requiring backtracking.

Story Diagrams support two kinds of activities, statement activities and Story Patterns. A statement activity consists of a chunk of Java code[1], e.g. used for I/O-operations, mathematical computations and method invocations, which are hard to express in graph rewrite rules. A Story Pattern is a graph rewrite rule showing left- and right-hand side within one picture. According to our experiences with 1000 pages of Progres specifications, graph rewrite rules often look-up quite complex patterns but usually they perform only small modifications. In these cases, the graph rewrite rule consists of quite large left- and right-hand sides which are very similar to each other and one has to compare the two sides thoroughly in order to determine the specified effects (modifications). Thus, the combination of left- and right-hand side into a single picture results in a more concise and readable notation.

Actually, we use a kind of UML object or collaboration diagram as notation for graph rewrite rules. This notation is quite familar to our target customers, i.e. software engineers. The depicted object structure defines the left-hand side of our rule. Creation of objects and links is shown by attaching a series of + symbols. Deletion is shown by canceling objects and links with two parallel lines.[2] Optionally, created objects and links

---

1. So far, this code is not checked by the Fujaba environment but just copied into the generated method implementation. Adequate compile time checks are current work.
2. In UML collaboration diagrams, one can already indicate creation and deletion using {new} and {destroyed} constraints, respectively. However, due to our experiences in teaching Story Diagrams to students, these constraints are easily overlooked. Thus, we employ the notational adaption possibilities of UML that allow to depict such constraints using specific icons.

are shown in green color, deleted objects and links are shown in red color. Thus, the left-hand side of the depicted rule consists of the normal objects and edges together with the (red) canceled objects and links. The right-hand side consists of the normal and created objects and links. This notation was inspired by the Sesam notation, cf. [LD96].

In object-oriented terms, a Story Pattern represents a complex boolean condition on a number of bound and unbound variables (like a term in first order logic). Unbound variables are shown as boxes containing name and type, e.g. $\boxed{\text{e1 :Elevator}}$. Bound variables are shown as boxes containing only their name, e.g. $\boxed{\text{e1}}$. Subsequent Story Patterns may use variables bound in previous Story Patterns (or statements) of the same Story Diagram. In addition, formal parameters and the self reference "this" may serve as bound variables. A link in a Story Pattern represents the boolean condition that the objects matched by the corresponding variables are connected by such a link.

Generally, a Story Pattern is executed by binding all its unbound variables to objects such that the represented condition evaluates to true. If this is possible, the specified modifications are performed and the Story Pattern succeeds, otherwise it fails. One may use the special keywords **success** and **failure** as guards for transitions leaving Story Patterns to branch on the **success** or **failure** execution of a Story Pattern. Generally, Story Patterns are restricted to isomorphic matches. One may state exceptions explicitly using {maybe v1 = v2} constraints. The identification condition is ensured statically.

For example, Story Pattern 1 of Figure 2 consists of one bound variable (this) and three unbound variables e1, l1, and l2. Elevator e1 needs to be connected to the current house (denoted by this) via a has link and to level l1 via an at link. Levels l1 and l2 are reached from this via levels links. The levels link leading to l2 is qualified by the number of level l1 plus the value of the direction attribute of this (this.direction may contain 1 or -1). If a match is found, Story Pattern 1 cancels the at link between e1 an l1 and adds a new at link between e1 and l2. This models a move of the elevator to the next level in the current direction. Story Pattern 1 may fail if no level l2 exists above or below the current level. In that case, execution follows the [failure] transition and Story Pattern 2 toggles the current direction and revisits Story Pattern 1. Note, attribute assignments are shown in the so-called attribute compartment of a variable. We use the := operator to distinguish attribute assignments from (equality) attribute conditions (shown in the same compartment). The latter employ the == operator.

After a successful move, branch activity 3 is reached. On every odd step, we execute Story Pattern 4. Story Pattern 4 looks for an arbitrary Person p3 staying at an arbitrary level l3 for at least 3 steps. The latter is required by the attribute condition arrival >= this.step+3 shown in the so-called attribute compartment of variable p3. In addition, the *negative* variable e2 *(crossed out by a big X)* requires that no elevator exists at the chosen level l3. If Story Pattern 4 finds a match, it activates person p3 via replacing the staysAt link by a waitsAt link.

Story Pattern 5 contains a set-valued variable p2 shown by two stacked dashed rectangles. Set-valued variables represent not just one node but the set of all nodes that match the depicted constraints. Thus, p2 models the set of all person(object)s in elevator e1 (cf. the isIn link) that want to go to level l2 (cf. the wantsTo link). For all persons in p2, Story Pattern 3 cancels the attached isIn and wantsTo links and creates staysAt links, modeling that these persons leave the elevator and stay at their target level. Note, that variable e1 and l2 have been bound during the execution of Story Pattern 1 and still refer to the same objects in Story Pattern 5. In addition to set-valued variables Story Pattern provide optional variables (shown as dashed rectangle). Optional variables are bound to nodes if possible and ignored otherwise.

Story Pattern 6 represents a so called for-all Pattern indicated by two stacked activity shapes. A for-all Pattern is executed for all possible bindings of its variables. Story Pattern 6 contains two unbound variables, p4 and l4. Thus, each person p4 waiting at the current level l2 stops waiting (cf. the canceled waitsAt link) and enters the elevator (cf. the added isIn link) and presses his personal target button. The latter is simulated by adding a wantsTo link to level l4 which is determined by the value of attribute this.choice.

A for-all pattern may have subsequent activities executed for each of its matches. Therefore, we provide the special transition guards [each time] and [end]. Each time Story Pattern 6 is applied, execution proceeds with activity 7. Activity 7 contains a simple pseudo code statement assigning a new random value to this.choice. Thus, for each Person waiting at the current level, Story Pattern 6 selects a new randomly chosen target level.

The remaining activities count the number of elevator steps and limit the number of simulation steps. Our example contains no ordered or sorted links. For the means provided to deal with ordered or sorted links see [FNT98].

3 Formal foundations

The semantics of Story Diagrams are based on a translation to Progres. Despite sorted, ordered, and qualified associations, the object oriented data model of Story Diagrams corresponds to the directed, attributed, node and edge labeled graphs of Progres. Thus, a Story Pattern corresponds directly to a Progres production where the left-hand side is derived from the normal and canceled Story Pattern parts and the right-hand side corresponds to normal and created parts. Within a Story Diagram, variables bound to objects in one Story Pattern may be used as bound variables in subsequent Story Patterns in the same diagram. This passing of Story Pattern variable values is simulated by Progres production parameters. Attribute handling is translated to Progres conditions and transfer clauses. Optional, set-valued and negative parts have direct correspondencies. For example, the Progres production simulating Story Pattern 4 is shown in Figure 3.

Story Diagrams are restricted to so-called well-formed control flow. Basically, the control flow built by the transitions of a Story Diagram must represent nested sequences, branches, and loops. This enables a translation into Progres sequence-, choose-, and loop-statements. However, much care has to be taken to avoid the special backtracking mechanisms of Progres, that are NOT part of the Story Diagram semantics. Basically, triggering backtracking is avoided by embedding the productions simulating single Story Patterns into a

   choose when Step_X then sdm__Success := true else sdm__Success := false end
construct. If the execution of Step_X fails, the second choose branch is executed (which

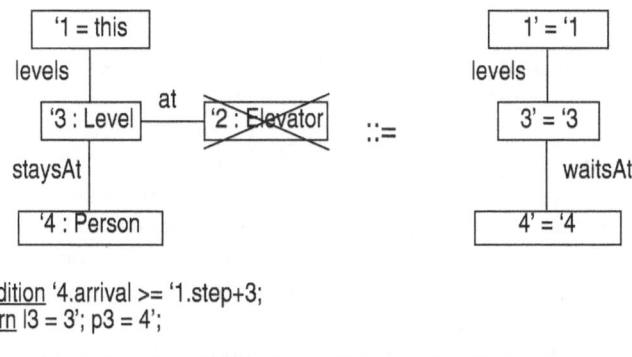

condition '4.arrival >= '1.step+3;
return l3 = '3'; p3 = '4';
end

Figure 3 : Simulating Story Pattern 4 with Progres

will succeed). After all, the special boolean variable sdm__Success signals success or failure of Step_X and may be used for subsequent branching.

Using such a translation, the well defined and elaborated semantics of Progres provides us with sound answers to a lot of detailed questions e.g. concerning the combination of multiple set-valued and optional and negative nodes within one rule or the handling of the identification condition in case of non-isomorphic matches. However, the translation of sorted, ordered, and qualified associations to Progres needs additional efforts since such elements are not provided by the Progres graph model. This translation is not yet formally specified. On the other hand, a semantic definition through a translation into Progres answers semantics questions to graph grammar people, only. It is of little help for our target customers, i.e. software engineers. Thus, a direct formalization staying close to object-oriented concepts is future work.

4 Translating Story Diagrams to Java

Story Diagrams drop the backtracking mechanisms of Progres. This enables the *Fujaba environment* (cf. [FNT98]) to generate standard object-oriented (Java) code from Story Diagrams. This Java code uses normal main-memory objects and does not rely on any huge library (like e.g. Progres generated code relies on GRAS, the Progres object database). Only some small helper functions are used to make the code more readable. Thus this code works seamlessly with other parts of a system (e.g. the graphical user interface). Due to the usage of Java, the code is platform independent.

Figure 4 to 7 show a cut-out of the Java code generated for class House of our example. First, the Fujaba generator translates UML class diagrams to Java. UML classes correspond directly to Java classes, cf. line 3. According to Java style guides, we translate UML class attributes to private Java attributes accessible via appropriate get- and set-methods. For example, attribute **direction** has methods **getDirection** and **setDirection**, cf. line 4 to 10.

According to UML semantics, associations are bi-directional. Thus, we implement associations by pairs of references in the respective classes. For example, association has

```
1: import java.util.*;
2: import com.objectspace.jgl.*;
3: public class House {
4: private int direction = 1;
5: private int getDirection ( ) {
6:    return this.direction;
7: } // getDirection
8: private int setDirection (int newDirection){
9:    return this.direction = newDirection;
10: } // setDirection
11: private OrderedMap levels
12:              = new OrderedMap ();
13: public void addToLevels (Level elem) {
14:    if (!this.hasInLevels (elem)){
15:       Integer k = new Integer (
16:                        elem.getNo ()));
17:       this.levels.add (k, elem);
18:       elem.setRevLevels (this);
19: }  } // addToLevels
20: public void removeFromLevels (...) { ... }
21: public boolean hasInLevels (Level elem) {
22:    if (elem == null) { return false;
23:    } else {
24:       Integer k = elem.getNo();
25:       return (this.levels.get(k) != null));
26: }  } // hasInLevels
27: public Enumeration elementsOfLevels ( )
28: {  return this.levels.elements (); }
29: public Level getFromLevels (int key) { ... }
30: private Elevator has;
31: public Elevator getHas () { ... }
32: public Elevator setHas () { ... }
33: ...
34: public void createObjects () { ... }
35: public void removeYou () { ... }
```

Figure 4 : Java class House, part a

of Figure 1 is implemented by an attribute has of type Elevator in class House (cf. line 30) and a reverse reference attribute revHas[1] of type House in class Elevator. Again, we provide appropriate access methods for these attributes (cf. line 31 and 32). For multi-valued associations we use standard container classes provided by the java generic library [JGL98]. We use class OrderedSet (balanced sorted trees) for normal (and sorted) associations and class DList (doubly linked lists) for ordered associations. We could implement qualified associations via class HashMap (hash tables). However, we use class OrderedMap (balanced binary tree of key-value pairs) since during pattern matching we frequently traverse the whole collection which is less efficient for hash tables. Thus, association levels of Figure 1 is implemented by attribute levels of type OrderedMap in class House (cf. line 11 and 12) and attribute revLevels of type House in class Level.

For container class attributes we provide appropriate access methods, that allow to add and remove elements (e.g. addToLevels line 13 and removeFromLevels line 20), to check the containment of an element (e.g. hasInLevels line 21), to visit all contained elements (e.g. elementsOfLevels line 27), and in case of a qualified association to retrieve an element via a key (e.g. getFromLevels line 29).

In order to guaranty the consistency of the pairs of references that implement the associations, the respective access methods for reference attributes call each others. For example, in line 18 method addToLevels calls method setRevLevels on the added level object elem, passing itself as parameter. This establishes the reverse reference. In the same way, removeFromLevels resets the revLevels reference at the corresponding Level object. The same holds for set methods of single valued reference attributes. Thus, to establish a pair of references one just calls the appropriate set or addTo method on

---

1. If no explicit role name is given, we use the prefix ´rev´ for the reverse reference.

```
36: public void doDemo () { ...
37:   while (!sdm__Success){
38:     // story pattern (1)
39:     sdm__Success = false;
40:     try{
41:       e1 = this.getHas(); // bind e1
42:       SDM.ensure (e1 != null);
43:       l1 = e1.getAt(); // bind l1
44:       SDM.ensure (l1 != null);
45:       SDM.ensure (l1.getRevLevels() ==
46:                                   this);
47:       l2 = this.getFromLevels (l1.getNo()
48:              + this.getDirection()); // bind l2
49:       SDM.ensure (l2 != null);
50:       // check isomorphic binding
51:       SDM.ensure (l1 != l2);
52:       e1.setAt (null); // delete link
53:       e1.setAt (l2); // create link
54:       sdm__Success = true;
55:     } catch (SDM.Exception e) { }
```

Figure 5 : Java class House, part b

one of the participating objects and passes the other as parameter. For example, the call e1.setAt (l2) in line 53 of Figure 5 creates an at link between elevator e1 and level l2. For the removal of links, again an appropriate method call on one of the objects suffices, cf. line 52.

Now, the Fujaba generator is able to translate Story Diagrams to standard Java code using the implementation of classes, attributes, and associations described above. Story Diagrams become methods of the depicted classes. The translation of Story Diagrams consists of two main tasks, translating the control flow of Story Diagrams and translating Story Patterns. Story Diagrams are restricted to so-called well-formed control flow. Basically, the control flow built by the transitions of a Story Diagram must represent nested sequences, branches, and loops. This enables a direct translation into Java block-, if-, and while statements. For example, the control flow of Figure 2 corresponds to a while statement (cf. line 37) containing the code of Story Pattern 1 (line 38 to 55) followed by an if statement (line 56, Figure 7) with a success and a failure branch (line 92). Within the success branch activity 3 corresponds to an if statement (line 57) with Story Pattern 4 in its true branch (line 58 to 80). This if statement is followed by Story Pattern 5 (line 82). For-all Story Pattern 6 corresponds to a while statement (line 84) containing activity 7 (line 87). This while is followed by activity 8 (line 89) and an if statement (line 90) for activity 9. The if statement for activity 9 contains a return statement in its else branch (line 91). The failure branch (line 92) of the if statement (line 56) following Story Pattern 1 contains just Story Pattern 2 (line 93).

Our translation of Story Patterns uses the same graph pattern matching optimisation strategies as Progres, cf. [Zün96, SZ94, FNT98]. However, the code of Progres prototypes relies on special control flow interpreters that guide the execution of (non-deterministic) search steps within a big switch statement. Dropping these backtracking mechanisms enabled us to generate more seamless Java code using normal control structures.

Generally, the code generated for a Story Pattern consist of nested loops for each unbound variable in the Story Pattern, cf. Figure 6. At each level, we choose an already bound variable v0 and a link l1 connecting it to a still unbound variable v1. Then, we generate a loop that iterates through the set of nodes reached via l1 and bind these candidates to v1, one after the other. Within the loop, we generate nested if-statements that check all conditions related to v1. Once all conditions related to v1 are considered, we proceed with the next unbound variable. Altogether, this generates nested search loops that iterate through all possible combinations of variable bindings. Once all variables are bound and all condition checks are passed, i.e. at the inner most nesting level, a match is found and code executing the graph rewrite step is reached. In addition, the successful execution is signaled by assigning true to the special boolean variable sdm__Success. At each search loop level this variable is then used to terminate the search process.

```
a: for all (v1 reached from a bound variable via a depicted link) do {
b:    if (condition 1 on v1 holds) {
c:        ...
d:        if (condition k on v1 holds) {
e:            ....
f:                for all (vn reached from a bound variable via a depicted link) do {
g:                    ...
h:                    if (condition vm holds von vn) {
i:                        /* found a match */
j:                        /* execute modifications */
k:                        /* terminate search */
l:                    } // if
m:                    ...
n:                } // for all vn
o:            ...
p:        } // if
q:    ...
r:    } // if
s: } // for all v1
```

Figure 6 : Nested loops and conditions for Story Pattern matching

Actually, we generate slightly different code for the checking of conditions. For large rewrite rules the nested if-statements soon reach an excessive nesting depth. To avoid this, we use the exception mechanism of Java. We replace nested if-statements by a single try-catch block and use a small helper function called SDM.ensure to check conditions. In case of a failure, SDM.ensure just throughs an exceptions and thus leaves the whole try-catch block.

In case of Story Pattern 4 variable this is the only bound variable. The first search loop iterates through all levels l3 that are reached from this via a levels link, cf. lines 60 and 61 to 63 in Figure 7. Additionaly, the try-catch block from lines 64 to 79 surrounds all conditions related to variable l3. For example, line 65 checks whether traversing an at link against its direction[1] reaches an elevator (which is excluded in Story Pattern 4). The second search loop (line 68) iterates through persons p3 that stay at level l3. The body of the second search loop is surrounded by its own try-catch block (cf. line 64 to 79 and 71 to 77). Thus, a search exception raised e.g. in line 72 is caught in line 77 and another

---

1. to indicate the reverse direction of an association, we use the name prefix 'rev'

```
56:     if (sdm__Success){
57:        if (Math.odd (this.getStep()){
58:           sdm__Success = false; // activity (4)
59:           Enumeration I3Enum =
60:                      this.elementsOflevels ();
61:           while ( ! sdm_Success
62:                    && I3Enum.hasMoreElements ()) {
63:              I3 = I3Enum.nextElement ();
64:              try {
65:                 SDM.ensure (I3.getRevAt () == null));
66:                 Enumeration p3Enum =
67:                         I3.elementsOfRevStaysAt ();
68:                 while ( ! sdm__Success &&
69:                         && p3Enum.hasMoreElements ()) {
70:                    p3 = p3Enum.nextElement ();
71:                    try {
72:                       SDM.ensure (p3.arrival >
73:                                   this.getStep () + 3);
74:                       p3.setStaysAt (null);
75:                       p3.setWaitsAt (I3);
76:                       sdm__Success = true;
77:                    } catch (SDM.exception e) { }
78:                 } // while p3Enum
79:              } catch (SDM.exception e) { }
80:           } // while I3Enum
81:        } // if
82:        ... // story pattern (5)
83:        sdm__Success = true;
84:        while (sdm__Success){
85:           ... // story pattern (6)
86:           if (sdm__Success) {
87:              this.choice=Math.random(0,3); //step 7
88:           } }
89:        ... // statement 8
90:        if (this.step < 100){ // step 9
91:           } else {  return; }
92:     } else {
93:        ... // story pattern (2)
94: } } } } // doDemo
95: } // House
```

Figure 7 : Java class House, part c

iteration of the second search loop starts at line 68. In that case, line 69 checks whether (still) another person stays at the current level. If iterator p3Enum contains more elements, line 70 binds variable p3 to the next person and the constraints are checked again. If iterator p3Enum runs out of elements, the inner search loop terminates and the outer search loop tries to bind I3 to the next level. If we run out of levels, we terminate and the Story Pattern has failed (sdm__Success is still false). Once a level without elevator is found and the constraint of line 72 to 73 is fulfilled, the modifications of the Story Pattern are executed (cf. line 74 to 75) and sdm__Success becomes true (cf. line 76) terminating the while statements and signaling success.

The code of Story Pattern 4 use while loops and elementsOf methods in order to traverse to-many associations. In case of to-one association, this is simplified to a association look-up using a get method and a simple check for a not-null value. For example, the code for Story Pattern 1 of Figure 2 first binds variable e1 by calling e1=this.getHas() (cf. line 41, Figure 5). Next, line 42 checks whether this was successful using the condition e1!=null. Once e1 is successfully bound, we call I1=e1.getAt() to determine the current level I1 (line 43). Line 47 to 48 bind variable I2 via the qualified levels association. Line 45 and 46 show a link-constraint example ensuring that I1 is a level of the current house.

If all variables are bound and each constraint is checked, we execute the modifications of the Story Pattern. In our example we cancel the old at link (cf. line 52) and create a new at link connecting e1 and l2 (cf. line 53). Finally, we assign true to variable sdm__Success signaling the successful Story Pattern execution (cf. line 54). This is used in line 56 to determine the next activity.

Note, our pattern matching strategy requires that each (connected component of a) Story Pattern contains at least one bound variable that allows to reach all its unbound variables by traversing links. Other approaches like Progres and AGG have no such limitation. However, those approaches require an explicit graph or at least explicit object extensions in order to be able to search for matches. Managing such an explicit extension in our approach would disable the Java garbage collector to discard objects, automatically. Thus, it would re-introduce the problem of memory leaks to Java and aggravate a seamless integration with other system parts. Thus we decided to abandon such an explicit graph. Due to our experiences with up to 1000 pages of Progres specification, requiring patterns to contain a bound variable allowing to compute all other objects is not a severe restriction. One always finds some kind of root objects that he or she can use to access the desired objects.

5 Conclusions

Story Diagrams aim to push graph grammars to a broader industrial usage. Therefore, Story Diagrams adapt standard object-oriented modeling languages i.e. UML class diagrams, activity diagrams, and collaboration diagrams. Generally, Story Diagrams enhance object-oriented software development methods by appropriate means for modeling the evolution and dynamic behavior of complex object structures. In [JZ98] we propose Story Driven Modeling as a new method for the software development based on Story Diagrams and show its application in a case study. Story Driven Modeling is an extension of [SWZ95a].

The semantics of Story Diagrams is based on its predecessor Progres. However, we enhanced the data model of graphs towards the object-oriented data model. In addition, we dropped the backtracking mechanisms of Progres since extensive experiences have shown that it is seldom used. Dropping the backtracking mechanisms enabled us to translate Story Diagrams and Story Patterns into standard object-oriented Java code. The generated code does not require an extensive library and may be integrated seamlessly with other system parts and is platform independent due to the usage of pure Java. Currently, the generator adapts the graph pattern matching algorithm of Progres. Incorporation of the back jumping techniques developed by [Rud98] is current work.

Currently, the Fujaba environment allows convenient editing of Story Diagrams and UML class diagrams, cf. [FNT98]. Fujaba provides a reasonable number of consistency checks integrating class diagrams and Story Diagrams. The Fujaba generator supports the translation of class diagrams and Story Diagrams and a reasonable subset of Story Patterns to Java code. In addition it comprises a simple object structure browser used to visualize and debug the execution of Story Diagrams. The Fujaba environment is programmed in pure Java and thus available on all standard platforms. The Fujaba environ-

ment underlies the GNU-Licence and is available via the internet under
http://www.uni-paderborn.de/cs/fujaba/index.html

References

[BFG96] D. Blostein, H. Fahmy, A. Grbavec: Issues in the Practical Use of Graph Rewriting.
In *Proc. 5th. Int. Workshop on Graph-Grammars and their Application to Computer
Science*; LNCS 1073, pp. 38-55, Springer

[FNT98] T. Fischer, J. Niere, L. Torunski: Design and Implementation of an integrated development environment for UML, Java, and Story Driven Modeling. Master Thesis, University of Paderborn (In German)

[JGL98] Technical reference of the generic collection library for Java http://
www.objectspace.com/jgl/

[JSZ96] J.-H. Jahnke, W. Schäfer, and A. Zündorf: A Design Environment for Migrating Relational to Object Oriented Database Systems. In *Proceedings of the International Conference on Software Maintenance* (ICSM '96), IEEE Computer Society 1996

[JZ98] J.-H. Jahnke and A. Zündorf: Specification and Implementation of a Distributed Planning and Information System for Courses based on Story Driven Modeling. In *Proceedings of the Ninth International Workshop on Software Specification and Design*
April 16-18, Ise-Shima, Japan, IEEE CS, pp. 77-86

[LS88] C. Lewerentz and A. Schürr. GRAS, a management system for graph-like documents.
In *Proc. of the 3rd Int. Conf. on Data and Knowledge Bases*. Morgan Kaufmann,
1988.

[Rud97] M. Rudolf: Design and Implementation of an Interpreter for attributed graph rewriting
rules. Master Thesis, Technical University of Berlin (In German)

[Roz97] G. Rozenberg (ed): Handbook of Graph Grammars and Computing by Graph Transformation, World Scientific, 1997.

[LD96] J. Ludewig, M. Deininger: Teaching Software Project Management by Simulation:
The SESAM Project; in Irish Quality Association (eds.): *5th European Conference on
Software Quality*, Dublin, pp. 417-426

[SWZ95a] A. Schürr, A. Winter, A. Zündorf: Graph Grammar Engineering with PROGRES; in:
Schäfer W. (ed.): Software Engineering - ESEC '95; LNCS 989, pp. 219-234;
Springer (1995)

[SWZ95b] A. Schürr, A. Winter, A. Zündorf: Visual Programming with Graph Rewriting Systems; In: Proc. VL'95 11th Int. IEEE Symp. on Visual Languages, Darmstadt, Sept.
1995, IEEE Computer Society Press (1995)

[UML97] UML Notation Guide version 1.1. Rational Software, http://www.rational.com/uml/

[Zün96] Zündorf A.: Graph Pattern Matching in PROGRES; In:J. Cuny,H. Ehrig,G. Engels,G.
Rozenberg (eds): In *Proc. 5th. Int. Workshop on Graph-Grammars and their Application to Computer Science*; LNCS 1073, pp. 454-468 Springer

A Fully Abstract Model for Graph-Interpreted Temporal Logic*

Fabio Gadducci[1], Reiko Heckel[2], and Manuel Koch[1]

[1] Technical University of Berlin, Fachbereich 13 · Informatik,
Franklinstraße 28/29, D-10587 Berlin, Germany
{gfabio,mlkoch}@cs.tu-berlin.de
[2] University of Paderborn, Fachbereich 17
Warburgerstraße 100, D-33098 Paderborn, Germany
reiko@uni-paderborn.de

Abstract. Graph-interpreted temporal logic is an extension of propositional temporal logic for specifying *graph transition systems* (i.e., transition systems whose states are graphs). Recently, this logic has been used for the specification and compositional verification of safety and liveness properties of rule-based graph transformation systems. However, no calculus or decision procedure for this logic has been provided, which is the purpose of this paper.

First we show that any sound and complete deduction calculus for propositional temporal logic is also sound and complete when interpreted on graph transition systems, that is, they have the same discriminating power like general transition systems. Then, structural properties of the state graphs are expressed by *graphical constraints* which interpret the propositional variables in the temporal formulas. For any such interpretation we construct a graph transition system which is *typical* and *fully abstract*. Typical here means that the constructed system satisfies a temporal formula if and only if the formula is true for all transition systems with this interpretation. By fully abstract we mean that any two states of the system that can not be distinguished by graphical constraints are equal. Thus, for a finite set of constraints we end up with a finite state transition system which is suitable for model checking.

1 Introduction

Graph transformations [15] are a graphical specification technique developed in the early seventies as generalization of Chomsky-grammars. This rule-based formalism provides an intuitive description for the manipulation of graph-like structures as they occur in databases, object-oriented systems, neural networks, software or distributed systems. In particular, in order to be more suitable for such kinds of applications, the algebraic approaches to graph transformation [4,2,3]

---

* Research partially supported by the EC TMR Network GETGRATS (General Theory of Graph Transformation Systems) through the Technical University of Berlin and the University of Pisa.

H. Ehrig et al. (Eds.): Graph Transformation, LNCS 1764, pp. 310–322, 2000.

have been extended by various techniques for expressing static and dynamic consistency conditions of graphs and graph transformations [11,19,12,1,10,7,8,13].

In [7] Heckel presented a compositional verification approach for safety and liveness properties of graph transformation systems. The main focus was on how to derive global properties from local ones, while the question of how to verify local properties w.r.t. certain parts of the system was not considered. In this paper we try to close this gap by investigate techniques and tools for checking the validity of temporal properties w.r.t. graph transition systems. Further investigations with repect to this topic can be found in [7,13], where more detailed case studies are presented to motivate the concepts.

A graph transition system is a transition system where the states are graphs equipped with variable assignments. The variables themselves form a graph as well, and the assignments are represented by (partial) graph homomorphisms. We are then able to specify properties of states by *graphical constraints*: Such constraint is just a pattern state (i.e., a partial graph morphism) which is satisfied by a second state if this provides at least the same structure of the first. This concept has originally been developed in [11] where graphical constraints were used in order to express static consistency properties. In [12,10] it was combined with propositional temporal logic, able to express also dynamic properties of systems. The long-term goal of this work is to develop analysis techniques for transition systems specified by graph grammars. A variety of such techniques already exists in the literature on temporal logic (see e.g. [17]). Thus, in order to be able to reuse these tools, in Section 4 we compare the graph-based temporal logic with the classical, propositional temporal logic and show that the notions of validity of formulas are the same in both cases.

Most of the automated techniques for verifying temporal properties of systems assume transition systems with finite sets of states. Hence, for applying such techniques it is necessary to collapse infinite transition systems to finite ones by identifying states which are logically indistinguishable. A transition system is called *fully abstract* if it is completely collapsed in this sense, i.e., if any two states which cannot be distinguished by temporal properties are identified in the system. Thus, in general, this problem is stated w.r.t. *all* temporal formulas for a given set of propositional variables (see e.g, [18]). In this paper, we use graphical constraints in order to define the evaluation of propositional variables. Therefore we are interested in *structural* equivalence of states, i.e., whether two states are distinguishable by means of a given set of graphical constraints. Consequently, the results in this paper are largely independent of the temporal logic in use and could be easily transfered to more sophisticated logics.

On the other hand, a transition system may represent a whole class of systems if it is *typical* for that class in the sense that it satisfies a formula if and only if this formula is satisfied by *all* transition systems of that class. Then, the typical system can be used to examine the validity (entailment, tautology, etc.) of formulas for the whole class of system by checking the corresponding property in the typical transition system only.

In Section 5 we present a construction that, for a given interpretation of the propositional variables as graphical constraints, produces a transition system which is both typical and fully abstract. We argue that these two properties correspond, respectively, to the existence and uniqueness of transition system morphisms from all transition systems of the class to the typical, fully abstract one. In other words, the constructed system is a final object in a suitable category of transition systems. Moreover, the system is finite provided that we only use finitely many different constraints in the temporal specification.

2   Some Background on Propositional Temporal Logic

In the first part of this section we review the classical syntax and semantics for the propositional fragment of linear temporal logic, according to [17].

Definition 1 (temporal formula). *Let* Q *be a (countable) set of propositional variables. A temporal formula (short formula) is a term generated by the following syntax*

$$\Phi ::= Q \mid \neg\Phi \mid \Phi_1 \wedge \Phi_2 \mid \Phi_1 \mathcal{U} \Phi_2 \mid \bigcirc \Phi$$

We let $\Phi, \Phi_1 \dots$ *range over the set* TF *of formulas.* △

The operators $\neg$ and $\wedge$ are the usual ones for negation and conjunction. The operators $\bigcirc$ and $\mathcal{U}$ constitute instead the temporal part of the logic. Roughly, since the semantics is given in terms of an evaluation over (sequences of) states of a *transition system*, the *next-time*-operator $\bigcirc$ demands that a formula holds in the immediate successor state, while the *until*-operator $\mathcal{U}$ requires that a formula eventually holds and until then a second formula is true. In the following, we apply the usual abbreviations for implication $\Longrightarrow$ and disjunction $\vee$. The boolean constant *true* can be defined as a logical tautology, for instance as $q \vee \neg q$ for propositional variables q in Q. The temporal *sometimes*-operator is defined by $\Diamond\Phi := true\mathcal{U}\Phi$, the *always*-operator by $\Box\Phi := \neg\Diamond\neg\Phi$.

As we mentioned, the classical semantics is based on transition systems. A transition system consists of a set of states, possibly infinite, and a relation on these states representing state transitions. The validity of a formula refers then to *runs* generated by such a transition system intended as infinite sequences of states. The set R of runs for examining validity is a designated subset of all possible runs through the transition system.

Definition 2 (transition system). *A* transition system *is a triple* $T = \langle S, \rightarrow, R \rangle$ *where* S *is a non-empty set of* states, $\rightarrow \ \subseteq S \times S$ *is a transition relation, and* R *is a suffix closed set of* runs. *A run* $\sigma = \sigma(0)\sigma(1)\dots$ *with* $\sigma(i) \in S$ *is a maximal length path through the transition system. Its i-th suffix is given by* $\sigma|_i = \sigma(i)\sigma(i+1)\dots$.

A transition system morphism $h : \langle S, \rightarrow, R \rangle \rightarrow \langle S', \rightarrow', R' \rangle$ *is a function* $h : S \rightarrow S'$ *such that* $h(R) \subseteq R'$, *where* $h(\sigma(0)\sigma(1)\dots) = h(\sigma(0)) \ h(\sigma(1))\dots$ *for all* $\sigma = \sigma(0)\sigma(1)\dots \in R$. △

We always assume that every state occurs in some run. Transition system morphisms allow to simulate the runs of the source system within the target system, in order to preserve, as shown later, the notion of *validity* of a formula. Together with an evaluation of the variables, the transition system constitutes a temporal model. The evaluation assigns to each variable a subset of S, where the variable is deemed to be true at each element of this subset.

Definition 3 (temporal model). *Given a transition system $T = \langle S, \rightarrow, R \rangle$ and a set of propositional variables Q, an* evaluation $\mathcal{V} : Q \rightarrow \mathcal{P}(S)$ *assigns to each variable a set of states $\mathcal{V}(q) \subseteq S$. The pair $\mathcal{M} = \langle T, \mathcal{V} \rangle$ constitutes a* temporal model.

A morphism $h : \mathcal{M} \rightarrow \mathcal{M}'$ *of temporal models $\mathcal{M} = \langle T, \mathcal{V} \rangle$ and $\mathcal{M}' = \langle T', \mathcal{V}' \rangle$ is a transition system morphism $h : T \rightarrow T'$ such that $s \in \mathcal{V}(q) \Leftrightarrow h(s) \in \mathcal{V}'(q)$ for all $q \in Q$, $s \in S$.*

The category of temporal models and their morphisms, with the obvious composition and identities, is denoted by $\mathbf{TM}_Q$. △

An evaluation denotes for each variable those states where the variable is deemed to be true. We have now all the components needed to inductively introduce the notion of *satisfaction* of temporal formulas Φ for a given computation σ.

Definition 4 (satisfaction of temporal formula). *Let $\mathcal{M} = \langle T, \mathcal{V} \rangle$ be a temporal model over Q, where $T = \langle S, \rightarrow, R \rangle$. A formula Φ is satisfied by a run $\sigma \in R$, denoted as $\sigma \models \Phi$, if one of the following cases is verified:*

- $\sigma \models q \Leftrightarrow \sigma(0) \in \mathcal{V}(q)$ *for all $q \in Q$*
- $\sigma \models \neg\Phi \Leftrightarrow \sigma \not\models \Phi$
- $\sigma \models \Phi_1 \wedge \Phi_2 \Leftrightarrow \sigma \models \Phi_1$ *and* $\sigma \models \Phi_2$
- $\sigma \models \bigcirc\Phi \Leftrightarrow \sigma|_1 \models \Phi$
- $\sigma \models \Phi_1 \mathcal{U} \Phi_2 \Leftrightarrow \exists k \in \mathbb{N}, \sigma|_k \models \Phi_2$ *and for each $j, 0 \leq j < k, \sigma|_j \models \Phi_1$*

The formula Φ is $\mathcal{M}$-true, short $\models_{\mathcal{M}} \Phi$, if $\sigma \models \Phi$ for all $\sigma \in R$. It is valid, *denoted by $\models \Phi$, if $\models_{\mathcal{M}} \Phi$ for all models $\mathcal{M}$.* △

The semantical idea behind morphisms of temporal models is invariance of satisfaction: For each temporal formula Φ and every run σ in $\mathcal{M}$, $\sigma \models \Phi \Leftrightarrow h(\sigma) \models \Phi$.

A proof system for a temporal language consists of a set of axioms schemes and inference rules. A *proof* of Φ is a finite sequence of formulae $\Phi_1, ..., \Phi_n$ with $\Phi = \Phi_n$ and where each Φ_i is either an axiom instance or the result of an application of a rule instance to premises belonging to the set $\{\Phi_1, .., \Phi_{i-1}\}$. Φ is a *theorem* of this proof system, denoted by $\vdash \Phi$, if there is a proof of Φ. A proof system is said to be *sound* w.r.t a satisfaction relation $\models$ if every provable formula is valid and it is said to be *complete* w.r.t. $\models$ if every valid formula is provable. A sound and complete proof system for the satisfaction relation $\models$ is given in [17].

Given an at least countable set S (i.e., such that there exists an injective function $c : \mathbb{N} \rightarrow S$), it is easy to understand that we could restrict our attention to the full sub-category $\mathbf{TM}_Q^S$ of $\mathbf{TM}_Q$, containing only those temporal models whose set of states is contained in S, without losing expressive power. In other words, a formula Φ is valid, that is $\models \Phi$, if and only if $\models_\mathcal{M} \Phi$ for all models $\mathcal{M} \in \mathbf{TM}_Q^S$, denoted as $\models_S \Phi$.

Of course, such a result simply says that we could restrict our attention to that family of models, whose set of elements is contained in $\mathbb{N}$ (and see again [17]). Nevertheless, such a formulation allows us to investigate the class of temporal models over a given interpretation; and to ask for the characterization of the *minimal model* satisfying a given formula under that interpretation. Such concerns will be the basis for our notion of *graph interpretation* of temporal logic, given in Section 4 and Section 5.

3   Putting Graphs into the View

This section introduces a toy example of a token ring algorithm for mutual exclusion (taken from [7]), to provide the basic ideas motivating the graph interpretation of temporal logic. This example is introduced more detailed in [8]. Another application of the concepts is given in [13], where they are applied to a case study of a distributed configuration management system.

The overall structure of the system is specified by a *type graph* (see e.g., [9]) as shown below on the left. It may be read like an entity/relationship schema specifying the node and edge types which may occur in the *instance graphs* modeling system states. *Processes* are drawn as black nodes and *resources* as light boxes. An edge from a process to a resource models a *request*. An edge in the opposite direction means that the resource is currently *held by* the process.

The token ring is a cyclic list of processes, where an edge between two processes points to the *next* process. For each resource there is a *token*, represented by an edge with a flag, which is passed from process to process along the ring. If a process wants to use a resource, it waits for the corresponding token. Mutual exclusion is ensured because there is only one token for each resource in the system.

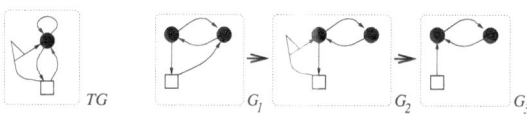

On the right, a sequence of graphs is depicted modeling a possible evolution of a system with two processes and one resource. Formally, an instance of a type graph TG is represented as *typed graph*, i.e., a graph G together with a *typing homomorphism* $t_G : G \to TG$ (indicated in the sample states by the symbols used for vertices). A morphism of typed graphs $\langle G, t_G \rangle$ and $\langle H, t_H \rangle$ is a graph homomorphism $f : G \to H$ which is compatible with the typing, that is, $t_H \circ f = t_G$. Thus, graphs and graph morphisms typed over TG form a comma category $\mathbf{Graph} \downarrow TG$. In the following all graphs and all morphisms are assumed to be typed over a fixed type graph TG.

The figure below shows an example of the temporal properties we would like to express. The left formula states for all processes p and resources r that always there is a future state where process p gets hold of the token belonging to resource r. The formula in the middle says that whenever p requests a resource r and gets the corresponding token, then r is held by p in the very next computation step. The right formula states the desired liveness property: Whenever a process p requests a resource r, eventually it will get it, or withdraw its request.

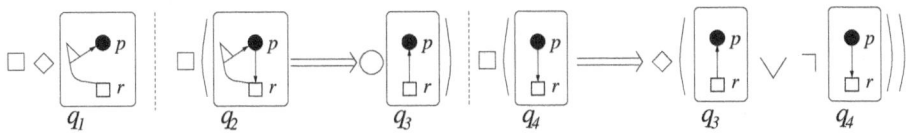

Such properties are not expressible in *propositional* temporal logic but require a combination of propositional formulas with so-called *graphical constraints* for expressing structural properties of system states. Graphical constraints are based on the idea of "specification by example": A constraint is a pattern for all states with a certain property (providing a certain structure). This approach is typical for rule-based systems, in particular rule-based graph transformation, where a rule is a minimal pattern of the transformations to be performed.

In order to keep track of the processes and resources during the computation, in the example we used graphical variables p and r to denote them. Thus, a state of our system is not just a graph but incorporates an assignment of the graphical variables. Since we have to handle the deletion and creation of graphical objects, such assignments are modeled as *partial* morphisms from a graph of variables to the state graphs. The graph of variables X in the example consists of the process node p and resource node r with corresponding type. The morphisms are indicated by labeling the nodes of the target graph with their pre-image in X, e.g., q_1 represents the (total) graph morphism from X to the depicted graph mapping p to the only process node and r to the only resource node.

4   An Interpretation over Graphs

We split this section in two distinct parts. The first one presents some (we believe) original results on the category of partial morphisms between graphs. The second applies these results, in order to provide a suitable notion of *graph model* for temporal formulas: the equivalence of such a model with the usual semantics presented in Section 2 is then proved.

4.1   A Pre-order Structure over Partial Morphisms

Formally, a *partial graph morphism* g from G to H is a total graph morphism $\bar{g} : dom(g) \to H$ from some subgraph $dom(g)$ of G, called the *domain* of g, to H. Hence, partial morphisms are often written as spans $G \hookleftarrow dom(g) \xrightarrow{\bar{g}} H$. The set of all partial morphisms with source X is denoted by $\mathcal{P}\mathcal{M}or(X)$.

Definition 5 (embedding relation). *Let X be a graph. The embedding relation $\lhd_X \subseteq \mathcal{P}\mathcal{M}or(X) \times \mathcal{P}\mathcal{M}or(X)$ is defined as follows. Let $a = (X \hookleftarrow$*

$X_a \xrightarrow{\bar{a}} Y_a$) and $b = (X \hookleftarrow X_b \xrightarrow{\bar{b}} Y_b)$, then $a \lhd_X b$ if and only if $X_a \hookrightarrow X_b$ and there is a total graph morphism $f : Y_a \to Y_b$ such that the diagram below commutes.

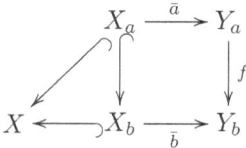

$\triangle$

Thus, $a \lhd_X b$ means that Y_a can be embedded in Y_b while respecting the "assignments" a and b, i.e., the second "state" provides at least all the structure of the first, and possibly some more. Notice that $\lhd_X$ is a pre-order, since it is reflexive and transitive, but does not satisfy asymmetry (not even up-to graph isomorphism)

The pre-order "$\lhd_X$" shall be used throughout the paper for modeling logical consequence (of constraints), satisfaction (of graphical constraints by states) and specialization (of states). When clear from the context, the index X will often be skipped.

Proposition 1 ($\lhd$ has lub). *Let X be a graph. The pre-order $\lhd_X$ on partial morphisms $\mathcal{P}\mathcal{M}or(X)$ has least upper bounds.*

Proof. If $A \subseteq \mathcal{P}\mathcal{M}or(X)$ is a set of partial graph morphisms $a = (X \hookleftarrow X_a \xrightarrow{\bar{a}} Y_a)$, its least upper bound is $lub(A) = (X \hookleftarrow \bigcup_{a \in A} X_a \xrightarrow{u} Y)$ where Y is the colimit object of the family of total morphisms $(\bigcap_{a \in A} X_a \hookrightarrow X_a \xrightarrow{\bar{a}} Y_a)_{a \in A}$ and $u = \bigcup_{a \in A} \bar{a}$ is the union of the total morphisms $\bar{a}$, characterized as the arrow induced by the universal property of the colimit (see the diagram below for the case of $A = \{a, b\}$ where the colimit is given by the outer pushout diagram).

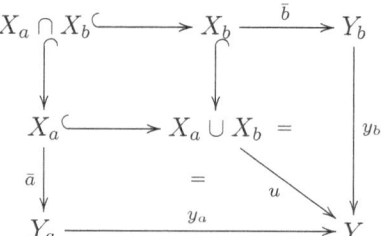

In order to see that $lub(A)$ is indeed an upper bound, observe that there is an inclusion $X_a \hookrightarrow \bigcup_{a' \in A} X_{a'}$ for each $a \in A$. Moreover, a morphism $y_a : Y_a \to Y$ commuting the newly formed diagrams is obtained as colimit injection. That $lub(A)$ is *least* upper bound follows by the fact that the union of subgraphs of X is least upper bound for inclusion of subgraphs, and by the universal property of colimits. $\qquad \square$

Notice that if $X_a = X_b$, the graph Y is just the pushout of $\bar{a}$ and $\bar{b}$. On the other hand, if X_a and X_b are disjoint, then $Y = Y_a + Y_b$ is the disjoint union. An example of the construction of a least upper bound can be found in Section 5.

4.2   Introducing Graph Models

Our *graph model* will be based on a suitable notion of transition system, whose states are assignments (that is, partial graph morphisms) from a fixed graph of variables X.

Definition 6 (temporal graph model). *A graph transition system under X is a transition system $GT = \langle S, \rightarrow, R \rangle$ where $S \subseteq \mathcal{P}\mathcal{M}or(X)$. A (temporal) graph model over X is a pair $\mathcal{G}\mathcal{M} = \langle GT, I \rangle$, where $I : Q \rightarrow \mathcal{P}\mathcal{M}or(X)$ is an interpretation of the propositional variables. A morphism $h : \mathcal{G}\mathcal{M} \rightarrow \mathcal{G}\mathcal{M}'$ of graph models $\mathcal{G}\mathcal{M} = \langle GT, \mathcal{I} \rangle$ and $\mathcal{G}\mathcal{M}' = \langle GT', \mathcal{I}' \rangle$ is a transition system morphism $h : GT \rightarrow GT'$ such that $I(q) \lhd s \Leftrightarrow I'(q) \lhd h(s)$ for all $q \in Q$, $s \in S$.*

The category of graph models over X and their morphisms, with the obvious composition and identities, is denoted by $\mathbf{GM}_Q^X$.   △

It is easy to define a functor $F : \mathbf{GM}_Q^X \rightarrow \mathbf{TM}_Q$: for a graph model $\mathcal{G}\mathcal{M} = \langle GT, I \rangle$, where $GT = \langle S, \rightarrow, R \rangle$, the induced temporal model $F(\mathcal{G}\mathcal{M})$ is the pair $\langle GT, \mathcal{V}_I \rangle$, where $\mathcal{V}_I(q) = \{ s \in S \mid I(q) \lhd s \}$.

The functor F is neither full, nor faithful (since the morphisms in $\mathbf{TM}_Q^{\mathcal{P}\mathcal{M}or(X)}$ need not to preserve the pre-order structure). Nevertheless, it can be used to lift the notions of satisfiability over graph models: given a graph model $\mathcal{G}\mathcal{M} = \langle GT, I \rangle$ a run σ of GT *satisfies* a temporal formula Φ, written $\sigma \models_{\mathcal{G}\mathcal{M}} \Phi$, if the formula is satisfied in the induced temporal model, that is, $\sigma \models_{F(\mathcal{G}\mathcal{M})} \Phi$. Similarly for the notion of $\mathcal{G}\mathcal{M}$-*truth*, $\models_{\mathcal{G}\mathcal{M}} \Phi$, or *G-validity*, $\models_G \Phi$.

As anticipated above, a propositional variable q (abstractly representing a state property) is interpreted by a pattern state $I(q)$ for the intended property. The structure of $I(q)$ is inherited to all states s in which $I(q)$ may be embedded, that is, where $I(q) \lhd s$. Thus, the evaluation $\mathcal{V}_I(q)$ consists of all those states of the system which are reachable from $I(q)$ via $\lhd$.

The following proposition states that the translation between graph and temporal model induced by the functor F is nevertheless surjective, up-to equivalence of systems.

Proposition 2 (temporal model vs. graph model). *For each temporal model $\mathcal{M} = \langle T, \mathcal{V} \rangle$ there exists a graph model $\mathcal{G}\mathcal{M} = \langle GT, I \rangle$ over X such that the induced temporal model $F(\mathcal{G}\mathcal{M}) = \langle GT, \mathcal{V}_I \rangle$ is equivalent to $\mathcal{M}$, in the sense that $\models_{\mathcal{M}} \Phi$ if and only if $\models_{F(\mathcal{G}\mathcal{M})} \Phi$ for all formulas Φ.*   △

As a consequence of Proposition 2, both notions of validity of formulas coincide.

Theorem 1 (validity). *A temporal formula Φ is valid, $\models \Phi$, if and only if it is G-valid, $\models_G \Phi$.*

Thus every sound and complete calculus for the satisfaction relation defined in Section 2 is also sound and complete for the class of temporal models induced by graph models.

Typically, graph transition systems are generated by *graph grammars* providing a set of rules together with a start graph. All notions and results of this paper are fairly independent of the particular graph transformation approach one may choose in order to generate these systems. Nevertheless we would like to give an idea about the relationship between graph transition systems and graph grammars (even if space limitations inhibit us to provide a full set of definitions). The basic effect of applying a transformation rule to a graph is to remove certain graphical elements and to create some new elements which are linked in a suitable way to elements which have been preserved. Starting from a given graph G_0, in this way we may generate derivations $G_0 \Longrightarrow G_1 \Longrightarrow G_2 \Longrightarrow \ldots$ by repeated application of rules. A sequence of partial graph morphisms $a_0 a_1 a_2 \ldots$ from a given graph of variables X into the graphs of the derivation is regarded as a run σ if the assignments are compatible with the deletion, preservation, and creation of items in the following sense. If $G_i \Longrightarrow G_{i+1}$ is a graph transformation step with assignments $a_i : X \to G_i$ and $a_{i+1} : X \to G_{i+1}$, then a_i and a_{i+1} have to agree on all elements that are preserved from G_i to G_{i+1}, only elements of G_i that are deleted by the step may be forgotten from a_i to a_{1+1}, and only those elements that are new in G_{i+1} may be used to extend the assignment a_{i+1} with respect to a_i.

5   A Typical Graph Transition System

In this section we construct for an interpretation I a graph transition system that is *typical for I* in the sense that its induced model satisfies exactly those formulas that are satisfied by all models induced by transition systems with that interpretation. This transition system is *minimal*, in the sense that it is distinguished by the property of being the final object in $\mathbf{GM}_{Q,I}^X$, the full subcategory of $\mathbf{GM}_Q^X$ containing all those models with interpretation I. The typical system is fully abstract with respect to satisfaction in the sense that any two states which are not distinguishable by constraints from the interpretation I are equal in the typical system.

The idea for this abstraction is to build for every possible configuration of truth values of propositional variables a state which implements this configuration, i.e., where only such constraints are satisfied that interpret a variable which is true in the given configuration. Such states are constructed as least upper bounds of sets of states (seen as constraints) with respect to the relation $\lhd$: from a logical point of view, we may think of $lub(C)$ as the conjunction of the set C of constraints.

The figure below shows the construction of the least upper bound for the two contstraints q_2 and q_3 used to interpret the propositional variables of the temporal formulas in Section 3. The graph X contains a process node p and a resource node r. Since both constraints are total the construction of the least upper bound is just their pushout (cf. the proof of Proposition 1). It can be easily checked out that a state satisfies the least upper bound of q_2 and q_3 if and only if both the constraint q_2 and the constraint q_3 are satisfied.

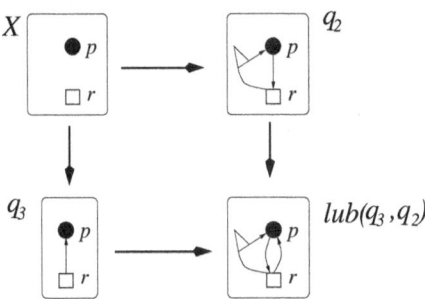

Construction 2 (typical graph transition system) *For an interpretation* $I : Q \to \mathcal{P}\mathcal{M}or(X)$, *let* $\mathcal{P}_I(Q)$ *be the set of all subsets* Q' *of* Q *which are closed under entailment, that is, where for all* $q \in Q$, $I(q) \lhd lub(Q')$ *implies* $q \in Q'$. *Then, the* typical graph transition system $GT_I = \langle S_I, \to_I, R_I \rangle$ *has as states all partial morphisms* $lub(I(Q'))$ *for* $Q' \in \mathcal{P}_I(Q)$, *the transition relation* $\to_I = S_I \times S_I$ *is the full cartesian product, and* R_I *is set of all paths through* $\to_I$. $\triangle$

Notice that the typical graph transition system has finitely many states whenever the set of constraints $I(Q)$ used for interpreting the propositional variables is finite.

The states of the typical transition system for the example of the previous section are shown below. The three graphs on the left in the upper row and the rightmost one in the lower row are the original constraints from the formula. The triangles among them indicate the "entailment" relation $\lhd$. Constructing all least upper bounds we obtain the remaining four constraints, where the empty graph is obtained as the colimit over the empty family of morphisms.

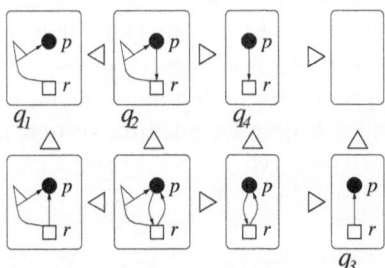

Due to the construction of the states of the typical transition system as least upper bounds, they satisfy exactly those constraints from which they are constructed, i.e. $I(q) \lhd lub(I(Q'))$ if and only if $q \in Q'$ for all q in Q and Q' in $\mathcal{P}_I(Q)$.

The following theorem characterizes the typical graph transition system for I as a final object in $\mathbf{GM}_{Q,I}^X$.

Theorem 3 (fully abstract transition system). *The typical graph transition system GT_I for an interpretation I is final in $\mathbf{GM}_{Q,I}^X$, that is, for each graph transition system GT there exists a unique morphisms $!_{GT} : GT \to GT_I$ in $\mathbf{GM}_{Q,I}^X$.*

Proof. For each graph transition system $GT = \langle S, \to, R \rangle$ in $|\mathbf{GM}_{Q,I}^X| \langle GT, \mathcal{V}_\mathcal{I} \rangle$ we define a morphism $!_{GT} : GT \to GT_I$. Let for a state $a \in S$ the set $Q(a) = \{q \in Q \mid I(q) \lhd a\}$. Then $Q(a)$ is closed under entailment, i.e., $Q(a) \in \mathcal{P}_I(Q)$, and we define $!_{GT}(a) = lub(I(Q(a)))$. $!_{GT}$ is a model morphisms from $\langle GT, \mathcal{V}_I \rangle$ to $\langle GT_I, \mathcal{V}_I \rangle$ since, by construction, $I(q) \lhd a$ iff $I(q) \lhd lub(I(Q(a)))$. Thus, $!_{GT}$ is a morphism of graph transition systems.

Its uniqueness follows from the fact that for a set of propositional variables Q', $lub(I(Q'))$ is the only state in GT_I that satisfies exactly the constraints of this set. Since morphisms of graph transition systems have to preserve and reflect the satisfaction of propositional variables, the mapping $a \mapsto lub(I(Q(a)))$ is forced by this condition. □

The unique morphism $!_{GT}$ collapses a potentially infinite graph transition system GT to a finite one, given by the image of GT under $!_{GT}$. Finality is a categorical way of saying that GT_I is both typical and fully abstract: The existence of the morphisms $!_{GT} : GT \to GT_I$ for all graph transition systems GT expresses the fact that GT_I is typical. This is made precise in the corollary below. The uniqueness of these morphisms implies that GT_I is fully abstract: If there are two different states indistinguishable by the constraints, there can be two different candidates for the definition of $!_{GT}$.

Corollary 1 (GT_I is typical for I). *The graph transition system GT_I is typical among the transition systems with interpretation I in the sense that for all temporal formulas Φ, $\models_{GT_I,I} \Phi$ if and only if $\models_I \Phi$, that is, $\models_{GT,I} \Phi$ for all graph transition systems GT.*

Proof. "$\Longleftarrow$": Since GT_I is in $\mathbf{GM}_{Q,I}^X$, the claim holds.

"$\Longrightarrow$": We show that a graph transition system $\langle S, \to, R \rangle \in \mathbf{GM}_{Q,I}^X$ with $\not\models_{GT,I} \Phi$ implies $\not\models_{GT_I,I} \Phi$. If $\not\models_{GT,I} \Phi$, there is a run $\sigma \in R$ such that $\sigma \not\models_I \Phi$. Since Theorem 3 provides a graph transition system morphism $!_{GT} : GT \to GT_I$, and such morphisms preserve and reflect satisfaction of temporal formulas, $!_{GT}(\sigma) \not\models_I \Phi$ implying that $\not\models_{GT_I,I} \Phi$. □

6   Conclusion

The paper contributes to the ongoing work to develop a visual design technique for software systems that additionally supports the designer by analysis techniques. Visual design techniques [6,14,16] have a growing influence on the design techniques for software systems. However, most of them do not provide concepts to model and verify dynamic system properties.

This paper introduced a graphical interpretation for the propositional fragment of linear temporal logic, in order to express dynamic properties of graph transition systems. We showed that the semantics based on graphical transition systems is compatible with the classical semantics based on (general) transition systems. This allows us to reuse the classical calculi and tools for examining the validity of uninterpreted propositional temporal formulas.

In order to take into account the interpretation by graphical constraints, we construct for every interpretation I a typical model which is finite if the interpretation uses only finitely many constraints.

Based on these two results, techniques developed in the theory of the classical semantics of temporal logic may be applied. In [5,13], for example, a model checker is applied to the (temporal model induced by the) typical graph transition system in order to check a temporal formula.

A further abstraction of the typical transition system can be achieved by analyzing the bisimilarity of states w.r.t. *temporal* properties in the sense of [18]. Moreover, it has to be investigated how the concepts and results of this paper can be used to analyze properties of a particular graph grammar, e.g., how to check the validity of a formula in the generated graphical transition system.

References

1. I. Claßen, M. Gogolla, and M. Löwe. Dynamics in information systems: Specification, construction, and correctness. Technical Report 96–01, Technische Universität Berlin, 1996. 311
2. A. Corradini, U. Montanari, F. Rossi, H. Ehrig, R. Heckel, and M. Löwe. Algebraic approaches to graph transformation, Part I: Basic concepts and double pushout approach. In Rozenberg [15], pages 163–245. 310
3. H. Ehrig, R. Heckel, M. Korff, M. Löwe, L. Ribeiro, A. Wagner, and A. Corradini. Algebraic approaches to graph transformation, Part II: Single pushout approach and comparison with double pushout approach. In Rozenberg [15], pages 247–312. 310
4. H. Ehrig, M. Pfender, and H.J. Schneider. Graph grammars: an algebraic approach. In *14th Annual IEEE Symposium on Switching and Automata Theory*, pages 167–180. IEEE, 1973. 310
5. F. Gadducci, R. Heckel, and M. Koch. Model checking graph-interpreted temporal formulas. In *Prelim. Proc. 6th Int. Workshop on Theory and Application of Graph Transformation (TAGT'98), Paderborn*, 1998. 321
6. The Object Management Group. OMG UML Specification, V. 1.3, 1999. 320
7. R. Heckel. Compositional verification of reactive systems specified by graph transformation. In *Fundamental Approaches to Software Engineering*, volume 1382 of *LNCS*, pages 138–153. Springer Verlag, 1998. 311, 314
8. R. Heckel. *Open Graph Transformation Systems: A New Approach to the Compositional Modelling of Concurrent and Reactive Systems*. PhD thesis, TU Berlin, 1998. 311, 314

9. R. Heckel, A. Corradini, H. Ehrig, and M. Löwe. Horizontal and vertical structuring of typed graph transformation systems. *Math. Struc. in Comp. Science*, 6(6):613–648, 1996. Also Tech. Rep. 96-22, TU Berlin. 314

10. R. Heckel, H. Ehrig, U. Wolter, and A. Corradini. Integrating the specification techniques of graph transformation and temporal logic. In *Proc. Mathematical Foundations of Computer Science (MFCS'97), Bratislava*, volume 1295 of *LNCS*, pages 219–228. Springer Verlag, 1997. 311

11. R. Heckel and A. Wagner. Ensuring consistency of conditional graph grammars – a constructive approach. *Proc. of SEGRAGRA'95 "Graph Rewriting and Computation", Electronic Notes of TCS*, 2, 1995. http://www.elsevier.nl/locate/entcs/volume2.html. 311

12. M. Koch. Modellierung und Nachweis der Konsistenz von verteilten Transaktionsmodellen für Datenbanksysteme mit algebraischen Graphgrammatiken. Technical Report 96-36, TU Berlin, 1996. Master's thesis. 311

13. M. Koch. *Integration of Graph Transformation and Temporal Logic for the Specification of Distributed Systems*. PhD thesis, TU Berlin, 1999. 311, 314, 321

14. G. Rasmussen, B. Henderson-Sellers, and G.C.Low. An object-oriented analysis and design notation for distributed systems. *Object Currents*, 1(10), 1996. 320

15. G. Rozenberg, editor. *Handbook of Graph Grammars and Computing by Graph Transformation, Volume 1: Foundations*. World Scientific, 1997. 310, 321

16. J. Rumbaugh, M. Blaha, W. Premerlani, E. Eddy, and W. Lorenson. *Object-Oriented Modeling and Design*. Prentice Hall International, 1991. 320

17. C. Stirling. Modal and temporal logics. In *Background: Computational structures*, volume 2 of *Handbook of Logic in Computer Science*, pages 477–563. Clarendon Press, Oxford, 1992. 311, 312, 313, 314

18. J. van Benthem. Correspondence theory. In D. Gabbay and F. Günther, editors, *Handbook of Philosophical Logic, Vol. II*, pages 167 – 248. Reidel, 1984. 311, 321

19. A. Wagner. *A Formal Object Specification Technique Using Rule-Based Transformation of Partial Algebras*. PhD thesis, TU Berlin, 1997. 311

More About Control Conditions
for Transformation Units*

Sabine Kuske

Universität Bremen, Fachbereich 3
Postfach 33 04 40, D-28334 Bremen, Germany
kuske@informatik.uni-bremen.de

Abstract. A transformation unit is a structuring principle for composing graph transformation systems from small units. One of the basic
components of a transformation unit is its control condition which allows
to restrict the non-determinism of graph transformation. The concept of
transformation units is generic in the sense that each formalism which
specifies a binary relation on graphs can be used as a control condition.
This paper discusses a selection of concrete classes of control conditions
which seem to provide reasonable expressive power for specifying and
programming with transformation units. These include regular expressions, once, as-long-as-possible, priorities, and conditionals; some of them
were already used in an ad hoc manner in earlier papers. It is shown which
classes of control conditions can be replaced by others without changing
the semantics of the corresponding transformation unit. Moreover, three
properties of control conditions are studied: minimality, invertibility and
continuity.

1 Introduction

Real applications of graph transformation may often consist of hundreds of rules
which can only be managed in a reasonable and transparent way with the help
of a structuring principle. Therefore, several modularization concepts for graph
transformation systems are currently under development (cf. also [HEET99]).
One of them is the transformation unit, a structuring principle which allows to
decompose large graph transformation systems into small reusable units that can
import each other (see [KK96,KKS97,AEH⁺99,KK99]). It transforms graphs by
interleaving rule applications with calls of imported transformation units. One
fundamental feature of the transformation unit is its independence of a particular
graph transformation approach, i.e. of a selected graph data model, a specific
type of rules, etc. (see [Roz97] for an overview of the main graph transformation
approaches).

Each transformation unit contains a control condition that allows to regulate
its graph transformation process. This is meaningful because formalisms which

---

* This work was partially supported by the ESPRIT Working Group Applications of
  Graph Transformation (APPLIGRAPH) and the EC TMR Network GETGRATS
  (General Theory of Graph Transformation Systems).

H. Ehrig et al. (Eds.): Graph Transformation, LNCS 1764, pp. 323–337, 2000.

allow to specify transformations of graphs in a rule-based way are usually non-deterministic for two reasons. First, there may be more than one rule which can be applied to a certain graph. Second, a rule can be applied to various parts of a graph. Hence, in order to program or specify with graph transformation, it is often desirable to regulate the graph transformation process, for example by choosing rules according to a priority, or by prescribing a certain sequence of steps (cf. e.g. [Sch75,Bun79,Nag79,MW96,Sch97,TER99], see also [DP89] for regulation concepts in string transformation).

The concept of transformation units is generic in the sense that each formalism which specifies a binary relation on graphs can be used as a control condition. In order to use transformation units for specifying and programming with graph transformation, concrete classes of control conditions must be employed.

Based on the variety of control mechanisms for graph transformation proposed in the literature, we discuss in this paper a selection of concrete classes of control conditions for transformation units which seem to provide reasonable expressive power for the practical use of transformation units. These include regular expressions, priorities, conditionals, once, and as-long-as-possible, where some of them were already used in an ad hoc manner in earlier papers [KK96,AEH$^+$99]. We define their semantics in the context of transformation units and show how some classes of control conditions can be expressed by others without changing the interleaving semantics of the corresponding transformation unit. Moreover, three properties of control conditions are studied: minimality, invertibility, and continuity. The first two are required to perform certain operations on transformation units like flattening and inversion. The third leads to a fixed point semantics in the case of recursive transformation units. Please note that, due to space limitations, proofs are only sketched in this paper.

2   Transformation Units

A transformation unit comprises descriptions of initial and terminal graphs, a set of rules, a control condition, and a set of imported transformation units. It transforms initial graphs to terminal ones by interleaving rule applications with calls to imported transformation units such that the control condition is obeyed. Transformation units are a fundamental concept of the graph- and rule-centered language GRACE, currently under development by researchers from various sites. In the following we briefly recall the definition of transformation units together with its interleaving semantics.

Since transformation units are independent of a particular graph data model or a specific type of graph transformation rules, we assume that the following are given: a class $\mathcal{G}$ of *graphs*, a class $\mathcal{R}$ of *rules*, and a *rule application operator* $\Rightarrow$ specifying a binary relation $\Rightarrow_r \subseteq \mathcal{G} \times \mathcal{G}$ for each $r \in \mathcal{R}$. Intuitively, $\Rightarrow_r$ yields all pairs (G, G') of graphs where G' is obtained from G by applying r. In order to specify the initial and terminal graphs of a transformation unit, we assume in addition that a class $\mathcal{E}$ of *graph class expressions* is given where each $X \in \mathcal{E}$ specifies a set $SEM(X) \subseteq \mathcal{G}$. The control condition of a transformation unit is taken from a class $\mathcal{C}$ of *control conditions* each of which specifies a binary relation on graphs. Since control conditions may contain identifiers (usually for imported transformation units or local rules), their semantics depends on their *environment*, a mapping which associates each identifier with a binary relation on graphs. This means more precisely that each $C \in \mathcal{C}$ specifies a binary relation $SEM_E(C) \subseteq \mathcal{G} \times \mathcal{G}$ for every mapping $E \colon ID \to 2^{\mathcal{G} \times \mathcal{G}}$ where ID is a set of names. All these components form a *graph transformation approach* $\mathcal{A} = (\mathcal{G}, \mathcal{R}, \Rightarrow, \mathcal{E}, \mathcal{C})$.

A *transformation unit* over $\mathcal{A}$ is a system $t = (I, U, R, C, T)$ where $I, T \in \mathcal{E}$, $R \subseteq \mathcal{R}$, $C \in \mathcal{C}$, and U is a set of (already defined) transformation units over $\mathcal{A}$. This should be taken as a recursive definition of the set $\mathcal{T}_\mathcal{A}$ of transformation units over $\mathcal{A}$. Hence, initially U is the empty set yielding unstructured transformation units which may be used in the next iteration, and so on. Note that, for reasons of space limitations, in this paper we consider only transformation units with an acyclic import structure. But the presented results (apart from Observation 2) can easily be transferred to the more general case of transformation units having a cyclic import structure which were studied in [KKS97].

The interleaving semantics of a transformation unit contains a pair (G, G') of graphs if G is an initial graph and G' is a terminal graph, G can be transformed into G' using the rules and the imported transformation units, and (G, G') is allowed by the control condition. Formally, let $t = (I, U, R, C, T) \in \mathcal{T}_\mathcal{A}$ and assume that every $t' \in U$ already defines a binary relation $SEM(t') \subseteq \mathcal{G} \times \mathcal{G}$ on graphs. Then the *environment* of t is the mapping $E(t) \colon ID \to 2^{\mathcal{G} \times \mathcal{G}}$ defined by $E(t)(id) = SEM(id)$ if $id \in U$, $E(t)(id) = \Rightarrow_{id}$ if $id \in R$, and $E(t)(id) = \emptyset$, otherwise.[1] The *interleaving semantics* of t is the relation $SEM(t) = (SEM(I) \times SEM(T)) \cap RIS(t) \cap SEM_{E(t)}(C)$ where $RIS(t) = (\bigcup_{id \in U \cup R} E(t)(id))^*$.[2] The name $RIS(t)$ stands for *relation induced by the interleaving sequences of t* because it consists of all pairs $(G, G') \in \mathcal{G} \times \mathcal{G}$ such that there is an interleaving sequence $G_0, \dots, G_n$ of t with $G_0 = G$, $G_n = G'$. A sequence $G_0, \dots, G_n$ $(n \geq 0)$ is called an *interleaving sequence of t* if, for $i = 1, \dots n$, $G_{i-1} \Rightarrow_r G_i$ for some $r \in R$ or $(G_{i-1}, G_i) \in SEM(t')$ for some $t' \in U$.

3 Control Conditions

With control conditions, the non-determinism of the graph transformation process can be restricted so that a more functional behaviour is achieved. A typical example is to allow only iterated rule applications where the sequences of applied rules belong to a particular control language. In general, every description of a binary relation on graphs may be used as a control condition. In the following, we discuss a selection of control conditions suitable to specify and program with

---

[1] For technical simplicity, we assume here that ID contains U and R as disjoint sets. Nevertheless, if transformation units are used in a specification or programming language, ID should provide a set of predefined identifiers out of which the elements of U and R are named. Note that such an explicit naming mechanism can be added to the presented concepts in a straighforward way.

[2] For a binary relation ρ its reflexive and transitive closure is denoted by ρ^*.

graph transformation. These conditions are regular expressions, once, as-long-as-possible, priorities, and conditionals.

3.1 Regular Expressions

Regular expressions – well known from the area of formal languages – are useful to prescribe the order in which rules or imported transformation units should be applied. Here, we consider the class REG of regular expressions over ID which is recursively given by $\epsilon, \emptyset \in REG$, $ID \subseteq REG$, and $(C_1 ; C_2)$, $(C_1 | C_2)$, $(C^*) \in REG$ if $C, C_1, C_2 \in REG$. In order to omit parentheses, we assume that * has a stronger binding than $;$ which in its turn has a stronger binding than $|$. Intuitively, the expression ϵ requires that no rule or imported unit is applied, $\emptyset$ specifies the empty set, and $id \in ID$ applies the rule or transformation unit id exactly once. Moreover, $C_1 ; C_2$ applies first C_1 and then C_2, $C_1 | C_2$ chooses non-deterministically between C_1 and C_2, and C^* iterates C arbitrarily often. Formally, this means that for each environment E we have $SEM_E(\epsilon) = \Delta\mathcal{G}$, $SEM_E(\emptyset) = \emptyset$, $SEM_E(id) = E(id)$ for each $id \in ID$, $SEM_E(C_1 ; C_2) = SEM_E(C_1) \circ SEM_E(C_2)$, $SEM_E(C_1 | C_2) = SEM_E(C_1) \cup SEM_E(C_2)$, and $SEM_E(C^*) = SEM_E(C)^*$.[3]

The following transformation unit $Eulerian\_test$ has a regular expression as its control condition. It imports the four units $relabel\_all\_nodes(*, even)$, $check\_degree$, $relabel\_all\_edges(ok, *)$, and $relabel\_all\_nodes(even, *)$, and applies them in this order exactly once. The initial graphs are all (undirected) unlabelled and connected graphs, whereas the terminal graphs are unlabelled.

$Eulerian\_test$	
initial:	*unlabelled, connected*
uses:	$relabel\_all\_nodes(*, even)$, $relabel\_all\_nodes(even, *)$,
	$relabel\_all\_edges(ok, *)$, $check\_degree$,
conds:	$relabel\_all\_nodes(*, even) ; check\_degree ;$
	$relabel\_all\_edges(ok, *) ; relabel\_all\_nodes(even, *)$
terminal:	*unlabelled*

Given an appropriate interleaving semantics for the imported units, $Eulerian\_test$ can be used to check for each unlabelled and connected graph G whether it is Eulerian, i.e. $(G, G) \in SEM(Eulerian\_test)$ iff G is Eulerian. For this, the unit $relabel\_all\_nodes(*, even)$ labels each unlabelled node of its input graph with $even$; the unit $check\_degree$ keeps the label of each node with an even degree, changes the label of each node with an odd degree into odd, and labels each edge with ok; $relabel\_all\_edges(ok, *)$ unlabels each ok-labelled edge; and the unit $relabel\_all\_nodes(even, *)$ unlabels each $even$-labelled node. (Note that the symbol $*$ stands for unlabelled.) The unit $check\_degree$, which does the main part of the algorithm, is presented in Sect. 3.3.

---

[3] $\Delta\mathcal{G}$ denotes the identity relation on $\mathcal{G}$ and $SEM_E(C_1) \circ SEM_E(C_2)$ the sequential composition of the relations $SEM_E(C_1)$ followed by $SEM_E(C_2)$.

Regular expressions can be extended in a straightforward way from the set *ID* to an arbitrary set M of control conditions of some other type. Hence, each regular expression over M also serves as a control condition the semantics of which is defined analogously to the one above. This "extended class" of regular expressions is useful for iterating the execution of or choosing non-deterministically between various control conditions in M.

3.2 Once

The control condition $once(id)$, where id is (the name of) a rule or an imported transformation unit, allows only interleaving sequences in which id is applied exactly once. All other rules or imported units can be applied arbitrarily often. Formally, for each environment E, $(G, G') \in SEM_E(once(id))$ if there exist $G_0, \ldots, G_n \in \mathcal{G}$ and $id_1 \cdots id_n \in ID^*$ such that (1) $G_0 = G$ and $G_n = G'$, (2) $(G_{i-1}, G_i) \in E(id_i)$ for $i = 1, \ldots, n$, and (3) there exists exactly one $i \in \{1, \ldots, n\}$ with $id_i = id$.

An example of a transformation unit using the control condition once and double pushout rules [CEH+97] is *Eulerian_cycle*, which searches for a Eulerian cycle in a connected unlabelled non-empty graph. The rule r_1 is applied exactly once and marks a node with a *start*-loop and a *cont*-loop, where *cont* stands for *continue*. With rule r_1 it is determined at which node the unit begins to traverse its input graph with r_2. The rule r_2 passes the *cont*-loop of a node v to a node v' provided that v and v' are connected via an unlabelled edge e, and labels e with E. After each application of r_2 the E-labelled edges form a path in the current graph from the node with the *start*-loop to the node with the *cont*-loop. The rule r_3 of *Eulerian_cycle* deletes the *cont*-loop and the *start*-loop provided that they are attached to the same node. The terminal component requires that all edges are labelled with E. The semantics of *Eulerian_cycle* consists of all pairs (G, G') where G is a non-empty Eulerian graph and G' is obtained from G by labelling all edges with E.

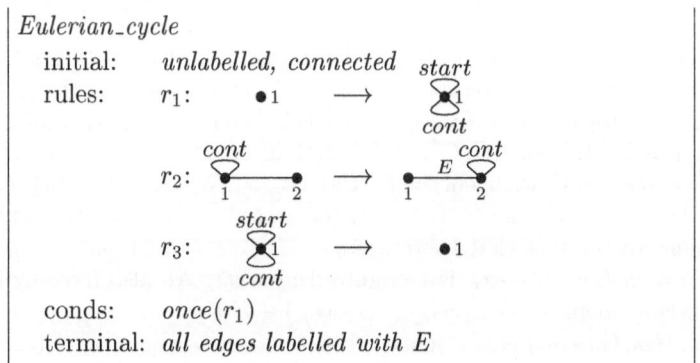

The next observation states that once can be replaced by a regular expression. This follows from the definitions of the semantics of regular expressions, once, and transformation units.

Observation 1 Let $t = (I, U, R, C, T)$ with $C = once(id)$ and let $C' = c\,;\, id\,;\, c$ where $c = (id_1 \mid \cdots \mid id_n)^*$ and $\{id_1, \dots, id_n\} = (U \cup R) \backslash \{id\}$. Then $SEM_{E(t)}(C) = SEM_{E(t)}(C')$.

If the underlying approach fulfills some weak assumptions, each transformation unit with a regular expression as its control condition can be transformed into a semantically equivalent one with the control condition $once(r)$ where r is a rule. We illustrate the construction for the case where the underlying rules are double pushout rules. (But note that in general it can be done for many other rule classes.) For this purpose, we consider an arbitrary but fixed graph transformation approach $\mathcal{A} = (\mathcal{G}, \mathcal{R}, \Rightarrow, \mathcal{E}, \mathcal{C})$ where $\mathcal{G}$, $\mathcal{R}$, and $\Rightarrow$ are defined as in the double pushout approach, and $REG \subseteq \mathcal{C}$.

Observation 2 Let $t = (I, U, R, C, T) \in \mathcal{T}_{\mathcal{A}}$ with $C \in REG$. Then we can construct $t' = (I, U', R', once(r), T)$ such that $r \in R$ and $SEM(t) = SEM(t')$.

Proof. (Sketch) The unit t' is obtained as follows: (1) Construct a finite deterministic automaton aut with state set Q, alphabet ID, transition function $\delta \colon Q \times ID \to Q$, initial state $s \in Q$, and final state $f \in Q$ such that aut recognizes the language specified by C, and the states of Q do not occur as node labels in $\mathcal{G}$. (Note that aut can always be constructed.) (2) For each $\bar{t} = (\bar{I}, \bar{U}, \bar{R}, \bar{C}, \bar{T}) \in U$ and $q, q' \in Q$ let $(\bar{t}, q, q')$ be the unit

$$
\begin{array}{|ll}
(\bar{t}, q, q') & \\
\quad \text{initial:} & \{G \uplus \bullet_q \mid G \in SEM(\bar{I})\} \\
\quad \text{uses:} & \bar{U} \\
\quad \text{rules:} & \bar{R} \cup \{rem(q), ins(q')\} \\
\quad \text{conds:} & rem(q)\,;\, C\,;\, ins(q') \\
\quad \text{terminal:} & \{G \uplus \bullet'_q \mid G \in SEM(\bar{T})\}
\end{array}
$$

where for each $q \in Q$ and each $G \in \mathcal{G}$, $G \uplus \bullet_q$ denotes the disjoint union of G and a q-labelled node, $rem(q)$ removes a q-labelled node, and $ins(q)$ inserts a q-labelled node (i.e. $rem(q) = (\bullet_q, \emptyset, \emptyset)$ and $ins(q) = (\emptyset, \emptyset, \bullet_q)$). For all $r = (L, K, R) \in R$ and $q, q' \in Q$ let (r, q, q') be the double pushout rule $(L \uplus \bullet_q, K, R \uplus \bullet_{q'})$. Then define $U' = \bigcup_{t \in U} \{(t, q, q') \mid q, q' \in Q, \delta(q, t) = q')\}$, $R' = \bigcup_{r \in R} \{(r, q, q') \mid q, q' \in Q, \delta(q, r) = q')\} \cup \{ins(s), rem(f)\}$, and $r = ins(s)$.

As shown in [KK96], $(G, G') \in SEM(t)$ iff there are $x_1 \cdots x_n \in (U \cup R)^*$ and $G_0, \dots, G_n \in \mathcal{G}$ with $G_0 = G$ and $G_n = G'$ such that (a) $x_1 \cdots x_n$ is specified by C and (b) for $i = 1, \dots, n$, $(G_{i-1}, G_i) \in E(t)(x_i)$. By definition (a) is equivalent to the fact that there are $q_0, \dots, q_n \in Q$ with $q_0 = s$, $q_n = f$ and $\delta(q_{i-1}, x_i) = q_i$ (i.e. $x_1 \cdots x_n$ is recognized by aut). We also have by definition that (b) is equivalent to $(G_{i-1} \uplus \bullet_{q_{i-1}}, G_i \uplus \bullet_{q_i}) \in E(t')(x_i, q_{i-1}, q_i)$ $(i = 1, \dots, n)$. Moreover, $(G_0, G_0 \uplus \bullet_s) \in E(t')(ins(s))$ and $(G_n \uplus \bullet_f, G_n) \in E(t')(rem(f))$.

Hence, we get on the one hand that $SEM(t) \subseteq SEM(t')$. Conversely, by definition of t' and aut the unit t' always applies $ins(s)$ in the first step and $rem(f)$ in the last (exactly once). Hence, on the other hand $SEM(t') \subseteq SEM(t)$.

$\square$

3.3 As-Long-As-Possible

In general, a control condition C specifies a binary relation on graphs. The control condition C! iterates this process as long as possible. For example, if C is a rule, the condition C! applies this rule to the current graph as long as possible. Formally, for each environment E the condition C! specifies the set of all pairs $(G, G') \in SEM_E(C)^*$ such that there is no $G'' \in \mathcal{G}$ with $(G', G'') \in SEM_E(C)$.

The control condition of the following transformation unit *check_degree* requires that the rules r_1, r_2, and r_3 are applied as long as possible. The rules are relabelling rules, i.e. they do not change the underlying structure of a graph but only its labelling. For example, r_1 relabels two *even*-labelled nodes into *odd*-labelled ones and an unlabelled edge connecting these nodes into an *ok*-labelled one.

check_degree
rules: r_1: $\overset{even}{\underset{1}{\bullet}}\,\overset{even}{\underset{2}{\bullet}} \longrightarrow \overset{odd}{\underset{1}{\bullet}}\,ok\,\overset{odd}{\underset{2}{\bullet}}$

r_2: $\overset{odd}{\underset{1}{\bullet}}\,\overset{odd}{\underset{2}{\bullet}} \longrightarrow \overset{even}{\underset{1}{\bullet}}\,ok\,\overset{even}{\underset{2}{\bullet}}$

r_3: $\overset{odd}{\underset{1}{\bullet}}\,\overset{even}{\underset{2}{\bullet}} \longrightarrow \overset{even}{\underset{1}{\bullet}}\,ok\,\overset{odd}{\underset{2}{\bullet}}$

conds: $(r_1 \mid r_2 \mid r_3)$!

When used within the unit *Eulerian_test* of Sect. 3.1 the input graph of *check_degree* is some non-empty connected graph where all edges are unlabelled and all nodes are labelled with *even*. Remember that for such a graph, the unit *check_degree* keeps the label of each node with an even degree, changes the label of each node with an odd degree into *odd*, and labels each edge with *ok*.

The next observation states that each condition of the form $(r_1 \mid \cdots \mid r_n)$! where $r_1, \dots, r_n$ are rules can be replaced by the expression $(r_1 \mid \cdots \mid r_n)^*$ if the underlying class $\mathcal{E}$ of graph class expressions fulfills the following requirements: (1) For each rule set P, the expression $reduced(P)$ is in $\mathcal{E}$ which specifies the set of all graphs that are reduced w.r.t. P, i.e. $G \in SEM(reduced(P))$ if and only if no rule in P is applicable to G; (2) $\mathcal{E}$ is closed under intersection, i.e. for all $X_1, X_2 \in \mathcal{E}$, $X_1 \wedge X_2 \in \mathcal{E}$ with $SEM(X_1 \wedge X_2) = SEM(X_1) \cap SEM(X_2)$. The proof is obtained in a straightforward way.

Observation 3 Let $t = (I, U, R, (r_1 \mid \cdots \mid r_n)!, T)$ with $r_i \in R$ $(i = 1, \dots, n)$. Let $t' = (I, U, R, (r_1 \mid \cdots \mid r_n)^*, reduced(\{r_1, \dots, r_n\}) \wedge T)$. Then $SEM(t) = SEM(t')$.

3.4 Priorities

A *priority* consists of a set $N \subseteq ID$ and a non-reflexive partial order $<$ on N. Intuitively, such a condition allows to apply a rule or a transformation unit $id \in N$ to the current graph G only if no rule or transformation unit $id' \in N$

of higher priority is applicable to G. Formally, let $C = (N, <)$ be a priority and let for each $id \in N$, $HP(id, C) = \{id' \mid id < id'\}$. Then for each environment E, $(G, G') \in SEM_E((N, <))$ if there exist $G_0, \ldots, G_n \in \mathcal{G}$ such that (1) $G_0 = G$ and $G_n = G'$ and (2) for $i = 1, \ldots, n$, $(G_{i-1}, G_i) \in E(id_i)$ for some $id_i \in N$ and for all $id \in HP(id_i, C)$ there is no $G \in \mathcal{G}$ with $(G_{i-1}, G) \in E(id)$.

The following transformation unit $test\_adjacency$ has a priority control condition. If the input graph is an undirected node-labelled graph with exactly one s-labelled node v, it tests whether v is adjacent to each c-labelled node. In this case v is labelled with c and all c-labelled nodes with ok. Otherwise v becomes unlabelled. The rules are again relabelling ones. The condition $r_2 < r_1$ guarantees that the label s is only deleted if there exists a c-labelled node not adjacent to v. Because of $r_3 < r_2$ the label of v can only be transformed into c if v is adjacent to all c-labelled nodes in the input graph. The terminal expression requires that no rule of $\{r_1, r_2, r_3\}$ is applicable to the output graph. (Note that the unit $test\_adjacency$ can be used for example to find a clique in an undirected unlabelled graph.)

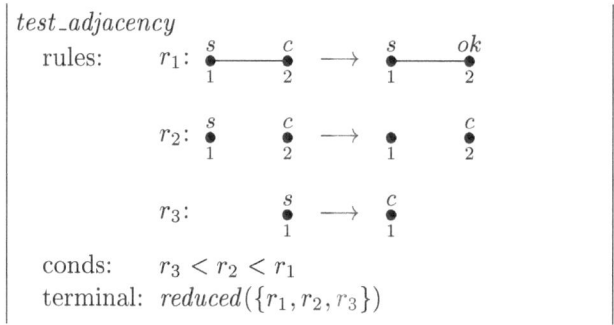

Each transformation unit t with a priority control condition can be transformed into a semantically equivalent unit t' the control condition of which is composed of regular expressions and as-long-as-possible. Roughly speaking, the construction is based on the fact that the control condition $(\{id_1, id_2\}, <)$, where $id_2 < id_1$ can be replaced by $(id_1 \mid (id_1 \,!)\,;\, id_2)^*$.

Observation 4 Let $t = (I, U, R, C, T)$ with $C = (N, <)$ and $N = \{id_1, \ldots, id_n\}$. Let $C' = (p(id_1) \mid \cdots \mid p(id_n))^*$, where for each $id \in N$, $p(id) = id$ if $HP(id, C) = \emptyset$, and $p(id) = ((p(id_1') \mid \cdots \mid p(id_m'))\,!)\,;\, id$ if $m > 0$ and $\{id_1', \ldots, id_m'\} = HP(id, C)$. Then $SEM(t) = SEM((I, U, R, C', T))$.

Proof. (Sketch) For each $id \in N$ let the *recursion depth* $rd(id)$ of id be defined by $rd(id) = 0$ if $HP(id, C) = \emptyset$, and $rd(id) = \max\{rd(id_1'), \ldots, rd(id_m')\} + 1$ if $\{id_1', \ldots, id_m'\} = HP(id, C)$ and $m > 0$. (Note that $rd(id)$ is well defined because $<$ is acyclic.) By induction on $rd(id_i)$ $(i = 1, \ldots, n)$ we get $SEM_{E(t)}(p(id_i)) \subseteq SEM_{E(t)}((N, <))$ which implies that $SEM_{E(t)}(C') \subseteq SEM_{E(t)}((N, <))$. Conversely, consider for $(G, G') \in SEM_{E(t)}((N, <))$ a sequence $G_0, \ldots, G_n$ of graphs satisfying the properties according to the definition of priorities. By induction on n we get $SEM_{E(t)}((N, <)) \subseteq SEM_{E(t)}(prior(id_1) \mid \cdots \mid p(id_n))^*)$. Hence, $SEM(t) = SEM((I, U, R, C', T))$.   □

Let $t = (I, U, R, C, T)$ be a transformation unit where C is a *rule priority* $(N, <)$, i.e. $<$ is defined only on R. Then under some assumptions t can be transformed into a semantically equivalent transformation unit with the default control condition *true* specifying all pairs of graphs.

The basic idea of the construction is to encode priorities into negative contexts of rules which forbid the application of the rules to a graph G if certain subgraphs – specified in the negative contexts – occur in G. Rules with negative contexts are a restricted version of the rules with negative application conditions studied within the single pushout approach in [HHT96].

In order to transform the rule priority of t into *true*, we assume that each rule p of the underlying approach $\mathcal{A}$ has a *left-hand side* $lhs(p)$ and is applicable to a graph G whenever there exists a graph morphism from $lhs(p)$ to G. (Note that for example the single pushout approach [EHK$^+$97] as well as the hyperedge replacement approach [DHK97] and the node replacement approach [ER97] fulfill these requirements.) A rule with negative contexts is of the form $r = (\mathcal{N}, p)$ where $\mathcal{N}$ is a set of graphs, p is a rule, and for each $G \in \mathcal{N}$ the left-hand side of p is a proper subgraph of G, i.e. $lhs(p) \subset G$. Such a rule r is applied to a graph G according to the following steps: (1) CHOOSE a graph morphism g from $lhs(p)$ to G. (2) CHECK the negative context condition: for each $G' \in \mathcal{N}$ there is no graph morphism h from G' to G such that h restricted to $lhs(p)$ is equal to g. (3) APPLY p to G in the same way as in $\mathcal{A}$.

Observation 5 Let $t = (I, U, R, C, T) \in \mathcal{T}_\mathcal{A}$ such that $C = (N, <)$ is a rule priority with $N = R \cup U$. For each $p \in \mathcal{R}$ let $neg(p, C) = (\mathcal{N}, p)$ such that $\mathcal{N} = \{lhs(p) \uplus lhs(p') \mid p' \in HP(p, C)\}$. Let $t' = (I, U, \{neg(p, C) \mid p \in R\}, true, T)$. Then $SEM(t) = SEM(t')$.

Proof. By definition we have $G \Rightarrow_{neg(p,C)} G'$ iff $G \Rightarrow_p G'$ and for each $p' \in HP(p, C)$ there is no graph G'' such that $G \Rightarrow_{p'} G''$. Based on this, we get by definition that $SEM_{E(t)}(C) = RIS(t')$. It follows that $SEM(t')=_{def} SEM(I) \times SEM(T) \cap RIS(t') = (SEM(I) \times SEM(T)) \cap SEM_{E(t)}(C)$. Moreover, by definition $SEM_{E(t)}(C) \subseteq RIS(t)$. Hence, $SEM(t) = SEM(t')$.   □

Remark. Note that the requirement $N = R \cup U$ can be made without loss of generality because $SEM_{E(t)}(C) \subseteq RIS(t)$. More precisely, t can be transformed into a semantically equivalent transformation unit $t' = (I, U', R', (N' <'), T) \in \mathcal{T}_\mathcal{A}$ where $U' = U \cap N$ and $R' = R \cap N$, $N' = U' \cup R'$, and $<'$ is the restriction of $<$ to N'.

3.5   Conditionals

A *conditional* is of the form IF X THEN C where X is a graph class expression and C a control condition (which is already defined). For each environment E

it allows all pairs (G, G') of graphs such that $G \in SEM(X)$ and $(G, G') \in SEM_E(C)$.

For each graph class expression X let $\neg X$ specify all graphs which are not in $SEM(X)$ (i.e. $SEM(\neg X) = \mathcal{G} - SEM(X)$). Then we can construct control conditions which iterate conditionals and branch in the control flow, if we combine conditionals with regular expressions (over control conditions). Some examples are:

- (IF X THEN C) | (IF $\neg X$ THEN C') executes C if X is fulfilled and C' otherwise.
- (IF X THEN C)! executes C as long as the current graph belongs to $SEM(X)$.
- (IF X THEN C)* executes C if X is fulfilled and iterates this arbitrarily often.

Obviously, each control condition C of the previous paragraphs can be expressed by the conditional IF all THEN C where all denotes a graph class expression specifying the set of all graphs. Moreover, conditionals do not increase the semantic power of transformation units, i.e. $(I, U, R, \text{IF } X \text{ THEN } C, T)$ can be replaced by the semantically equivalent unit $(I \wedge X, U, R, C, T)$.

3.6   Further Control Conditions

There are some further control conditions worth to be mentioned. First, each transformation unit t can serve as a control condition because semantically it specifies a binary relation on graphs. For each environment E, the semantics of the control condition t is given by $SEM(t)$. Second, each pair $(X_1, X_2) \in \mathcal{E} \times \mathcal{E}$ defines a binary relation on graphs by $SEM((X_1, X_2)) = SEM(X_1) \times SEM(X_2)$ and, therefore, it can be used as a control condition which is independent of the choice of an environment, i.e. $SEM_E((X_1, X_2)) = SEM((X_1, X_2))$ for all environments E. Third, another kind of rule priorities is based on graph transformation approaches where a rule r is applied to a graph G at a so-called *occurrence* which is a subgraph of G. (This is the case in all major graph transformation approaches.) Such a rule priority allows to apply a rule at an occurrence occ only if no occurrence of a rule with higher priority overlaps with occ (cf. [LM93]). Fourth, BCF expressions [Sch97] can be used as control conditions. Since they provide a certain kind of implicit import which in general may be cyclic, they should be used for transformation units with non-hierarchical import structure. In the case of acyclic BCF expressions, it can be shown that they can be directly translated into semantically equivalent transformation units with non-hierarchical import structure using only control conditions of the types presented in the previous subsections.

4   Properties of Control Conditions

In the following, we present some features of control conditions. First, we introduce minimal control conditions which are used for the flattening of transformation units in [KK96] and show that for each transformation unit there is a semantically equivalent one whose control condition is minimal. The semantics of a transformation unit with a minimal control condition can be computed without constructing the interleaving sequences. It is just obtained from the control condition and the initial and terminal expressions. Second, invertible control conditions are considered. Invertibility of operations is a useful property for systems where previous states should be reconstructible. In this sense, an interesting question is under which conditions a transformation unit t can be transformed into a transformation unit t' which specifies the inverted relation of t. It turns out that this is only possible if the control condition of t is invertible (cf. [KK96]). Finally, we consider continuity of control conditions, a property which leads to a fixed point semantics in the case of cyclic import structures (cf. [KKS97]).

4.1   Minimality

A control condition C of a transformation unit $t = (I, U, R, C, T)$ is *minimal* (with respect to t) if C only specifies pairs of graphs which occur in the relation induced by the interleaving sequences of t, i.e. $SEM_{E(t)}(C) \subseteq RIS(t)$. In a transformation unit t with a minimal control condition the explicit computation of $RIS(t)$ is no longer necessary, i.e. $SEM(t) = (SEM(I) \times SEM(T)) \cap SEM_{E(t)}(C)$. As mentioned above, minimality of control conditions is one of the features required for flattening a transformation unit, i.e. for constructing a semantically equivalent one with empty import.

It can be shown by induction on the structure of regular expressions that each $C \in REG$ is minimal. Moreover, by definition priorities are minimal and the control conditions of the form C! are minimal if C is minimal.

Moreover, for each transformation unit there is a semantically equivalent one with a minimal control condition if the underlying approach provides regular expressions and intersection of control conditions. The proof is straightforward.

Observation 6 Let $t = (I, U, R, C, T)$ and let $C' = C \wedge (id_1 | \cdots | id_n)^*$ where $\{id_1, \ldots, id_n\} = U \cup R$ and $SEM_E(C') = SEM_E(C) \cap SEM_E((id_1 | \cdots | id_n)^*)$ for each environment E. Then C' is minimal with respect to t, and $SEM(t) = SEM((I, U, R, C', T))$.

4.2   Invertibility

An *invertible* class $\mathcal{C}$ of control conditions contains for each $C \in \mathcal{C}$ a condition C^{-1} which, roughly speaking, specifies the inverse of the relation specified by C. More precisely, if C specifies in some environment E the relation ρ, the condition C^{-1} specifies the inverse of ρ in the inverted environment by associating with each identifier id the inverse of $E(id)$, i.e. $SEM_{inv(E)}(C^{-1}) = \rho^{-1}$ where $inv(E)(id) = E(id)^{-1}$ for each $id \in ID$. Invertibility of control conditions is required for the inversion of transformation units, a useful operation for reconstructing previous states of graph transformation systems when programming with transformation units.

It can be shown that the class REG and the class of all boolean expressions over REG are invertible (cf. [KK96]). Hence, $once(id)$ is also invertible (see Observation 1). Moreover, under some assumptions we can construct an inverse control condition for each condition of the types introduced in Sects. 3.3–3.5.

Concretely, for each conditional IF X THEN C we can construct its inverse control condition if C can be inverted. The proof follows directly from the definitions.

Observation 7 Let $\mathcal{C}$ be an invertible class of control conditions and let $C \in \mathcal{C}$. Then $SEM_E(\text{IF } X \text{ THEN } C)^{-1} = SEM_{inv(E)}(C^{-1}; \text{IF } X \text{ THEN } \epsilon)$ for each environment E and each graph class expression X.

To invert as-long-as-possible and priorities, we require that for certain control conditions C and each environment E there exists a graph class expression (C, E) specifying all graphs G such that there is no G' with $(G, G') \in SEM_E(C)$. The proof of Observation 8 follows directly from the definitions. For Observation 9 let $p(id)$ be defined as in Observation 4.

Observation 8 Let $\mathcal{C}$ be an invertible class of control conditions and let $C \in \mathcal{C}$. Then $SEM_E(C!)^{-1} = SEM_{inv(E)}(\text{IF } (C, E) \text{ THEN } (C^{-1})^*)$ for each environment E.

Observation 9 Let $C = (N, <)$ be a priority with $N = \{id_1, \dots, id_n\}$. For each environment E and each $id \in N$ let $p(id, E) = id$ if $HP(id, C) = \emptyset$, and $p(id, E) = id$; IF $((p(id_1') | \cdots | p(id_m')), E)$ THEN $(p(id_1', E) | \cdots | p(id_m', E))^*$ if $HP(id, C) = \{id_1', \dots, id_m'\}$ and $m > 0$. Then $SEM_E((N, <))^{-1} = SEM_{inv(E)}((p(id_1, E) | \cdots | p(id_n, E))^*)$.

Proof. (Sketch) By Observation 4 $SEM_E((N, <))^{-1} = SEM_E((p(id_1) | \cdots | p(id_n))^*)^{-1}$, which is equal to $(SEM_E(p(id_1))^{-1} \cup \cdots \cup SEM_E(p(id_n))^{-1})^*$. By induction on the recursion depth $rd(id)$ of $id \in N$, we get $SEM_E(p(id))^{-1} = SEM_{inv(E)}(p(id, E))$. Hence, $SEM_E((N, <))^{-1} = (SEM_{inv(E)}(p(id_1, E)) \cup \cdots \cup SEM_{inv(E)}(p(id_n, E)))^* = SEM_{inv(E)}((p(id_1, E) | \cdots | p(id_n, E))^*)$. □

Remark. The constructions for inverting priorities can be simplified by distributing them over several transformation units. The idea is that for each environment E each control condition of the form id; IF $((p(id_1) | \cdots | p(id_n)), E)$ THEN $(p(id_1, E) | \cdots | p(id_m, E))^*$ can be expressed for example by id; id' where id' is an imported unit with control condition IF $((p(id_1) | \cdots | p(id_n)), E)$ THEN $(p(id_1, E) | \cdots | p(id_m, E))^*$. Analogously, the control condition of id' can be simplified into IF (id'', E) THEN id''', etc.

4.3   Continuity

A control condition C is *continuous* if for each chain of environments $E_1 \subseteq E_2 \subseteq \cdots$ the following holds: Computing first the semantics of the control condition with respect to each environment in the chain and then taking the union of the resulting relations yields the same result as computing the semantics of the control condition with respect to the union of the environments in the chain, i.e. $\bigcup_{i\in\mathbb{N}} SEM_{E_i}(C) = SEM_{\bigcup_{i\in\mathbb{N}} E_i}(C)$.[4]

Observation 10 Regular expressions are continuous. Priorities and as-long-as-possible are non-continuous.

Proof. (Sketch) The first statement follows by induction on the structure of regular expressions. Now, let $p = (\{id_1, id_2\}, <)$ be the priority control condition with $id_2 < id_1$. Let $G_0, \ldots, G_4$ be non-isomorphic graphs. Let $E_0 \subseteq E_1 \subseteq \cdots$ be such that for each $i > 1$, $E_i = E_{i+1}$, and $E_0(id_1) = \{(G_1, G_2)\}$, $E_0(id_2) = \{(G_2, G_3)\}$, $E_1(id_1) = \{(G_1, G_2), (G_2, G_4)\}$, and $E_1(id_2) = E_0(id_2)$. Then by definition $(G_1, G_3) \in SEM_{E_0}(p)$, but $(G_1, G_3) \notin SEM_{E_1}(p)$. Hence, $(G_1, G_3) \in \bigcup_{i\in\mathbb{N}} SEM_{E_i}(p)$, but $(G_1, G_3) \notin SEM_{\bigcup_{i\in\mathbb{N}} E_i}(p)$. This implies that priorities are not continuous. The proof of the non-continuity of the condition as-long-as-possible is very similar. □

Remark. In the case of a transformation unit t with a possibly cyclic import structure one gets a fixed point semantics if the control conditions are continuous w.r.t. imported transformation units (cf. [KKS97]). This means that a fixed point also exists in the cases of $(N, <)$ and $C!$ provided that $<$ and C only refer to local rules.

5   Conclusion

We have proposed classes of control conditions for transformation units and shown interrelations between them. Moreover, some properties of control conditions were given. There remains some further work to be done:

- The practical usability of the proposed classes of control conditions should be examined with the help of realistic case studies.
- As mentioned before, for each BCF expression with acyclic import structure there is a semantically equivalent transformation unit with the presented classes of control conditions. We conjecture that the set of all BCF expressions can be translated into transformation units with arbitrary import structure and the control conditions discussed here. This should be thoroughly worked out, since a large number of regulation mechanisms proposed for transformation systems can be defined with BCF expressions.
- In the area of term graph rewriting, reduction strategies like *innermost*, *leftmost*, etc. are used that cannot be expressed with the conditions presented in Sect. 3 because their semantics depend on the specific structure of the intermediate term graphs in interleaving sequences. In general, there may

---

[4] For two environments E_1 and E_2, $E_1 \subseteq E_2$ if for each $id \in ID$, $E_1(id) \subseteq E_2(id)$; and $E_1 \cup E_2$ is the environment defined by $E_1 \cup E_2(id) = E_1(id) \cup E_2(id)$ for all $id \in ID$.

be various application areas of graph transformation which demand such a particular type of control conditions. Those classes of conditions should be formulated and studied within the framework of transformation units.

- In [EH86], a general notion of application conditions for graph transformation rules is introduced. Special classes of this notion are studied in [LM93] or [HHT96], for example. The relations between such application conditions and control conditions of transformation units should be studied systematically. Note that Observation 5 is a first step in this direction, because it relates rule priorities with negative application conditions.
- The presented control conditions should be compared with those employed in rule-based systems which are not based on graph transformation (like, for example, expert systems).
- In Sect. 4, some properties of control conditions were presented. Probably, there are further interesting properties in the context of regulating the graph transformation process in a modular way. This should be worked out.

Acknowledgement

I thank Hans-Jörg Kreowski, Frank Drewes, Renate Klempien-Hinrichs, and the anonymous referees for their helpful comments.

References

AEH+99. Marc Andries, Gregor Engels, Annegret Habel, Berthold Hoffmann, Hans-Jörg Kreowski, Sabine Kuske, Detlef Plump, Andy Schürr, and Gabriele Taentzer. Graph transformation for specification and programming. *Science of Computer Programming*, 34(1):1–54, 1999. 323, 324

Bun79. Horst Bunke. Programmed graph grammars. In Volker Claus, Hartmut Ehrig, and Grzegorz Rozenberg, editors, *Proc. Graph Grammars and Their Application to Computer Science and Biology*, volume 73 of *Lecture Notes in Computer Science*, pages 155–166, 1979. 324

CEH+97. Andrea Corradini, Hartmut Ehrig, Reiko Heckel, Michael Löwe, Ugo Montanari, and Francesca Rossi. Algebraic approaches to graph transformation part I: Basic concepts and double pushout approach. In Rozenberg [Roz97]. 327

DHK97. Frank Drewes, Annegret Habel, and Hans-Jörg Kreowski. Hyperedge replacement graph grammars. In Rozenberg [Roz97], pages 95–162. 331

DP89. Jürgen Dassow and Gheorghe Păun. *Regulated Rewriting in Formal Language Theory*, volume 18 of *EATCS Monographs on Theoretical Computer Science*. Springer-Verlag, 1989. 324

EEKR99. Hartmut Ehrig, Gregor Engels, Hans-Jörg Kreowski, and Grzegorz Rozenberg, editors. *Handbook of Graph Grammars and Computing by Graph Transformation, Vol. II: Applications, Languages and Tools*. World Scientific, Singapore, 1999. To appear. 337

EH86. Hartmut Ehrig and Annegret Habel. Graph grammars with application conditions. In Grzegorz Rozenberg and Arto Salomaa, editors, *The Book of L*, pages 87–100. Springer-Verlag, Berlin, 1986. 336

EHK⁺97.   Hartmut Ehrig, Reiko Heckel, Martin Korff, Michael Löwe, Leila Ribeiro, Annika Wagner, and Andrea Corradini. Algebraic approaches to graph transformation II: Single pushout approach and comparison with double pushout approach. In Rozenberg [Roz97], pages 247–312. 331

ER97.   Joost Engelfriet and Grzegorz Rozenberg. Node replacement graph grammars. In Rozenberg [Roz97], pages 1–94. 331

HEET99.   Reiko Heckel, Gregor Engels, Hartmut Ehrig, and Gabriele Taentzer. Classification and comparison of modul concepts for graph transformation systems. In Ehrig et al. [EEKR99]. To appear. 323

HHT96.   Annegret Habel, Reiko Heckel, and Gabriele Taentzer. Graph grammars with negative application conditions. *Fundamenta Informaticae*, XXVI(3,4):287–313, 1996. 331, 336

KK96.   Hans-Jörg Kreowski and Sabine Kuske. On the interleaving semantics of transformation units — a step into GRACE. In Janice E. Cuny, Hartmut Ehrig, Gregor Engels, and Grzegorz Rozenberg, editors, *Proc. Graph Grammars and Their Application to Computer Science*, volume 1073 of *Lecture Notes in Computer Science*, pages 89–108, 1996. 323, 324, 328, 332, 333, 334

KK99.   Hans-Jörg Kreowski and Sabine Kuske. Graph transformation units and modules. In Ehrig et al. [EEKR99]. To appear. 323

KKS97.   Hans-Jörg Kreowski, Sabine Kuske, and Andy Schürr. Nested graph transformation units. *International Journal on Software Engineering and Knowledge Engineering*, 7(4):479–502, 1997. 323, 325, 333, 335

LM93.   Igor Litovsky and Yves Métivier. Computing with graph rewriting systems with priorities. *Theoretical Computer Science*, 115:191–224, 1993. 332, 336

MW96.   Andrea Maggiolo-Schettini and Józef Winkowski. A kernel language for programmed rewriting of (hyper)graphs. *Acta Informatica*, 33(6):523–546, 1996. 324

Nag79.   Manfred Nagl. *Graph-Grammatiken: Theorie, Anwendungen, Implementierungen*. Vieweg, Braunschweig, 1979. 324

Roz97.   Grzegorz Rozenberg, editor. *Handbook of Graph Grammars and Computing by Graph Transformation, Vol. I: Foundations*. World Scientific, Singapore, 1997. 323, 336, 337

Sch75.   Hans-Jürgen Schneider. Syntax-directed description of incremental compilers. In D. Siefkes, editor, *GI — 4. Jahrestagung*, volume 26 of *Lecture Notes in Computer Science*, pages 192–201, 1975. 324

Sch97.   Andy Schürr. Programmed graph replacement systems. In Rozenberg [Roz97], pages 479–546. 324, 332

TER99.   Gabriele Taentzer, C. Ermel, and Michael Rudolf. The AGG-approach: Language and tool environment. In Ehrig et al. [EEKR99]. To appear. 324

Integrity Constraints in the Multi-paradigm Language PROGRES

Manfred Münch[1], Andy Schürr[2], and Andreas J. Winter[1]

[1] Lehrstuhl für Informatik III, RWTH Aachen
Ahornstr. 55, D-52074 Aachen, Germany
{muench,winter}@i3.informatik.rwth-aachen.de
[2] Software Engineering Institute, University BW München
Werner-Heisenberg-Weg 39, D-85577 Neubiberg, Germany
schuerr@informatik.unibw-muenchen.de

Abstract. PROGRES is a multi-paradigm visual programming or exe-cutable specification language, which has a well-defined static type con-cept. It supports programming with graph rewriting systems. An inte-grated type-checker is able to check the static semantics of a specification. This paper presents the integration of static integrity constraints to the language which allow to check a specification's semantics at run-time. It discusses a number of important design decisions such as when constraints have to be checked and what happens if constraint violations are detected.

1 Introduction

The idea to use graph transformations for specification and very high level programming purposes is now about thirty years old. Some early visual pro-gramming languages such as Plan2D [2] are indeed graph rewriting languages. Graph grammars e.g. are used for the definition of the syntax and semantics of visual languages like in DiaGen [11]. Beyond that there is a still growing number of visual languages that rely directly on the graph rewriting paradigm, such as GOOD [13]. Also, process modelling systems were described with the help of graph grammars and realized by graph rewriting systems. An example for that is DYNAMITE [6] which implements dynamic task nets. Other applications of graph grammars and graph rewriting systems can be found in Computer Integrated Manufacturing systems (see [27]).

In this paper we will focus on PROGRES (PROgramming with Graph REwriting System) [18]. It is a visual programming language in the sense that it has a graph-oriented data model and a graphical syntax for its most im-portant language constructs. We distinguish between data definition and data manipulation and do not rely on the rule-oriented programming paradigm for all purposes.

Most rule-based visual programming languages are untyped languages, i.e. processed data is only implicitly defined by the set of their rules. An example for these languages is KidSim [22]. Even those languages which combine rules

H. Ehrig et al. (Eds.): Graph Transformation, LNCS 1764, pp. 338–352, 2000.
© Springer-Verlag Berlin Heidelberg 2000

with class diagrams (see [9])do not check at compile-time whether rules produce inconsistent data configurations or not. In contrast to rule-based languages there are some visual dataflow languages which have a static type concept. Such a static type concept enables the system to validate a program at compile-time.

PROGRES combines the well-known class diagrams of object-oriented languages with graph rewriting rules. Its language is statically typed and we have a two-level polymorphic type system with multiple inheritance. With the help of type information an integrated analyzer can assert the correctness of a specification regarding its static semantics. That means that rules produce only graphs which fulfil the requirements of the specified class diagrams.

The integrated analyzer is a powerful tool. However, most (static) integrity constraints are not definable as a class diagram. Therefore we need a separate constraint definition language. OOA/OOD languages usually either use informal constraint definitions or first order logic for this purpose. In PROGRES we have integrated an approach which is similar to the Object Constraint Language OCL of the Unified Modelling Language UML (see [15] for further details concerning UML and OCL).

This paper presents the integration of integrity constraints into the PROGRES programming language. The main design goal was to support the user in debugging a specification at run-time. We have also adopted the idea of active integrity constraints as they are realized in e.g. active database systems [28]. This allows the user to specify a repair action if a constraint is not met.

In the next section related work will be discussed. Section 3 introduces the reader to the PROGRES language. Section 4 discusses the integrity constraints. Section 5 deals with active constraints, an extension of the static integrity constraints. In section 6 the evaluation strategy of constraints will be discussed. Section 7 deals with the results of these new features and shows the possibilities for future research.

2   Related Work

The semantics of PROGRES is developed on the basis of the logic-based graph grammar theory (see [12]). One of the main purposes of the system is to generate visual programming environments. PROGRES itself is a visual programming language that has been enriched by some textual language constructs.

Constraint languages are well-known e.g. in database systems. Since many years constraints are used in active database systems in order to ensure the integrity of a database ([10] and [23]). C. Bauzer Medeiros gives a definition of those integrity constraints in [10]:

An integrity constraint in a database environment is a statement of a condition that must be met in order to maintain data consistency.

Bauzer Medeiros presents a language called ALICE/RL (Assertion Language for Integrity Constraint Expressions/Rule Language). This language (or a variation of it) is used in active database environments and consists of Event-

Condition-Action triples where an action is triggered if the condition holds after the occurence of a specified event. This action can be seen as a sort of repair action. The ideas of defining a condition which has to be met and otherwise executing a repair action has directly been transferred to PROGRES. The disadvantage of ALICE/RL is that it allows its users only a textual definition of first order logic expressions instead of a visual representation for its constraint expressions.

Another already existing constraint language is UML's (graphical) OCL [15]. However, a realization of UML's OCL in PROGRES does not make any sense since PROGRES has more precise type-checking rules than OCL. Furthermore OCL does not fit to the historically grown subgraph pattern matching of PROGRES or the concept of derived attributes, since OCL is mainly based on SmallTalk.

A visual representation of constraints is supported by the language G-Log (see [14]). G-Log constraints can be rewritten in equivalent horn clause logic expressions. However, these constraints are only a subset of the constraints we want to be able to define in PROGRES. Another serious disadvantage of G-Log is that there is no implementation yet.

Courcelle shows in [1] that verification of arbitrary monadic second order logic formulae is theoretically feasible for a restricted class of context-free algebraic graph grammars. More precisely he proves that it is decidable whether the generated languages of certain context-free graph grammars fulfill a given set of formulae. Despite of Courcelle's research results a verification is not possible in PROGRES. We will discuss that briefly in section 4.

Another inspiring approach for this work was the integration of graphical constraints to graph grammars based on category theory [5]. This approach presents the integration of pre- and postconditions to graph rewriting rules. Although some ideas could be taken over this approach aims for different goals such as verification of graph grammars. However, currently there is no implementation available.

Last but not least we have to mention the rather general graph transformation framework in [8]. It proposes the usage of graph expressions for the definition of static integrity constraints (in the sense of this paper) as well as for the definition of pre- and postconditions to graph transformations. PROGRES constraints presented here may be seen as one concrete instantiation and implementation of the general framework with one exception: constraint-violating PROGRES graph transformations are not handled like graph transformations with unsatisfied application conditions (as we will see later on).

In this paper we will not only discuss the visual representation of constraints w.r.t. visual programming with the graph rewriting system PROGRES but also design decisions as which kinds of constraints we have implemented and when they are checked. We also address the topic of how to react to a constraint violation.

3   The PROGRES Language

A PROGRES specification is composed of two parts. The first part describes the static properties of a class of graphs. The second part defines the operations that can be applied to the graph. Operations are graph manipulations which are denoted as graph rewriting rules. In the following we will present both parts of a specification.

The first part describes the class diagram of a graph consisting of the definitions for nodes and edges of this graph. Nodes are considered to be independent objects. These nodes are defined by classes. Classes can be either abstract, i.e. without any instances, or concrete. It is also possible to construct an inheritance hierarchy over classes.

Node classes can contain attribute definitions. These attribute definitions can also be redefined in subclasses (which has not been done in the example of fig. 1). PROGRES supports the definition of attributes at node classes as well.

Fig. 1 shows an example of a node class definition with attribute definitions. The node class *PERSON* has an attribute *Age* and the node class *AIRCRAFT* has the attribute *Seats*. Furthermore this graph schema contains several edge-type definitions. E.g. a node of the class *PERSON* may be connected to a node of the class *AIRCRAFT* (which can also be a node of a subclass of *AIRCRAFT*, i.e. *Boeing*737 or *Boeing*747, of course) by the edge sits_in. This edge describes that a person has entered an aircraft. *Boeing*737 e.g. is a concrete class which inherits all properties of the abstract class *AIRCRAFT* (denoted by the arrow with a hollow arrowhead). Concrete classes, i.e. classes from which instances may appear in the host graph, are represented by rectangular boxes with rounded edges.

Edges represent binary relationships between nodes. Their definition contains the type of the edge and the two node classes that denote the type of the source and target class, resp., of this edge. Furthermore the cardinalities of the edges are annotated at each edge. In fig. 1 the edge *holds* describes that a person can hold a set of flight tickets but on the other hand one flight ticket can only be held by one single person. In contrast to nodes, edges cannot be considered as independent objects. They do not possess attributes or identifiers and they are implicitly deleted if their source or target will be deleted.

After the class diagram is defined in the first part of a specification, in the second part of a specification the operations on a graph are described. This graph is called the host graph. The operations can be divided into two different language constructs: productions and transactions. Modifications of a graph are realized by productions. These are graph rewriting rules which are denoted diagrammatically. All productions have a left-hand side and a right-hand side. Fig. 2 shows an example of a production.

The production in fig. 2 describes the usual check-in procedure at an airport. A person holding (a valid) flight ticket and probably one or more suitcases waits at the counter (clerk) to check in. If this configuration can be found in the host graph the production can be executed. That means that the person gets

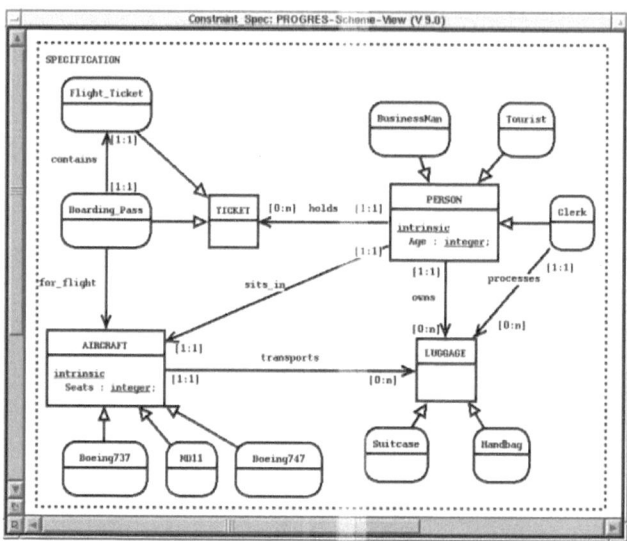

Fig. 1. Definition of a class diagram

a boarding pass that contains the person's flight ticket and leaves his or her luggage at the counter.

The rectangular solid boxes denote nodes in the host graph which have to be found. Nodes with dashed lines are optional, i.e. the specified production can be executed although these nodes have not been found. Double rectangles match a set of nodes.

In the production shown in fig. 2 the node with identifier '3 on the left-hand side, which is of the class *LUGGAGE*, matches a set of nodes of maximum size (which may also be empty - indicated by a dashed border of the rectangle). The arrows represent edges in the graph. On the left-hand side of the production a graph pattern is identified and replaced by the graph pattern which is specified on the right-hand side of our production. The notation $x' = `x$ means that these nodes are replaced identically. Nodes which appear on the left-hand side only will be deleted, even if they are shared, and nodes that appear on the right-hand side only are newly created (e.g. the boarding pass in fig. 2). The same happens to edges. The new subgraph is embedded automatically, unless otherwise stated (see [21]). Identically replaced nodes will be embedded into the graph preserving any edges to nodes which were not matched by the production. Dangling edges which can occur if a node is deleted will be removed. However, there is no garbage collector which deletes 'unconnected' nodes implicitly.

Another language construct that is available in the operational part of a specification is a transaction. Transactions are written down textually and provide control structures such as loops, if-then-else constructs, deterministic as well as nondeterministic operation sequences, and guarded conditional statements.

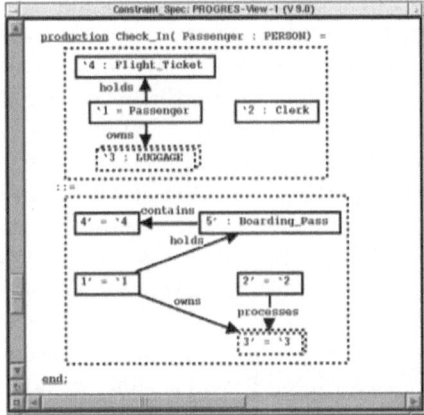

Fig. 2. The production Check_In

These are the most important features of the PROGRES language to understand the explanations in the following.

4 Integrity Constraints

The PROGRES environment supports the user with many useful tools. One of the most powerful ones is the analysis tool. It works incrementally and because of the strict type concept it is capable of checking the static semantics with the help of more than 300 consistency rules. Although this tool covers many sources of errors one of the most important cannot be checked: the "integrity" of a host graph which is produced by executing a PROGRES-specification. That means that the user cannot ensure that the graph always meets a set of user-specified conditions which must be met beyond the type consistency. This is a major deficiency for large specifications. For this reason a couple of different constraint expression have been added to PROGRES.

In principle integrity constraints may be used to check e.g. correct cardinalities of edges (although in PROGRES, this is already done by the type checker), the applicability or definedness of graph transformations, or if a class of graphs described by a specification can be reduced to a class of graphs with certain properties. Those formulae are static integrity constraints in the classical sense. Their expressiveness is mainly based on first order logics, horn-clause logics or monadic second order logics. Therefore it is also possible to describe existentially or universally quantified constraints. It is also conceivable to write down formulae in temporal logics. However, we only support static integrity constraints without temporal aspects.

Constraints can be used for different purposes. In [4], Heckel describes a constraint system for graph grammars based on the algebraic approach with the purpose to verify a set of graph transformation rules. Because of PROGRES' derived attributes which can influence the execution of a specification and its non-determinism a verification is only possible for a small subset of PROGRES specifications. We use constraints for runtime-checking of our specification. Of course this is always very time-intensive, strategies to gain a better efficiency are known. We make use of an incremental attribute evaluation mechanism and an efficient subgraph pattern matching algorithm (see section 5 for further information).

In this section we discuss the constraint language we have integrated in PROGRES. First the different forms of passive constraints will be explained which allow for an integrity check of a host graph a specification produces. After that active constraints will be presented in the next section. Active constraints are an extension of (passive) integrity constraints by adding an action that can be triggered if a specified condition does not hold on the host graph.

Integrity constraints can be partitioned into four groups:

- First, we implemented graphical global graph consistency conditions based on horn-clause logics.
- Another group are textual global graph consistency conditions, which are not addressed in this paper. They contain first order logics and allow for definedness checks and support the control structures of PROGRES' transactions.
- Furthermore, we discuss textual (graphical) pre- and postconditions to productions and transactions. Their expressiveness is like textual global graph consistency conditions.
- Finally, we introduce local (textual) constraints of node classes, which are called constraint attributes. They can use PROGRES' path expressions which are similar to first order logics, limited to coherent subgraphs.

In the following subsections the use of constraints will be demonstrated by some examples. We will not define the syntax of our constraint language formally. The PROGRES language manual [17] gives a thorough overview of the syntax of constraints.

4.1   Global Graph Consistency

Graphs can contain semantical faults made by the specifier although the graph was built up conforming to the class diagram (see fig. 1). This means that the host graph does not meet the requirements, i.e. conditions, which are explicitly defined in the form of constraints. With the help of these constraints the user is able to find those errors and correct them in the specification.

Global graph consistency conditions (which will be called global constraints in the following) enable the user of the PROGRES system to specify graph patterns which have to be present in a host graph at any time or which may never appear. This has the same semantics as an existentially quantified expression. Certainly, this kind of specifying constraints is not enough to express all possible conditions which must hold on a graph. Particularly with global constraints it is very helpful to be able to postulate universally quantified expressions.

Consider this formula:

$$\forall Aircraft : transports(Aircraft, Luggage)$$
$$\rightarrow (\exists Person : owns(Person, Luggage) \land$$
$$sits\_in(Person, Aircraft))$$

This formula claims that for all aircrafts that transport luggage there must be a person owning this luggage and sitting in this aircraft. The constraint in fig. 3 is constructed according to the formula. In PROGRES the predicates of the formula above are represented by edges and the variables by nodes (independent objects). Thereby every match of the for-part of the constraint in fig. 3 must be extendible to a match of the $ensure$-part of that constraint. This is denoted by the nodes '1 and '2 in the $ensure$-part of the constraint. These are the same nodes which have been matched in the for-part. (Note that the owns-edge was never affected, i.e. deleted or changed, by the production $Check\_In$). More formally, assume that P is the subgraph of the for-part of our constraint, Q is the subgraph of the $ensure$-part. Then a constraint is satisfied in a graph G if and only if for all matches m of P in G there exists a match n of Q in G such that n extends m.

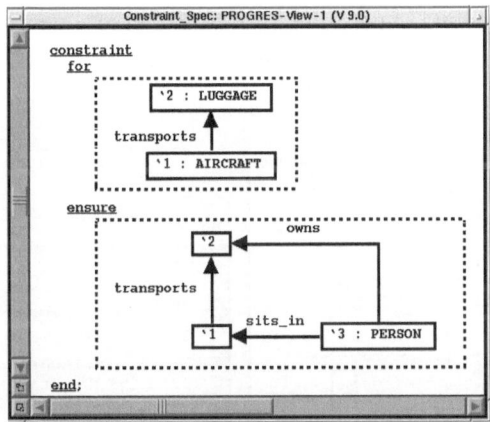

Fig. 3. Universally quantified constraint

A very important topic is the time when global constraints should be checked. The PROGRES system always starts execution with an empty initial host graph. Therefore it does not make sense to check constraints all the time since existentially quantified constraints cannot hold at the very beginning. Note that it does not matter for universally quantified constraints. If these constraints do not find a proper match m in the graph G for the specified subgraph P in the for-part of the constraint they will not be checked and thus cannot fail. Another idea is to check a constraint after every operation. However, existentially quantified constraints still make problems then. Furthermore this behaviour can already be modelled by postconditions to those operations (see subsection 4.2). We have decided to check constraints on a more course level, i.e. after the execution of every operation which is exported by a package (see [20] for further information about PROGRES packages). We think that we have tackled all problems mentioned before by this solution. Constraints can be seen as contracts packages have to obey. However, due to the fact that the integration of the module concept is new to PROGRES we currently discuss about the best solution of the triggering mechanism of constraints.

4.2   Pre- and Postconditions

Global constraints are a general instrument to check the integrity of a host graph. A disadvantage of global constraints is that all of the specified constraints have to be met when they are checked. For many purposes it would be very helpful to have only a small set of conditions which are valid for only one operation. An example is the application of an operation which requires a certain state of the host graph. This is very difficult to express in terms of global constraints.

This is the reason why we have integrated preconditions to operations into the PROGRES language. Consider the example of a transaction in fig. 4. This operation may only be applied to a host graph if there were not too many boarding passes handed out at the check-in. Otherwise people run into problems when they board the aircraft nevertheless.

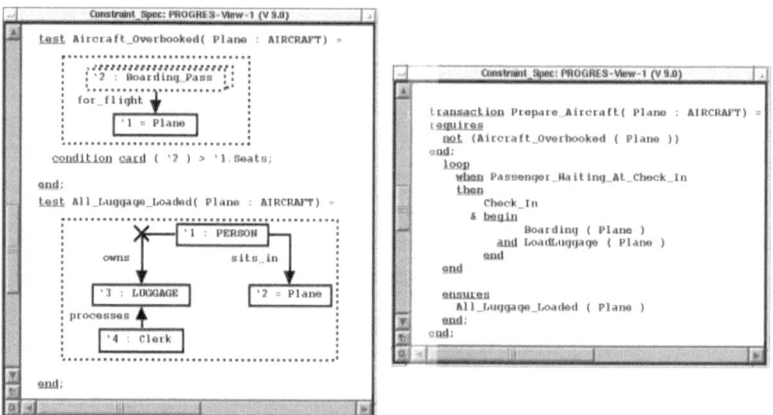

Fig. 4. Precondition to a transaction

To check the integrity of a single operation it is necessary to have postconditions as well. With postconditions the user is able to check if this operation has modified the host graph in the way he wanted it to be changed. If we consider the same transaction it should be made sure that every piece of luggage which is owned by a passenger is in the aircraft and not left behind at the check-in office. The *ensure*-part of the transaction in fig. 4 models this situation. This postcondition expresses that there may be no owns-edge (crossed-out edge) from any passenger sitting in the aircraft to a piece of luggage which is still processed by the clerk at the check-in counter.

Both, pre- and postcondition call a test which may not fail. Otherwise the pre- or postcondition is not met. That means that there must be a match of the specified graph pattern in the host graph. Note that tests are existentially quantified.

The transaction and the tests shown in fig. 4 make use of parameters which can be passed to every operation in PROGRES. An expression like '$x = Plane$ (as shown in both tests) claims that the match of this node must be exactly the node given in the parameter[1].

4.3   Constraint Attributes

Until now all constraints were associated with either the host graph or operations (pre- and postconditions). Beyond that it is also possible to bind constraints to node classes. Therefore another kind of attribute, the constraint attribute, has been added to the language. Of course all constraint attributes have to be of boolean type. As with every other attribute, constraint attributes can be redefined as well. It is possible to define abstract constraint attributes, i.e. without an evaluation function, at node classes. The function will be bound to a subclass to this attribute. Beyond that it is also possible to attach a repair action to a constraint attribute (refer to subsection 4.4 fo further information about repair actions). Fig. 5 shows an example of a constraint attribute definition. The constraint is actually the same as the precondition in fig. 4. The only difference is that this constraint attribute is checked for all instances of the *AIRCRAFT* node class while the precondition is only checked for the specific aircraft which is dealt with in the transaction.

4.4   Active Constraints

Active constraints are an extension of integrity (or passive) constraints. We have adopted the idea to model active constraints similar to Event-Condition-Action triples as to find in e.g. ALICE/RL. Passive constraints can be compared with a combination of an event and a condition only. The action gives the user the chance to specify an operation which repairs the violated condition. This is sensible because this mechanism can be exploited to help the specifier debugging a specification automatically or to enable the user to specify an event-controlled

---

[1] Appropriate values for those variables can be obtained by calling tests or productions with out-parameters; i.e. tests or productions can assign a node or a set of nodes to a variable and return it to the calling operation for further use.

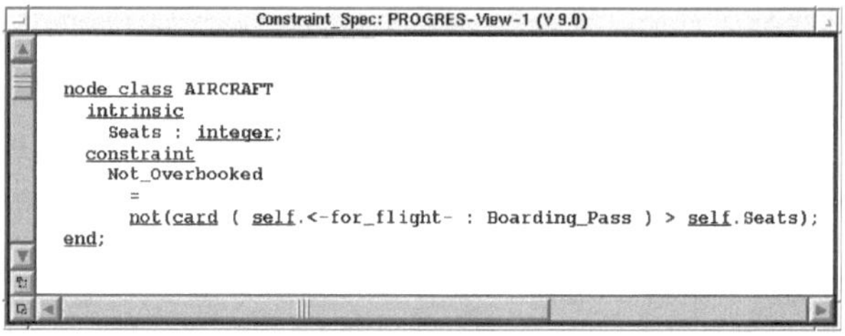

Fig. 5. Defining constraint attributes

behaviour conveniently. In PROGRES we have added such repair actions to global and local constraints at node classes.

Active constraints are said to be violated if the constraint is not met and the repair action cannot be executed successfully, i.e. the repair action cannot repair anything. Of course, such a repair action can violate another (active) constraint again. The task to care about the termination problem is left to the user of the PROGRES system.

Finally we have to discuss what happens if a constraint is violated. There are several ways how the system can react. The first and most restrictive reaction would be to stop the execution immediately and force the user to repair the specification. Only then the system allows to restart the execution. Another idea is to weaken the last condition and stop the execution, roll back the last (and eventually violating) operation so that the user can modify this operation and resume the execution. An even weaker reaction would be to mark the erroneous subgraph and go on nevertheless. However, we have decided to choose a variant of the second solution, i.e. to stop the execution and mark the constraint which has been violated. PROGRES offers an *Undo*-operation which allows the user to roll back every single step to find where the violating operation is located.

Active constraints behave similarly if their repair action cannot be applied anymore or if its application does not yield the desired result. However, the system tries to repair the erroneous subgraph first without notifying the user. We are aware that the price for this convenience is a different semantics of the specification when the constraint checking is switched on.

5   Implementation Aspects

The PROGRES programming environment currently consists of about 750,000 lines of code, including the DBMS GRAS and the graphical user interface. It is available as free software on Solaris2 for Sun Workstations and on Linux for PC.

PROGRES is used for different purposes around the world. Meanwhile there are very large specifications which consist of about 200 pages and rewrite rules with more than 70 nodes on the left- and right-hand side.

An integrated interpreter enables PROGRES to animate a specification. For this reason the system translates the specification into intermediate abstract graph rewriting code. This code may be executed directly by an integrated abstract graph rewriting machine, which is a conventional stack machine enriched by graph rewriting operations and backtracking capabilities. It is also possible to store and resume debugging sessions, to undo or redo sequences of graph modifications and to recompile and execute modified parts of the specification incrementally.

Beyond that the PROGRES system can compile the abstract graph rewriting code to C source code. The biggest specifications generate about 100,000 lines of C code. The C code accompanied by the environment's graph-oriented DBMS GRAS [7] and a Tcl/Tk-based user interface may be used as a rapid prototype (see [21] for details). We could already develop many useful visual environments for CIM [27], software engineering systems [12], and visual programming languages like BDL [19].

One of the most important issues is the efficiency of the constraint checking algorithm. As we have already discussed, global constraints are only checked against public operations in the according module. The main defining quantity for the run-time behaviour of the constraint checking algorithm is the graph-pattern search algorithm. This algorithm has a complexity of $O(n^k)$ where n is the number of nodes in the host graph and k the number of nodes to be matched. Therefore, checking universally quantified constraints is the most expensive operation because they have to find every match of the given graph pattern in the for-part of the constraint. Existentially quantified constraints just need to find one match which is a bit cheaper. The overall complexity of constraint checking is approximately (#constraints) * $O(n^k)$. However, an integrated search planner often reduces the worst-case complexity considerably [29]. It computes a partial order for searching nodes in a graph and proposes the cheapest evaluation order. Therefore all presented examples in this paper have a complexity of $O(n)$. Nevertheless, constraint checking is a very expensive operation regarding time and for that reason it can be switched off by the user of the PROGRES system.

In PROGRES local constraint attributes at node classes have to be defined textually and can be evaluated much more efficiently using an incremental attribute evaluation algorithm. The complexity is $O(\#$ maybe affected nodes). These nodes are determined by an attribute dependency graph. The attribute values are only computed for those nodes where the value of relevant attributes was changed.

6   Conclusion and Future Work

In this paper we presented the integration of constraints, pre- and postconditions to the PROGRES language. There are two main classes of constraints: integrity constraints and active constraints that allow for event-controlled programming.

We have shown what constraints are good for and how they can be used. Furthermore we have discussed several different approaches and reasoned about the solutions we have favoured.

All kinds of discussed constraints are available in the current implementation of the PROGRES system. It supports validation of specification in several ways:

- The syntactical correctness of a specification is guaranteed by a syntax-directed editor and parser
- An integrated analyzer can guarantee that all parts of a specification are well-typed
- Integrity constraints as well as pre- and postconditions of rules allow a run-time checking of the specification
- An automatic verification of a subset of PROGRES specifications is feasible but its description is out of the scope of this paper (cf. [5]).

Constraints are also interesting for generating a prototype from a specification. With the help of these constraints the user can make sure that the prototype will not produce an erroneous host graph. This leads to much safer generated visual programming environments. Active (attribute) constraints can be used for defining views on graphs which is also a very desirable feature when host graphs are growing larger.

Currently we are integrating a UML-like module concept into the PROGRES language (see [20]). In connection with modules constraints will be a crucial means for programming by contract. The constraints will play the role of UML's OCL expressions (cf. section 1). The main future goal is to blend the standard object-oriented modelling notations with visual programming with graph rewriting rules.

References

1. B. Courcelle. The expression of graph properties and graph transformations in monadic second-order logic. In *chapter 5 in [16]*, pages 313–400. 1997. 340
2. E. Denert, R. Franck, and W. Streng. PLAN2D - towards a two-dimensional programming language. LNCS 26, pages 202–213. Springer Verlag, Berlin, 1974. 338
3. G. Engels and H. Schneider, editors. *Int. Journal of Software Engineering and Knowledge Engineering: Special Issue on Graph Grammar-Based Specifications*, volume 7. World Scientific, Singapore, 1997. 351
4. A. Habel, R. Heckel, and G. Taentzer. Graph grammars with negative application conditions. *Fundamenta Informaticae*, XXVII, pages 1/2, 1996. 343
5. R. Heckel and A. Wagner. Ensuring consistency of conditional graph grammars - a constructive approach. In *SEGRAGRA '95 "Graph Rewriting and Computation"*. Elsevier, Electronic Notes of TCS, 2, 1995. 340, 350
6. P. Heimann, G. Joeris, C.-A. Krapp, and B. Westfechtel. DYNAMITE: Dynamic task nets for software process management. In *Proceedings of the 18th International Conference on Software Engineering*, pages 331–341, Los Alamitos (CA), 1996. IEEE Computer Society Press. 338

7. N. Kiesel, A. Schürr, and B. Westfechtel. Gras, a graph-oriented database system for (software) engineering applications. *Information Systems*, 20(1):21–51, 1995. 349

8. H.-J. Kreowski, S. Kuske, and A. Schürr. Nested graph transformation units. In *[3]*, pages 479–502, 1997. 340

9. D. W. McIntyre and G. E. P. Visual tools for generating iconic programming environments. In *[24]*, pages 162–168, 1992. 339

10. C. B. Medeiros and M. J. Andrade. Implementing integrity control in active data bases. *Journal of Systems Software*, 27:171–181, 1994. 339

11. M. Minas and G. Viehstaedt. Diagen: A generator for diagram editors providing direct manipulation and execution of diagrams. In *[25]*, pages 203–210, 1995. 338

12. M. Nagl, editor. *Building Thightly Integrated (Software) Development Environments: The IPSEN Approach*, volume 1170 of *Lecture Notes in Computer Science*. Springer Verlag, Berlin, 1996. 339, 349

13. J. Paredaens, J. Bussche, M. Andries, M. Gemis, M. Gyssens, I. Thyssens, D. van Gucht, V. Sarathy, and L. Saxton. An overview of GOOD. *ACM SIGMOD Record*, 21(1):25–31, 1992. 338

14. J. Paredaens and P. Peelman. G-log, a declarative graphical query specification language. Technical Report TR 91-16, Dept. Wiskunde en Informatica, UIA, Antwerpen, 1991. 340

15. Rational Software Corporation. UML semantics, version 1.1. http://www.rational.com, 1997. 339, 340

16. G. Rozenberg, editor. *Handbook on Graph Grammars: Foundations*, volume 1. World Scientific, Singapore, 1997. 350

17. A. Schürr. *The PROGRES Language Manual*. RWTH Aachen, D-52056 Aachen, Germany. http://www-i3.informatik.rwth-aachen.de/research/progres/ProgresSyntax.html. 344

18. A. Schürr. PROGRES, a visual language and environment for programming with graph rewrite systems. Technical Report AIB 94-11, 1994. 338

19. A. Schürr. BDL - a nondeterministic data flow programming language with backtracking. In *[26]*, pages 394–401, 1997. 349

20. A. Schürr and A. Winter. UML packages for PROGRES. In *TAGT'98 — 6th International Workshop On Theory And Application Of Graph Transformation*, pages 132–139, Paderborn, 1998. 346, 350

21. A. Schürr, A. Winter, and A. Zündorf. Visual programming with graph rewriting systems. In *[25]*, pages 195–202, 1995. 342, 349

22. D. C. Smith, A. Cypher, and J. Spohrer. Kidsim: Programming agents without a programming language. *Communications of the ACM*, 37(7):54–67, 1994. 338

23. S. D. Urban and A. M. Wang. The design of a constraint/rule language for an object-oriented data model. *Journal of Systems Software*, 28:203–224, 1995. 339

24. *Proc. IEEE Workshop on Visual Languages (VL'92)*, Los Alamitos, CA, 1992. IEEE Computer Society Press. 351

25. *Proc. IEEE Symposium on Visual Languages (VL'95)*, Los Alamitos, CA, 1995. IEEE Computer Society Press. 351

26. *Proc. IEEE Symposium on Visual Languages (VL'97)*, Los Alamitos, CA, 1997. IEEE Computer Society Press. 351

27. B. Westfechtel. A graph-based system for managing configurations of engineering design documents. *International Journal of Software Engineering and Knowledge Engineering*, 6(4):549–583, 1996. 338, 349

28. J. Widom and S. Ceri, editors. *Active Database Systems: Triggers and Rules For Advanced Database Processing*. Morgan Kaufmann Publishers, Inc, 1996. 339

29. A. Zündorf. Graph pattern matching in PROGRES. In *5th International Workshop on Graph Grammars and Their Application to Computer Science*, Lecture Notes in Computer Science 1073, pages 454–468, Berlin, 1996. Springer Verlag. 349

A Framework for Adding Packages to Graph Transformation Approaches

Giorgio Busatto[1], Gregor Engels[2], Katharina Mehner[2], and Annika Wagner[2]

[1] Department of Computer Science, Leiden University
P.O. Box 9512, The Netherlands
busatto@wi.leidenuniv.nl
[2] Department of Computer Science
Paderborn University, D-33095, Germany
{engels,mehner,awa}@uni-paderborn.de

Abstract. Graphs are a commonly used formalism for modeling many different kinds of static and dynamic data. In many applications, data modeling can be improved by using hierarchically structured graphs. But, while there already exist hierarchical graph data models, no general-purpose hierarchical graph data model exists yet, which unifies common features of these domain-specific models. In this paper, we present *graph packages*, a general formalism for defining hierarchical graphs, supporting the most important features found in known applications.

Because of the dynamic nature of graphs, hierarchical graph transformation is also an important issue to be dealt with when using hierarchical graphs. Motivated by the successful application of graph grammars to the specification of graph transformations, we also introduce a framework that allows to specify hierarchical graph transformations by combining existing graph grammar approaches and our graph package concept. These concepts are a step towards the definition of a general-purpose hierarchical graph data model.

1 Introduction

Graphs are a very natural and intuitive approach for modeling realistic problems and have been successfully used in different fields of computer science to model, for example, database structures in information systems, data structures in software specification and in programming language semantics, and hypertext structure in hypermedia systems.

A graph traditionally consists of a set of nodes and a set of edges, the nodes representing entities of some kind, and the edges representing relations between them. A natural example of graph modeling is the World-Wide Web (WWW), the well-known distributed information system (see [W3]), wherein documents (called *pages*) contain the actual information, and references between pages (called *hyperlinks*) represent navigational paths between pages. It is natural to model pages as nodes of a graph, and hyperlinks as edges, thus visualizing the structure of the WWW.

H. Ehrig et al. (Eds.): Graph Transformation, LNCS 1764, pp. 352–367, 2000.

Graphs, however, are difficult to manage and to understand if they contain too many nodes or edges. For example, a graph-like presentation of a large information system like the web is of little use when trying to find some specific page, and it only leads to the well known "lost in hyper-space" problem.

In such situations, it is a common solution to adopt a "divide and conquer" approach: A graph is divided into smaller, more manageable components. If a two-layer decomposition is not sufficient, the same process is applied to the subcomponents as many times as needed. The result is a tree-like or dag-like (if sharing of subcomponents is allowed) decomposition of a graph, which we call a *hierarchical graph* (HG for short).

If we look again at the web, we notice that this way of organizing information is commonly used. In fact, large collections of pages are grouped into *web sites*. The pages of a web site can be further grouped into more fine grained sub-collections, e.g. according to their content, or their owner. Note that this kind of structuring is not reflected by the underlying flat data model of the WWW. However, more recent hypermedia information systems integrate similar ideas into their hypermedia model, for example, HyperWave (see [Hyp]) allows for a hierarchical structuring of documents through the concept of a *collection*.

The advantage of using a hierarchical structuring of information is that we can have different views of the data, according to the point in the hierarchy from which we look at it. For example, knowing that there is a web site about a certain topic helps us to abstract from the actual pages contained in that site, and concentrate on the overall content of the site. We will consider those pages individually only when visiting that site, i.e. when we decide to descend one level in the hierarchy.

In some cases the concept of hierarchical structuring needs additional features: If a graph is decomposed in a hierarchical fashion, it is, in general, not advisable to give direct access to components that are found in the inner levels of the hierarchy. For example, on the web not all the pages contained in a certain site have the same importance for an external user: A homepage is likely to be interesting, and therefore it makes sense to have direct access to it, whereas a page dealing with a very specific topic will only be interesting once the user has decided to visit the site. These features, though not currently supported by the data model of the WWW, are often implemented via ad-hoc solutions (homepages, "cookies", CGI scripts, etc). What is interesting for us, is that once we have a hierarchical decomposition of a graph, we often need to hide graph elements inside the hierarchy. This is the well-known idea of *encapsulation*, or *information hiding*, as found in programming languages (e.g. in Modula2, see [Wir85]), and software engineering (e.g. in UML, see [UML]).

We know numerous applications of hierarchical graphs, e.g. the definition of programming language semantics in [Pra79], the visualization and design of hypertexts in [BRS92], graph visualization in software systems (see e.g. [LB93], [Him94], and several commercial software or API's), and the representation of hypermedia document structures (see again our web example, and [Hyp]). Hierarchical structuring is used in all these examples, whereas encapsulation is used in HyperWave and in many graph visualization systems.

Although hierarchical graphs are often used, and in spite of the fact that there is an overall agreement on basic concepts like the use of hierarchies and of encapsulation, there does not exist a general approach to hierarchical graphs, nor any general-purpose hierarchical graph data model. Therefore, in different applications, ad-hoc notions of hierarchical graphs are introduced, and the term *hierarchical graph* is applied to data models that can differ a lot from each other. Due to this variety, we feel the need for a standard hierarchical graph notion and data model, possibly integrating the (most important) concepts of known approaches. This should both help to compare the features and expressive power of different approaches to hierarchical graphs, and provide a reference data model that allows to avoid developing new data models around the same concepts.

One of the main reasons for the lack of generality of many hierarchical graph data models is, in our opinion, the fact that the hierarchy is strictly coupled with the underlying graph. Such approaches, which we call *coupling*, cannot be general, because they use special features of the graph to build up the hierarchy. For example, the *encapsulated hierarchical graph* approach of [ES95], where the hierarchy is defined by using *complex nodes* which can be nested in each other, is not convenient in applications where we do not want to associate the elements of the hierarchy to any graph element at all (see e.g. [BBH94]), unless we introduce artificial nodes, the only purpose of which is to model the hierarchy information.

We have tackled these problems by adopting the alternative *decoupling* approach, which consists of building the hierarchy as a structure separate from the graph, and then assigning the various elements (nodes and/or edges) of the graph to the elements of the hierarchy. As a result, we define the concepts of a *graph package* (which models an element of a hierarchy), and of a *graph package system* (GPS, which models a complete hierarchy), which allow to build a hierarchical decomposition of a generic graph, and to encapsulate graph elements according to such decomposition. The visibility information is also added as separate information.

After discussing the concepts of hierarchy and encapsulation, we want to consider the dynamic aspects of hierarchical graph data modeling. If we look back at our example, we see that pages and links between them, as well as the page organization (e.g. the site structure) are changing all the time. This is a very special example of the general fact that graphs and hierarchical graphs are often used to model dynamic structures. Therefore, in a fully fledged hierarchical graph data model, a language for the specification of hierarchical graph transformation plays a very important role.

In this respect, it is natural to describe the dynamic aspects of graphs by means of graph grammars and graph transformation systems, since these formalisms have been successfully used for the specification of complex graph transformations, both in theoretical computer science (e.g. for studying the properties of sets of graphs), and in practical applications (e.g. as a software specification language). An overview on graph grammars, as well as an introduction to the literature on the subject can be found in [Roz97].

There are already examples of graph transformation approaches using hierarchical graphs, for example [Pra79] and [LB93]. These do not offer a general approach to hierarchical graph transformation, since they use a very special notion of hierarchical graphs. In order to tackle this problem, we again use the decoupling idea, and study hierarchical graph transformation as a combination of graph transformation and hierarchy transformation. In this case, the graph part of the transformation can be specified by a graph transformation system, while hierarchy transformation is handled separately.

In this scenario, we have identified two possible approaches. The first is the *descriptive* approach, in which the package hierarchy is described in terms of the underlying graph, i.e. there is some general mechanism, that allows to determine a hierarchical graph, starting from a given flat graph. In this case, graph transformation automatically induces transformations on induced hierarchical graph, and graph transformation rules can be re-interpreted as hierarchical graph transformation rules. The second is the *constructive* approach, where, besides graph transformation rules, there exist hierarchy transformation rules, which affect the graph and the packages directly. Therefore, the package structure is *constructed* through repeated transformation rule applications.

In this paper, we adopt the descriptive approach, which is suggested by applications where the hierarchical structure is computed from and added to a given graph, and it is not meant to be directly accessed by hierarchical graph transformations (see e.g. [BRS92]). The descriptive approach is appealing, in that it allows to reinterpret traditional graph-rewrite rules as hierarchical graph rewrite rules. Furthermore, we have developed a method (called the *qualification mechanism*), which allows to lift a given graph to a hierarchical graph in a general way. The qualification mechanism has the additional requirement that the underlying graphs are typed.

As hinted above, the descriptive approach allows to interpret existing graph transformation rules as hierarchical graph transformation rules. We have formalized this idea of lifting existing graph transformation rules to hierarchical graph transformation rules, and defined an abstract *framework* that identifies some minimal requirements of a graph transformation system for the lifting to be applicable.

In this paper we proceed as follows: Section 2 presents a running example, which will be used in the paper to illustrate and motivate our approach. Section 3 introduces a formal definition of the basic features of graph packages. Section 4 deals with a descriptive way of defining graph packages, and with the qualification mechanism. Section 5 deals with graph package system transformation. Section 6 gives a summary and discusses possible future work.

2   An Illustrating Example

In order to illustrate our definitions, we will introduce more details about the web example, that we have mentioned in the introduction. The WWW is a distributed information system based on the Internet. The information is represented as *hyper-documents*, i.e. as a collection of documents containing cross-references (or *hyperlinks*) between them. The single documents, called *pages*, are provided by different *web sites*. The sources and targets of hyperlinks within pages are called *anchors*. The WWW can be seen as a large graph, the pages and anchors being the nodes and the hyperlinks being the edges. Insertion and deletion of pages and links can be modeled as graph transformations.

In our example, we assume to have web pages documenting two European research projects, APPLIGRAPH and GETGRATS. Figure 1 shows a graphical representation of the relevant part of the web. Projects, sites (such as Pisa and Bremen), pages and anchors are represented as nodes, where Pi stands for page i and small filled squares denote anchors. Each project is documented by three pages which are not necessarily located at the same web site. The fact that a certain page documents a project and that a page is provided by a site as well as the fact that an anchor belongs to a page is represented by solid arrow-headed edges in the graph. Hyperlinks are depicted as dashed arrow-headed edges.

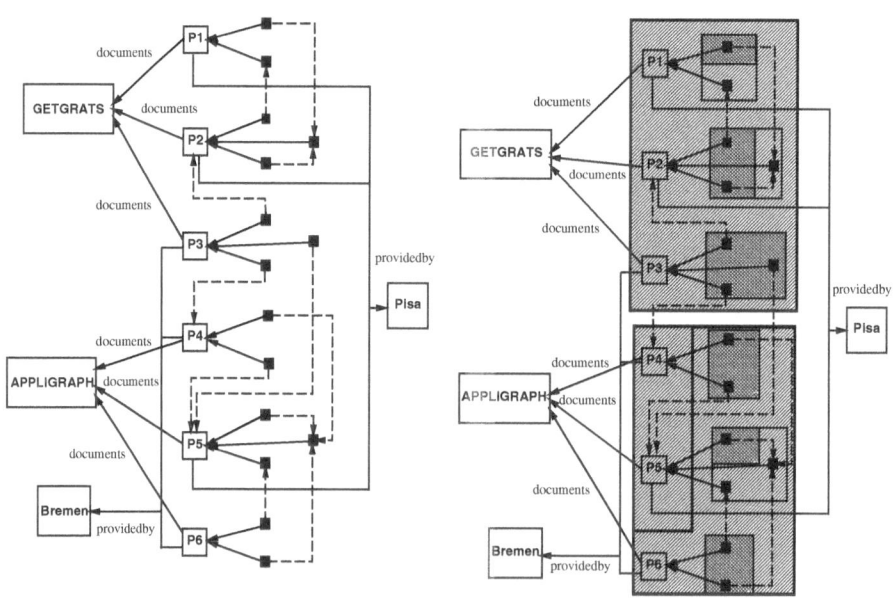

Fig. 1. Graph package system for the WWW example.

The basic model of the web uses only a few different node and edge types as can be seen in the example. Hyperlinks are allowed between any pages. This unrestricted use of hyperlinks often results in a complex net of pages and links. As a result the managing of web sites becomes very difficult and authors and developers often develop additional strategies to handle the inherent complexity, taking into account additional information about the web structure. For example, very often pages are grouped according to their contents and such a group of pages has a so-called homepage at the top of the hierarchy, i.e. a page through which all other pages in the group can be accessed. An arbitrary user of the WWW can create hyperlinks to any page inside such a group, but the homepage is the most reliable page to which one can refer, as it is left to the maintainer of the homepage to add the necessary hyperlinks to the other pages in the group. Hence, only a homepage should be accessible from outside.

Such ideas concerned with grouping and hierarchical structuring of pages or hiding of pages are not captured by the basic model of the WWW, which shows that the graph model in its present fashion is insufficient. However, in the web example we repeatedly find the pattern that a project is documented by several pages. The pages related to the same project naturally form a group. The same holds for the anchors of a page. The right hand side of figure 1 suggests how we can add the missing information. In the example, we have identified groups of pages and anchors belonging together. Such a grouping is called a package and denoted as a rectangle drawn around them, e.g. there is a package around the pages P4, P5, and P6. The example also shows how to build hierarchical structures. Packages can be nested, for example the contents of each page is organized in a package. Notice also that we are considering a very generic kind of grouping, that does not interfere with node typing, i.e. we can have nodes of the same type in different groupings, and nodes of different types in the same grouping.

In the above discussion, we suggested that only certain pages, like homepages, should be accessible from outside the project and access to other pages should be restricted. We can achieve this by means of packages, simply by hiding inside a package the pages to which we want to forbid access. Notice again that the encapsulation concept is independent from typing, since different pages (and thus entities with the same type) can have different visibilities with respect to a certain package.

In our example, we display the information about visibility as follows. The elements which are not visible outside a package are shaded, e.g. P6 is not visible outside the package, i.e. it can not be accessed via an hyperlink from a page outside the package. Hidden pages and anchors can only be accessed by anchors of the same package.

3 Graph Packages

In this section, we formalize and complete the ideas about graph packages already introduced in section 2. For the definition of our graph package concept we use typed directed graphs. Graph packages can actually be defined on simpler kinds of graphs (untyped, unlabelled directed graphs), but we need typed graphs in section 4 for defining graph packages in a descriptive way.

Definition 1 (Graphs and typed graphs). *A directed graph (in the sequel also called a* graph*) is a quadruple* $G = (N_G, E_G, s_G, t_G)$*, where* N_G *is the set of nodes of* G*,* E_G *is the set of edges of* G*, and* $s_G, t_G : E_G \to N_G$ *are two functions mapping an edge to its source (resp. target) node.*

Given two graphs G*,* H*, we say that* G *is an* instance *of the graph schema* H *iff there exist two functions* $\tau_N : N_G \to N_H$*, and* $\tau_E : E_G \to E_H$*, such that, for all* $u, v \in N_G$*,* $e \in E_G$*, with* $u = s_G(e)$*, and* $v = t_G(e)$*, we have* $\tau_N(u) = s_H(\tau_E(e))$*, and* $\tau_N(v) = t_H(\tau_E(e)))$*. If a graph* G *is an instance of a schema* H*, then we say that* G *is a* (H-)typed *graph. Given a graph schema* H*, we indicate the set of all its instances with* $\mathcal{G}(H)$*.*

For our running example, we have the graph schema in figure 2. Hyperlink edges can be of two types: from an anchor to an anchor, or from an anchor to a page (we have kept the dashed notation to relate these edge types to hyperlinks in figure 1 more easily). `belongsto` edges relate anchors to the owning page.

We are now ready to introduce our formalization of the package concept. Remember that the basic ideas of graph packages are hierarchical node grouping, and node visibility. Therefore, each node is assigned to exactly one package. In-

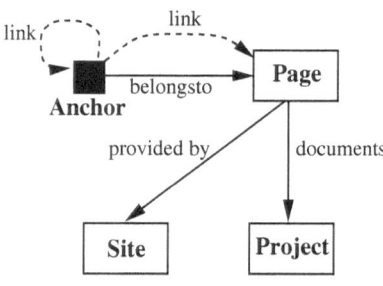

Fig. 2. Graph schema.

stead of directly defining the visibility of a node with respect to each package we use the idea of import and export interfaces for packages, as known from modularization concepts in software engineering. Furthermore, we want node visibility to restrict the possibility of drawing edges across the hierarchy (i.e. from one package to another).

Definition 2 (Graph package systems). *A* graph package system *(GPS) is a tuple* $S = (G, P, \prec_P, \text{Owner}, \text{Exp}, \text{Imp})$*, where*

- G *is a typed graph,*
- P *is a finite set, whose elements we call* graph packages*,*
- $\prec_P \subseteq P \times P$ *is a nesting relation between packages, which induces a tree structure on* P*, whose root we indicate as* $\rho(P)$ *(the root package),*
- $\text{Owner} : N_G \to P$ *is a function, mapping each node of* G *to the package that owns it,*
- $\text{Exp}, \text{Imp} : P \to 2^{N_G}$ *are two functions, defining the sets of* exported *resp.* imported *nodes of each package,*

such that S satisfies the package visibility conditions *and the* forbidden edges condition *(see definitions 3, 4 in the sequel).*

S is called a graph package system over *G. For all* $p \in P$, *we let* $\mathrm{Own}(p) := \{n|n \in \mathrm{N}_G \wedge \mathrm{Owner}(n) = p\}$, *and we call it the* set of own nodes *of p. Given a set of typed graphs* $\mathcal{G}$, *the set of all graph package systems over graphs belonging to* $\mathcal{G}$ *is denoted by* $\mathcal{P}(\mathcal{G})$, *i.e.* $\mathcal{P}(\mathcal{G}) := \{S|S = (G, P, \prec_P, \mathrm{Owner}, \mathrm{Exp}, \mathrm{Imp}) \wedge G \in \mathcal{G} \wedge$ "*S is a GPS*"$\}$.

In our web example, the root package $\rho(P)$ is not depicted in figure 1. The nodes owned by $\rho(P)$ are GETGRATS, APPLIGRAPH, Bremen and Pisa and we have e.g. Owner(GETGRATS) $= \rho(P)$. Let GGP and AGP represent the GETGRATS and the APPLIGRAPH package, respectively. We have $GGP, AGP \prec_P \rho(P)$, i.e. they are the packages nested in the root package. Let PP_1 through PP_6 be the six page packages, then $PP_1, PP_2, PP_3 \prec_P GGP$ and $PP_4, PP_5, PP_6 \prec_P AGP$. An example for ownership is Owner(P1) = Owner(P2) = GGP. As far as the visibility is concerned, we have, for example, $\mathrm{Exp}(AGP) = \{$P4, P5$\}$ and $\mathrm{Imp}(GGP) = \{$P4, P5, GETGRATS, Bremen, Pisa, A12, A22$\}$, where A12 indicates the second anchor of page P1, and A22 indicates the second anchor of page P2. Notice that imported nodes are not depicted explicitly in figure 1.

For the definition of graph package systems we have used two conditions that we are going to define now. The first states constraints on owned, imported and exported nodes, according to the hierarchical structure of the packages.

Definition 3 (Package visibility conditions). *Let us consider a tuple* $S = (G, P, \prec_P, \mathrm{Owner}, \mathrm{Exp}, \mathrm{Imp})$ *as in 2. Then the* package visibility conditions *for S state that:*

1. *For all* $q \in P$, *if* $\exists p \in P$ *such that* $q \prec_P p$ *we have* $\mathrm{Imp}(q) \subseteq \mathrm{Own}(p) \cup [\mathrm{Imp}(p) \setminus \mathrm{Exp}(q)] \cup \bigcup_{r \prec_P q} \mathrm{Exp}(r)$ *and for* $\rho(P)$ *we have* $\mathrm{Imp}(\rho(P)) \subseteq \bigcup_{r \prec_P \rho(P)} \mathrm{Exp}(r)$.

2. *For all* $q \in P$ *we have* $\mathrm{Exp}(q) \subseteq \mathrm{Own}(q) \cup \left[\bigcup_{r \prec_P q} \mathrm{Exp}(r) \right]$.

3. $\mathrm{Exp}(\rho(P)) = \emptyset$.

Intuitively, definition 3 states that: (1) The imported nodes of a package q must either be owned nodes of its parent package p, or imported into p (but not imported from q itself), or be exported by one of the subpackages $r_1, \ldots, r_n$ of q. (2) The exported nodes of a package q must either be owned nodes of q or be imported from one of the subpackages $r_1, \ldots, r_n$ of q. (3) The root package has no exported nodes.

The visibility conditions state that the imported nodes of PP_3 can only be nodes that are imported into or owned by the parent package GGP. We assume that $\mathrm{Imp}(PP_3) = \{$P2, P3, P4, P5, Bremen, GETGRATS$\}$, which is possible by assuming that the last four nodes are imported into GGP (whereas P2 and P3 are owned by GGP). As far as exported nodes are concerned, we assume (see figure 1) that $\mathrm{Exp}(PP_3) = \emptyset$.

The following condition forbids certain edges across different packages, by imposing that the target of an edge be at least imported into the package that owns the source of the edge.

Definition 4 (Forbidden edges condition). *Let us consider a tuple* $S =$ $(G, P, \prec_P, \mathrm{Owner}, \mathrm{Exp}, \mathrm{Imp})$ *as in 2. Then the* forbidden edges condition *for* S *states that: for all edges* $e \in \mathrm{E}_G$, *if* $s_G(e) \in \mathrm{Own}(p)$ *(for some package p), then* $t_G(e) \in \mathrm{Own}(p) \cup \mathrm{Imp}(p)$.

In our example, an edge from an anchor that is owned by the page package PP_3 to the page P4 is allowed, because P4 belongs to $\mathrm{Imp}(PP_3) =$ {P2, P3, P4, P5, Bremen, GETGRATS}, whereas an edge to P6 is not allowed.

Condition 4 can be easily extended to undirected graphs, if we consider that an undirected edge can be represented by a couple of directed edges. As a result, if an undirected edge exists between two nodes u, v, then u must be imported into or owned by $\mathrm{Owner}(v)$, and v must be imported into or owned by $\mathrm{Owner}(u)$.

4   Descriptive Definition of Graph Packages

Now that we have described all the features of our structuring concept, we want to study possible ways to define the structuring on a given graph. We follow a so-called *descriptive* approach, meaning that the structuring of the graph is described via a general mechanism on the schema level, using some properties of the graph. We extend graph schemata to *qualifying schemata*, which allow to associate packages to certain nodes, via the so-called *qualification* mechanism. As a result of this construction, it is possible to associate a unique graph package system S to a graph G, if G is compatible with the qualification described by the qualifying schema.

Definition 5 (Qualifying schema). *Let* $H = (\mathrm{N}_H, \mathrm{E}_H, s_H, t_H)$ *be a graph schema. Then* H *can be extended to a* qualifying schema $\Sigma = (H, \mathcal{QE})$, *where* $\mathcal{QE} \subseteq \mathrm{E}_H$ *is a set of* qualifying edge types, *such that,* $\forall \varepsilon_1, \varepsilon_2 \in \mathcal{QE}$, $s_H(\varepsilon_1) =$ $s_H(\varepsilon_2) \Rightarrow \varepsilon_1 = \varepsilon_2$, *i.e. every node type is the source of at most one qualifying edge type, and there are no loops in the graph* $H' = (\mathrm{N}_H, \mathcal{QE}, s_{H'}, t_{H'})$, *where, for all* $e \in \mathcal{EQ}$, $s_{H'}(e) = s_H(e)$ *and* $t_{H'}(e) = t_H(e)$.

We call $\mathcal{QN} := \{n \in \mathrm{N}_H | \exists \varepsilon \in \mathcal{QE}.n = t_H(\varepsilon)\}$ *the set of* qualifying node types. *We call* $\{n \in \mathrm{N}_H | \exists \varepsilon \in \mathcal{QE}.n = s_H(\varepsilon)\}$ *the set of* qualified node types.

The idea behind qualifying node types is that their instances (called *qualifying nodes*, as explained below) are used to determine which packages are to be put in the graph package system. For each qualifying edge type, there is exactly one target node type (for example, documents has the target Project), and that node type is a qualifying node type. Forbidding loops of qualifying edge types ensures that the subgraph of H induced by them is a tree. The use of

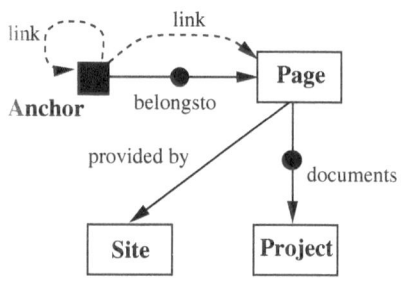

Fig. 3. Qualifying schema.

this will be clarified in construction 1.

In our example, we use the qualifying schema Σ shown in figure 3. The qualifying edge types (e.g. the one between `Page` and `Project`) are indicated graphically by drawing a black bullet on them. Therefore, we have $\mathcal{QN} = \{\texttt{Project}, \texttt{Page}\}$, $\mathcal{QE} = \{\texttt{Page} -\texttt{documents} \rightarrow \texttt{Project}, \texttt{Anchor} -\texttt{belongsto} \rightarrow \texttt{Page}\}$.

The next definitions show how we can use a qualifying schema for constructing a graph package system on top of a given graph G, by looking at the nodes and edges of the graph G and at their properties (i.e. in a descriptive way).

Definition 6 (Qualified instance). *Let $\Sigma = (H, \mathcal{QE})$ be a qualifying schema, and let G be an instance of the graph schema H, via the typing functions τ_N, τ_E. Then (G, QE), where $QE := \{e \in E_G | \tau_E(e) \in \mathcal{QE}\}$, is called a qualified instance of Σ, if $\forall e_1, e_2 \in QE.s_G(e_1) = s_G(e_2) \Rightarrow e_1 = e_2$.*

QE is called the set of qualifying edges. *We call $QN = \{n \in N_G | \exists e \in QE.n = t_G(e)\}$ the set of* qualifying nodes *of (G, QE). We call $\{n \in N_G | \exists e \in QE.n = s_G(e)\}$ the set of* qualified nodes.

Note that a given graph G of a graph schema H cannot always be instantiated to a qualified instance (G, QE) of a given qualifying schema $(H, \mathcal{QE})$, since we require the uniqueness of the qualifying edge for each qualified node. This property will later on guarantee that each node is owned by exactly one package.

Construction 1 (GPS via qualified instance).
Given a qualifying schema $\Sigma = (H, \mathcal{QE})$, and a graph G, instance of H, we define the following construction of a GPS $S = (G, P, \prec_P, \text{Owner}, \text{Exp}, \text{Imp})$:

1. *If possible, extend G to a qualified instance (G, QE) of Σ. Let QN be the set of qualifying nodes for (G, QE).*
2. *Let P be a finite set (of packages) such that: There exists a total injective mapping $m : QN \rightarrow P$. For all $n \in QN$, we indicate $m(n)$ by p_n, and we say that p_n is a* qualified package *or the* package qualified by n. *Furthermore, there exists exactly one element $\rho \in P \setminus \{p_n | n \in QN\}$, which we call the* non-qualified package. *Then let $\rho(P) := \rho$.*
3. *Let $\text{Owner}(m) := p_n$ for $m \in N_G$, if there exists $e \in QE$ with $s_G(e) = m$ and $t_G(e) = n$, where $n \in QN$ is the qualifying node of package $p_n \in P$. Let $\text{Owner}(m) := \rho(P)$ otherwise.*
4. *For all $p, q \in P$, define the nesting relation as*

$$p \prec_P q \equiv \exists n \in QN.p = p_n \wedge \text{Owner}(n) = q$$

5. *For all $p \in P$, define $\text{Exp}(p) := \emptyset$*
6. *Let $\text{Imp}(\rho(P)) := \emptyset$, and, for all $n \in QN$, let $\text{Imp}(p_n) := \{n\}$.*

If such a construction succeeds, we call $(G, P, \prec_P, \text{Own}, \text{Exp}, \text{Imp})$ a graph package system induced by Σ and G.

It is easy to prove the following proposition.

Proposition 1. *For each qualifying schema $(H, \mathcal{QE})$, and each graph G, instance of H, and construction 1 succeeds, then the resulting graph-package system is unique up to isomorphism.*

This means that the hierarchical structure and the distribution of nodes among the packages is fully determined by the qualification schema and the underlying graph. Notice also that at least the qualifying node of each qualified package must be imported into the package, otherwise the edge restriction condition would forbid the edges linking qualified nodes to corresponding qualifying nodes.

Notice that the visibility information (i.e. Exp, Imp) was defined in a trivial way during this construction. In a real case, like in the web example, the visibility should be defined in a more refined way (notice, in fact, that the visibility of nodes in the web example cannot be expressed using construction 1 as it is). This would involve modifying steps 5 and 6 by using some more refined method of defining exported and imported nodes. For example, one could use attributed graphs, and define the exported nodes of a package as those nodes that satisfy some fixed condition on their attributes. Such methods are very much dependent on the kind of application to which hierarchical graphs are applied, and providing a thoroughly worked out example thereof would be outside the scope of this paper.

Abstracting from the actual method used, we are interested in situations where the GPS can be built only looking at the underlying graph G. This leads to the following definition.

Definition 7 (Induced GPS's). *Let $\mathcal{G}$ be a set of typed graphs, and let $\mathcal{P}(\mathcal{G})$ be the set of GPS's over $\mathcal{G}$. If there exists a partial function $f : \mathcal{G} \to \mathcal{P}(\mathcal{G})$, then, for all $G \in \mathcal{G}$, if $f(G)$ is defined we say that f induces the GPS $f(G)$ on G.*

The function f is partial because we do not require that the used structuring algorithms always succeed. In the case of qualification, it may happen that the condition to be checked on a graph G, when building a qualification instance, fails, or that the forbidden edges condition is not satisfied, therefore making the whole construction fail.

5   Graph Package Transformation

In this section, we discuss transformations on hierarchically structured graphs. Our main goal is to reuse graph rewrite rules and their semantics according to an existing graph rewrite approach in order to define the semantics of rewrite steps on a hierarchically structured graph. We present a general framework that supports these goals and show how to use it with the GPS's presented in section 3 as one special kind of hierarchically structured graphs.

Consider the following general situation, where a rewrite relation $\Rightarrow \subseteq \mathcal{G} \times \mathcal{G}$ on graphs shall be extended to a rewrite relation $\Rightarrow_{\mathcal{S}}$ on structures $\mathcal{S}$ *depending* on the graphs. The dependency can be expressed by a function $f : \mathcal{G} \to \mathcal{S}$. Then we can define $\Rightarrow_{\mathcal{S}} \subseteq \mathcal{S} \times \mathcal{S}$ by $f(G) \Rightarrow_{\mathcal{S}} f(G')$, if $G \Rightarrow G'$.

Now we apply this idea to the set of typed graphs $\mathcal{G}$ and the GPS's induced by them. Here the function f is given by the qualification mechanism described in section 4 and $\mathcal{S}$ is instantiated to $\mathcal{S} := \mathcal{P}(\mathcal{G})$. In this case, f is a partial function, as the induced GPS does not exist for all graphs.

Additional restrictions on the application of rewrite rules on structured graphs can emerge from constraints of the structuring concept used to instantiate the framework. We want to define restrictions on the allowed transformations, that take into account the hierarchical structuring and information hiding as given by the GPS approach. This restricts the rewrite relation on GPS's, i.e. $\Rightarrow_{GPS} \subseteq \Rightarrow_{\mathcal{S}}$. Since information hiding is defined w.r.t. packages, we decided that also transformations must be applied with respect to a package, i.e. rule applications must be local to a package, which we call the *affected package*. This has the following consequences:

- The search for a match of a rule is restricted to elements visible to a package.
- The affected package has to be preserved, i.e. it cannot be deleted or moved.
- Only nodes internal to a package can be deleted or created.

In order to check these conditions, it is not sufficient that the graph transformation approach that we want to extend provides a direct derivation relation. We also need:

1. A way to speak about "matched nodes", e.g. a match could be represented by a mapping from the nodes in the left-hand side of the rule to nodes in the host graph.
2. A way to speak about preserved/deleted/created nodes, e.g. every direct derivation step $G \Rightarrow G'$ should allow to define a (partial) tracking function $nt : N_G \rightarrow N_{G'}$.
3. A way to speak about preserved/deleted/created packages, e.g. an analogous tracking function $pt : P \rightarrow P'$ (where P, P' are the sets of packages for $f(G)$, and $f(P')$ respectively) for packages.

For example, in the single pushout (SPO) approach to graph rewriting (see e.g. [Roz97, p. 247ff.]), finding a mapping as in point 1, as well as a node tracking function, is already part of the rewrite process. The existence of a node tracking function, together with the qualification mechanism described in section 4, allows to define a package tracking function in a straightforward way.

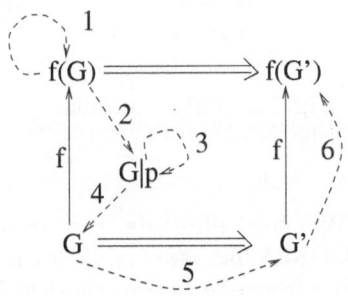

Based on this idea, we obtain the schema for graph package transformation illustrated in figure 4, where a GPS

Fig. 4. GPS transformation step.

$f(G)$ is obtained via a function f from an underlying graph G (induced package idea) and the transformation is obtained by applying a rule r (with extended semantics) to $f(G)$ and a chosen package p of $f(G)$ (local package transformation idea). The transformation involves the steps below, where the numbers of the steps correspond to the numbers at the dashed lines in figure 4.

1. Select the rewrite rule r and the affected package p.
2. Find a match for r in the projection $G|_p$ of G to the affected package p.
3. Check application conditions ensuring that the transformation will be local to p. If these conditions are not satisfied the application fails.
4. Extend the match for r in $G|_p$ to a match for r in G. If this extension is impossible the application fails.
5. Apply rule r to G obtaining G'.
6. Build $S' = f(G')$. If $f(G')$ is not defined, the GPS transformation step fails.

If step 6 is completed successfully, then we say that $S' = f(G')$ is obtained from $S = f(G)$ through a GPS derivation step via the rule r and package p, and we write this as $S \Rightarrow_{GPS} S'$.

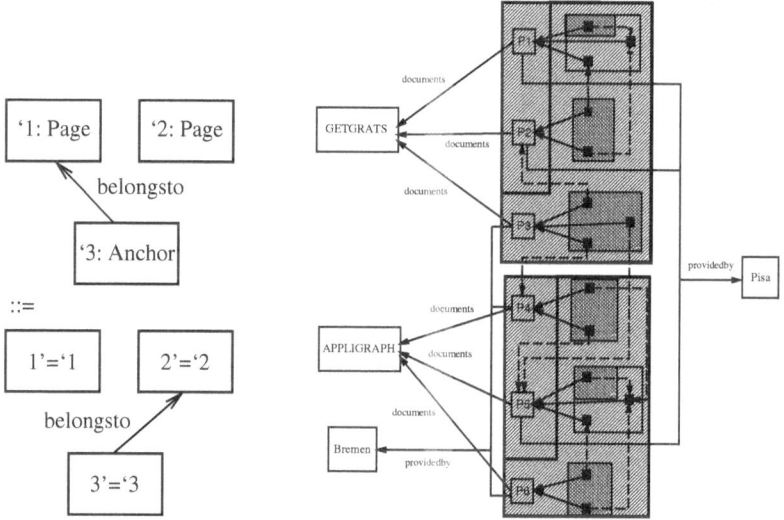

Fig. 5. Move anchor rule.   New graph package system.

We now illustrate the transformation with an example. Consider a PROGRES-like (see e.g. [Sch]) rule r as in figure 5, which allows to move an anchor from one page to another. Let us also consider the graph package system depicted in figure 1, and the package $p = $ GETGRATS in it, thus completing step 1 above.

We can find a match for r in the graph restricted to GETGRATS, for example '1 → P2, '2 → P1, '3 → A4, thus completing step 2.

Applying this rule (with the current match) preserves package p (its qualifying node is preserved), and it does not change the contents of any package other than p. If another match were used, for example '2 → P4, the application of the rule would insert the matched anchor into package APPLIGRAPH, thus violating the locality. After verifying the locality condition we have completed step 3.

The shown match is also a match for the whole graph underlying the graph package system in figure 1. We can then apply r, moving the matched anchor from P2 to P1. This completes steps 4 and 5.

We apply the function f (in the example the f is defined through the qualification mechanism) to build a new graph package system on top of the new graph. This completes step 6, and the result is shown in figure 5. Notice that the anchor has moved to a new sub-package of p, since its qualifying node has changed (from P2 to P1). Notice also that the visibility information is not what we would obtained directly from construction 1. As discussed in section 4, our web example would need a more refined method for defining visibility, which is out of the scope of this paper to discuss.

6   Conclusions

In this paper, we have presented a graph structuring concept called graph package, that provides hierarchical structuring of generic graphs and encapsulation of graph elements. This concept is inspired by existing hierarchical graph data models in the literature, as well as by modularization concepts from the software engineering and programming language areas. Thanks to the principle of decoupling the hierarchy and visibility information from the graph itself, graph packages can be applied to different kinds of graphs, and are therefore a step towards the definition of a general-purpose hierarchical graph data model.

We have also investigated the idea of defining hierarchical graphs in a descriptive way, i.e. by deriving the package structure from the information contained in the underlying graph. Combining the descriptive approach and the idea of decoupling, we can reuse existing graph transformation approaches for specifying HG transformation. These ideas have been formalized in a general framework, which allows to lift a graph rewrite step in some existing graph transformation approach, to a rewrite step on a hierarchical graph.

Although simple and attractive from a theoretical point of view, we need more experience with this framework, in order to evaluate its applicability. In this respect, we would like to instantiate it to one (or more) existing graph transformation approach(es). We would then apply HG as a data model for a concrete application, and specify graph and HG transformation using the chosen graph transformation approach and the framework. A good candidate application is the specification of a web authoring tool, to be used both for re-engineering (restructure existing hypertexts), and for forward engineering (support development of new hypertexts) of web documents.

Another open issue is that of visibility/encapsulation. Although useful for certain applications, it is still an open question whether a generic HG data model should include encapsulation. In fact there are several cases where encapsulation is not used (see e.g. [Pra79], [BRS92]), mainly because it is not needed, and/or because it does not fit very smoothly with the use of dag-like hierarchies . For these reasons, we would like to distinguish between a proper HG data model, *without* encapsulation, and an *encapsulated* HG (EHG) data model, supporting encapsulation. Thus we can say that the graph package systems presented in this paper are more in the direction of EHG. Future research should be aimed at developing a theory of HG and HG transformation without visibility, and at adding the visibility concept as an orthogonal concept, making EHG an extension of pure HG. Pure hierarchical graphs, combined with the decoupling idea, would also offer the interesting advantage of allowing multiple hierarchies built on top of the same graph.

As a final topic, we would like to spend a few words about the constructive approach. In the constructive case we keep the decoupling idea (i.e. we use graph packages), but we drop the descriptive idea (we do not require that a GPS is induced by the underlying graph). Then we have to deal with the problem of defining transformation rules on GPS's, that act on the package structure, besides affecting the graph structure. We would like to transfer to GPS's the graph grammar idea of substituting substructures, which requires a concept of substructure in a package hierarchy, a sensible embedding mechanism, and to find a way to coordinate the package transformation with the graph transformation. The constructive approach and the related issues mentioned above are the subject of ongoing research.

References

BBH94. W. Bachl, F.-J. Brandenburg, and T. Hickl. Hierarchical graph design using higraD. Technical report, Passau, 1994. 354

BRS92. R. A. Botafogo, E. Rivlin, and B. Shneiderman. Structural analysis of hypertext: identifying hierarchies and useful metrics. *ACM Transactions on Information Systems*, 10:142–180, Apr 1992. 353, 355, 366

ES95. Gregor Engels and Andy Schürr. Encapsulated hierachical graphs, graph types, and meta types. In *SEGRAGRA'95, Joint COMPU-GRAPH/SEMAGRAPH Workshop on Graph Rewriting and Computation*, volume 2 of *Electronic Notes in Theoretical Computer Science*. Elsevier, 1995. 354

Him94. Michael Himsolt. Hierarchical graphs for graph grammars. In *Pre-proceedings of the Fifth International Workshop on Graph Grammars and Their Application to Computer Science*, 1994. 353

Hyp. Hyperwave www homepage. URL: http://www.hyperwave.de. 353

LB93. Michael Löwe and Martin Beyer. AGG — an implementation of algebraic graph rewriting. In *Rewriting Techniques and Applications*, volume 690 of *Lecture Notes in Computer Science*, pages 451–456. Springer-Verlag, 1993. 353, 355

Pra79. Terrence W. Pratt. Definition of programming language semantics using grammars for hierarchical graphs. In V. Claus, H. Ehrig, and G. Rozenberg, editors, *Graph-Grammars and Their Application to Computer Science* and Biology, volume 73 of *Lecture Notes in Computer Science*, pages 389–400, 1979. 353, 355, 366

Roz97.   Grzegorz Rozenberg, editor. *Handbook of Graph Grammars and Computing by Graph Transformations. Vol. I: Foundations.* World Scientific, 1997. 355, 363

Sch.   A. Schürr. *A Guided Tour through the PROGRES Environnement.* RWTH Aachen, D-52056 Aachen, Germany. 364

UML.   Uml documentation. Published on the World-Wide Web at the URL: `http://www.rational.com/uml/index.jtmpl`. 353

W3.   Homepage of the world wide web consortium. URL: `http://www.w3.org`. 352

Wir85.   Niklaus Wirth. *Programming in Modula2.* Springer, 1985. 353

Refinements of Graph Transformation Systems via Rule Expressions[*]

Martin Große–Rhode[1], Francesco Parisi Presicce[2], and Marta Simeoni[2]

[1] TU Berlin, Germany
mgr@cs.tu-berlin.de
[2] Università di Roma *La Sapienza*, Italy
{parisi,simeoni}@dsi.uniroma1.it

Abstract. Graph transformation systems are formal models of computational systems, specified by rules that describe the atomic steps of the system. A refinement of a graph transformation system is given by associating with each of its rules a composition of rules of a refining system, that has the same visible effect as the original rule. The basic composition operations on graph transformation rules are sequential and parallel composition, corresponding to temporal and spatial refinements respectively. Syntactically refinements are represented by rule expressions that describe how the refining rules shall be composed.

1 Introduction

Refinements are the basic steps in the development of system specifications. Starting from an abstract description of the system's behaviour stepwise refinements yield more and more concrete specifications, that should finally be directly implementable on a machine. In this paper we consider refinements of graph transformation systems, that are formal models of systems whose states are graphs and whose transitions are given by graph transformations. A graph transformation system is specified by a set of graph transformation rules that define its basic steps. A refinement of a more abstract specification by a more concrete one can then be given by associating with each rule of the more abstract specification a composition of rules of the more concrete specification, such that the composed rule coincides with the translation of the abstract rule to the finer type system. Hiding of the internal implementation parts is thereby obtained by restriction of the refinement to the types of the abstract specification, that are supposed to form a subsystem of the types of the more concrete specification. It can be shown then, that all derivations of the abstract system yield implementations in the concrete system, whose visible parts coincide with the given abstract ones, i.e., refinement preserves behaviour. The basic composition operations for

---

[*] This work has been supported by the EEC TMR network GETGRATS (General Theory of Graph Transformation Systems) ERB-FMRX-CT960061 and the ESPRIT Basic Working Group APPLIGRAPH (Applications of Graph Transformation Systems) 22565

H. Ehrig et al. (Eds.): Graph Transformation, LNCS 1764, pp. 368–382, 2000.

graph transformation rules are sequential composition by concatenation and parallel composition by amalgamation, i.e., refinement in space and time, that can be combined arbitrarily. For the concrete definition of a refinement we introduce a language of expressions that can be evaluated via the above mentioned rule operations. Refinement of graph transformation systems is transitive, and substitution of rule expressions as syntactical composition represents the semantical composition of refinements by iterated rule composition.

The paper is organized as follows. In the next section we review the basic notions of typed graph transformation systems that we need for the presentation of refinements here. Then we introduce the syntax, i.e., the rule expressions, and the evaluation of expressions by concrete semantical operations on rules. In Section 4 we introduce refinements and their most important properties, such as transitivity and preservation of behaviour. In Section 5 we briefly discuss a module concept for graph transformation systems, based on refinements as relation between export interface and implementation body. A conclusion, with a discussion on related works and further developments is given in Section 6.

2   Typed Graph Transformation Systems

In this section we briefly review the standard definitions and facts of graph transformation systems. For a detailed introduction and survey for the untyped case see [3]. Types are presented e.g. in [2,8].

A graph $G = (G_E, G_N, G_s, G_t)$ is given by sets G_E and G_N of edges and nodes respectively, and functions $G_s, G_t : G_E \rightarrow G_N$ that assign source and target nodes to the edges. A graph morphism $k = (k_E, k_N) : G \rightarrow H$ is a pair of functions $k_E : G_E \rightarrow H_E$ and $k_N : G_N \rightarrow H_N$ on edges and nodes respectively that are compatible with the source and target functions, i.e., $H_s \circ k_E = k_N \circ G_s$ and $H_t \circ k_E = k_N \circ G_t$. This defines the category **Graph**.

Let TG be a fixed graph, called type graph. A TG–typed graph $(t_G : G \rightarrow TG)$ is given by a graph G and a graph morphism $t_G : G \rightarrow TG$ that expresses the typing of the nodes and edges of G in TG. A morphism $k : (t_G : G \rightarrow TG) \rightarrow (t_H : H \rightarrow TG)$ of TG–typed graphs is a graph morphism $k : G \rightarrow H$ that is compatible with the typing morphisms, i.e., $t_H \circ k = t_G$. This defines the category **Graph**$_{TG}$.

Each type graph morphism $f : TG \rightarrow TG'$ induces a pair of adjoint functors between the corresponding categories of typed graphs. The free functor (forward retyping) $f^> : \mathbf{Graph}_{TG} \rightarrow \mathbf{Graph}_{TG'}$ is given by the composition of a typed graph $t_G : G \rightarrow TG$ with f and the identity on morphisms. The forgetful functor (backward retyping) $f^< : \mathbf{Graph}_{TG'} \rightarrow \mathbf{Graph}_{TG}$ is given by pullbacks, as depicted in Figure 1, where $f^<(G')$ is a pullback of f and $t_{G'}$, $f^<(H')$ is a pullback of f and $t_{H'}$, and $f^<(k')$ is the morphisms induced by $f^<(G') \rightarrow TG \xrightarrow{f} TG'$ and $f^<(G') \rightarrow G' \xrightarrow{k'} H' \xrightarrow{t_{H'}} TG'$.

A TG–typed graph transformation rule $L \xleftarrow{l} K \xrightarrow{r} R$ is given by TG–typed graphs L, K, and R, the left hand side, gluing graph, and right hand side respectively, and TG–typed graph morphisms $l : K \rightarrow L$ and $r : K \rightarrow R$.

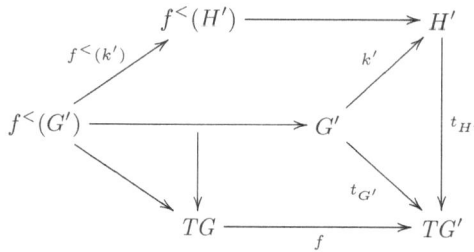

Fig. 1. Backward Retyping by Pullbacks

Throughout this paper we assume that l is an inclusion, i.e., $K \subseteq L$, and r is injective. Let $Rule_{TG}$ denote the set of all such rules. A *typed graph transformation system specification (tgts–specification)* $\mathbf{G} = (TG, N, \pi)$ is given by a type graph TG, a set N of names, and a mapping $\pi : N \to Rule_{TG}$ that associates a TG–typed rule to each name.

A *morphism of tgts–specifications* $f = (f_{TG}, f_N) : \mathbf{G} \to \mathbf{G}'$ is given by a type graph morphism $f_{TG} : TG \to TG'$ and a mapping $f_N : N \to N'$ between the sets of rule names, such that $f_{TG}^{>}(\pi(n)) = \pi'(f_N(n))$ for each $n \in N$. This defines the category **TGTS**.

Given a tgts–specification $\mathbf{G} = (TG, N, \pi)$ a *direct derivation* $n/m : G \Rightarrow H$ w.r.t. $\mathbf{G}$ from a TG–typed graph G via a rule with name n and a matching morphism $m : L \to G$ is a pair (n, S), where $n \in N$, S is a double pushout diagram

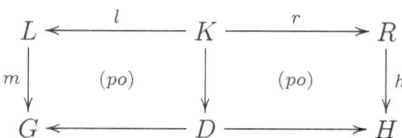

in $\mathbf{Graph}_{TG}$, and $\pi(n) = (L \xleftarrow{l} K \xrightarrow{r} R)$. G is called the *input*, H the *output*, and $h : R \to H$ the comatch of $n/m : G \Rightarrow H$. The sequential semantics $Seq(\mathbf{G})$ of $\mathbf{G}$ is the labeled transition system given by all TG–typed graphs as states and all direct derivations as labeled transitions. Later on we will also consider a more general semantics with amalgamated (parallel) derivations.

3   An Algebra of Rules

The composition operations for graph transformation rules we consider here are sequential and parallel composition for refinement in time and space respectively (cf. [3,6]). Below we give concrete set theoretic definitions of these operations, based on union and renaming of rules (see also [5]). Since union identifies items of different rules that happen to have the same names we introduce renaming as an auxiliary operation to control the identifications, as well as expressions for identity rules on arbitrary typed graphs. As a formal prerequisite for the renaming operation we assume a global set $\mathcal{N}$ of names to be given, such that all nodes and edges that occur in some rule are taken from $\mathcal{N}$. Substitution of expressions is defined as usual.

Definition 1 (Rule Expressions). *Given a type graph TG, a set N of rule names, and a countably infinite set $\mathcal{N}$ of names, the set $E = Exp(TG, N, \mathcal{N})$ of rule expressions w.r.t. TG, N, and $\mathcal{N}$ is given by the grammar*

$$E ::= N \mid E; E \mid E \cup E \mid E[\nu] \mid id_G$$

where a renaming ν of $\mathcal{N}$ is given by $\nu = x_1/y_1, \ldots, x_n/y_n$ with $x_i, y_i \in \mathcal{N}$ and $x_i \neq x_j$ for all $i \neq j$, and the index G in an identity expression id_G is a (syntactic expression for a) TG–typed graph.

Given another set N' of rule names a substitution σ is a function $\sigma : N \to Exp(TG, N', \mathcal{N})$. It extends to a function $\sigma^* : Exp(TG, N, \mathcal{N}) \to Exp(TG, N', \mathcal{N})$ by

$$
\begin{aligned}
\sigma^*(n) &= \sigma(n) & \sigma^*(e[\nu]) &= (\sigma^*(e))[\nu] \\
\sigma^*(e; e') &= \sigma^*(e); \sigma^*(e') & \sigma^*(id_G) &= id_G \\
\sigma^*(e \cup e') &= \sigma^*(e) \cup \sigma^*(e')
\end{aligned}
$$

Corresponding to these syntactic operations we introduce now the semantic operations on rules.

Sequential Composition Given TG–typed rules $p = L \xleftarrow{l} K \xrightarrow{r} R$ and $p' = L' \xleftarrow{l'} K' \xrightarrow{r'} R'$ their sequential composition $p; p' = L'' \xleftarrow{l''} K'' \xrightarrow{r''} R''$ is defined if $R = L'$.

In this case $L'' = L$, $R'' = R'$, and the gluing graph K'' is given by the preimage $K'' = r^{-1}(K') \subseteq K$ of $K' \subseteq L' = R$ under $r : K \to L' = R$. (See Figure 2.) The morphisms are given by the compositions $l'' = l \circ l'|_{K''} : K'' \to L$ and $r'' = r' \circ r|_{K''} : K'' \to R'$.

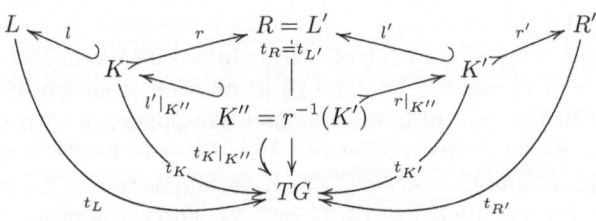

Fig. 2. Sequential Composition of Typed Rules

Then l'' is an inclusion and r'' is injective, because r and r' are injective. The typings $t_{K''} = t_K|_{K''} : K'' \rightarrow TG$, $t_{L''} = t_L : L \rightarrow TG$, and $t_{R''} = t_{R'} : R' \rightarrow TG$ are compatible with the inclusion l'' (obviously) and r'', because for all $x \in K''_E \cup K''_N$

$$t_{K'}(r|_{K''}(x)) = t_{L'}(l'(r|_{K''}(x))) = t_{L'}(r(l'|_{K''}(x))) = t_R(r(l'|_{K''}(x))) = t_K(l'|_{K''}(x)) = t_K|_{K''}(x) ,$$

i.e., $t_{K'} \circ r|_{K''} = t_K|_{K''}$, and $t_{R'} \circ r' = t_{K'}$. That means, $p; p'$ is a TG–typed rule.

As shown in [3] sequential composition of rules is in one-to-one correspondence with the sequential composition of derivations, i.e., given rules p and p' with names n and n' respectively and $R = L'$ there are direct derivations

$$n/m : G \Rightarrow H \text{ and } n'/h : H \Rightarrow K ,$$

where $h : R \rightarrow H$ is the comatch of n/m, if and only if there is a direct derivation

$$n; n'/m : G \Rightarrow K .$$

Union The union of two functions $f : M \rightarrow N$ and $f' : M' \rightarrow N'$ is a relation $f \cup f' \subseteq (M \cup M') \times (N \cup N')$. We say f *and* f' *are* $\cup$–*compatible* if $f \cup f'$ is a function $M \cup M' \rightarrow N \cup N'$, i.e.

$$\forall x \in M \cap M' \quad f(x) = f'(x) .$$

The union of two TG–typed rules $p = L \xleftarrow{l} K \xrightarrow{r} R$ and $p' = L' \xleftarrow{l'} K' \xrightarrow{r'} R'$ is defined if all functions concerned are $\cup$–compatible. That means, for $X \in \{L, K, R\}$, the source and target functions X_s, X'_s and X_t, X'_t, the node and edge components of the typings $t_X, t_{X'}$, and the morphisms l, l' and r, r' must be $\cup$–compatible. Moreover, the union $r \cup r'$ is required to be injective. Then the union $p \cup p'$ is defined by

$$p \cup p' = (L \cup L' \xleftarrow{l \cup l'} K \cup K' \xrightarrow{r \cup r'} R \cup R') .$$

If the requirements stated above are satisfied the union $p \cup p'$ is a TG–typed rule again.

Operationally the union of two (composed) rules can be understood as the execution of two threads of control, whose initial and final state are compatible. For a pair of simple rules p_1 and p_2 as above it is shown in [3] that direct derivations with $p_1 \cup p_2$ are in one-to-one correspondence with unions of compatible local direct derivations. Let $m : L_1 \cup L_2 \rightarrow G$ be a match of $p_1 \cup p_2$ to an input graph G and $G = G_1 \cup G_2$ be a decomposition of G, which need not be disjoint, such that $m(L_i) \subseteq G_i$ $(i = 1, 2)$. Furthermore let $L_0 := L_1 \cap L_2$ and m_i $(i = 0, 1, 2)$ be the restriction $m_i = m|_{L_i} : L_i \rightarrow G$. Suppose m satisfies the compatibility conditions

   – $m_i : L_i \rightarrow m(L_1 \cup L_2) \cap G_i$ is a match of p_i , i.e., it satisfies the gluing condition locally,

   – m_1 deletes an item $x \in G_1 \cap G_2$ if and only if m_2 deletes x.

Under these conditions there is a direct derivation $p_1 \cup p_2/m : G \Rightarrow H$ if and only there are direct derivations $p_i/m_i : G_i \Rightarrow H_i$ $(i = 1, 2)$ and $H = H_1 \cup H_2$. (More precisely, H is the pushout of H_1 and H_2 w.r.t. the output H_0 of p_0/m_0 : $G_0 \Rightarrow H_0$.) That means the input graph G is decomposed into two parts G_1 and G_2 with intersection G_0, each rule receives its part including a copy of G_0, and after the rule application the results are glued together again. In the extreme case, both threads receive a copy of all of G, i.e., $G_1 = G_2 = G$, as in the usual implementation of threads. Analogously, if $p_1 = p'_1; \dots ; p'_k$ and $p_2 = p''_1; \dots ; p''_m$ are sequential compositions each thread receives its part G_i of the input graph G, including a copy of G_0, and works on it with $p_i/m_i : G_i \Rightarrow H_i$. The overall compatibility requirement for the well definedness of the union $p_1 \cup p_2$ ensures that the outputs H_1 and H_2 of these local transformations can be glued together again and yield a final result $H = H_1 \cup H_2$. The intermediate states, however, might be inconsistent.

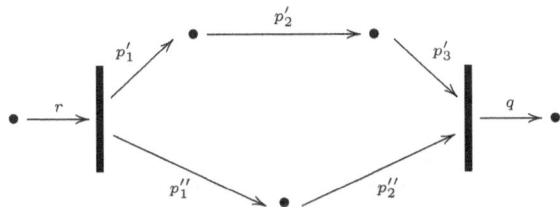

Fig. 3. The composition $r; (p'_1; p'_2; p'_3) \cup (p''_1; p''_2); q$

Renaming Each renaming $\nu = x_1/y_1, \dots , x_n/y_n$ of $\mathcal{N}$ induces an endo–function $\nu : \mathcal{N} \rightarrow \mathcal{N}$ by $\nu(x_i) = y_i$ and $\nu(x) = x$ for all $x \in \mathcal{N} - \{x_1, \dots , x_n\}$. A function $f : N \rightarrow N'$ with $N, N' \subseteq \mathcal{N}$ is called *compatible with ν in $\mathcal{N}$*, if

$$\forall x, x' \in N \quad \nu(x) = \nu(x') \Rightarrow \nu(f(x)) = \nu(f(x')) .$$

In this case the renamed function $f[\nu] : \nu(N) \rightarrow \nu(N')$ is defined by $f[\nu](\nu(x)) = \nu(f(x))$. A function $f : N \rightarrow Z$ with $N \subseteq \mathcal{N}$ is *compatible with ν*, if

$$\forall x, x' \in N \quad \nu(x) = \nu(x') \Rightarrow f(x) = f(x') ,$$

and the renamed function $f[\nu] : \nu(N) \rightarrow Z$ is defined by $f[\nu](\nu(x)) = f(x)$ in this case.

    Given now a TG–typed graph $(t_G : G \rightarrow TG)$ its renaming $(t_G : G \rightarrow TG)[\nu] = (t_G[\nu] : G[\nu] \rightarrow TG)$ is defined if the source and target functions G_s

and G_t are compatible with ν in $\mathcal{N}$, and the typing t_G is compatible with ν. In this case

$$G[\nu] = (\nu(G_E), \nu(G_N), G_s[\nu], G_t[\nu])$$

and $t_G[\nu]$ is the renaming of t_G by ν.

Finally the renaming $p[\nu]$ of a TG–typed rule $p = L \xleftarrow{l} K \xrightarrow{r} R$ by ν is defined if all renamings $L[\nu]$, $K[\nu]$, and $R[\nu]$ are defined, and l and r are compatible with ν in $\mathcal{N}$. Then

$$p[\nu] = L[\nu] \xleftarrow{l[\nu]} K[\nu] \xrightarrow{r[\nu]} R[\nu] \ .$$

A bijective renaming does not change the effect of a rule, because isomorphic rules yield isomorphic sets of derivations. Bijective renamings will only be used inside parallel and sequential composition to identify or distinguish items of different rules. A non–injective renaming however reduces the possible applications or effects of a rule. If, for instance, an edge $e : n_1 \to n_2$ is renamed by $[n_1/n, n_2/n]$ it becomes a loop. Thus if it occurs in a left hand side of a rule the renamed rule can only be applied at loops, if it occurs in a right hand side the renamed rule can only create loops.

This restriction of applications/effects is also the aim of *amalgamation* of rules, that is used for synchronization (see [1]). Categorically an amalgamation is a pushout w.r.t. a common subrule in $Rule_{TG}$. Set theoretically a pushout can be obtained by union with appropriate renaming. With the latter, synchronization of parts of the rules is expressed by giving equal names to items that shall be shared and distinct names to items that shall be kept distinct.

Identity Rules For each TG–typed graph $t_G : G \to TG$ the *identity rule* id_G is defined by

$$id_G = G \xleftarrow{id_G} G \xrightarrow{id_G} G$$

with each component typed by t_G .

Sequential composition is defined for pairs of rules where the right hand side of the first rule coincides with the left hand side of the second one. This condition can now be obtained by an appropriate renaming and/or a union with identity rules (cf. Example 1).

We summarize these constructions in the following definition of rule operations, where also the evaluation of rule expressions is introduced.

Definition 2 (Rule Operations). *Let* $p = L \xleftarrow{l} K \xrightarrow{r} R$ *and* $p' = L' \xleftarrow{l'} K' \xrightarrow{r'} R'$ *be* TG–*typed rules, and* ν *a renaming of* $\mathcal{N}$*. Then*

$$p; p' \;=\; (L \xleftarrow{\;lol'|_{K''}\;} r^{-1}(K') \xrightarrow{\;r'or|_{K''}\;} R') \;, \qquad \text{if } R = L' \,,$$

$$p \cup p' = (L \cup L' \xleftarrow{\;l \cup l'\;} K \cup K' \xrightarrow{\;r \cup r'\;} R \cup R') \,,$$

$$\text{if the unions of all functions are well defined,}$$

$$p[\nu] \;=\; (L[\nu] \xleftarrow{\;l[\nu]\;} K[\nu] \xrightarrow{\;r[\nu]\;} R[\nu]) \;, \text{ if all renamings are well defined,}$$

$$id_G \;=\; (G \xleftarrow{\;id_G\;} G \xrightarrow{\;id_G\;} G) \,.$$

Given a function $\pi : N \to Rule_{TG}$ the corresponding partial evaluation function $\pi^* : Exp(TG, N, \mathcal{N}) \dashrightarrow Rule_{TG}$ is inductively defined by

$$\begin{array}{ll} \pi^*(n) & = \pi(n) \qquad\qquad\qquad \pi^*(e[\nu]) = \pi^*(e)[\nu] \\ \pi^*(e; e') & = \pi^*(e); \pi^*(e') \qquad\quad \pi^*(id_G) = id_G \\ \pi^*(e \cup e') & = \pi^*(e) \cup \pi^*(e') \end{array}$$

whenever the right hand side of the equation is well–defined.

4   Refinements

Having introduced the syntactic expressions for rule compositions the definition of a refinement of a tgts–specification $\mathbf{G}_1$ by a tgts–specification $\mathbf{G}_2$ is now straightforward. It is given by a graph morphism to relate the type graphs and a mapping that associates with each rule name of the first specification an evaluable expression over the names of the second one. For that purpose we first introduce the closure of a tgts–specification w.r.t. rule compositions.

Definition 3 (Compositional Semantics). *The* compositional closure $\mathbf{G}^* = (TG, N^*, \pi^*)$ *of a tgts–specification* $\mathbf{G} = (TG, N, \pi)$ *is given by the same type graph, the set* $N^* \subseteq Exp(TG, N, \mathcal{N})$ *of rule expressions that are evaluable w.r.t.* π*, and the extension* π^* *of* π *to* N^* *. The* compositional semantics $Com(\mathbf{G})$ *of* $\mathbf{G}$ *is then defined by* $Com(\mathbf{G}) = Seq(\mathbf{G}^*)$ *.*

The free functor associated with the type graph morphism yields the second correctness criterion for the refinement: The forward retyping of each original rule must coincide with the correponding composed rule of the refining system. Translation in this direction is necessary in order to ensure preservation of behaviour (see Theorem 2).

Definition 4 (Refinement). *Given tgts-specifications* $\mathbf{G}_i = (TG_i, N_i, \pi_i)$ $(i = 1, 2)$ *a* refinement $r : \mathbf{G}_1 \to \mathbf{G}_2$ *is a tgts-specification morphism* $r : \mathbf{G}_1 \to \mathbf{G}_2^*$ *. That means,* $r = (r_{TG}, r_N)$ *is given by a graph morphism* $r_{TG} : TG_1 \to TG_2$ *and a function* $r_N : N_1 \to Exp(TG_2, N_2, \mathcal{N})$ *, such that for each* $n \in N_1$ *the rule* $\pi_2^*(r_N(n))$ *is well defined and* $r_{TG}^>(\pi_1(n)) = \pi_2^*(r_N(n))$ *.*

Since a single rule name is an expression, too, refinements subsume tgts–specification morphisms as defined in Section 2 as special cases.

The refinement of tgts–specifications is transitive. Composition of refinements is given by the composition of the type graph morphisms and substitution of rule expressions. Moreover, tgts–specifications and refinements yield a category.

Theorem 1 (Transitivity). *Given refinements $r = (r_{TG}, r_N) : \mathbf{G}_1 \to \mathbf{G}_2$ and $r' = (r'_{TG}, r'_N) : \mathbf{G}_2 \to \mathbf{G}_3$ their composition $r' \circ r : \mathbf{G}_1 \to \mathbf{G}_3$, defined by*

$$r' \circ r = (r'_{TG} \circ r_{TG}, r'_N{}^* \circ r_N) : \mathbf{G}_1 \to \mathbf{G}_3 \ ,$$

is a refinement. Composition is associative and there are neutral (identity) refinements. That means, tgts–specifications and refinements define a category, denoted by **Ref***.*

Proof. (Sketch) Composition of graph morphisms and the name components is associative and has identities. The correctness of the composition can be shown by induction on $r_N(n)$.

Example 1 (Communicating Agents). Consider two agents P and Q. Agent P sends a message a to Q. The abstract behaviour of this system is specified by the one rule

$$send : \quad \boxed{a} \leftarrow \!-\!- \!\!\!\!\!\! \bigcirc\!P \qquad \bigcirc\!Q \qquad \Longrightarrow \qquad \bigcirc\!P \qquad \bigcirc\!Q -\!-\!\!\to \boxed{a}$$

over the type graph ⊟ $-\!-\!-$ ◯. Circles denote agents, shaded boxes messages, and a dashed arrow indicates that an agent holds a message. The typing of the graphs is indicated by their graphical representation. The gluing graph of the rule is given implicitly here as in all following rules by the intersection of the left and right hand side.

Now this specification is refined by one that also models the channel over which the message is transmitted. The type graph of the second specification is given by

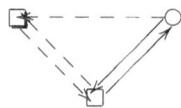

Solid boxes denote channels, agents may have writing or reading access to a channel, and a message may be at the input or the output port of a channel. The refining system has rules to connect and disconnect agents and channels, to submit a message to a channel and receive a message from a channel, and to transmit a rule through a channel

$$w\_connect : \quad \bigcirc\!A \qquad \Longrightarrow \qquad \bigcirc\!A -\!\!\to \boxed{c}$$

$$r\_connect : \quad \bigcirc\!A \qquad \Longrightarrow \qquad \bigcirc\!A \leftarrow\!- \boxed{c}$$

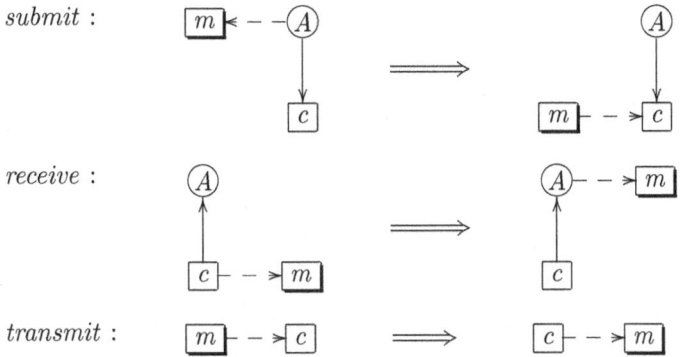

submit :

receive :

transmit :

The rules *w_disconnect* and *r_disconnect* are the inverses of the corresponding connection rules. For the refinement we need furthermore identity rules $id_{a \leftarrow P}$, $id_{P \rightarrow l}$, $id_{Q \leftarrow l}$, and $id_{Q \rightarrow a}$ on the graphs $\boxed{a} \leftarrow - - \textcircled{P}$, $\textcircled{P} \longrightarrow \boxed{l}$, $\textcircled{Q} \leftarrow \boxed{l}$, and $\textcircled{Q} - - \ast \boxed{a}$. The refinement is then given by the inclusion of the type graphs and the mapping

$$send \mapsto (id_{a \leftarrow P} \cup w\_connect[A/P, c/l] \cup r\_connect[A/Q, c/l]);$$
$$(submit[m/a, A/P, c/l] \cup id_{Q \leftarrow l});$$
$$(id_{P \rightarrow l} \cup transmit[m/a, c/l] \cup id_{Q \leftarrow l});$$
$$(id_{P \rightarrow l} \cup receive[m/a, A/Q, c/l]);$$
$$(w\_disconnect[A/P, c/l] \cup r\_disconnect[A/Q, c/l] \cup id_{Q \rightarrow a})$$

Note how the nodes of the refining rules are used as parameters here, that are instantiated via the renamings. The renaming $[c/l]$ is formally not necessary, but it documents that P and Q have to be connected to *the same* channel in order to communicate with each other.

It is easy to check that the rule composition is well defined, and that the forward retyping of the *send* rule coincides with the result of the composition. Thus the indicated mapping is in fact a refinement.

The most important property of refinements is of course preservation of behaviour. That means, each path in $Seq(\mathbf{G_1})$ should be mapped to a path in $Com(\mathbf{G_2})$ such that its restriction to the smaller type graph coincides with the given path. To support the transitivity of refinements we state this property immediately for composed derivations, that contain direct derivations and their sequential compositions as special cases.

Theorem 2 (Preservation of Behaviour). *Let $r = (r_{TG}, r_N) : \mathbf{G_1} \rightarrow \mathbf{G_2}$ be a refinement of tgts–specifications. For each composed derivation $e/m : G \Rightarrow H$ in $Com(\mathbf{G_1})$ there is a composed derivation $r_N^*(e)/r_{TG}^>(m) : r_{TG}^>(G) \Rightarrow r_{TG}^>(H)$ in $Com(\mathbf{G_2})$. Moreover, if r_{TG} is injective, then*

$$r_{TG}^< \left(r_N^*(e)/r_{TG}^>(m) : r_{TG}^>(G) \Rightarrow r_{TG}^>(H) \right) \cong (e/m : G \Rightarrow H) .$$

Proof. (Sketch) Free functors preserve pushouts, and if r_{TG} is injective then $r_{TG}^{<} \circ r_{TG}^{>} \cong Id_{\mathbf{Graph}_{TG_2}}$.

The refinement category **Ref** does not have all colimits. For instance, there is in general no way to obtain a smallest common refinement of two independent refinements $r_P(n)$ and $r'_P(n)$ of a rule name n. However, if one of the refinements is just an inclusion of names a pushout exists. This special case is already sufficient for the module concept for graph transformation systems that we discuss below.

Theorem 3 (Pushouts with Inclusions). *Given refinements* $r = (r_{TG}, r_N)$: $\mathbf{G}_1 \to \mathbf{G}_2$ *and* $i = (i_{TG}, i_N) : \mathbf{G}_1 \to \mathbf{G}_3$ *with* i_N *injective and* $i(n) \in N_3$ *for all* $n \in N_1$ *there is a pushout of* r *and* i *in* **Ref** .

Proof. (Sketch) To obtain a pushout first the pushout of the type graphs is constructed in **Graph**. Then in the name set N_3 the image $i_N(N_1)$ is replaced by N_2 with the corresponding parts of the rule mappings π_3 and π_2 adjusted to the new type graph.

5   Modules

In [7] we have introduced a module concept for typed graph transformation systems, based on the notions of spatial or temporal decomposition of rules, considered as separate kinds of refinements. In this section we formulate the same concept in terms of refinements via rule expressions introduced above. Since the module concept is rather independent of the kind of refinements the improved version of combined spatial and temporal refinements introduced here can be plugged in easily. The required properties are exactly the ones shown in Theorems 1 and 3.

A module for typed graph transformation systems (tgts–module) is composed of three tgts–specifications: two interfaces, import and export, and a body that implements the services offered by the export interface, possibly using the features required by the import interface. The export is related to the body by a refinement that defines the implementation, whereas the import interface is simply a subsystem of the body, meaning that the external features required by the import interface are simply included in the body.

Definition 5 (Module). *A tgts–module* $MOD = (IMP \xrightarrow{m} BOD \xleftarrow{r} EXP)$ *is given by tgts–specifications* IMP, BOD, *and* EXP, *an injective tgts–morphism* $m : IMP \to BOD$, *and a refinement* $r : EXP \to BOD$. *It can be visualized by:*

$$
\begin{array}{ccc}
& & EXP \\
& & \downarrow r \\
IMP & \xrightarrow{\ m\ } & BOD
\end{array}
$$

The syntax of a tgts–module is quite similar to the one of an algebraic specification module (see [4]). The only difference is the lack of the parameter part. However, the semantics of an algebraic specification module is a functor from the algebras of the import interface to the algebras of the export interface, $SEM : \mathcal{A}lg(IMP) \rightarrow \mathcal{A}lg(EXP)$, while the semantics of a tgts–module is simply given by the semantics of the export interface, which becomes the only visible part of the whole module from the outside environment (hence hiding of information is supported). The internal components, import and body, together with the export interface, are needed to define a structuring means which supports both the notions of implementation (via refinement morphisms) and information hiding.

In the communicating agents example for instance one could designate the *transmit* rule, concerning the transmission of the messages through the channel, as import interface, which is then further refined by another module. This defines a communicating agents module, whose semantics is the exchange of messages between two agents. The communication is internally implemented by sending the messages over a channel, whose internal behaviour is arbitrary, i.e., deferred to further refinement.

Given two tgts–modules $MOD = (IMP \rightarrow BOD \leftarrow EXP)$ and $MOD' = (IMP' \rightarrow BOD' \leftarrow EXP')$, a module morphism $mod : MOD \rightarrow MOD'$ is defined as a triple of tgts–morphisms $(mod_I : IMP \rightarrow IMP', mod_B : BOD \rightarrow BOD', mod_E : EXP \rightarrow EXP')$ such that following diagrams (in **TGTS** and **Ref**, respectively) commute.

$$
\begin{array}{ccc}
IMP & \xrightarrow{\; m \;} & BOD \\
{\scriptstyle mod_I}\downarrow & = & \downarrow{\scriptstyle mod_B} \\
IMP' & \xrightarrow[\; m' \;]{} & BOD'
\end{array}
\qquad\qquad
\begin{array}{ccc}
BOD & \xleftarrow{\; r \;} & EXP \\
{\scriptstyle mod_B}\downarrow & = & \downarrow{\scriptstyle mod_E} \\
BOD' & \xleftarrow[\; r' \;]{} & EXP'
\end{array}
$$

Following the line of proofs in [7], it can be shown that modules and module morphisms form a category **MOD**. This allow us to define, analogously to algebraic specification modules, the operations of *union* and *composition* of modules.

The *union* of two modules MOD_1 and MOD_2 w.r.t. a common submodule MOD_0 is a module MOD_3 where each component is obtained by taking first the disjoint union of the corresponding components of MOD_1 and MOD_2 and then identifying the parts in common contained in MOD_0. This operation is well–defined if the result of this construction is a module again, i.e., if the induced morphism from the resulting export to the body is a refinement. In categorical terms, MOD_3 is a pushout object (that is determined up to isomorphism) of MOD_0 and the module morphisms $mod_1 : MOD_0 \rightarrow MOD_1$ and $mod_2 : MOD_0 \rightarrow MOD_2$.

The idea for the *composition* operation of two modules $MOD = (IMP \xrightarrow{m} BOD \xleftarrow{r} EXP)$ and $MOD' = (IMP' \xrightarrow{m'} BOD' \xleftarrow{r'} EXP')$, is to relate them via a tgts–morphism h, called *interface morphism*, from the import interface of MOD to the export interface of MOD', i.e $h : IMP \rightarrow EXP'$, and to create a new module having the import interface of MOD', the export interface of MOD and a body implementing the features of both MOD and MOD'. Precisely:

$$MOD'' = (IMP'' \xrightarrow{m^* \circ m'} BOD'' \xleftarrow{r^* \circ r} EXP'')$$

where $IMP'' = IMP'$, $EXP'' = EXP$ and BOD'' is the pushout object (and m^* and r^* are the induced morphisms) of the following diagram in **Ref**:

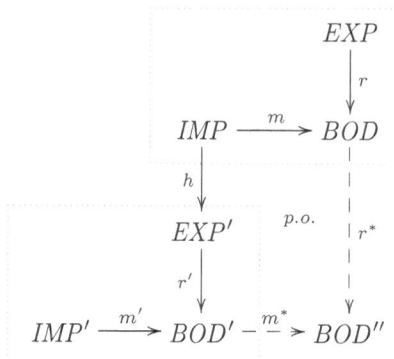

Even if the definition of a tgts–module and the operations of union and composition of tgts–modules have been inspired by the algebraic specification modules, there is a main difference between the two kinds of specifications: tgts–specifications are used for formally specifying dynamically evolving systems where graphs are states and graph transformations are state transitions. Hence, it is very important to preserve the dynamical behaviours of the systems whenever their corresponding specifications are related via tgts–morphisms or refinement morphisms. This is ensured by the Preservation of Behaviour properties stated by Theorem 2. For a tgts–module, this means that each derivation of the export (resp. import) interface can be translated along the refinement (resp. injective tgts–morphism) into a derivation of the body.

This dynamical aspect is not present in algebraic specifications since each of them specifies an algebra (initial semantics) or a class of algebras (loose semantics), that model only the static functional view of the components.

6   Conclusion

We have introduced syntax and semantics of refinements of typed graph transformation systems. On the semantical side we have used set theoretic operations to define sequential and parallel compositions of rules. With appropriate renamings also all amalgamations of rules can be obtained. These concrete operations are partial, because independent renaming of nodes and edges does not necessarily define a graph, and union of typed rules does not necessarily yield a typed rule. Correspondingly the evaluation of rule expressions is partial, and the definedness of the evaluation becomes one of the correctness criteria of refinements. The most important properties of refinement are its transitivity, resp. composability (see Theorem 1), and the preservation of behaviour (Theorem 2).

Transitivity of refinements and the Pushouts with Inclusions property (see Theorem 3) are moreover fundamental for defining a module concept for typed graph transformation systems: the module concept we have already introduced in [7] has been rephrased here in terms of refinements via rule expressions.

Other notions of refinement have been suggested before. The spatial and temporal refinements introduced in [6] can be considered as a preliminary version of the general notion defined here. In [8] the type graphs of the systems are related by a partial graph morphism instead of a total one, which induces a combination of backward and forward retyping of the rules. Each rule is refined by a sequential derivation of the refining system, which includes both a specialization (see below) and a sequential composition of rules; parallel composition is not considered. The retyped rule is allowed to be an instance of the refining rule, thus the refining rule may only be applied to special cases. (In this sense this definition of refinement could be called a specialization instead.) This leads to a much weaker comparison of behaviours: The visible part of the refining rule need not coincide with the original rule.

Further research is devoted to an extension of the format of rules and composition operations. For example, with conditional rules and a *case*–composition an abstract rule could be refined by a list of alternatives that implement the abstract rule in different ways, depending on some context information. Or rules could be refined by instances of the rules of another system, thus specializing their applications. A more difficult topic is to allow iteration in the refining rules, i.e., an implementation where the number of steps is not fixed in advance, but depends on the specific context.

References

1. P. Böhm, H.-R. Fonio, and A. Habel. Amalgamation of graph transformations: a synchronization mechanism. *Journal of Computer and System Science*, 34:377–408, 1987. 374
2. A. Corradini, U. Montanari, and F. Rossi. Graph processes. *Special issue of Fundamenta Informaticae*, 26(3,4):241–266, 1996. 369
3. A. Corradini, U. Montanari, F. Rossi, H. Ehrig, R. Heckel, and M. Löwe. Algebraic approaches to graph transformation part I: Basic concepts and double pushout approach. In G. Rozenberg, editor, *Handbook of Graph Grammars and Computing by Graph transformation, Volume 1: Foundations*, pages 163–246. World Scientific, 1997. 369, 370, 372
4. H. Ehrig and B. Mahr. *Fundamentals of Algebraic Specification 2: Module Specifications and Constraints*, volume 21 of *EATCS Monographs on Theoretical Computer Science*. Springer Verlag, Berlin, 1990. 379
5. G. Engels, H. Ehrig, R. Heckel, and G. Taentzer. A combined reference model- and view-based approach to system specification. *Int. Journal of Software and Knowledge Engeneering*, 7(4):457–477, 1997. 370

6. M. Große–Rhode, F. Parisi Presicce, and M. Simeoni. Spatial and temporal refinement of typed graph transformation systems. In L. Brim, J. Gruska, and J. Zlatuška, editors, *Proc. Mathematical Foundations of Computer Science 1998*, pages 553–561. Springer LNCS 1450, 1998. 370, 381

7. M. Große–Rhode, F. Parisi Presicce, and M. Simeoni. Refinements and modules for typed graph transformation systems. In J.L. Fiadeiro, editor, *Workshop on Algebraic Development Techniques (WADT'98), at ETAPS'98, Lisbon, April 1998*, pages 137–151. Springer LNCS 1589, 1999. 378, 379, 381

8. R. Heckel, A. Corradini, H. Ehrig, and M. Löwe. Horizontal and vertical structuring of typed graph transformation systems. *Math. Struc. in Comp. Science*, 6(6):613–648, 1996. 369, 381

Simple Modules for GRACE*

Reiko Heckel[1], Berthold Hoffmann[2], Peter Knirsch[2], and Sabine Kuske[2]

[1] Fachbereich 17 Informatik, Universität-GH Paderborn
Warburger Str. 100, D-33098 Paderborn, Germany
reiko@uni-paderborn.de
[2] Fachbereich 3 Mathematik – Informatik, Universität Bremen
Postfach 33 04 40, D-28334 Bremen, Germany
{hof,knirsch,kuske}@informatik.uni-bremen.de

Abstract. The language GRACE is being proposed for specifying and programming in a graph-centered environment based on graph transformation. Emphasis in the design of GRACE is laid on modular structuring of programs. In this paper we present a simple kind of modules for the language which encapsulate rules and graph transformation units, and export some of them. The local transformation units define compound graph transformations that call rules and other transformation units which are either defined by the surrounding module, or imported from other modules. The interleaving semantics of modules specifies a binary relation on graphs for each exported item.

1 Introduction

GRACE is a graph transformation language for specifying and programming in a graph-centered environment that is being developed by researchers at TU Berlin, Universität Bremen, Universität Erlangen, Universität der Bundeswehr München, Universität Oldenburg, and Universität Paderborn. In contrast to other graph transformation languages like PROGRES [Sch94] and AGG [TB94], GRACE is *approach-independent* in that it may host different notions of graph transformation – in a single program eventually.

GRACE shall extend the abstract notion of graph transformation by concepts for programming and specification that support the development of large-scale systems. As a first step, *transformation units* were introduced in [KK96] to specify control, functional composition, and encapsulation (see also [KKS98,AEH+99]). In this paper, we relieve transformation units from their interfaces and local rules, and introduce *modules* with export and import interfaces that encapsulate transformation units and rules. The resulting structure is similar to that of procedures and packages in Ada [Ada95], or of functions and modules in Haskell [PHA+97].

---

* This work was partially supported by the ESPRIT Working Group Applications of Graph Transformation (APPLIGRAPH) and the EC TMR Network GETGRATS (General Theory of Graph Transformation Systems).

H. Ehrig et al. (Eds.): Graph Transformation, LNCS 1764, pp. 383–395, 2000.
© Springer-Verlag Berlin Heidelberg 2000

This kind of modules will not suffice alone: we also need a concept similar to classes in object-oriented languages, for *data-driven encapsulation*. However, a simple module concept will also be useful on top of such a rather "fine-grained" class concept, see Java [AG95].

The paper is organized as follows: the approach-independence of GRACE is explained in the following section. Afterwards, we introduce graph transformation units which, together with graph transformation rules, constitute the basic components of modules. In section 4 graph transformation modules are presented together with their operational semantics. Section 5 shows how structured graph transformation systems can be built up from a set of modules resulting in a module system. The usefulness of this concept is illustrated with a small example from the area of graph algorithms. We conclude by presenting related work and outlining how graph transformation modules can be extended with respect to various aspects.

Throughout this paper, *ID* shall denote a vocabulary of *identifiers* that is used to name entities like rules, transformations, and modules. If $N \subseteq ID$ is a subset of names, a *named set* (over N) is a mapping $NS : N \rightarrow S$ into some given set S. We write $n : s \in NS$ to refer to some $n \in N$ with $NS(n) = s$.

2   Approach-Independence

There are many ways to define graph transformation (see [Roz97] for a survey). All of them allow some class of graphs to be transformed by applying some kind of rules. In most cases, these rules can be applied in arbitrary order and frequency; some notions of *programmed* graph transformation also allow to specify conditions on graphs, and to control the application of rules.

Since GRACE is *approach-independent*, we just specify the requirements for graph transformation approaches and make the language generic with respect to this specification.

Definition 1. ([KK96]) A *graph transformation approach* is a system

$$\mathcal{A} = \langle \mathcal{G}, \mathcal{R}, \Rightarrow, \mathcal{E}, \mathcal{C} \rangle$$

with a class $\mathcal{G}$ of *graphs*, a class $\mathcal{R}$ of *graph transformation rules*, a *rule application operator* $\Rightarrow$, a class $\mathcal{E}$ of *graph class expressions*, and a class $\mathcal{C}$ of *control conditions*, where

- $\Rightarrow$ defines a *graph transformation relation* $\Rightarrow_r \subseteq \mathcal{G} \times \mathcal{G}$ for every rule $r \in \mathcal{R}$,
- each $X \in \mathcal{E}$ specifies a *graph language* $\mathcal{L}(X) \subseteq \mathcal{G}$, and
- each $C \in \mathcal{C}$ specifies a *graph transformation relation* $SEM_E(C) \subseteq \mathcal{G} \times \mathcal{G}$, for every *graph transformation environment* $E : ID \rightarrow 2^{\mathcal{G} \times \mathcal{G}}$, a mapping that associates transformation relations with names which may occur in C. [1]

For each control condition $C \in \mathcal{C}$, $names(C) \subseteq ID$ denotes the set of names occurring in C.

---

[1] For a set A, 2^A denotes its power set.

Example 1. As an example we define a simple graph transformation approach which will be used throughout this paper.

- The class of graphs consists of all undirected graphs without multiple edges labeled over an alphabet Σ with $* \in \Sigma$. The symbol $*$ is a special one standing for *unlabeled*.

  Formally, each graph $G \in \mathcal{G}$ is a system (V, E, l, m) where V is a finite set of *nodes*, $E \subseteq V \times V$ is a finite set of *edges*, $l \colon V \to \Sigma$ and $m \colon E \to \Sigma$ are mappings associating a *label* to each node and each edge in G.

  A graph $G' = (V', E', l', m') \in \mathcal{G}$ is a *subgraph* of G if $V' \subseteq V$, $E' \subseteq E$, $l' = l|V'$, and $m' = m|E'$.[2]

  An *occurrence* g of G' in G consists of two mappings $g_V \colon V' \to V$ and $g_E \colon E' \to E$ such that the label of each node and edge in G' is preserved, i.e. $g_V \circ l = l'$ and $g_E \circ m = m'$.

- Each rule is of the form $r = (\mathcal{N}, L \to R)$ where L and R are graphs with equal sets of nodes and $\mathcal{N}$ is a set of graphs such that for each $M \in \mathcal{N}$, L is a proper subgraph of M. It is known as rule with negative context condition as described in [HHT96]. An example of such a rule is depicted below in pseudo code notation.

  The rule *mark* consists of a left- and a right-hand side separated by an arrow. The left-hand side shows an abbreviated notation of a negative context condition meaning that the rule *mark* can be applied if there is neither an unlabeled nor a m-labeled edge between the nodes. This directly corresponds to the above introduced set $\mathcal{N} = \{$ ●———■ , ●——$\overset{m}{}$——■ $\}$. If the negative context condition is satisfied an m-labeled edge is inserted.

- A rule $r = (\mathcal{N}, L \to R)$ is applied to a graph G resulting in graph G' according to the following steps : (1) CHOOSE an occurrence g of L in G. (2) CHECK the negative context condition: for each $M \in \mathcal{N}$ there is no occurrence h of M in G such that h restricted to L is equal to g. (3) REMOVE the image of each edge in L, i.e. for each $e \in E_L$ remove $g_E(e)$ (4) ADD R to the resulting graph by gluing each node v in R with its image $g_{V_L}(v)$ in G. Hence, the rule *mark* inserts an m-labeled edge between two existing nodes if they are neither connected by an unlabeled nor by an m-labeled edge. figure 1 shows an application of the rule *mark*.

- As graph class expressions we use the default expression *all* specifying the set of all graphs, as well as F and $\neg F$ for all $F \subseteq \Sigma$. The expression F specifies all graphs labeled over F whereas $\neg F$ admits only graphs without labels from F.

- Typical examples of control conditions are regular expressions over *ID* and as-long-as-possible [Kus98]. Here, we combine both in the following way:

---

[2] For a function $f \colon A \to B$ and a set $C \subseteq A$ $f|C$ denotes the restriction of f to C.

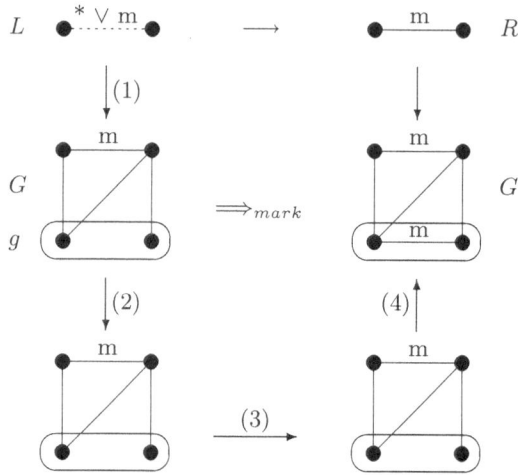

Fig. 1. An application of the rule *mark*

for each $n \in ID$, $n!$ is a control condition which applies the rule or transformation unit named n as long as possible, i.e. for each environment E,
$SEM_E(n!) = \{(G, G') \in E(n)^* \mid \neg \exists G'' \in \mathcal{G} : (G', G'') \in E(n)\}$. Let $\mathcal{C}' = \{n! \mid n \in ID\}$. Then the class of control conditions of our example approach
consists of all regular expressions over $\mathcal{C}' \cup ID$, which are recursively defined
as follows: $\mathcal{C}' \cup ID \in REG$, and $(C_1 ; C_2) \in REG$ if $C_1, C_2 \in REG$. Semantically, $n \in ID$ applies the rule or transformation unit n exactly once,
i.e. $SEM_E(n) = E(n)$ for each environment E; $(C_1 ; C_2)$ applies first C_1 and
then C_2, i.e. $SEM_E((C_1 ; C_2)) = SEM_E(C_1) \circ SEM_E(C_2)$. For example, the
condition $(mark! ; erase)$ where *erase* is a rule applies first *mark* as long as
possible and then *erase* once.

Eventually, it shall be possible to define *heterogeneous* graph transformation
systems, composed from modules defined in different approaches. For now, we
stick to homogeneous systems specified in a single approach.

3   Graph Transformation Units

In a simple graph transformation system, rules define basic graph transformations that may be applied in arbitrary order and frequency. We introduce graph
transformation units for defining compound graph transformations that call
rules and other transformation units, by control conditions of the underlying
approach. This provides for *functional composition*. Also, they allow pre- and
post-conditions to be specified, by graph class expressions of the underlying approach. This allows to *type* units with respect to the graphs they transform. The
rules and transformation units used in a graph transformation unit are defined
or imported by the surrounding module, and are used via the transformation
environment in which the unit is interpreted.

Definition 2. A *graph transformation unit* over $\mathcal{A} = (\mathcal{G}, \mathcal{R}, \Rightarrow, \mathcal{E}, \mathcal{C})$ is a tuple $trafo = \langle I, C, T \rangle$ where

- $I \in \mathcal{E}$ is an *initial graph class expression*,
- $C \in \mathcal{C}$ is a *control condition*, and
- $T \in \mathcal{E}$ is a *terminal graph class expression*.

Let $\mathcal{T}_\mathcal{A}$ denote the set of graph transformation units over $\mathcal{A}$.

The semantics of each transformation unit is a binary relation on graphs. It consists of each pair (G, G') of graphs where G is initial, G' is terminal, (G, G') is specified by the control condition, and G' can be obtained from transforming G with the rules and transformation units to which the names in the control condition refer. Hence, the semantics of a transformation unit is computed with respect to an environment which associates a semantics to each name occurring in the control condition.

Definition 3 (semantics of graph transformation units). Consider a graph transformation unit $trafo = \langle I, C, T \rangle$ over $\mathcal{A} = (\mathcal{G}, \mathcal{R}, \Rightarrow, \mathcal{E}, \mathcal{C})$ and a graph transformation environment $E \colon ID \to 2^{\mathcal{G} \times \mathcal{G}}$. Then the *semantics* of $trafo$ in E, denoted by $SEM_E(trafo)$, consists of all pairs (G, G') of graphs such that

- $(G, G') \in \mathcal{L}(I) \times \mathcal{L}(T)$,
- $(G, G') \in SEM_E(C)$, and
- $(G, G') \in (\bigcup_{n \in names(C)} E(n))^*$.

Remarks.

1. The last point of the preceding definition means that for each $(G, G') \in SEM_E(trafo)$ there is a sequence $G_0, \ldots, G_j$ of graphs such that (1) $G_0 = G$ and $G_j = G'$, and (2) for $i = 1, \ldots, j$ $(G_{i-1}, G_i) \in E(n)$ for some $n \in names(C)$. Hence, if E associates to each name in C the semantics of the rule or transformation unit it refers to, G' is obtained from G by interleaving rule applications with applications of transformation units.
2. Note that all control conditions of the class $\mathcal{C}$ in our example approach fulfil the property of the last item in the definition. But in general one may have other types of control conditions. A simple example is a pair of graphs without names, specifying itself.
3. The graph transformation units of [KK96] contain also a set U of used (or imported) graph transformation units, and a set R of local graph transformation rules. In our definition of graph transformation units, imports and rules belong to the surrounding graph transformation module, and are used via the environment.

Example 2. An example of a transformation unit is *invert* where *unmark* and *erase* are names of rules.

> *invert*
>   initial: $\{*\}$
>   conds: *mark!* ; *erase!* ; *unmark!*
>   terminal: *all*

The unit *invert* takes an unlabeled graph as its input, applies first the rule *mark* as long as possible, then the rule *erase* as long as possible, and finally the rule *unmark* as long as possible. As terminal graphs all graphs are allowed. This transformation unit adds m-labeled edges until the graph is complete. After that it deletes previously existing unlabeled edge and then finally unmarks the m-labeled edges. Therefore *invert* returns the complement of its input graph.

4   Graph Transformation Modules

A graph transformation module encapsulates a named set of graph transformation rules and graph transformation units, and exports some of them. It imports transformation units and rules from other modules. The semantics of a module is a set of graph transformations, i.e. a set of binary relations on graphs.

More precisely, a graph transformation module consists of an import part, a body, and an export part. The body defines a set of named local rules and transformation units. The export part consists of a subset of the names of local rules and transformation units. It determines which local units and rules can be imported by other modules. Hence, a local unit or rule cannot be used from the outside if its name is not contained in the export. The import part consists of a set of names (refering to rules and transformation units of other modules) which can be used in local transformation units. This means that each local unit may contain names from the import part in its control condition. All other names of the control condition must refer to other local rules or transformation units.

Definition 4. Let $\mathcal{A} = (\mathcal{G}, \mathcal{R}, \Rightarrow, \mathcal{E}, \mathcal{C})$. Then a triple $mod = \langle IM, DEF, EX \rangle$ where

- $IM \subseteq ID$ is a set of *imported names*,
- $DEF : N \rightarrow \mathcal{R} \cup \mathcal{T}_{\mathcal{A}}$ is a finite named set of *rules and transformation units* over some set $N \subseteq ID$ of names,
- $EX \subseteq N$ is a set of *exported names*,

is called a *graph transformation module* over $\mathcal{A}$ if the following conditions are satisfied:

- No imported name is re-introduced for a local rule or transformation unit, i.e. $N \cap IM = \emptyset$.
- Transformation units use only names that are imported, or locally defined. More precisely, for all $n : \langle I, C, T \rangle \in DEF$, $names(C) \subseteq N \cup IM$.

Let $\mathcal{M}_{\mathcal{A}}$ denote the set of graph transformation modules over $\mathcal{A}$.

For technical simplicity, we define the semantics for non-recursive modules in the following, wherein graph transformation units do not call themselves recursively. Formally, a module M is *non-recursive*, if the relation $\leadsto^M$ on the set of transformation units in M is acyclic, where $t\colon \langle I, C, T\rangle \leadsto^M t'\colon \langle I', C', T'\rangle$ if $t' \in names(C)$. Note, however, that the definition of the semantics below can be generalized to the case of recursive modules analogously to the generalization of the interleaving semantics of transformation units to the nested case (see [KKS98]).

The semantics of a module *mod* associates a binary relation on graphs to each exported name. It is defined as the restriction of the so-called environment of *mod* to its export part. This module environment maps a binary relation on graphs to each locally defined transformation unit and rule depending on the semantics of the imported names. More precisely, it associates to each imported name its semantics, to each (name of a) local rule r its corresponding binary relation $\Rightarrow_r$, to each (name of a) local transformation unit t the semantics of t within the environment of *mod*, and the empty set to each other name.

Definition 5 (semantics of graph transformation modules). Consider a graph transformation module $mod = \langle IM, DEF, EX\rangle$ over $\mathcal{A}$, and a graph transformation environment $E\colon ID \rightarrow 2^{\mathcal{G}\times\mathcal{G}}$ defining the imports. Then the *module environment* of *mod* with respect to E is the mapping $E_{mod}\colon ID \rightarrow 2^{\mathcal{G}\times\mathcal{G}}$ defined by

$$E_{mod}(n) = \begin{cases} E(n) & \text{if } n \in IM, \\ \Rightarrow_r & \text{if } n : r \in DEF, \\ SEM_{E_{mod}}(\langle I, C, T\rangle) & \text{if } n : \langle I, C, T\rangle \in DEF, \\ \emptyset & \text{otherwise.} \end{cases}$$

The *export semantics* is obtained as the restriction of the module environment to exported names:

$$EXP_{mod,E}(n) = \begin{cases} E_{mod}(n) & \text{if } n \in EX, \\ \emptyset & \text{otherwise.} \end{cases}$$

Note that the mapping E_{mod} is well-defined because the dependency relation $\leadsto^M$ is acyclic.

Remarks

1. A conventional graph transformation system (consisting of a set R of graph transformation rules) corresponds to a module $mod = \langle \emptyset, DEF, N\rangle$ where $DEF : N \rightarrow \mathcal{R}$ defines rules only.
2. A transformation unit $trut = \langle I, U, R, C, T\rangle$ as in [KK96] corresponds to a graph transformation module $mod = \langle IM, DEF, EX\rangle$ with $IM = U$, $DEF = R \cup \{t : \langle I, C, T\rangle\}$, and a single export $EX = \{t\}$.
3. If a graph transformation module $mod = \langle IM, DEF, EX\rangle$ exports all defined transformation units and no rule, every unit $t : \langle I, C, T\rangle$ in *mod* corresponds to a "[KK96]-transformation unit" $trut(t) = \langle I, IM \cup U, R, C, T\rangle$ where R are

the rules defined in DEF, and U are the transformation unit names defined in DEF. Hence, in this case a module corresponds to a set of "[KK96]-transformation units".

Notice that all these equivalences only hold up to the handling of names, which is not explicitly treated in the original concept of transformation units.

An example of a module is *invert_graph* in the upper frame of figure 2. It consists of the rules *mark*, *erase*, and *unmark*. The rule *erase* deletes an unlabeled edge and *unmark* deletes the label m of an arbitrary m-labeled edge. The module *invert_graph* contains also the local transformation unit *invert*. The control condition *mark!* ; *erase!* ; *unmark!* requires that first, an m-labeled edge between every pair of non-adjacent nodes is inserted. Second, all unlabeled original edges are deleted. Third, *unmark* deletes all edge labels m. Hence, the resulting terminal graph of *invert* is the complement of its input graph. The transformation unit *invert* is exported by the module *invert_graph*.

5 Module Systems

A GRACE program, or *system* is just a (named) collection of graph transformation modules wherein all imports correspond to some export.

Definition 6. A named set MS of graph transformation modules is called a *graph transformation module system*, shortly module system, if the following conditions hold true:

- The names exported by the modules are mutually distinct, i.e. for all modules $m : \langle IM, DEF, EX \rangle$ and $m' : \langle IM', DEF', EX' \rangle$ in MS, $EX \cap EX' \neq \emptyset$ implies $m = m'$.
- All transformation units or rules imported by a module are exported by another module, i.e. for all $m : \langle IM, DEF, EX \rangle$ in MS and all imported names $n \in IM$, there is a module $m' : \langle IM', DEF', EX' \rangle$ so that $n \in EX'$.

The modules composing a GRACE program are *simple* in that they only restrict the use of names for the programmers. GRACE programs have a flat semantics, given by the export semantics of the monolithic collection of all definitions of all its modules.

Definition 7 (flattening of a graph transformation module system).
Let S be a graph transformation system.

The *flattening* of MS is the module $\widetilde{MS} = \langle \emptyset, \widetilde{DEF}, \widetilde{EX} \rangle$ given by the empty import, the disjoint union $\widetilde{DEF}$ of all definitions, and the (non-disjoint) union $\widetilde{EX}$ of all exports of modules in MS.

The *flat semantics* of MS is given by the export semantics $EXP_{\widetilde{MS}, \emptyset}$ where $\emptyset$ is the empty environment.

A structured definition of the semantics is based on the additional assumption that the *uses relation* between modules is well-founded, i.e., descending the chain of imports we are always able to find a module with empty import. Then, the semantics of the modules in the system can be computed *bottom-up*, starting from these basic modules.

Definition 8 (uses relation, structured semantics). Let, in a module system MS, the relation $\mathcal{U} \subseteq MS \times MS$ be defined by

$$m : \langle IM, DEF, EX \rangle \; \mathcal{U} \; m' : \langle IM', DEF', EX' \rangle \;\; \text{iff} \;\; IM \cap EX' \neq \emptyset$$

Assume that MS is finite, and that the relation $\mathcal{U}$ is acyclic. Then, the semantics of a module mod within MS is defined by $SEM_{MS}(mod) = EXP_{mod,E(mod)}$, where the environment $E(mod)$ is given by

$$E(mod) = \bigcup_{mod \; \mathcal{U} \; mod'} SEM_{MS}(mod')$$

Note that the union is disjoint as exports of modules are disjoint. If mod has no import that means $SEM_{MS}(mod) = EXP_{mod,\emptyset}$.

Since we require a hierarchical import structure, the structured and the flattened semantics are equivalent:

Proposition 1. *For every graph transformation module system MS with acyclic use relation:*

$$Sem_{MS}(mod) = EXP_{\widetilde{MS},\emptyset}$$

Example 3. In the following, we define a graph transformation module system that computes maximal independent sets of nodes and cliques in graphs. It implements algorithms given in [Chr75]. The underlying graph transformation approach has been described previously. A set of nodes in a graph is said to be *independent* if it does not contain two nodes which are connected via an edge. An independent set is *maximal independent* if the addition of any other node would yield a non-independent set. In contrast with independent sets, a clique is a complete subgraph. To facilitate the computation, we use an interesting relation between those two kinds of sets: cliques of a graph are maximal independent sets of its complement and vice versa.

The module system of figure 2 realizes both algorithms —computing maximal independent sets and cliques— using two modules. The module system is shown as a surrounding box, modules are boxes situated inside.

The semantics of the module *invert_graph* was already discussed in section 4. The module *sets* imports the transformation unit *invert* from the module *invert_graph*. It contains the local rule *select* which labels an unlabeled node with s provided that it has no adjacent s-labeled node. The local transformation unit *maxiset* applies *select* as long as possible to an unlabeled input graph, i.e. it computes a maximal independent set. Finally, the transformation unit *clique* computes first the complement graph of its unlabeled input graph and determines then a maximal independent set of it. Hence, the s-labeled nodes of the

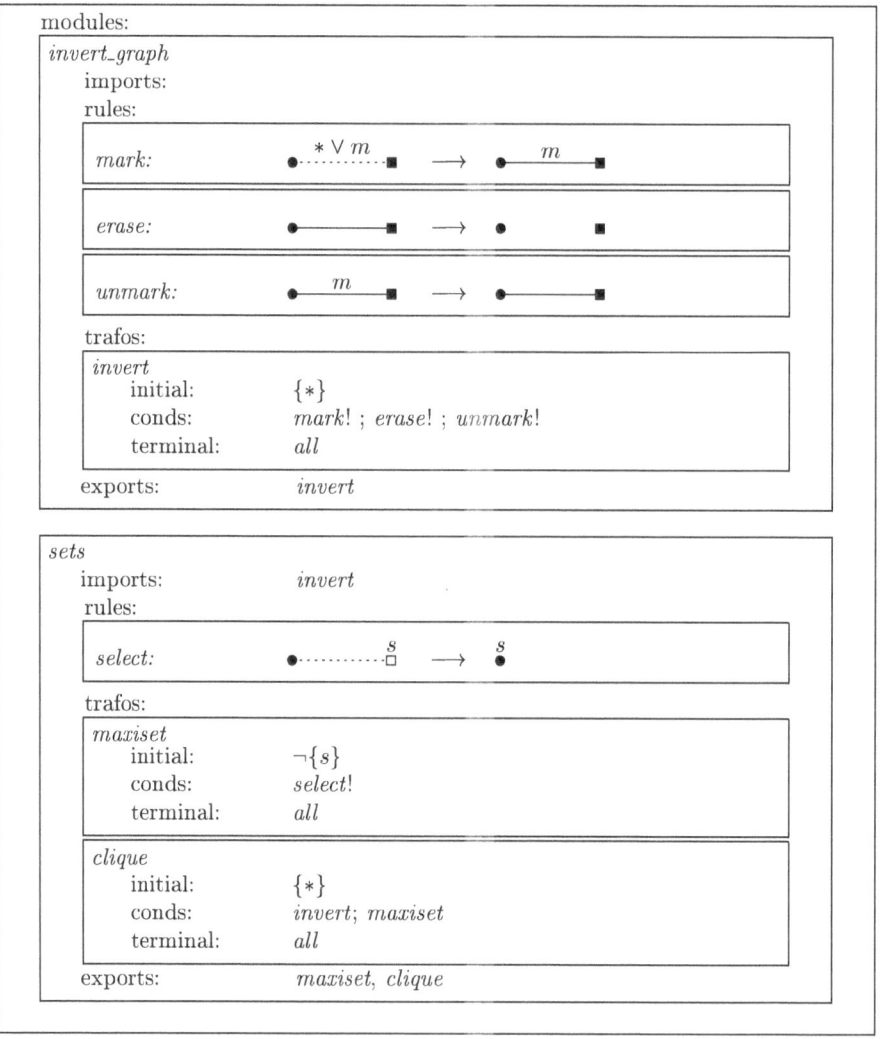

Fig. 2. A module system

result of *clique* form a clique of the input graph. Both local transformation units of the module *sets* are exported.

Of course such algorithms can be easily implemented using only transformation units in the sense of [KK96]. But here modules group rules and transformation units according to their use. The first module is graph oriented. It is responsible for inverting a graph. The second module focusses on set theoretic aspects. The transformation unit *invert* which is exported by the module *invert_graph* is imported and used by the module *sets*. The module *sets* provides two transformation units that can be used by other modules. The export of more than one transformation unit was not possible in GRACE, yet.

6   Conclusion

In this paper we have devised a module concept that allows transformations and rules to be encapsulated so that only some of them are exported, and the rest is hidden. The concept is a straight-forward extension of the transformation units defined in [KK96]; it is simple as it just provides *information hiding* for operations, without making the semantics more complicated.

In this section we relate our work to other module concepts for graph transformation in order to identify the issues for further investigations. (See [HEET99] for a more detailed comparison of module concepts for graph transformation.)

Related Work

The module concept recently proposed for PROGRES [Sch97] is based on the package concept of the object-oriented modeling language UML. Similar to our proposal its emphasis is on information hiding for definitions. However, this includes also the encapsulation of graph types which is not (yet) supported in our approach.

In [EE96], distributed states, inheritance, and import-export interfaces are investigated in an approach-independent axiomatic style inspired by modularity concepts for algebraic specifications. However, these concepts remain on an abstract level and are not formally integrated.

The module concept for typed graph transformation systems [CMR96,GPS98] applies the import-export architecture of [EE96] to the DPO approach and uses refinement relations between graph transformation systems in order to model the implementation of exported operations in the body. Also here, graph types may be defined locally, and abstract views of these graphs for other modules can be specified.

DIEGO [TS95] proposes distributed encapsulated graph objects (in the DPO approach) that communicate (and synchronize) via export-import interfaces. As concurrency and distribution is not—yet!—considered for GRACE, these concepts do not occur in our proposal too.

Future Work

The main advantage of our approach over the surveyed concepts is that it is completely defined, and still approach-independent. Thus it can be applied to practically every graph transformation approach with only little effort. However, the comparison indicates several aspects by which the presented model has to be extended and refined:

- Functional abstraction (by transformation units) has to be paired with *data abstraction* for structuring the graphs. If transformation units are tied to such substructures, we get graph *classes* and *objects* like in object-oriented languages such as Java [AG95]. (See [Hof99] for a proposal.)
- Given graph objects, *concurrency* and *distribution* can be introduced.
- Finally, we can study *heterogeneous systems* which are composed from modules specified in different graph transformation approaches.

Let's do it soon, and let's do it with grace!

References

Ada95.   *The Programming Language Ada.* ISO standard 8652, 1995. See also at URL `http://lglwww.epfl.ch/Ada/`. 383

AEH+99.   Marc Andries, Gregor Engels, Annegret Habel, Berthold Hoffmann, Hans-Jörg Kreowski, Sabine Kuske, Detlef Plump, Andy Schürr, and Gabriele Taentzer. Graph transformation for specification and programming. *Science of Computer Programming* 34:1–54, 1999. 383

AG95.   Ken Arnold, and James Gosling. *The Java Programming Language (2nd edition).* Addison Wesley's Java Series, 1998. 384, 394

Chr75.   N. Christofides. *Graph Theory: An Algorithmic Approach.* Academic Press, Inc., London, 1975. 391

CMR96.   A. Corradini, U. Montanari, and F. Rossi. Graph processes. *Fundamenta Informaticae*, 26(3,4):241–266, 1996. 393

EE96.   H. Ehrig and G. Engels. Pragmatic and semantic aspects of a module concept for graph transformation systems. In *5th Int. Workshop on Graph Grammars and their Application to Computer Science, Williamsburg '94, LNCS 1073*, pages 137–154. Springer Verlag, 1996. 393

GPS98.   M. Große–Rhode, F. Parisi-Presicce, and M. Simeoni. Spatial and temporal refinement of typed graph transformation systems. In *Proc. Mathematical Foundations of Computer Science (MFCS'98), LNCS 1450*, pages 553–561, Springer Verlag, 1998. 393

HHT96.   Annegret Habel, Reiko Heckel, and Gabriele Taentzer. Graph grammars with negative application conditions. *Fundamenta Informaticae*, XXVII:1/2, 1996. 385

HEET99.   Reiko Heckel and H. Ehrig and G. Engels and Gabriele Taentzer. Classification and comparison of module concepts for graph transformation systems. In G. Engels, H. Ehrig, H.-J. Kreowski, and G. Rozenberg (eds.): *Handbook of Graph Grammars and Computing by Graph Transformation, Vol. II: Specification and Programming*, to appear, 1999. 393

Hof99.   Berthold Hoffmann. From graph transformation to rule-based programming
            with diagrams. In M. Nagl, and A. Schürr, editors, *Proc. Int'l Workshop
            and Symposium on Applications of Graph Transformation with Industrial
            Relevance (AGTIVE)*, to appear in Lecture Notes in Computer Science,
            1999. 394

KK96.   Hans-Jörg Kreowski and Sabine Kuske. On the interleaving semantics of
            transformation units — a step into GRACE. In Janice E. Cuny, Hartmut
            Ehrig, Gregor Engels, and Grzegorz Rozenberg, editors, *Proc. Graph Gram-
            mars and Their Application to Computer Science*, number 1073 in Lecture
            Notes in Computer Science, pages 89–108, 1996. 383, 384, 387, 389, 390,
            393

KKS98.   Hans-Jörg Kreowski, Sabine Kuske, and Andy Schürr. Nested graph trans-
            formation units. *International Journal on Software Engineering and Knowl-
            edge Engineering*, 7(4):479–502, 1998. 383, 389

Kus98.   Sabine Kuske. More about control conditions for transformation units. In
            *Prelim. Proc. 6th Int. Workshop on Theory and Application of Graph Trans-
            formation (TAGT'98), Paderborn*, pages 150–157, 1998. 385

PHA+97.   John Peterson, Kevin Hammond, Lennart Augustsson, Brian Boutel, War-
            ren Burton, Joseph Fasel, Andrew D. Gordon, John Hughes, Paul Hu-
            dak, Thomas Johnsson, Mark Jones, Erik Meijer, Simon Peyton Jones,
            Alistair Reid, and Philip Wadler. *Report on the Programming Lan-
            guage Haskell, Version 1.4.* Yale University, April 1997. See URL
            http://www.haskell.org/report/index.html. 383

Roz97.   Grzegorz Rozenberg, editor. *Handbook of Graph Grammars and Computing
            by Graph Transformation, Vol. I: Foundations.* World Scientific, Singapore,
            1997. 384, 395

Sch94.   Andy Schürr. PROGRES: A VHL-language based on graph grammars. In
            Hans-Jürgen Schneider and Hartmut Ehrig, editors, *Proc. Graph Transfor-
            mations in Computer Science*, number 776 in Lecture Notes in Computer
            Science, pages 641–659, 1994. 383

Sch97.   A. Schürr. Programmed graph replacement systems. In Rozenberg [Roz97],
            pages 479 – 546. 393

TB94.   Gabriele Taentzer and M. Beyer. Amalgamated graph transformation sys-
            tems and their use for specifying AGG – an algebraic graph grammar system.
            In Hans-Jürgen Schneider and Hartmut Ehrig, editors, *Proc. Graph Trans-
            formations in Computer Science*, number 776 in Lecture Notes in Computer
            Science, pages 380–394, 1994. 383

TS95.   G. Taentzer and A. Schürr. DIEGO, another step towards a mod-
            ule concept for graph transformation systems. *Proc. of SEGRAGRA'95
            "Graph Rewriting and Computation"*, Electronic Notes of TCS, 2, 1995.
            http://www.elsevier.nl/locate/entcs/volume2.html. 393

UML Packages for PROgrammed Graph REwriting Systems

Andy Schürr[1] and Andreas J. Winter[2]

[1] Software Engineering Institute, University BW München
Werner-Heisenberg-Weg 39, D-85577 Neubiberg, Germany
schuerr@informatik.unibw-muenchen.de
[2] Lehrstuhl für Informatik III, RWTH Aachen
Ahornstr. 55, D-52074 Aachen, Germany
winter@i3.informatik.rwth-aachen.de

Abstract. Specification and rapid prototyping of graph manipulation software by means of PROgrammed Graph REwriting Systems (PRO-GRES) is a paradigm attracting more and more interest in various fields of computer science. Specifications for process modeling tools, database query languages, etc. suffer from the lack of any module concept. This paper introduces a module concept for the graph rewriting (transformation) language PROGRES that is closely related to the package concept of the standardized OO modeling language UML. It supports a variety of software design patterns including the construction of "Abstract Graph Types" and "Updatable Graph Views".

1 Introduction

Visual programming languages and tools based on the concept of *graph transformations* (GTs) are attracting more and more interest in various fields of applied computer science as well as in related engineering disciplines. They combine two successful principles in one formalism: (1) *graphs* as a well-understood and popular data model and (2) *rules* as declarative means for the description of complex transformation and inference processes on complex data structures. Nevertheless, the graph transformation community is still waiting for the breakthrough of their languages and tools in the real (industrial) world.

The GT language *PROGRES* [Sch91,Zün96] is used at various sites for specification and rapid prototyping activities. Specifications of real configuration management, process modeling, reverse engineering, and distributed system analyzing tools usually have a size of more than 100 printed pages. Keeping these specifications in a *consistent* state and *reusing* generic parts of one specification within another specification is a nightmare without the existence of any *module concept*. Thus, the lack of any (implemented) GT module concept is one of the main hindrances for a wider distribution of GT specification or programming languages in general and the language PROGRES in particular.

This problem should be familiar for software developers of the late 60ies, expert system developers of the late 70ies, and object-oriented (OO) modelers of the 80ies. Well-known software engineering concepts like "abstract data

H. Ehrig et al. (Eds.): Graph Transformation, LNCS 1764, pp. 396–410, 2000.
© Springer-Verlag Berlin Heidelberg 2000

types" [Par72] and "programming–in–the–large" [DK76] have been invented and
have lead to the development of modular programming languages like Modula-
2 or Ada [WS84] and software design languages like HOOD [Rob72]. The OO
modeling community made significant progress introducing a module concept
that allows the distributed development of large software analysis and design
models and the reuse of existing submodels. The *Unified Modeling Language*
UML [Rat97] developed by Booch, Rumbaugh, and Jacobson is the first OO
modeling notation addressing all facets of a state-of-the-art module concept.

In the meantime a number of graph transformation researchers have joined
their efforts in the so-called GRACE initiative [AEH+96]. Their common goal is
the development of a GRAph Centered Environment supporting modular programming with various forms of graph data models, rewrite rules, and rule regulation mechanisms. Related publications present the first attempts to introduce
import/export as well as inheritance/refinement relationships into the world
of GT languages [EE96,HCE96], to encapsulate GT algorithms by means of
so-called transformation units [KK96,Sch96], to adapt the concept of database
views to graph types [NS96], to apply transformation rules on hierarchical graph
packages [BEM+99], and to develop new concepts for the design of distributed
systems of graph objects [TS95]. An overview and comparison of GT module
concepts can be found in [EEH99].

Taking all these sources of inspiration into account, we have developed a
module concept for PROGRES which is kept as simple and flexible as possible.
The version presented here is based on the *formal definition* of a refined UML
package concept as introduced in [SW97].

The rest of this paper is organized as follows: very brief introductions to
PROGRES and UML's package concept are followed by a short presentation of
the PROGRES package concept and supported software design styles (patterns).
The paper concludes with a discussion of related work and the summary.

2   The Language PROGRES

In this section we give a very brief introduction to the PROGRES language. For
a detailed discussion of its features and the integrated development environment
the reader is referred to [SWZ98,Sch91,Zün96,Nag96].

PROGRES is based on the data model of *directed attributed graphs* and allows
to describe transformations on graphs by rules declaratively. It was originally designed for describing internal data structures, operations, and relations of tightly
integrated Software Engineering tools. Its semantics is formally defined and it is
supported by an integrated set of tools which allow for specifications to be edited
graphically or textually, to be analyzed, and to be executed. Therefore, PRO-
GRES constitutes an *executable visual language* for specifying systems based on
internal graph structures. Together with the ability to generate stand-alone prototypes from specifications the suitability of our *Graph Grammar Engineering*
approach [SWZ95] has been demonstrated for some application areas.

PROGRES specifications consist of two parts: the *static graph schema* and the schema-consistent *graph transformations*. The schema allows to express static properties of the regarded class of graph. Its visual notation has the form of EER diagrams with *entities* called node types and *binary relationships* called edge types. Properties of node types, which are not relationships between nodes, are modeled as attributes. A *type hierarchy*, similarly to object-oriented approaches, can be established on these node types. Defining edge types in the schema allows to check statically whether edges are used in a correct context (source and target nodes) in graph transformations.

The screenshot of the schema editor in Fig. 1 shows the graph schema of an Airline Reservation System. *Node types* are represented by solid boxes. Arrows with triangle heads depict the is a-hierarchy and the other arrows indicate *edge types* representing binary relationships between nodes of the corresponding types.

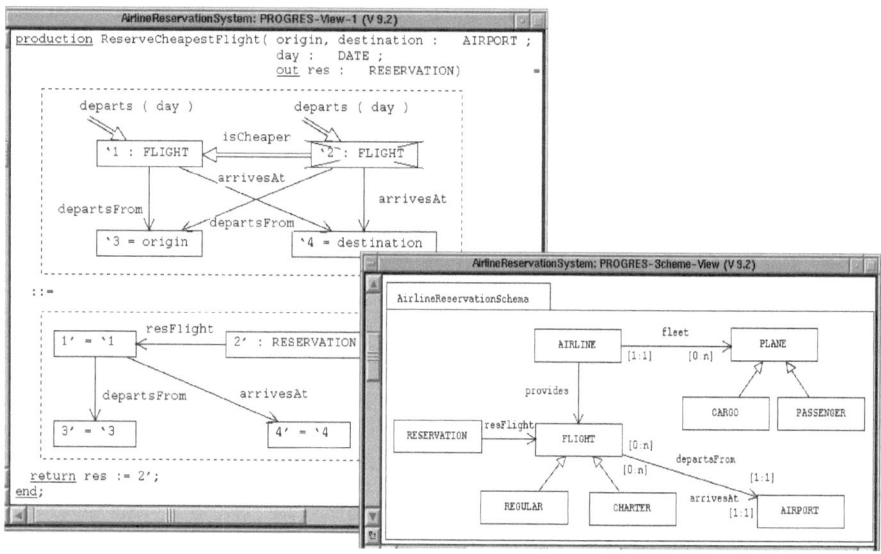

Fig. 1. PROGRES graph schema and production rule

Having defined the static graph schema, *production rules* can be developed allowing schema-consistent transformations only. The rule ReserveCheapestFlight in Fig. 1 creates a reservation node for a flight from the given origin to the given arrival airport provided it is the cheapest available flight on the intended date. It consists of a left and a right hand side. The left hand side describes the graph pattern which is matched and replaced by the pattern on the right hand side when the production rule is applied. The rule's left hand side also refers to forbidden context (crossed-out node) and uses derived graph properties like value restrictions and path navigations (double arrows).

To summarize, PROGRES and its integrated environment define and implement a graph-transformation-based, strongly typed programming language with well-defined syntax, static and dynamic semantics. Being a mixed textual and diagrammatic language, it permits different styles of programming, and supports

- graphical and textual definition of schemas with derived graph properties,
- rule-oriented and diagrammatic specification of graph transformations with parameterized graph production rules, and
- imperative programming of composite graph transformation processes by means of deterministic and nondeterministic control structures.

3   The UML Package Concept

The information hiding and modularization concept of the *UML packages* was strongly influenced by the design of the OO programming languages C++ and Java. It has the following properties:

- A *package* builds a shell around a group of closely related declarations; usually these declarations have the form of a diagram and define the data structures and the operations of a modeled (software) system.
- It is the single purpose of a package to regulate the *visibility* of its declarations, i.e. to restrict the usage of its declarations to well-defined parts of a system model.
- As a consequence, packages have *no run-time semantics* at all, i.e. the semantics of a system model is not changed if all its declarations are put into a single package.
- A *dependency* from a client package to a server package reveals some of the server's declarations (resources) while others remain hidden.
- *Import* dependencies may be used to access a package's public resources, whereas *refinement* relationships provide access to its protected resources.
- Furthermore, UML supports *nesting* of packages with inverse scoping rules of block-structured languages, i.e. a parent package has an implicit import dependency to any child package, but not the other way round.
- A package can be developed, compiled, and tested independently from others and it may be replaced by another package with the same export interface.

Table 1 explains the interaction between the visibility tag of a package resource having a value from the ordered set { public > protected > private } and the two types of package dependencies in more detail.

The UML visibility rules as presented here are simplified due to the fact that they do not take implicit import dependencies as well as visibility tags of package dependencies themselves into account. Please note that UML version 1.1 neither explains how visibility tags of dependencies affect the visibility of imported resources nor their purpose. For further details concerning still existing

		public	protected	private
---------		--------	-----------	---------
import		visible	invisible	invisible
refine		visible	visible	invisible

Table 1. Simplified visibility rules of UML

design flaws of the UML package concept, proposals how to remove them, and a precise formal definition of visibility rules the reader is referred to [SW97].

In the following section we will explain our interpretation of the visibility tags for package dependencies. Furthermore, we will introduce a number of graph transformation-specific, but not PROGRES-specific package concept extensions.

4   The PROGRES Package Concept

Adopting the package concept of UML as the *PROGRES modularization concept* we explain our interpretation of the visibility tags of package dependencies. Later on we introduce two additional tags for exported package resources. They allow to restrict the usage of certain kinds of exported resources in client packages having either an import or a refinement dependency to the exporting package. Furthermore, we explain the "semantics" of import and refinement dependencies between packages more precisely than in the UML standard and to fine-tune the standard UML semantics for the purposes of a GT-based language.

Combining the two types of package dependencies with three different visibility tags we are able to distinguish six different types of package dependencies:

Interface import: a public import dependency corresponds to the concept of definition module imports in Modula-2. It allows to import those (node) types from server packages which are used as parameter types of public graph transformation operations of the client package.

Protected import: a protected import dependency has to be used, whenever certain implementation details based on imported resources have to be hidden from regular clients, but must be revealed to refining packages for redefinition purposes.

Implementation import: a private import dependency corresponds to the concept of implementation module imports in Modula-2. It allows to import all those resources used for implementing the given package without revealing the dependency to the package's clients.

Interface inheritance: a public refinement dependency defines a kind of subtype relationship between two packages. It is an assertion for all regular clients of these packages that the interface of the refining package is an extension of the interface of the refined package[1].

---

[1] A refining package may redefine inherited operations as long as parameter lists of these operations are not modified and all inherited integrity constraints as well as pre- and postconditions are still observed [WS97].

Protected inheritance: a protected refinement dependency allows the construction of a kind of subtype relationship between two packages that is invisible to regular clients, but visible for refining packages.

Implementation inheritance: a private refinement dependency is closely related to the well-known concept of implementation inheritance. It allows to implement a package as a refinement of another package without any restrictions concerning the export interface of the refining package. It is a matter of debate in the software engineering community whether implementation inheritance is a dangerous or a useful concept.

Based on the explanation of six different types of dependencies we are now able to discuss their semantics in more detail. Remember that packages do not have a run-time (dynamic) semantics at all, i.e. we are talking about compile-time (static) semantics only. All static semantic rules discussed below restrict the way how exported declarations of a server package may be used in its client packages and how declarations may be decorated with visibility tag values. There are no restrictions at all (except several hundreds of "programming–in–the–small" static semantics rules documented in [Sch91]) concerning the usage of declarations as long as the using declaration is part of the same package.

Before introducing static semantics rules restricting the usage of declarations across package boundaries, we have to summarize the available categories of declarations in PROGRES and the ways how these declarations may be used for building new declarations (cf. Sec. 2)[2]:

1. **Operation call:** once defined graph tests or transformations may be used to construct new more complex graph tests or transformations; the new operation calls the already defined graph tests and transformations.
2. **Parameter type:** attribute types and node types may be used as formal parameter types of graph tests and graph transformations.
3. **Schema extension:** attribute types, e.g. other node types and built-in types, are used for defining attributes of new node types. Node types are also used as source and target types for the declaration of new edge types (horizontal schema extension).
4. **Graph query:** attributes, node types as well as edge types are needed to define the graph patterns of subgraph tests or application conditions of graph transformations. In both cases these elements are used for inspecting but not for modifying a given (sub-)graph.
5. **Graph modification:** attributes, node types, and edge types are used for constructing the left- and right-hand sides of graph transformations, i.e. for defining necessary graph modifications.
6. **Schema refinement:** new node types may be defined as subtypes of already existing ones. They may redefine inherited attributes of their supertypes (vertical schema extension).

---

[2] Due to lack of space, we have omitted the discussion of the following categories of declarations: functions, path expressions, restrictions, and integrity constraints.

7. **Operation refinement:** a forthcoming real object-oriented version of PRO-GRES will not only support the redefinition of attribute evaluation functions, but also the redefinition of graph transformations that are associated with (complex) node types.

The first four items of the list allow to build new *abstract graph types* on top of existing abstract graph types. The notion of abstract graph types is adopted from abstract data types [Par72]. It refers to a certain kind of package design where the internal graph structure is hidden and merely a family of interface operations is exported. A client graph type usually defines its own graph structure and creates edges between its own nodes and the nodes of the underlying abstract graph type. Nodes and edges of the underlying abstract graph type are manipulated by calling the associated operations only[3]. This leads to a procedural and nonvisual programming style discussed in [WS97].

In order to preserve the visual flavor of PROGRES specifications it is sometimes useful to permit the graph transformations of one package to create and destroy instances of imported node and edge types directly. This corresponds to item 5 above, where we allow the unrestricted usage of (imported) node and edge types in the left- and right-hand side of graph transformation rules (productions). It is an important decision for the design of a package whether its export interface supports direct manipulation of exported node and edge types. Its implementation must be prepared for keeping the exported and externally modifiable parts of its graph structure consistent with its hidden parts. The reader is referred to [WS97] for a detailed discussion of how this kind of *view update problem* may be solved using active integrity constraints.

The discussion above shows that we need additional means to distinguish between the export of node types, edge types, and attributes for the definition of parameters, schema extensions, and graph queries on one hand, and the export of these resources for direct graph modification purposes on the other hand. As a consequence we are introducing an additional tag for exported resources (declarations) with two possible values: **modifiable** (the default) and **immutable**.

Modifiable resources may be used without any restrictions, whereas imported immutable resources may not be used for direct graph manipulation purposes in client packages (the tag does not impose any restrictions inside its own package).

Similarly it is not always useful to allow the redefinition of exported attributes and graph transformations and the definition of new subtypes of an exported node type. Consequently, we introduce another tag for exported resources, which has the two possible values: **redefinable** (the default) and **final**.

Redefinable resources may be used without any restrictions, whereas imported final resources may not be used for redefinition purposes in client packages (the tag does not impose any restrictions inside its own package).

Table 2 summarizes the interaction between the visible, the modifiable, and the redefinable tag of exported package declarations and the two package dependencies. It was constructed as follows: Table 1 was used to determine whether a

---

[3] Nodes and Edges of the underlying graph structure may neither be created nor deleted directly.

resource of one package is visible for its dependent packages. Private resources of a package are for instance never visible for the outside world. This has the consequence that all entries of the private column of Table 1 and thus also of Table 2 are blank. All cells of Table 1 with the value visible had to be split in four subcases in accordance to the four possible combinations of the previously introduced modifiable and redefinable tags.

export tags	public				protected				private
	mr	mf	ir	if	mr	mf	ir	if	-
import	mf	mf	if	if	-	-	-	-	-
refine	mr	mf	mr	mf	mr	mf	ir	if	-

Table 2. Visibility rules of PROGRES

A public export which has e.g. a tag m = modifiable and a tag r = redefinable may be used by its client packages with import dependencies for m = modification, i.e. direct manipulation purposes. Furthermore its r = redefinable tag is downgraded to f = final due to the fact that regular client packages never have the permission to refine the imported graph schema or the imported operations. All remaining cells of Table 2 have to be interpreted accordingly.

The definition of the entries of Table 2 was rather straightforward, with one exception: The immutable tag of redefinable as well as final public exports is only valid for regular import dependencies. It is upgraded to modifiable in the case of a refinement dependency. Otherwise it would be necessary to introduce a fourth export tag that allows to distinguish between immutable for import and refinement dependency and immutable for import but modifiable for refinement.

Due to the fact that refining packages usually have to manipulate instances of inherited node and edge types directly and the fact that immutable final resources may be moved to a separate package, we made the decision that the immutable flag of public resources is ignored by refinement dependencies between packages. It is a matter of debate whether the immutable flag of protected export resources should be ignored, too. In this case the entries ir and if in the fields determined by row refine and columns protected should be changed from the presented values ir = immutable redefinable and if = immutable final to the new values mr = modifiable redefinable and modifiable final.

Having explained the interaction between the three different tags of exported package resources and the two different types of package dependencies we will now discuss the restrictions concerning the selection of certain tag values. We cannot give a complete list of all related static semantics rules here, but the following list items present the most important rules:

– The visibility tag value of an export operation may not be higher than the visibility tag value of its parameter types. A public operation with a private parameter type is one example of a forbidden combination of values.

- The visibility tag value of an edge type or attribute declaration may not be higher than the visibility tag value of the referenced node types. It is for instance useless to define a public attribute for a protected node type or a protected edge type between private node types.
- The supertypes of a node type may not have a lower visibility tag value than the regarded node type itself. Otherwise, it would be possible to export type hierarchies with "holes", i.e. to hide subtype relationships between externally visible node types.
- Mutable supertypes may not have immutable subtypes. Otherwise, it would be possible to destroy (create) instances of immutable node types by handling them as indirect members of the mutable supertype.

The presented static semantics rules are not able to prohibit *unresolvable inheritance conflicts*: Let us assume that a package P declares a public node type A with a private attribute a := e as well as two public subtypes A1 and A2 redefining the initial value of attribute a from e to e1 and e2, respectively. Furthermore, assume that another package Q imports the node types A1 and A2 for constructing a type B is_a A1, A2. The new node type B inherits the attribute a with initial value e1 from A1 and with the initial value e2 from A2. The only way to resolve the inheritance conflict, the introduction of a new initial value for attribute a in subtype B, is blocked due to the fact that the private attribute a of package P is invisible to package Q.

Currently, we have no good idea how to avoid this problem without introducing very restrictive rules such as *"public or protected node types may not have private redefinable attributes"* or *"the concept of multiple inheritance may not be used across package boundaries"*. Please note that e.g. Java avoids this problem by disallowing multiple inheritance, whereas C++ simply ignores the discussed problem due to the lack of a properly defined package concept.

5   Specification-in-the-Large Patterns

Having presented the most important "screws" of our package concept and their meaning we will show how they should be used in practice. It is the purpose of this section to demonstrate that the introduced concepts allow the application of rather different *software design styles* for structuring large specifications. We will explain some *important design patterns* on an abstract level. Please note that it may be necessary to use combinations of the presented patterns for establishing more complex dependencies between client and server packages.

First of all we have to prove that PROGRES packages are the appropriate means for constructing *abstract graph types* in the usual sense [Par72]. Fig. 2 sketches the relationships between a server package that realizes an abstract graph type and a client package that constructs a new (abstract) graph type using the imported graph type.

The server package exports a number of public, immutable, and final declarations. Some declarations define the constructed graph type's schema, others

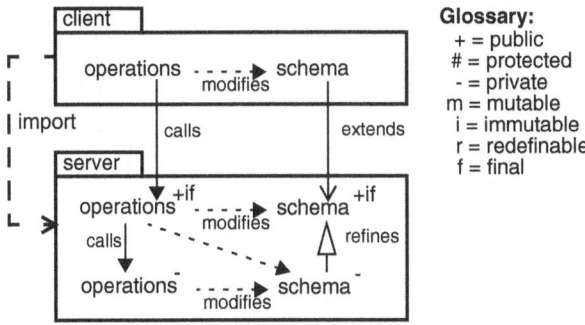

Fig. 2. Full data abstraction pattern

the accompanying graph transformation operations. The schema exported by the server package may consist of a number of node type and edge type declarations. The classical abstract graph type consists of a single immutable node type declaration, though. This type declaration as well as the associated interface operations have a final tag if the construction of subtypes should be prevented. These subtypes would have the permission to redefine and thereby to alter the exported operations and the underlying graph structure.

The revelation of the hidden complex graph structure required for implementing the functional behavior of the package is done in the private part of the abstract graph type. It extends and refines the public part of the schema. That means, the concept of aggregation is simulated by choosing one node type as the only visible representative for the underlying realization. The server's exported operations invoke the hidden operations in the private part. They perform the transformation on the internal graph structure. That means, access to the internal graph structure is only possible via exported interface operations of the abstract graph type. In this way the package can ensure that the internal graph structure remains consistent because clients do not have any possibility to circumvent the prepared interface operations performing well-defined transformations only. Consequently, the operations in the client package essentially consist of calls of interface operations exported by other packages.

Fig. 3 explains the definition of *"semi-abstract" graph types*. It exports a larger part of its graph schema. Marking the public schema with the modifiable flag allows its clients to manipulate instances of visible node and edge type declarations directly. The depicted server package constructs a kind of updatable graph view without any exported operations. The client package uses its own graph transformations to manipulate the visible part of the server's graph structure. In contrast to the encapsulated abstract graph type in Fig. 2 the client is not restricted to call exported operations of the server package only. That means, the client is able to define its graph transformations in a visual way by accessing instances of the server's schema. These graph transformations are even allowed to create and destroy instances of the imported schema.

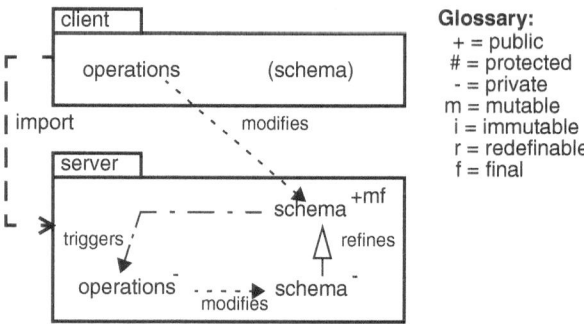

Fig. 3. Updatable view definition pattern

Graph manipulations of the client may introduce inconsistencies between the visible and the hidden part of the server's graph structure because the client does not see the server package's internals. Thus, it is not possible to ensure the preservation of the consistent context for changes made by the client directly.

Active integrity constraints known from active database systems [WC96], are responsible for detecting inconsistencies and for taking repair actions [MSW99]. If a constraint detects a graph pattern that corresponds to an inconsistent execution state, a hidden operation is triggered which is responsible for reestablishing consistency between the visible and externally modifiable part of the implemented graph structure and the associated hidden parts [WS97].

The last pattern presented in Fig. 4 sketches the construction of *subtypes of abstract graph types*. It is more or less a new variant of the pattern for abstract graph types in Fig. 2. In addition the client is allowed to refine the schema and the inherited operations. Please note that it is a matter of debate whether the operations of the client package should be allowed to manipulate instances of the inherited type definitions of the server package as depicted in Fig. 4. The main problem is that the client package is always able to circumvent the restriction to manipulate instances of inherited node types directly by deriving a new locally modifiable node type from the inherited unmodifiable node type. This is one of the reasons to upgrade the immutable flag to modifiable for inherited public resources also marked with the redefinable flag (cf. Table 2 of Sec. 4).

The design pattern of Fig. 2 with a public import dependency to a refinable server package is closely related to the introduction of a *parameter part* shared by the import and the export part of a given package in [EE96]. And, the design pattern of Fig. 4 with a private refinement dependency corresponds to what is usually called *implementation inheritance*. Taking into account

- that packages can be nested in order build complex subsystems with their own subarchitectures
- that a client package is implemented on top of more than one server package
- that a server package is used by different client packages in different ways

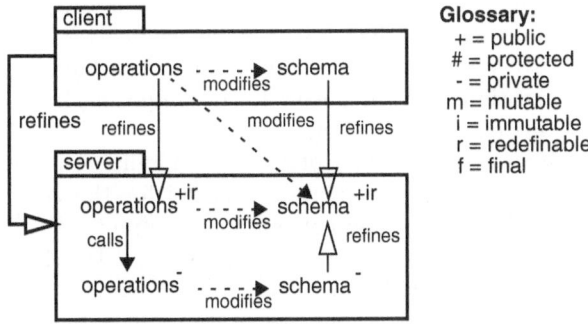

Fig. 4. Subtype definition pattern

it should be rather obvious that the PROGRES package concept supports an overwhelming variety of different software design styles and patterns. It is the subject of future research activities to identify useful design patterns (cf. [GHJ95] or [Fow97]) for the construction of large graph transformation system and to invent new patterns for more specific purposes.

6 Summary

The language PROGRES and its environment are the results of many years of application-oriented graph grammar or graph transformation research activities. Nowadays they are used at various sites for specifying and prototyping software, database, and knowledge-engineering tools.

Despite our success in demonstrating the usefulness of graph grammar engineering concepts and tools for software development, the currently available PROGRES release[4] is not yet ripe for real industrial software development projects. Compared with the history of imperative programming languages, PROGRES has reached the state of languages like Algol-68 or Pascal. It has a sophisticated type system and enforces a well-structured style of programming, but gives not yet any support for "specification–in–the–large" activities.

Considering the pile of module concept papers, it was and is clearly not our goal to start from scratch and to reinvent well-known "programming–in–the–large" concepts. On the contrary, we started the development of the PROGRES module concept based on our experiences with the design of a module interconnection language MIL [Nag90] and efforts in the object-oriented world to introduce a package concept for the Unified Modeling Language UML [Rat97]. Modules in the sense of MIL and packages in the sense of UML give the flexibility to adhere to quite different popular *software design styles*:

---

[4] For further details concerning PROGRES release version 9 cf.: http://www-i3.informatik.rwth-aachen.de/research/progres

- Packages that export a single "main" node type and a number of related operations define abstract data types in the classical sense [Par72].
- Packages with type declarations only in their interfaces and without any associated operations are our means for decomposing large graph schemas in the same way as in PCTE [Ecm90].
- Packages that export a number of directly modifiable types together with carefully defined active constraints are useful for the construction of updatable graph views, that may not only be manipulated using a number of predefined graph transformations as suggested in [EHT97].
- Packages that export a single transformation at their interfaces together with an appropriate set of graph schema declarations allow the definition of graph transformation units as suggested in [KK96].
- Packages that summarize and propagate the exported resources of a number of imported packages to their own interfaces offer more or less the concept of horizontally structured graph types as presented in [HCE96].
- Packages which refine the graph schemas and operations of a number of related packages offer more or less the concept of vertically structured graph types as presented in [HCE96].

Efforts are still necessary to develop all the details of a more ambitious package concept that supports the construction of hierarchical and distributed graph transformation systems. Finally, we are assembling more knowledge about useful graph transformation system design patterns and their relationships to pattern catalogues as presented in object-oriented design books.

References

AEH+96. M. Andries, G. Engels, A. Habel, B. Hoffmann, H. Kreowski, S. Kuske, D. Plump, A. Schürr, G. Taentzer: *Graph Transformation for Specification and Programming*. Science of Computer Programming, 34(1):1–54, April 1999. 397

Bei95.   C. Beierle: *Concepts, Implementation, and Applications of a Typed Logic Programming Language*. In Beierle, Plümer (eds.): Logic Programming: Formal Methods and Practical Applications. Elsevier Science B.V., 1995.

BEM+99. G. Busatto, G. Engels, K. Mehner, A. Wagner: *A Framework for Adding Packages to Graph Transformation Approaches*. In G. Engels, G. Rozenberg (eds.): Proc. TAGT '98. This volume. 397

CEE96.   J. Cuny, H. Ehrig, G. Engels, G. Rozenberg (eds.): *Proc. 5th Int. Workshop on Graph Grammars and Their Application to Computer Science*. LNCS 1073, Springer, 1996.

DK76.   F. DeRemer, H. Kron: *Programming–in–the–large versus Programming–in–the–small*. IEEE Trans. on Software Engineering, SE-2(2):80–86, 1976. 397

Ecm90.   European Computer Manufacturers Association: *ECMA Standard 149: Portable Common Tool Environment (PCTE)*. Abstract Specification, 1990. 408

EE96.   H. Ehrig, G. Engels: *Pragmatic and Semantic Aspects of a Module Concept for Graph Transformation Systems*. In Cuny et al. [CEE96], 137–154. 397, 406

EEH99.   H. Ehrig, G. Engels, R. Heckel, G. Taentzer: *Classification and Comparison of Modularity Concepts for Graph Tranformation Systems*. In G. Engels, G. Rozenberg (eds.): Proc. TAGT '98. This volume. 397

EHT97.   G. Engels, R. Heckel, G. Taentzer, H. Ehrig: *A View-Oriented Approach to System Modelling Using Graph Transformations*. In M. Jazayeri, H. Schauer (eds.): Proc. ESEC '97. LNCS 1301, 327-343, Springer, 1997. 408

Fow97.   M. Fowler: *Analysis Patterns*. Addison Wesley, 1997. 407

GHJ95.   E. Gamma, R. Helm, R. Johnson, J. Vlissides: *Design Patterns*. Addison Wesley, 1995. 407

HCE96.   R. Heckel, A. Corradini, H. Ehrig, M. Löwe: *Horizontal and Vertical Structuring in Typed Graph Transformation Systems*. Mathematic Structures in Computer Science, 6(6):613–648, December 1996. 397, 408

KK96.   H.-J. Kreowski, S. Kuske: *On the Interleaving Semantics of Transformation Units — A Step into GRACE*. In Cuny et al. [CEE96], 89–106. 397, 408

MSW99.   M. Münch, A.J. Winter, A. Schürr: *Integrity Constraints in the Multi-Paradigm Language PROGRES*. In G. Engels, G. Rozenberg (eds.): Proc. TAGT '98. This volume. 406

Nag90.   M. Nagl: Software Engineering: *Methodological Programming in the Large*. Springer, 1990 (in German). 407

Nag96.   M. Nagl (ed.): *Building Tightly Integrated Software Development Environments: The IPSEN Approach*. LNCS 1170, Springer, 1996. 397

NS96.   M. Nagl, A. Schürr: *Software Integration Problems and Coupling of Graph Grammar Specifications*. In Cuny et al. [CEE96], 155–169. 397

Par72.   D. Parnas: *A Technique for Software Module Specifications with Examples*. In Comm. of the ACM, vol. 15, 330–336. ACM Press, 1972. 397, 402, 404, 408

Rat97.   RATIONAL SOFTWARE CORPORATION: *UML Semantics*. http://www.rational.com/uml. 397, 407

Rob72.   P. Robinson: *Hierarchical Object-Oriented Design*. Prentice Hall, 1992. 397

Sch91.   A. Schürr: *Operational Specification with PROgrammed Graph REwriting Systems*. Deutscher Universitäts-Verlag, 1991 (in German). 396, 397, 401

Sch96.   A. Schürr: *Programmed Graph Transformations and Graph Transformation Units in GRACE*. In Cuny et al. [CEE96], 122–136. 397

SW97.   A. Schürr, A.J. Winter: *Formal Definition and Refinement of UML's Package Concept*. In J. Bosch, S. Mitchell (eds.): Object-Oriented Technology — ECOOP '97 Workshop Reader. LNCS 1357, 211-215, Springer, 1997. 397, 400

SWZ95.   A. Schürr, A.J. Winter, A. Zündorf: *Graph Grammar Engineering with PROGRES*. In W. Schäfer, P. Botella (eds.): Proc. ESEC '95. LNCS 989, 219–234, Springer, 1995. 397

SWZ98.   A. Schürr, A.J. Winter, A. Zündorf: *PROGRES*. In G. Rozenberg (ed.): Handbook of Graph Grammars and Computing by Graph Transformation, vol. 2: Specification and Programming. World Scientific, 1997. 397

TS95.   G. Taentzer, A. Schürr: *DIEGO, another step towards a module concept for graph transformation systems*. In A. Corradini, U. Montanari (eds.): Proc. SEGRAGRA '95. Vol. 2 of ENTSC. Elsevier Science B.V., 1995. 397

WC96.   J. Widom, S. Ceri (eds.): *Active Database Systems: Triggers and Rules for Advanced Database Processing*. Morgan Kaufmann, 1996. 406

WS84.   R. Wiener, R. Sincovec: *Software Engineering with Modula-2 and Ada*. Wiley, 1984. 397

WS97.   A.J. Winter, A. Schürr: *Modules and Updatable Graph Views for PRO-grammed Graph REwriting Systems.* TR 97-3, RWTH Aachen, 1997. 402, 406

Zün96.   A. Zündorf: *A Development Environment for PROgrammed Graph REwriting Systems.* Deutscher Universitäts-Verlag, 1996 (in German). 396, 397

Incremental Development of Safety Properties in Petri Net Transformations*

Julia Padberg, Maike Gajewsky, and Kathrin Hoffmann

Technical University of Berlin
{padberg,gajewsky,hoffmann}@cs.tu-berlin.de

Abstract. The application of the general theory of high-level replacement systems has proven to be most rewarding in many different areas, especially in Petri nets [EGPP99]. In this paper the extension of high-level replacement systems to refinement morphisms [Pad99] is applied to place/transition nets. The combination of morphisms, that preserve safety properties, with transformations of place/transition nets leads to rules and transformations, that preserve safety properties. Moreover, we extend our approach so that rules can introduce new invariant formulas, that is new safety properties.

Keywords: High-Level Replacement System, Petri Net Transformation, Safety Property, Temporal Logic, Rule-Based Refinement

1 Introduction

The theory of high-level replacement systems can successfully be employed not only to graph transformations, but also in other areas, as Petri nets, to obtain results relevant for software engineering. In this context the notion of safety is of central importance. Intuitively, a safety property in the sense of [MP92] means that "nothing bad" can happen in the system, and is expressed by a temporal logic formula. In this paper we apply an extension of high-level replacement systems to Petri nets in order to achieve an integration of transformations, and hence all its advantages, with the preservation of safety properties.

The combination of graph grammars and Petri nets is usually done by simulating the behaviour of nets by graph rules, see e. g. [Kre81,Sch94,Cor95,Sch99]. In that approach the semantics of both formalisms and their comparison is of central interest. We follow a different approach using rules and transformations to represent stepwise modification of nets. This kind of transformation for Petri nets is considered to be a vertical structuring technique in software engineering, known as rule-based modification.

The combination of transformations with the preservation of safety properties is very relevant in software engineering as the verification of large and

* This work is partly founded by the joint research project "DFG-Forschergruppe PETRINETZ-TECHNOLOGIE" between H. Weber (Coordinator), H. Ehrig (both from the Technical University Berlin) and W. Reisig (Humboldt-Universität zu Berlin), supported by the German Research Council (DFG).

H. Ehrig et al. (Eds.): Graph Transformation, LNCS 1764, pp. 410–425, 2000.

complex systems is often necessary, but very complicated (if possible at all) and thus expensive. The aim of this paper is to present a method how to derive safety properties step-by-step while developing a system. We suggest to develop safety properties hand in hand with the model. Usually, when we enhance a model we implicitly introduce new safety properties. Our approach is to make new safety properties explicit and to preserve them during the subsequent development steps. The formal basis of this method are on the one hand rules and transformations that preserve safety properties. On the other hand we use rules that introduce new safety properties.

The notion of rules for Petri nets is based on transformations within the abstract frame of high-level replacement (HLR) systems. HLR systems have been introduced in [EHKP91] as a categorical generalization of the double pushout approach to graph transformations. The application of high-level replacement systems to a special domain as place/transition nets, algebraic specification etc. requires a suitable category, a class of distinguished morphisms, and the satisfaction of the HLR-conditions. For place/transition nets with an initial marking we introduce a new class of distinguished morphisms compatible with the HLR conditions on the one hand and initial marking on the other hand.

In [Pad96] high-level replacement systems have been extended in order to allow different kinds of morphisms within the transformation step. This enables the expression of refinement using a morphism, called Q-morphism, directly from the left-hand side of the rule to the right-hand side. Based on this extension we can integrate net transformations with place preserving morphisms [Peu97], leading to place preserving rules. Moreover, we introduce transition gluing morphisms. These may identify transitions but nevertheless preserve safety properties as well. Their integration with net transformation yields transition gluing rules. The fact that place preserving as well as transition gluing rules are safety preserving requires three nontrivial conditions, called Q-conditions, see Theorem 1 (PP-Rules are Safety Preserving) and Theorem 2 (TG-Rules are Safety Preserving). In addition to preservation of safety properties we develop a concept of introducing new safety properties by net transformations. Our last main theorem (see Theorem 3 (SI-Rules are Safety Introducing and Preserving)) allows to iteratively formulate additional requirements to the model. The proofs to these theorems require detailed information about the involved categories, morphisms and constructions, and are – due to space limitations – omitted in this paper. For full proofs we refer to [GHP99].

These technical results yield an important outcome for software engineering. The preservation of safety properties as well as the propagation of newly introduced safety properties can be ensured on the level of rules. Thus, all the safety properties are propagated to the resulting net that are either in the start net or added during a transformation sequence. In [PGE98] we consider a similar problem, where we investigate place preserving morphisms and the corresponding rules and transformations of algebraic high-level nets [PER95], but no transition gluing rules, no rules introducing invariant formulas, and no markings of the nets.

The paper is organized the following way: Section 2 illustrates the main ideas in terms of a small example. Then we review high-level replacement systems in Section 3. Place/transition nets together with morphisms are given in Section 4. We show that place/transition nets transformations via place preserving rules as well as transition gluing rules preserve safety properties (see Theorems 1 and 2). Furthermore we present rules for introducing new safety properties and prove them to be safety preserving as well (see Theorem 3). A conclusion and discussion of future work can be found in Section 5.

2   Example: Developing the Model of an Elevator and Its Safety Properties

In this section we stepwise develop a Petri net model of an elevator. The development of the model goes along with the development of safety properties for the model. These safety properties have to be proven only when introducing them, because they are preserved by all further modifications of the model. Some basic notions of the models (place/transition nets), temporal logic formulas, and refinement of models by transformations are given on an intuitive level in order to explain the example. In the subsequent Sections 3 and 4 these notions will be formalized. We distinguish two major steps in the modelling of the elevator. First we derive a simple elevator which can arbitrarily move up and down. This model is equipped in a second step with a simple control mechanism to call the elevator. The first floor of the elevator is modelled by the net given in Figure 1.

Analogously to graph grammars we call this initial model *start net*. There is a floor denoted by **f** and two states of the door. The places **dc** and **do** denote a closed, respectively opened door. The state of doors can be changed by the transitions *o* and *c* meaning opening and closing. The elevator can either go up, modelled by transition *u*, or come down, by transition *d*.

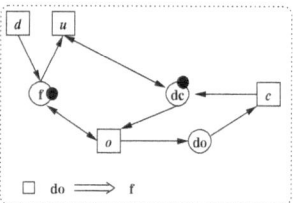

Figure 1: Start Net E_0

From the viewpoint of the first floor, the elevator vanishes by going up. Analogously, it appears by coming down in an unpredictable way, that is the pre domain of *d* is empty. The initial marking $M_0 = \mathbf{f} \oplus \mathbf{dc}$ denoted by black dots expresses that there is an elevator and the door is closed. Together with the start net E_0 there is given a safety property. For security reasons it should always be guaranteed that if the door stands open the elevator is on the floor. This is expressed by the temporal logic formula at the bottom line of Figure 1. Intuitively, a temporal logic formula, formalized in Definition 10, states facts about the markings and is given in terms of numbers of tokens on places. That is, the static formula $5\mathbf{a} \wedge 2\mathbf{b}$ is true for a marking M where at least 5 tokens are on place **a** and at least 2 tokens are on place **b**. The always operator in an invariant formula $\Box(5\mathbf{a} \wedge 2\mathbf{b})$ states that this is true for all reachable markings from M. In our case the safety property $\Box \mathbf{do} \Rightarrow \mathbf{f}$, meaning that "At any time, if the door is open then the elevator is on the floor" is satisfied. In fact, we can argue as follows: The formula **do** $\Rightarrow$ **f** is satisfied in the initial state. Moreover, u is the only transition which deletes the token on **f** and therefore may violate the formula. After its firing, o — the only transition to change the state of the door — is not enabled. Therefore, the door stays closed. Summarizing, the formula **do** $\Rightarrow$ **f** is always satisfied.

We are now going to enhance the model with further floors and requests bottoms. This will be done by adding floors to the start net, i. e. the application of the rules r_{int} and r_{fin} depicted in Figures 2 and 3. Application of a rule to a place/transition net (see also Definition 1) informally means replacing a subnet specified by the left-hand side of the rule with the net specified by the right-hand side. As the left-hand sides of r_{int} and r_{fin} are empty, we simply add the right-hand side to the (start) net. The property which should hold for each floor separately is again that the doors must be closed if the elevator is not in that floor. Corresponding-ly, the rules are *introducing new safety properties* (see Definition 7) depicted at the bottom of the net in the right-hand side. The formula $\Box$**do** $\Rightarrow$ **f** is satisfied for the net in the right-hand side of the rule, which can be seen analogously to the start net E_0.

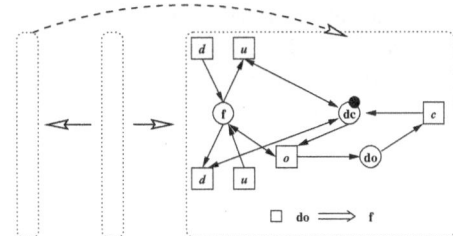

Figure 2: Rule r_{int} for Introducing an Intermediate Floor

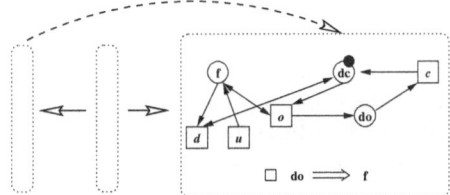

Figure 3: Rule r_{fin} for Introducing the Final Floor

Applying these rules, of course, we do not want to lose the safety property, which we already proved for the start net. Moreover, the introduced safety prop-erties should be propagated to the resulting net. The preservation of old safety properties and the satisfaction of the newly introduced safety properties is stated in Theorem 3. This means that the resulting net satisfies all the safety properties introduced by the rules and also all originally stated safety properties in the start net. The application of the rule r_{fin} depicted in Figure 3 and iterated applica-tion of the intermediate rule r_{int} yields an elevator with many (disconnected) floors and corresponding safety conditions.

In order to connect the floors we have to identify the up-going of the ele-vator from one floor with the coming-from-below from the next floor. This is achieved by the rule r_{glu} in Figure 4 that *glues the corresponding transitions*. It is compatible with the safety properties, stated in Theorem 2. This means, that in the derived net still all safety properties hold. A sample transformation sequence $E_0 \xrightarrow{r_{int}+r_{fin}} E_1 \xrightarrow{r_{glu}+r_{glu}} E_2$ yields the model E_2 of a simple elevator depicted in Figure 5, where $r_i + r_j$ designates the parallel application of rules r_i and r_j.

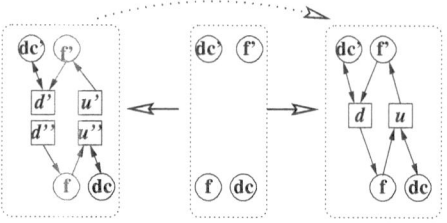

Figure 4: Rule r_{glu} for Gluing Transitions

The set of safety properties satisfied by E_2 is given by $\{\Box\textbf{do1} \Rightarrow \textbf{f1}, \Box\textbf{do2} \Rightarrow \textbf{f2}, \Box\textbf{do3} \Rightarrow \textbf{f3}\}$. This means that for all floors the safety property "At any time, if the door is open then the elevator is on the floor" holds. For enhancing this simple model, we want to add a simple control mechanism for calling the elevator. Three rules r_{rq_fin}, r_{rq_int}, and r_{rq_exc} (see Figures 6 here and Figures 7 and 8 in the Appendix), introduce exclusive requests to the elevator E_2. If there is a request at a floor, the elevator may not leave that floor, unless the door has been opened and subsequently closed. The insertion of requests to floors where only one direction of movement is possible is described by rule r_{rq_fin}. The marked place **nrq** designating no request and transition **r** (requesting) are added. Furthermore, the elevator may only move if there is no request on this floor which is captured by an additional arc to the transition **m**. By closing of the door the request is cleared which is modelled by the additional arc from **c** to **nrq**.

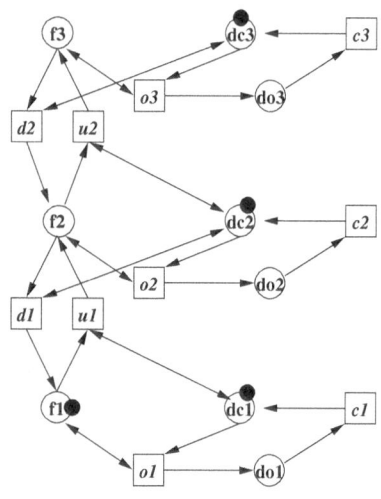

Figure 5: Simple Elevator E_2

Similarly, rule r_{rq_int} in the Appendix describes the insertion of requests to intermediate floors. The last rule r_{rq_exc} describes the mutual exclusion of the requests. All the rules $r_{rq...}$ do not change the environment of places, i. e. they are *place preserving*. By Theorem 1 they also preserve safety properties. Applying them to our simple elevator E_2 via $E_2 \xrightarrow{r_{rq\_fin}+r_{rq\_fin}+r_{rq\_int}} E_3 \xrightarrow{r_{rq\_exc}} E_4$ results in an elevator E_4 with a request mechanism (see Figure 9 in the Appendix), which still satisfies all the safety properties "At any time, if the door is open then the elevator is on the floor" for each floor.

The main advantage of our approach is that we do not have to prove the safety property in the net E_4 but just for the start net and for the rule introducing new safety properties. By these we could add further safety properties, which were preserved by transition gluing as well as place preserving rules. For software development this significantly decreases the cost of proving safety properties.

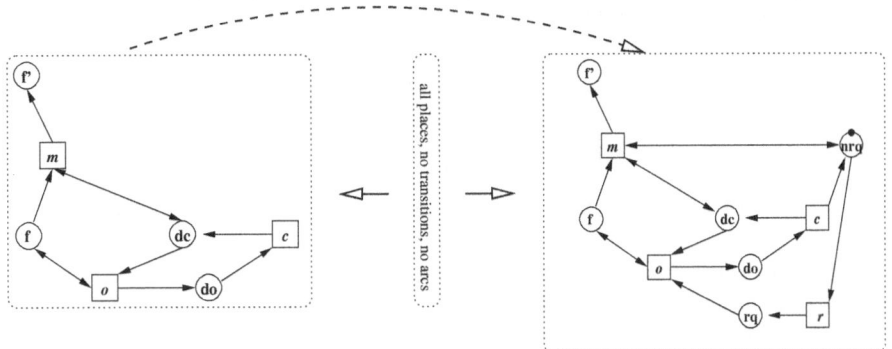

Figure 6: Rule r_{rq_fin}

3   Review of High-Level Replacement Systems

In this section we briefly review the concepts of high-level replacement (HLR) systems in the sense of [EHKP91], a categorical generalization of graph grammars. These concepts are mainly needed for our proofs in Section 4 and may be skipped by readers familiar with the theory of HLR and $\mathcal{Q}$-transformations.

High-level replacement systems are formulated for an arbitrary category $\mathcal{C}$ with a distinguished class $\mathcal{M}$ of morphisms. High-level replacement systems allow a great variety of interpretations of the concept of rules. Whereas in graph grammars rules describe the dynamic changes of graphs, in high-level replacement systems they define any kind of evolution of the system. In [EGPP99] they have been instantiated by algebraic specifications and Petri nets and used in the context of software engineering. Transformations of place/transition nets are used for the development of the system. In general, rules are splitted into a deleting part L, an adding part R and an interface K which is preserved, such that the rule r is given by $r = (L \xleftarrow{u} K \xrightarrow{v} R)$ where u and v are morphisms in the class $\mathcal{M}$.

Definition 1 (Rules and Transformations). *A rule* $r = (L \xleftarrow{u} K \xrightarrow{v} R)$ *in* $\mathcal{C}$ *consists of the objects* L, K *and* R, *called left-hand side, interface (or gluing object) and right-hand side respectively, and two*

morphisms $K \xrightarrow{u} L$ *and* $K \xrightarrow{v} R$ *with both morphisms* $u, v \in \mathcal{M}$, *a distinguished class of morphisms in* $\mathcal{C}$. *Given a rule* $r = (L \xleftarrow{u} K \xrightarrow{v} R)$ *a direct transformation* $G \xRightarrow{r} H$, *from an object* G *to an object* H *is given by two pushout diagrams*

$$
\begin{array}{ccccc}
L & \xleftarrow{u} & K & \xrightarrow{v} & R \\
{\scriptstyle g_1}\downarrow & {\scriptstyle (1)} & {\scriptstyle g_2}\downarrow & {\scriptstyle (2)} & \downarrow{\scriptstyle g_3} \\
G & \xleftarrow{c_1} & C & \xrightarrow{c_2} & H
\end{array}
$$

(1) *and* **(2)** *in the category* $\mathcal{C}$. *The morphisms* $L \xrightarrow{g_1} G$ *and* $R \xrightarrow{g_3} H$ *are called occurrences of* L *in* G *and* R *in* H, *respectively. By an occurrence of rule* $r =$

$(L \xleftarrow{u} K \xrightarrow{v} R)$ in a structure G we mean an occurrence of the left-hand side L in G.

A transformation sequence $G \overset{*}{\Longrightarrow} H$, short transformation, between objects G and H means G is isomorphic to H or there is a sequence of $n \geq 1$ direct transformations: $G = G0 \overset{r1}{\Longrightarrow} G1 \overset{r2}{\Longrightarrow} \ldots \overset{rn}{\Longrightarrow} Gn = H$.

Definition 2 (High-Level Replacement System). *Given a category $\mathcal{C}$ together with a distinguished class of morphisms $\mathcal{M}$ then $(\mathcal{C}, \mathcal{M})$ is called a HLR-category if $(\mathcal{C}, \mathcal{M})$ satisfies the HLR-Conditions (see [EGPP99]).*

In [Pad96,Pad99] we have introduced the notions of $\mathcal{Q}$-morphisms and $\mathcal{Q}$-rules, which are motivated by different kinds of refinement for Petri nets. Mostly refinement notions in Petri net literature concern the refinement of single net elements like transitions or places, for a survey see [BGV90]. Additionally, refinement of behavior — as considered here — is often given informally, or in terms of morphisms [Des91,DM90,BG92] that preserve or reflect the behaviour of a net. $\mathcal{Q}$-rules provide the formal frame for integration of such morphisms with rules and thus all their advantages like locality and genericity.

The main idea is to enlarge the given HLR-category in order to include morphisms, that are adequate for the refinement. The $\mathcal{Q}$-conditions [Pad99] state the additional requirements an HLR-category has to satisfy for the extension to refinement morphisms.

Definition 3 ($\mathcal{Q}$: Refinement Morphism [Pad99]). *Let $\mathcal{QC}$ be a category, so that $\mathcal{C}$ is a subcategory $\mathcal{C} \subseteq \mathcal{QC}$ and $\mathcal{Q}$ a class of morphisms in $\mathcal{QC}$.*

1. *The morphisms in $\mathcal{Q}$ are called $\mathcal{Q}$-morphisms, or refinement morphisms.*
2. *Then we have the following $\mathcal{Q}$-conditions:*
   Closedness: *$\mathcal{Q}$ has to be closed under composition.*
   Preservation of Pushouts: *The inclusion functor $I : \mathcal{C} \to \mathcal{QC}$ preserves pushouts, that is, given $C \xrightarrow{f'} D \xleftarrow{g'} B$ a pushout of $B \xleftarrow{f} A \xrightarrow{g} C$ in $\mathcal{C}$, then $I(C) \xrightarrow{I(f')} I(D) \xleftarrow{I(g')} I(B)$ is a pushout of $I(B) \xleftarrow{I(f)} I(A) \xrightarrow{I(g)} I(C)$ in $\mathcal{QC}$.*
   Inheritance of $\mathcal{Q}$-morphisms under Pushouts: *The class $\mathcal{Q}$ in $\mathcal{QC}$ is closed under the construction of pushouts in $\mathcal{QC}$, that is, given $C \xrightarrow{f'} D \xleftarrow{g'} B$ a pushout of $B \xleftarrow{f} A \xrightarrow{g} C$ in $\mathcal{QC}$, then $f \in \mathcal{Q} \Rightarrow f' \in \mathcal{Q}$.*
   Inheritance of $\mathcal{Q}$-morphisms under Coproducts: *The class $\mathcal{Q}$ in $\mathcal{QC}$ is closed under the construction of coproducts in $\mathcal{QC}$, that is, for $A \xrightarrow{f} B$ and $A' \xrightarrow{f'} B'$ we have $f, f' \in \mathcal{Q} \Rightarrow f + f' \in \mathcal{Q}$ provided the coproduct $A + A' \xrightarrow{f+f'} B + B'$ of f and f' exists in $\mathcal{QC}$.*
3. *A $\mathcal{Q}$-rule (r, q) is given by a rule $r = L \xleftarrow{u} K \xrightarrow{v} R$ in $\mathcal{C}$ and a $\mathcal{Q}$-morphism $q : L \to R$, so that $K \xrightarrow{u} L \xrightarrow{q} R = K \xrightarrow{v} R$ in $\mathcal{QC}$.*

The next fact states the class $\mathcal{Q}$ is also preserved under transformations.

Fact 4 ($\mathcal{Q}$-Transformations [Pad99]). *Let $\mathcal{C}$, $\mathcal{QC}$, $\mathcal{Q}$, and $I : \mathcal{C} \to \mathcal{QC}$ satisfy the $\mathcal{Q}$-conditions. Given a $\mathcal{Q}$-rule (r, q) and a transformation $G \overset{r}{\Longrightarrow} H$ in $\mathcal{C}$*

defined by the pushouts (1) and (2), then there is a unique $q' \in \mathcal{Q}$, such that $q' \circ c_1 = c_2$ and $q' \circ g_1 = g_3 \circ q$ in $\mathcal{QC}$. The transformation $(G \overset{r}{\Longrightarrow} H, q' : G \to H)$, or short $G \overset{(r,q')}{\Longrightarrow} H$, is called $\mathcal{Q}$-transformation. $R \overset{g_3}{\to} H \overset{q'}{\leftarrow} G$ is pushout of $G \overset{g_1}{\leftarrow} L \overset{q}{\to} R$ in $\mathcal{QC}$.

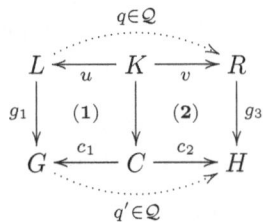

Proof Idea: Due to Definition 3, item 3 and the pushout properties of **(1)** the outer square commutes. Thus, by rearranging the diagram and decomposition of one pushout square into two pushout squares, the outer square is a pushout.

In [Pad99] it is shown, that well-known results from the double pushout approach to graph transformation and from high-level replacement systems, as Parallelism Theorem and Church-Rosser Theorems can be adapted to $\mathcal{Q}$-transformations. Especially we use the Parallelism Theorem in our ongoing example (see Section 2).

4   Rule-Based Refinement Preserving Safety Properties

In this section we are going to present the technical results leading to preservation of safety properties in place/transition nets by transformations. We define place/transition nets and special kinds of morphisms, namely place preserving and transition gluing morphisms. The integration of place preserving morphisms and rules in the context of the theory of $\mathcal{Q}$-transformations [Pad96] leads to place preserving rules. Analogously, we obtain transition gluing rules using transition gluing morphisms. Furthermore, we define formulas expressing safety properties and their preservation by special kinds of rules, the safety preserving rules. More precisely, preservation of safety properties always is intended to be "up to renaming of places" because of the use of morphisms and transformations. In our main results — Theorems 1 (PP-Rules are Safety Preserving) and Theorem 2 (TG-Rules are Safety Preserving) — we show that these rules preserve safety properties via transformations. Furthermore, we define rules introducing new safety properties and show in Theorem 3 (SI-Rules are Safety Introducing and Preserving), that they are also safety preserving. Due to space limitations we can only sketch the proofs here (detailed proofs can be found in [GHP99]). First, we recall place/transition nets and different kinds of morphisms. Note, that most notions are especially designed to preserve safety properties. Except for transition preserving morphisms they do not preserve the behaviour of nets unlike the majority of notions in the literature, for a survey see [Hof98].

Definition 5 (Place/Transition Nets). *A place/transition net is given by $N = (P, T, pre, post, M^0)$ with P the set of places, T the set of transitions,*

*pre, post : $T \rightarrow P^{\oplus}$ the pre- and postdomain[1] of transitions, and $M^0 \in P^{\oplus}$
describes the initial marking.*

The definition of morphisms is based on functions between the sets of places
and transitions. These functions have to be compatible with pre- and postfunc-
tions and with the initial marking. We introduce new notions of morphisms
that employ placewise comparison of markings (see condition (2) and marking
strictness of Definition 7). This comparison is based on the notion of restriction.

Definition 6 (Restriction). *Given $M \in P^{\oplus}$ and $P' \subseteq P$, then there is $M' \in P'^{\oplus}$ and $M'' \in (P \setminus P')^{\oplus}$ such that $M = M' \oplus M''$. We define $M_{|P'} = M'$
For $P' = \{p\}$ we denote $M_{|p} = M_{|P'}$. For a function $f_P : P_0 \rightarrow P$ we
write $M_{|f_P} = M_{|f_P(P_0)}$*

Definition 7 (PT-Morphisms and Categories). *Given $N_i = (P_i, T_i, pre_i, post_i, M_i^0)$, $i \in \{1, 2\}$ two place/transition nets. A morphism $f = (f_P, f_T)$:
$N_1 \rightarrow N_2$ with functions $f_P : P_1 \rightarrow P_2$, and $f_T : T_1 \rightarrow T_2$ is called*

loose *if the following embedding condition holds for all $t \in T_1$ and $p \in P_1$:*

(1) $f_P^{\oplus}(pre_1(t)) \leq pre_2(f_T(t))$ and $f_P^{\oplus}(post_1(t)) \leq post_2(f_T(t))$
(2) $f_P^{\oplus}(M_{1|p}^0) \leq M_{2|f_P(p)}^0$

This gives rise to the category **QPT**.
transition preserving *if it is loose and the following condition holds:*

(3) $f_P^{\oplus} \circ pre_1 = pre_2 \circ f_T$ and $f_P^{\oplus} \circ post_1 = post_2 \circ f_T$

This gives rise to the category **PT**. *Moreover, a transition preserving morphism is called:*

(marking strict)   *if $f_P^{\oplus}(M_{1|p}^0) = M_{2|f_P(p)}^0$*
(strict)   *if additionally f_P and f_T are injective*

place preserving [Peu97] *if it is a loose morphism and the following place
preserving conditions hold:*

(4) $\bullet(f_P(p)) = f_T(\bullet p)$ and $(f_P(p))\bullet = f_T(p\bullet)$ for all $p \in P_1$
       *where $\bullet p = \Sigma_{t \in T} post(t)(p) \cdot t$ and $p\bullet = \Sigma_{t \in T} pre(t)(p) \cdot t$
       define the pre and post sums of p with
       coefficients[1] $post(t)(p)$, $pre(t)(p)$.*
(5) $f_P^{\oplus}(M_1^0) = M_{2|f_P}^0$
(6) f_P and f_T are injective

transition gluing *if it is a loose morphism and the following holds:*

(7) $f_P : P_1 \xrightarrow{\sim} P_2$ is an isomorphism
(8) $f_P^{\oplus}(M_1^0) = M_2^0$
(9) f_T is surjective and $pre_2(t_2) = \Sigma_{t_1 \in f_T^{-1}(t_2)} f_P^{\oplus}(pre_1(t_1))$
       with $t_2 \in T_2$ and similarily for the post function.

---

[1] Elements w of the free commutative monoid $X^{\oplus}$ for some set X can be represented as
$w = \Sigma_{x \in X} \lambda_x x$ with coefficients $\lambda_x \in \mathbb{N}$. They can be considered as finite multisets.
Free commutative monoids imply the operations $\oplus, \ominus, \leq$ on linear sums.

Remark: Note, that these notions are novel and especially devised for the purpose of safety preservation for nets with initial marking. Particularly, place preserving morphisms preserve safety mainly due to condition (4) meaning, that the environment of places is preserved, s.t. no "new" arcs to or from transitions are added. Similarily condition (9) for transition gluing morphisms means, that no "new" arcs occur in the target net, and is thus mainly responsible for preserving safety.

The integration of different kinds of morphisms with rules leads to Definitions 8 and 9. In our main Theorems 1 and 2 we will show that these rules preserve safety properties. Note, that we demand strictness in Definitions 8 and 9 as these morphisms give rise to a HLR system.

Definition 8 (PP-Rules). *A pair (r, f) is called pp-rule, if $r = (L \xleftarrow{u} K \xrightarrow{v} R)$ is a rule with strict transition preserving morphisms u, v and a place preserving morphism $f : L \to R$ with $f \circ u = v$.*

Definition 9 (TG-Rules). *A pair (r, f) is called tg-rule, if $r = (L \xleftarrow{u} K \xrightarrow{v} R)$ is a rule with strict transition preserving morphisms u, v and a transition gluing morphism $f : L \to R$ with $f \circ u = v$.*

Example 1 (Morphisms and Rules in Elevator Example). In Figures 2, 3, 4, 6, 7, and 8 the morphisms denoted by $\longrightarrow$ are transition preserving, as there are no transitions in the interfaces of the rules. The morphisms denoted by $\longrightarrow$ are place preserving, as in the right-hand sides of the corresponding rules all new arcs are connected only to new places. Consequently, the rules in Figures 2, 3, 6, 7, and 8 are pp-rules. In Figure 4 the morphism denoted by $\cdots\!\!\!>$ is a transition gluing morphism and thus it denotes a tg-rule.

Next, we formalize safety properties in order to formulate our theorems concerning their preservation. We recall formulas over markings and their translations via morphisms. This allows expressing safety properties and their preservation via morphisms (see [Peu97]). The invariant formula $\Box\varphi$ expresses safety properties in the sense of [MP92], in e.g. [CM88,Sti92,BGV90,Rei98] further notions can be found.

Definition 10 (Formulas, Translations). *λp is a static formula for $\lambda \in \mathbb{N}$ and $p \in P$, static formulas are built up using the logical operators $\wedge$ and $\neg$. Let φ be a static formula over N. Then $\Box\varphi$ is an invariant formula.*
The validity of formulas is given w.r.t. the marking of a net. Let $M \in P^\oplus$ be an arbitrary marking of N then $M \models_N \lambda p$ iff $\lambda p \leq M$. For $M \models_N \neg\varphi_1$ iff $\neg(M \models_N \varphi_1)$ and $M \models_N \varphi_1 \wedge \varphi_2$ iff $(M \models_N \varphi_1) \wedge (M \models_N \varphi_2)$. The invariant formula $\Box\varphi$ holds in N under M iff φ holds in all states reachable from M: $M \models_N \Box\varphi \iff \forall M' \in [M\rangle : M' \models_N \varphi$. We also write $N \models \Box\varphi$ instead of $M^0 \models_N \Box\varphi$.
The translation of formulas T_f over N_1 along a morphism $f = (f_P, f_T) : N_1 \to N_2$ to formulas over N_2 is given for atoms by $T_f(\lambda p) = \lambda f_P(p)$. The translation

420 Julia Padberg et al.

of formulas is given recursively by $T_f(\neg\varphi) = \neg T_f(\varphi)$, and $T_f(\varphi_1 \wedge \varphi_2) = T_f(\varphi_1) \wedge T_f(\varphi_2)$, and $T_f(\Box\varphi) = \sqcup T_f(\varphi)$

Definition 11 (Safety Preserving Morphism). *A morphism $f : N_1 \rightarrow N_2$ is called safety preserving, iff for all invariant formulas $\Box\varphi$ we have $N_1 \models \Box\varphi \Longrightarrow N_2 \models T_f(\Box\varphi)$.*

Corollary 1. *Safety preserving morphisms are closed under composition.*

Based on the notion of safety preserving morphisms we can now define safety preserving rules and transformations. The general idea is that the application of a rule that preserves safety properties leads to a net transformation that also preserves safety properties.

Definition 12 (Safety Preserving Rule and Transformation). *Given a rule $r = (L \leftarrow K \rightarrow R)$ in an HLR system and an arbitrary morphism $f : L \rightarrow R$. The pair (r, f) is a safety preserving rule, iff*

1. *f is a safety preserving morphism*
2. *given a place/transition net N_1, and an occurrence $g_1 : L \rightarrow N_1$, the direct transformation $N_1 \overset{r}{\Longrightarrow} N_2$ yields a safety preserving morphism $\bar{f} : N_1 \rightarrow N_2$.*

We denote the direct transformation by $N_1 \overset{(r,f)}{\Longrightarrow} N_2$ and call it safety preserving transformation, short sp-transformation.

The following Theorems 1 and 2 provide sufficient conditions for safety preserving rules. In fact, these rules that have place preserving morphisms or transition gluing morphisms from the left to the right-hand side, preserve safety properties.

Theorem 1 (PP-Rules are Safety Preserving).

The subsequent proof idea merely sketches the most important steps of the corresponding proof (see [GHP99] on pages 25 and 31–41). **Proof Idea:** Place preserving morphism are safety preserving since both the environment and the marking of places are preserved. For condition 2 in Definition 12 we show that a place preserving morphism is a Q-morphism according to Definition 3. We use the category **QPT** of Definition 7 with loose morphisms as supercategory of **PT**. The class Q of morphisms is given by place preserving morphisms. Obviously the class Q of place preserving morphisms is closed under composition. Preservation of pushouts is due to the componentwise construction of pushout objects in both **QPT** and **PT**. Inheritance of place preserving morphisms under pushouts can be shown as follows: The conditions for place preserving morphisms in Definition 7 can be shown as follows: Condition 4 follows from injectivity, condition 2, and the construction of the pushout object in the category **QPT**. Condition 5 is due to the construction of pushouts. Condition 6 immediately follows by preservation of monomorphisms under pushouts. Finally, the inheritance of place preserving morphisms under coproducts is due to the componentwise construction of colimits. Thus, we can apply the theory of Q-transformation which exactly yields condition 2 of Definition 12 and thus the stated proposition.

Theorem 2 (TG-Rules are Safety Preserving).

The proof corresponding the ideas below can be found again in [GHP99] pages 27, and 31–34, and 41–46. **Proof Idea:** Transition gluing morphisms are safety preserving (Condition 1 in Definition 12) can be seen as follows: Due to condition (7) and (8) in Definition 7 each static formula is preserved and reflected. Especially, invariant formulas are preserved, which uses the fact that transition gluing morphisms reflect firing by condition (9). For Condition 2 in Definition 12 we again prove the Q-conditions where **PT** and **QPT** are given as in the proof of Theorem 1 and thus pushouts are preserved. The composition of two transition gluing morphisms trivially yields a morphism of the same kind. For inheritance of transition gluing morphisms under pushouts we argue as follows: Due to the fact that pushouts in the category **QPT** imply pushouts of the components we have the conditions (7) and (8) of Definition 7 for transition gluing morphisms. Condition (9) of Definition 7 follows directly as pushouts in **SETS** preserve surjective morphisms and due to the construction of the marking of the pushout object. Inheritance under coproducts is guaranteed as a special case of the fact that pushouts in the category **QPT** imply pushouts of the components.

Apart from preserving safety properties, we now present a concept for introducing new safety properties by rules. In Theorem 3 we state, that these rules additionally propagate safety properties via application.

Definition 13 (SI-Rule). *Given a place/transition net R and an invariant formula $\Box\varphi$ over R s. t. $R \models \Box\varphi$. An si-rule $r = (\emptyset \leftarrow \emptyset \rightarrow R)$ introduces the invariant formula $\Box\varphi$.*

Example 2 (SI-Rule). In our example in Section 2 the rules r_{int} and r_{fin} depicted in Figures 2 and 3 are introducing an invariant formula, i. e. si-rules.

Theorem 3 (SI-Rules are Safety Introducing and Preserving).

Proof Idea: (see again [GHP99] for details) The application of an si-rule to a net N yields the coproduct $N + R$ and an occurence $o : R \rightarrow N + R$ which is a coproduct injection and therefore place preserving. These morphisms preserve safety properties (see proof to Theorem 1). Furthermore, the unique morphism $\emptyset : \emptyset \rightarrow R$ (from the left-hand side to the right-hand side) is place preserving. Thus si-rules are a special case of pp-rules and Theorem 1 yields the stated proposition.
Remark: The parallel rule $r_{si} + r_{pp}$ of a si-rule r_{si} and a pp-rule r_{pp} is also place preserving because si-rules are special pp-rules. Furthermore, it can be considered as a more general safety introducing rule.

Example 3 (SP-Transformations for the Elevator). Again we consider the example given in Section 2 with respect to the safety property.
In the start net E_0 we have the safety property, that "At any time, if the door is open, then the elevator is on the floor". In the first transformation from E_0

to E_1 we apply rules r_{int} and r_{fin} in parallel (see Figures 2 and 3). These rules are introducing new safety properties for each additional floor. As r_{int} and r_{fin} are si-rules, the parallel rule $r_{int} + r_{fin}$ defined by $L_{floor} \leftarrow K_{floor} \rightarrow R_{floor}$ is also a si-rule. Its application transforms $E_0 \dashrightarrow E_1$ and the safety property for the start net is also preserved due to Theorem 3. With the newly introduced safety properties we have the safety property "At any time, if the door is open, then the elevator is on the floor" for all floors in E_1. Analogously, the application of $r_{glu} + r_{glu}$ leads to a transition gluing transformation $E_1 \dashrightarrow E_2$ which also preserves safety properties (Theorem 2). Similarly, application of $r_{rq\_int} + r_{rq\_fin}$ and r_{rq_exc} preserves the safety property by Theorem 1. We have the following sequence of transformations, where we skip the interfaces due to space limitations:

$$
\begin{array}{ccccc}
L_{floor} \dashrightarrow R_{floor} & L_{conn} \dashrightarrow R_{conn} & L_{req} \dashrightarrow R_{req} & L_{r_{rq\_exc}} \dashrightarrow R_{r_{rq\_exc}} \\
\downarrow \qquad\qquad\quad & \searrow \quad \downarrow \qquad\quad & \searrow \quad \downarrow \qquad\quad & \searrow \quad \downarrow \qquad\qquad \downarrow \\
E_0 \dashrightarrow\dashrightarrow E_1 & \dashrightarrow\dashrightarrow E_2 & \dashrightarrow\dashrightarrow E_3 & \dashrightarrow E_4
\end{array}
$$

where $X_{floor} = X_{r_{int}} + X_{r_{fin}}$, $X_{conn} = X_{r_{glu}} + X_{r_{glu}}$, and $X_{req} = X_{r_{rq\_int}} + X_{r_{rq\_fin}}$ for $X \in \{L, R\}$.

Due to Corollary 1 the concatenated morphisms in the bottom line of the diagram preserve the safety property "At any time, if the door is open, then the elevator is on the floor" is propagated to E_4.

5   Future Work

In the context of Petri nets rules and their application yield a vertical structuring technique for modification of nets. Based on the theory of $\mathcal{Q}$-transformations [Pad99] we have enhanced rule-based modification with the preservation of safety properties in the sense of [MP92]. Rules are equipped with an additional morphism from the left-hand side to the right-hand side that preserves safety properties. We have presented two kinds of rules, namely pp-rules (with a place preserving morphism from the left to the right) and tg-rules (with a transition gluing morphism from left to right). The Theorems 1 and 2 state that these rules preserve safety properties. In the corresponding proofs the theory of $\mathcal{Q}$-transformations is exploited to show that the application of such a rule yields a safety preserving morphism from the source to the resulting net. Furthermore, we presented the concept of introducing new safety properties by rules. Theorem 3 states that these rules do not only introduce new safety properties, but preserve (old) safety properties as well. For software engineering purposes these are important results as they allow an incremental development of safety properties in place/transition net transformations. Summarizing, applying safety preserving rules yields a system with all those safety properties that were true for a subsystem at a certain point of time. The simplest form is the simple propagation of newly introduced safety properties by si-rules. More complex properties which exploit the interaction between different components of the system cannot yet be handled by rules. Those properties have to be explicitly proven once and are subsequently propagated. Nevertheless, even in this case our approach simplifies verification as the subsystem in consideration might be much smaller than the resulting system. But this certainly should be an interesting topic of future research.

Furthermore, we want to rise the open question, what applications of these refinement morphisms can be found in other instances of high-level replacement systems, as graphs, algebraic specifications, etc. Because safety as well as liveness properties require a dynamic behavior, these obviously cannot be transferred from one graph or algebraic specification to another. But certainly the examination of other properties, that can be propagated in the frame of Q-morphisms, as graph theoretic properties or persistency for algebraic specifications, is an interesting and fruitful task.

References

BG92.   C. Brown and D. Gurr, *Refinement and Simulation of Nets – a categorical characterization*, 13th International Conference on Application and Theory of Petri Nets, LNCS **616** (K. Jensen, ed.), Springer Verlag, 1992, pp. 76–92. 416

BGV90.   W. Brauer, R. Gold, and W. Vogler, *A Survey of Behaviour and Equivalence Preserving Refinements of Petri Nets*, Advances in Petri Nets, LNCS **483** (1990), pp. 1–46. 416, 419

CM88.   K. M. Chandy and J. Misra, *Parallel program design: A foundation*, Addison-Wesley, 1988. 419

Cor95.   A. Corradini, *Concurrent Computing: From Petri Nets to Graph Grammars*, Proc. of the Joint COMPUGRAPH/SEMAGRAPH Workshop on Graph Rewriting and Computation (SEGRAGRA'95), Volume 2 of Electronic Notes in Theoretical Computer Science, 1995. 410

Des91.   J. Desel, *On Abstraction of Nets*, Advances in Petri Nets, LNCS **524**, Springer Verlag , 1991, pp. 78–92. 416

DM90.   J. Desel and A. Merceron, *Vicinity Respecting Net Morphisms*, Advances in Petri Nets, LNCS **483**, Springer Verlag, 1990, pp. 165–185. 416

EGPP99.   H. Ehrig, M. Gajewsky, and F. Parisi Presicce, High-Level Replacement Systems with Applications to Algebraic Specifications and Petri Nets, vol. 3: Concurrency, Parallelism, and Distribution, ch. 4, pp. 341–399, World Scientific, Handbook of Graph Grammars and Computing by Graph Transformations, 1999, pp. 341–399, to appear. 410, 415, 416

EHKP91.   H. Ehrig, A. Habel, H.-J. Kreowski, and F. Parisi-Presicce, *Parallelism and concurrency in high-level replacement systems*, Math. Struct. in Comp. Science **1** (1991), pp. 361–404. 411, 415

GHP99.   M. Gajewsky, K. Hoffmann, and J. Padberg, *Place Preserving and Transition Gluing Morphisms in Rule-Based Refinement of Place/Transition Systems*, Tech. Report 99/14, Technical University Berlin, 1999, to appear. 411, 417, 420, 421

Hof98.   Kathrin Hoffmann, *Structural Compatibility in Petri Nets: Morphisms and Categories of Nets*, Master's thesis, Technical University Berlin, 1998. 417

Kre81.   H.-J. Kreowski, *A comparison between Petri-nets and graph grammars*, LNCS **100**, Springer Verlag, 1981, pp. 1–19. 410

MP92.   Zohar Manna and Amir Pnueli, *The temporal logic of reactive and concurrent systems, specification*, Springer Verlag, 1992. 410, 419, 422

Pad96.   J. Padberg, *Abstract Petri Nets: A Uniform Approach and Rule-Based Refinement*, Ph.D. thesis, Technical University Berlin, 1996, Shaker Verlag. 411, 416, 417

Pad99.   Julia Padberg, *Categorical Approach to Horizontal Structuring and Refinement of High-Level Replacement Systems*, Applied Categorical Structures (1999), accepted. 410, 416, 417, 422

PER95.   J. Padberg, H. Ehrig, and L. Ribeiro, *Algebraic high-level net transformation systems*, Math. Struct. in Comp. Science **5** (1995), pp. 217–256. 411

Peu97.   S. Peuker, *Invariant Property Preserving Extensions of Elementary Petri Nets*, Tech. Report 97/21, Technical University Berlin (1997). 411, 418, 419

PGE98.   J. Padberg, M. Gajewsky, and C. Ermel, *Rule-Based Refinement of High-Level Nets Preserving Safety Properties*, Fundamental approaches to Software Engineering , LNCS **1382**, (E. Astesiano, ed.), Springer Verlag, 1998, pp. 221–238. 411

Rei98.   Wolfgang Reisig, *Elements of Distributed Algorithms. Modeling and Analysis with Petri Nets*, Springer Verlag, 1998. 419

Sch94.   H.-J. Schneider, *Graph grammars as a tool to define the behaviour of process systems: From Petri nets to Linda*, Preliminary Proc. of 5th International Workshop on Graph Grammars and their Application to Computer Science (G. Engels and G. Rozenberg, eds.), 1994. 410

Sch99.   H.-J. Schneider, Describing Systems of Processes by means of High-Level Replacement, vol. 3: Concurrency, Parallelism, and Distribution, ch. 7, World Scientific, Handbook of Graph Grammars and Computing by Graph Transformations, 1999, to appear. 410

Sti92.   C. Stirling, *Modal and temproal logics*, Handbook of Logic in Computer Science, vol. 2, Background: Computational structures, Clarendon Press, Oxford, 1992. 419

6 Appendix

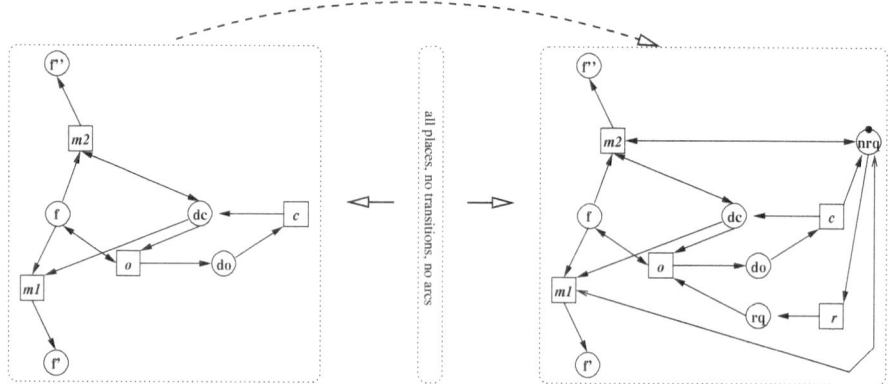

Figure 7: Rule r_{rq_int}

The difference of rule r_{rq_int} to rule r_{rq_fin} in Figure 6 is that the elevator may not move in neither direction if there is a request. This is achieved by the additional arcs both to transitions **m1** and **m2**.

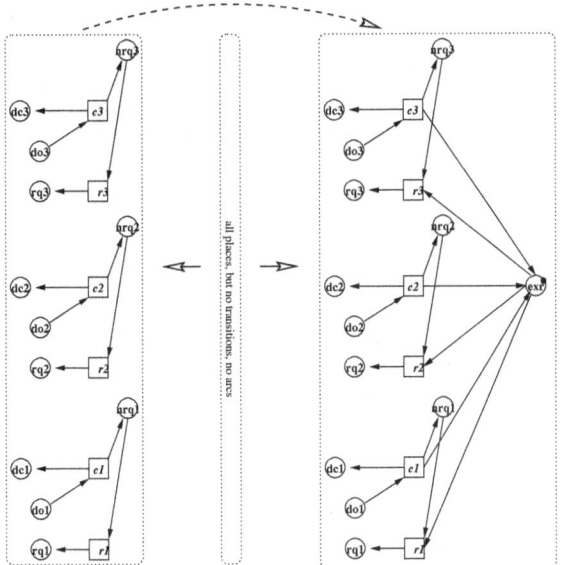

Figure 8: Rule r_{rq_exc}

Rule r_{rq_exc} adds the exclusion of requests. We insert the marked place **exr** and connect it to the net in such a way that it realizes the mutual exclusion of the transitions **r1**, **r2** and **r3** for calling the elevator. Once the token is withdrawn by any of these transitions the others cannot fire and it is only released when the request is cleared on that floor.

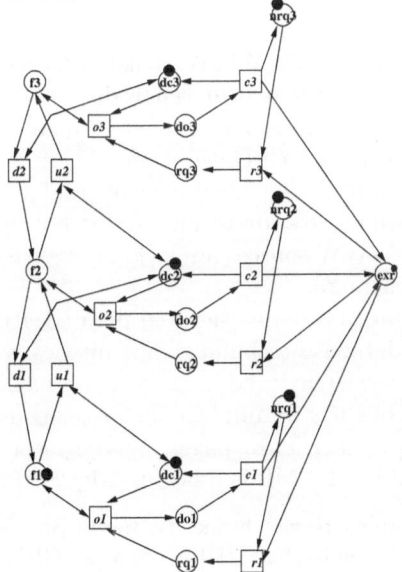

Figure 9: Elevator E_4 with request mechanism

Using Graph Transformation Techniques for Integrating Information from the WWW

Lukas C. Faulstich[1*]

Institut für Informatik, Freie Universität Berlin,
http://www.inf.fu-berlin.de/~faulstic

Abstract. The advent of the WWW has led to an abundance of information which is distributed on numerous Web sites of heterogeneous structure and coverage. It is therefore important to extract, combine and restructure these distributed data in order to facilitate the access to it. The HyperView methodology models Web sites and their HTML pages as graphs from which information is extracted by a hierarchy of views. We present the clustered graph data model (CGDM) used in the Hyper-View system. Views are defined using graph transformation rules. We use typed attributed Single Pushout graph transformation with application conditions on attributes.

The main contribution of this paper is a new rule activation mechanism that supports the incremental computation of views, a crucial requirement in the context of information extraction from Web sites. The HyperView prototype is currently used in the field of Digital Libraries. In this paper, we demonstrate our methodology in the domain of town information.

1 Introduction

The main problem with today's WWW is not a lack of information, but the diversity of sources on which the information is dispersed. This makes the WWW hard to use for planning purposes, for instance to plan how to spend an evening in a city like Berlin with a a rich cultural life. There is an increasing need for integration of WWW sources. The integration of databases is a well-studied field. Unfortunately, database integration techniques do not apply directly to WWW data integration since WWW sources are lacking essential database features like query processing or schemata.

The new research area on semistructured data ([1,4]) deals with approaches to adapt well-known database techniques like queries, views, and schemata to semistructured data. Semistructured data lacks an explicit structure to which it conforms perfectly, but has an implicit, heterogeneous structure with a high degree of variation instead. HTML pages are the most prominent example of semistructured data. Since HTML is a layout-oriented format, the semantics of

---

[*] Supported by the German Research Society, Berlin-Brandenburg Graduate School on Distributed Information Systems (DFG grant no. GRK 316)

H. Ehrig et al. (Eds.): Graph Transformation, LNCS 1764, pp. 426–441, 2000.

a page are usually hidden in its HTML code. Fortunately, often classes of pages sharing the same structure can be identified which are suitable automatic data extraction. This holds in particular for generated pages.

Running Example: Integration of town information. We define "Town Information" as information related to events, infrastructure and services of some urban community. In particular we focus on cultural and entertainment related information. Hence, events describe concerts, movie shows, street markets and other more, infrastructure means "venues" (locations of events, i.e., theaters, concert halls) as well as restaurants, pubs, bars, etc. Services include public transportation (e.g., bus and subway schedules), ticket reservation, weather forecasting, shopping, and so on.

To a large extent, town information is already available at various Web sites. Even though a few servers of city magazines specialize on this domain, it is still necessary to consult additional servers. For instance, detailed concert announcements can often only be found at the Web sites of concert halls or clubs, e.g., of the jazz club "A-Trane"[1] which serves as running example in this paper. Also, town maps and public transportation information are typically available from dedicated servers only.

Using the sources mentioned above, it is feasible to plan for instance how to spend an evening in Berlin. However, when trying to find and combine relevant pieces of information, the user is confronted with various problems. The main problem is the *missing link*: related information on different servers is not connected by hyper-links. If the user tries to find related information by consulting another server, he often encounters the problem that data from one server is of little use in another server. For instance, arbitrary street names can not be used in a public transportation server that requires the user to enter names of bus stops. Another problem is that for instance no server covers all concerts. To get the full picture, information has to be collected from Web sites, each of which is organized differently. Updates in the sources are another problem. To become aware of changes, the user has to check relevant servers on a regular basis. Finally, new Web sites are set up and existing ones are reorganized or even closed down quite often. This makes it necessary to deal with these changes.

The HyperView approach supports the building of integrated "virtual" Web sites that add missing links, enhance the coverage, and resolve the structural heterogeneity of the underlying sources. By using robust and high-level view definitions, it also tries to reduce the costs for setting up and maintaining such virtual Web sites. The HyperView approach does not attempt to automate the discovery and integration of new sources since this requires the experience of a human domain expert.

Organization of the paper: In Section 2 we give an overview of the HyperView approach. In Section 3, we demonstrate the application of the HyperView approach in the field of town information. A formal presentation of our view concept based on graph transformation techniques is given in Section 4. In Sec-

---

[1] <http://www.a-trane.de/>

tion 5, related work is reviewed, and Section 6 conludes this paper and gives a short outlook.

2   The **HyperView** Approach

The HyperView approach intends to couple several well-known Web sites relevant to a particular application domain to collect, restructure and connect information from different sources in a "virtual Web site". Virtual Web sites appear to the user just like conventional Web sites. The only difference is that the information presented to the user forms a view of the underlying Web sites from which data is extracted on demand. Virtual Web sites hide the structural heterogeneity of their sources, provide missing cross-links and achieve a broader coverage. For instance, in the domain of cultural town information it can combine event calendars from different Web sites and crosslink them with city maps, public transportation schedules from other servers.

In order to abstract from the concrete layout and notation used for HTML pages, the relevant data items are recognized in the syntax graphs of HTML pages and used to build up for each source an intermediary data layer termed *Abstract Content Representation* (ACR). The information stored in these ACRs is then integrated on a third layer into a common *database* (DB) graph where it can be queried by the user. The WWW user interface of the HyperView system supports a combination of browsing, canned queries in HTML forms and ad-hoc queries. The result pages returned to the user are represented as graphs in the *User Interface* (UI) layer. Each layer forms a view of the preceding layer that is implemented by a so-called *hyperview*. This conceptual structure is reflected in the layers of the HyperView architecture depicted in Figure 1.

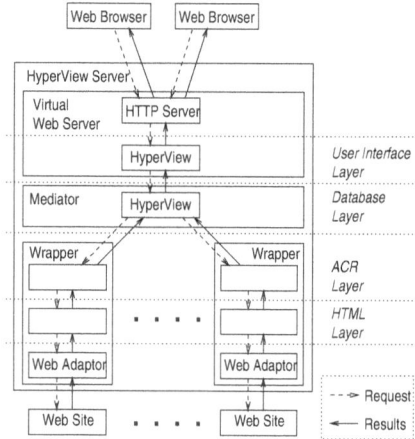

Fig. 1. Abstract architecture of the HyperView system

Since we do not want to maintain a huge data warehouse, information is loaded *on demand only* from the Web sources and the graphs at the higher layers are built incrementally triggered by requests from the succeeding layers. Thus, when the user is navigating the pages created by the WWW user interface, the HyperView system is actually navigating the underlying Web sites. To save WWW access costs, the HyperView caches the extracted information during a user session in order to avoid frequent reloading of pages for subsequent queries. This approach is reasonable in many application areas where sources are likely not to change within a typical user session.

To cope with the semistructured nature of WWW data, we introduce the CGDM graph data model presented in Section 4.1. CGDM supports the modularization of a graph into subgraphs called *clusters*. So-called *dependencies* between clusters control the existence of edges between vertices belonging to different clusters. We use CGDM to represent HTML syntax graphs, ACRs, DB, and UI graphs as clusters of a global data graph. We use CGDM also for schemata describing the structure of these layers. The HyperViews that map between the different layers are defined by sets of graph-transformation rules (cf. Section 4.2). The novelty of this approach lies in the demand-driven rule activation mechanism that is formally defined in the remainder of Section 4.

The graph representation of HTML pages and the graph pattern matching approach of HyperView make the HyperView system more robust against small structural changes in the sources and facilitate the rapid specification of ACR-views over HTML pages, compared to other character-based approaches (cf. Section 5).

Using a HyperView-based town information server, we could plan an evening in Berlin as follows: Find interesting entertainment events by either stating preferences in a HTML form or browsing the event database by categories like event type, town district, venue, time, etc. Follow links to external pages by the event promoters for background information. Follow a link from the address of a venue to the relevant part of the city map. Search/browse restaurants and pubs by their category and location Ask for a public transportation connection between two locations. As mentioned before, this town information server looks like a conventional Web site, but provides a value-added content compared to each of its sources.

3   Example: Integration of Town Information Web Sites

Before going into the underlying theory of the HyperView approach, we demonstrate an application in the field of town information.

Modeling. An event calendar of concerts, theater performances, movie shows, etc. is a central part of a HyperView server on (cultural) town information. In this example we first model the database schema describing this calendar, as depicted in Figure 4. The main concepts occuring in this schema are events and locations. As described in Def 1 in Section 4.1, a data graph conforms to a schema if it is mapped into the schema graph by an *interpretation* graph morphism.

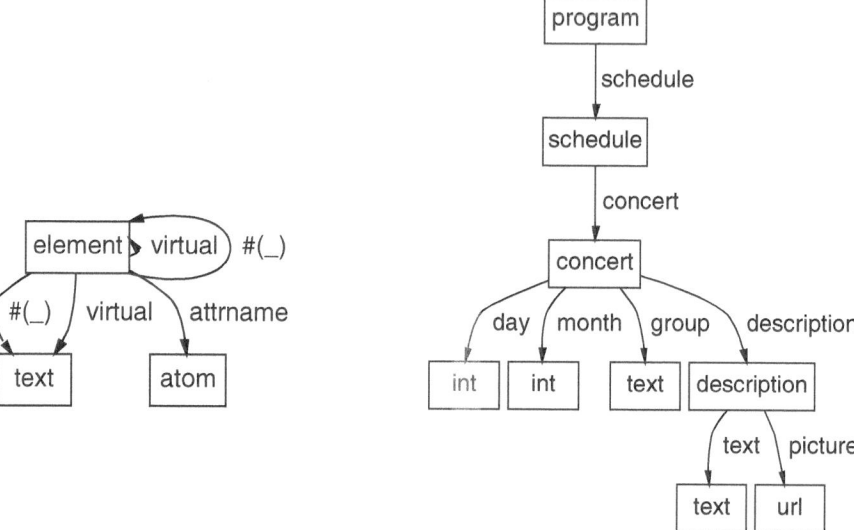

Fig. 2. Schema for HTML documents

Fig. 3. ACR schema for "A-Trane" program

Suppose that we want to integrate the monthly program[2] of the "A-Trane" jazz club into our **HyperView** event data base. We model the structure of *relevant parts* of the "A-Trane" Web site by the ACR schema shown in Figure 3. The list of all concert entries is represented by the **schedule** node. Each concert entry states the day of month, and points to a page describing the concert in detail. There can also be pictures of the band, which is represented by its URL.

The ACR graph conforming to the A-Trane ACR schema has to be extracted from HTML pages. We use the generic schema depicted in Figure 2 to model HTML pages as graphs. A HTML page consists of document elements that are linked to their children by numbered edges (labeled #(i), $i \in \mathbb{N}$) in order to form a parse tree. Text nodes form the leaves of these parse trees. Document elements may have attributes with atomic values. In order to implement robust pattern matching, several virtual edges are computed by the **HyperView** System, modeling e.g. subelement relationships or hyper-links.

Views. In order to set up a virtual Web site, several hyperviews have to be defined. For each source, a hyperview for extracting data from the pages of this Web site into the ACR graph, and a hyperview mapping this ACR graph into the DB graph has to be specified. Finally, a hyperview mapping the DB graph into the result pages of the UI graph has to be established. Each hyperview consists of a set of graph-transformation rules.

---

[2] <http://www.a-trane.de/Programm/programm.html>

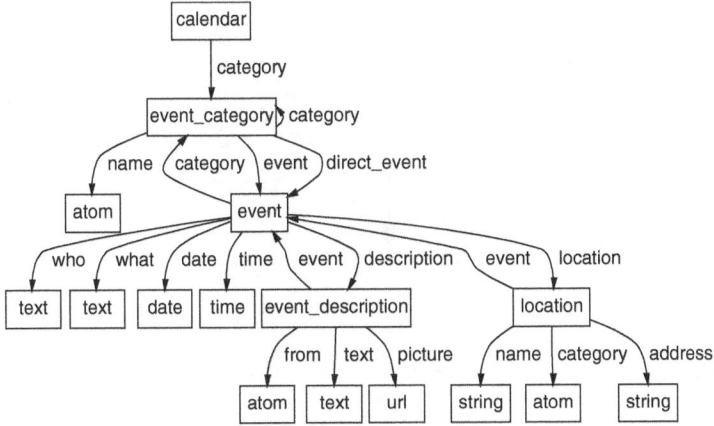

Fig. 4. DB schema for event calendar

In this subsection we give an example for one of the rules comprising the view specifications for the mapping between A-Trane ACR graph and the database graph. These rules are formulated against the schemata shown in Fig. 4 and Fig. 3.

In order to materialize the event edges emanating from the calendar node in the database graph, several alternative rules (one for each source of cultural events) may be activated. The rule atrane_event shown in Figure 5 extracts events from the "A-Trane" ACR. This rule matches a concert node in the "A-Trane" ACR and the location node describing the "A-Trane" in the database graph, if the values of the MONTH and DAY variables satisfy the constraint stated in the bottom line. This constraint also binds the YEAR variable to the current year and DATE to a date string concatenating the date components. For each successful match the rule creates an event node in the database graph. This node will be assigned the extracted date, the time 22:00 (the default time of concerts in A-Trane), and a link to the location node of the club.

4   Rule-Based View Mechanism

We now describe the formal framework of the HyperView system. For a more detailed presentation we refer to [9].

4.1   Data Model, Schemata, and Conformance

The WWW can be interpreted as a large graph of pages connected by hyper-links. Each of these pages is a structured document that can be represented by its parse tree. By combining these two representations we can view the WWW as a large graph of document elements that are clustered into pages. Hyper-links are edges that connect document elements of different clusters.

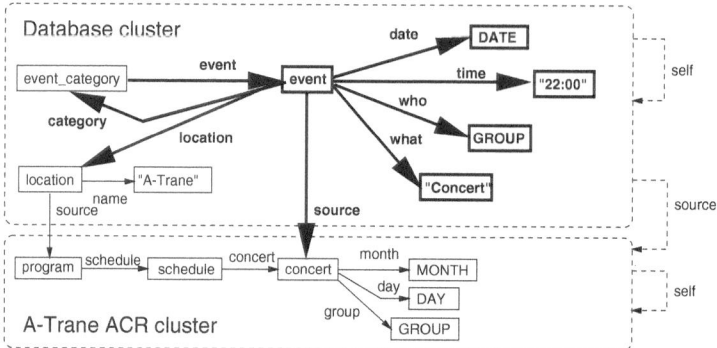

0 < MONTH =< 12 and 0 < DAY =< 31 and current_year(YEAR) and concat(YEAR,"/",MONTH,"/",DAY,DATE)

Fig. 5. Rule `atrane_event`. Notation: graph elements added by the rule are shown in **boldface**.

This motivates our *clustered* graph data model called CGDM. Large graphs are modularized by introducing clusters and dependencies which group vertices and edges, respectively. To each edge there exists a corresponding dependency which connects the cluster of its source with the cluster of its target. Hence, clusters and dependencies form a graph describing the structure of the clustered graph in the large.

A **clustered graph** is a triple $G = (G_{base}, G_{struct}, c)$ where the so-called **base graph** G_{base} and the **structure graph** G_{struct} are directed attributed graphs and c is a total morphism from G_{base} to G_{struct}. G_{base} represents the vertices and edges and G_{struct} the clusters and dependencies. The **clustering morphism** c assigns each vertex (edge) in G_{base} a corresponding cluster (dependency) in G_{struct}. We use the notation $G = (V, E, C, D, A, s, t, a, c)$ where V, E, C, D are disjoint carrier sets of vertices, edges, clusters, and dependencies, respectively. s and t assign source and target to edges and dependencies, $a : V \uplus E \uplus C \uplus D \longrightarrow A$ maps graph elements to the attribute algebra A. In the following we call clustered graphs simply **graphs**.

A graph morphism $f : G \longrightarrow H$ between clustered graphs is defined in the usual way by a family $(f_{vertex}, f_{edge}, f_{cluster}, f_{dep}, f_{attr})$ of *total* functions which map each of the carrier sets V_G, E_G, C_G, D_G, A_G of G to the corresponding sets of H. This implies that the category of clustered graphs with graph morphisms forms an instance of an **attributed graph structure** with total morphisms as defined in [16,14].

Pattern graphs (with variable set $\mathbb{V}$) are $T_\Sigma(\mathbb{V})$-labeled clustered graphs, i.e., their attribute algebra is formed by terms over a signature $\Sigma = (\mathbb{T}, \mathbb{O})$ with variables from $\mathbb{V}$. **Data graphs** are pattern graph whose labels are ground. We call $\mathbb{U}$ the set of all ground terms over Σ. The set $\mathbb{T}$ of sorts includes base types such as string, integer etc. **Schema graphs** are modeled as clustered $\mathbb{T}$-labeled graphs. A data graph is an instance of a given schema graph if there exists an *"interpretation"* graph morphism. The label of a schema element indicates the *type* of the labels of conforming instance elements.

Definition 1 (Interpretation, Conformance, Instance). *Let I a pattern graph and S a schema. A graph morphism $\iota : I \longrightarrow S$ is an* **interpretation** *of I w.r.t. S, if its attribute part $\iota_{attr} : T_\Sigma(\mathbb{V}) \longrightarrow \mathbb{T}$ assigns each label the type $T \in \mathbb{T}$ to which it belongs. A pattern graph I **conforms** to a schema S, if there exists an interpretation $\iota : I \longrightarrow S$. In this case we call I an* **instance** *of S.*

This definition is similar to the typing concept of [6], but supports clustered instances and schemata. It implies the following: (i) there may be several interpretations for I w.r.t. S, (ii) several parts of the instance graph may be interpreted by the same part of the schema, (iii) an interpretation must cover all elements of the instance graph, (iv) not all schema elements need to have corresponding data elements. In particular, the empty data graph conforms to any schema.

A morphism $f : G \longrightarrow H$ for clustered graphs with interpretations $\iota : G \longrightarrow S$ and $\kappa : H \longrightarrow S$ is called **type-compatible** if it satisfies $\kappa \circ f = \iota$.

Application: In the HyperView modeling technique, all schemata (HTML schema, ACR schemata for each source, and the database schema) form clusters of a global schema graph. In the same manner, HTML syntax graphs, ACR, DB, and result graphs form clusters of a global, clustered data graph. Edges between different clusters model hyper-links or point from a derived element to its source.

4.2 Rules

The concept of clustered graph is an instance of the general notion of attributed graph structures with additional equational constraints induced by the clustering morphism. In order to extend algebraic Single Push Out graph transformation on attributed graph structures ([16,14]) to clustered graphs, it is proven in [9] that a graph rewriting step preserves these equational constraints.

Since we need only non-deleting injective rules for our view definitions, we restrict the definition of Single Push Out rules $r : L \longrightarrow R$ by requiring that L is a subgraph of R, and r the inclusion morphism from L to R. In particular, the attribute part r_{attr} is the identity. R is labeled with typed terms from $T_\Sigma(\mathbb{V})$. To ensure that all rules conform to the global schema S, we introduce a typing morphism, the interpretation $\tau : R \longrightarrow S$. Moreover, we use a boolean term Γ as an application condition which has to be fulfilled by the variable binding induced by a match m. In summary, we use *typed attributed Single Push Out graph transformations with application conditions on attributes.*

Definition 2 (Rule). *A* **rule** *$p = (L, R, \Gamma, \tau)$ for a schema S consists of: (i) a clustered pattern graph R, called the* **right hand side** *(RHS) graph, (ii) a subgraph L of R, called the* **left hand side** *(LHS) graph, (iii) a typing morphism $\tau : R \longrightarrow S$, (iv) a boolean term Γ from $T_\Sigma(\mathbb{V})$ interpreted as an* **application constraint.**

A **substitution** is a function σ mapping a set of variables to terms. We denote the application of σ to a pattern graph G by $G\sigma$ and the application to a rule p by $p\sigma$. A **match** of a pattern graph Q in a data graph G is a graph morphism $m : Q \longrightarrow G$ that is type-compatible with given interpretations of Q and G in a schema graph S. m induces a substitution on the variables in Q. We denote the set of all matches for Q in G by $Matches(Q,G)$. A match of a rule p is a match for its LHS L_p. A match $\bar{m} : R_p \longrightarrow \bar{G}$ for its RHS satisfying Γ_p in a data graph $\bar{G}$ is called a **full match** of p.

The application constraint Γ_p can be used to compute bindings for variables occurring in R_p (cf. Figure 5). If Γ_p has multiple solutions for the same m, then the rule can be applied multiple times. We denote by $Apply(p|m)$ the set of all full matches $\bar{m} : R_p \longrightarrow \bar{G}$ resulting from all possible applications of p to the match $m : L_p \longrightarrow G$ and the resulting extensions of G to data graphs $\bar{G}$.

4.3   Oracles and Queries

In the last section we have defined the application of single rules. In this section, we define how a set of rules is used for answering *queries*. Roughly speaking, a query is a pattern graph Q and its solutions are matches of Q in data graphs which result from *dynamically expanding* a given initial data graph G_0, typically by applying rules to it.

Definition 3 (Query and Solution). *Let S be a schema and G_0 be a data graph with interpretation $\iota_0 : G_0 \longrightarrow S$. A **query** on G_0 is a tuple (Q,Q_0,Γ,τ) consisting of: (i) a pattern graph Q over a variable set $\mathbb{V}$, the **query graph**, (ii) a subquery $Q_0 \sqsubseteq Q$, the **anchor**, (iii) a constraint Γ being a boolean term from $T_\Sigma(\mathbb{V})$, (iv) a typing $\tau : Q \longrightarrow S$ of Q A **solution** for q is a triple (G,ι,m). It consists of a supergraph G of G_0 that has the extension $\iota : G \longrightarrow S$ of ι_0 as interpretation, and the match $m : Q \longrightarrow G$ of Q in G that induces a substitution for the variables of Q satisfying the query constraint Γ.*

An *oracle* can be seen as a "black box" which computes for a given query and an initial data graph a solution set satisfying all these requirements by dynamically extending the data graph. The result of such an oracle for a query q against a data graph G is shown schematically in Figure 6. More formally:

Definition 4 (Oracle). *An **oracle** Φ is a functor which takes a query $q = (Q,Q_0,\Gamma,\tau)$ and a data graph G_0 (with interpretation $\iota_0 : G_0 \longrightarrow S$) and returns a set $\Phi(q,G_0)$ of solutions (G,ι,m) for q w.r.t. G_0. For notational ease we write $(m : Q \longrightarrow G) \in \Phi(q,G_0)$ instead of $(G,\iota,m) \in \Phi(q,G_0)$. We use the abbreviation $\Phi(q|m_0) := \{m \in \Phi(q,G_0) \mid m|_{G_0} = m_0\}$ to express a call of an oracle with a fixed initial match $m_0 : Q_0 \longrightarrow G_0$.*

Example of an oracle. The HyperView System provides one builtin oracle, the **WWW oracle** for the HTML schema depicted in Fig. 2. A query anchor Q_0 has to match an existing part of a HTML cluster. The WWW oracle will try

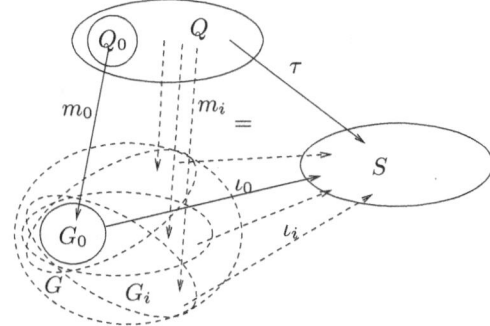

Fig. 6. A query $q = (Q, Q_0, \Gamma, \tau)$ against an oracle Φ w.r.t. an initial data graph G_0. The solution set is indicated by dashed lines.

to complete each match for Q_0 in the already materialized HTML clusters to matches for the whole query graph Q. The attempt to match href_target edges representing hyper-links will load HTML pages from the WWW. The WWW oracle supports virtual sub edges to denote a subelement relation among document elements.

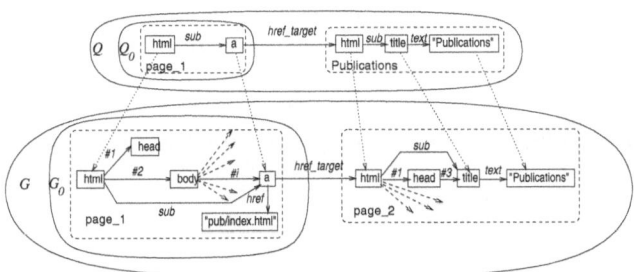

Fig. 7. Query against the WWW oracle matching the hyper-link from the author's home page to his publications page. The home page is assumed to be already materialized in a cluster page_1.

Applying a rule to a virtual data graph. If there is an oracle Φ available for queries against a certain cluster of the data graph G_0, we can use this oracle to apply a rule $p = (L, R, \Gamma, \tau)$ to G_0 even though we cannot find matches for p in G_0 itself. To do so, we have to formulate a query q for Φ which will return matches for L. It follows immediately that L should be contained in the query graph; in fact, we choose $Q = L$. Furthermore, the application constraint is used as a constraint for q and the restriction $\tau|_L$ of the typing τ as typing of Q. The anchor graph Q_0 of q is specified by a graph $A \sqsubseteq L$ that is added to the definition of p to indicate the portion of L for which a match in the already materialized data graph is required. Thus a rule gets the form $p = (A, L, R, \Gamma, \tau)$ and the query associated with this rule becomes $q = (L, A, \Gamma, \tau|_L)$. Using the oracle Φ

we obtain a set $\Phi(q, G_0)$ of matches $m_L : L \longrightarrow G \in \Phi(q, G_0)$ each of which is an extension of some match $m_A : A \longrightarrow G_0$.

To each $m_L \in \Phi(q, G_0)$, the rule can be applied in the usual way, producing full matches $m_R : R \longrightarrow \bar{G}$. This application of p against the oracle Φ over the initial data graph G_0 is depicted in Figure 8.

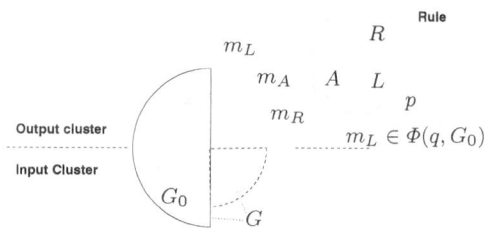

Fig. 8. Applying a rule to a *virtual* data graph with oracle Φ. The match for A is $m_A : A \longrightarrow G_0$, the match for the left hand side $m_L : L \longrightarrow G$, and the full match for the right hand side is $m_R : R \longrightarrow \bar{G}$.

Let $p = (A, L, R, \Gamma, \tau)$ be a production, Φ an oracle, and $m_A : A \longrightarrow G_0$ a partial match for p. Then we define $Apply^\Phi(p|m_A) := \{\bar{m} \in Apply(p|m) m \in \Phi(q|m_A)\}$ where $q = (L, A, \Gamma, \tau|_L)$, called the **rule application functor** for oracle Φ. This functor $Apply^\Phi(.|.)$ uses an oracle Φ to extend a partial match m_A for a rule p to a total match m and then applies p to this match using the functor $Apply(.|.)$ for conventional rule application without oracle as defined in Section 4.2.

4.4 Hyperviews

A hyperview defines an oracle specified by a set of rules. These rules are applied against a set of input clusters in order to answer queries posed against the oracle. By responding to these queries, the oracle materializes its output cluster as a function of its input clusters. As we have seen in the last section, the input clusters do not have to be materialized to apply rules against them: it is sufficient to have some oracle Φ for these clusters.

Thus we require all rules of a hyperview Π to be typed by a subschema that describes only the input and output clusters. Moreover, these rules are not allowed to add new elements to any of the *input* clusters.

Using a rule to answer a subquery. In order to use the rule set of a hyperview Π as an oracle, we have to specify how to use them to answer a query. We start with the problem of how to use a *single rule* p to answer a subquery.

Let Π be a hyperview and Φ an oracle for data graph clusters described by the input clusters of Π. Let $p \in \Pi$ one of its rules. Let $q = (Q, Q_0, \Gamma, \tau)$ be a query and $B \sqsubseteq Q$. We can use p to find a match for B if there is a suitable mapping $b : B \longrightarrow R_p$ called a **binding morphism**. The **binding region** B can be compared to the call site of a procedure in an imperative program or a Prolog goal. Applying rule p to a match $m : L \longrightarrow G$ yields a full match $\bar{m} : R \longrightarrow \bar{G}$. Figure 9 illustrates how this match can be lifted to a match $m_B : B \longrightarrow \bar{G}$ for B by defining $m_B = \bar{m} \circ b$.

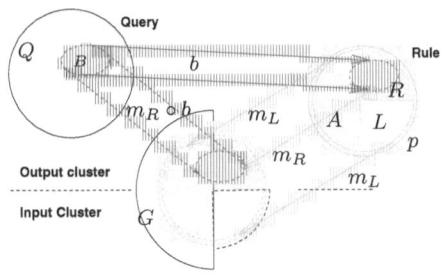

Fig. 9. Using a binding morphism to obtain a match for a subquery B.

Chaining rules to answer a query. To answer a whole query using a hyperview Π, we introduce the notion of a **query execution plan**. Such a plan consists essentially of a set of binding morphisms b_i for rules $p_i \in \Pi$ which cover (together with the anchor graph of the query) the whole query graph. If we can apply the rules p_i in such a way that the matches induced by the binding morphisms are compatible with each other and with a match for the anchor graph, we yield a match for the whole query graph being the union of all these matches.

Since a rule can be applied only if a match for its anchor graph can be found in the available data graph, care must be taken to activate rules in the right order. The query can be answered only if for each rule a subgraph matching its anchor graph already either exists in the initial data graph or is materialized by a preceding rule.

Our plan concept ensures statically that rules are executed in the right order. The key idea is to require that the anchor graph of a rule is either to be matched against the initial data graph or against the right hand side of a rule which is to be executed before. In the latter case, the match is defined by a so-called **rule dependency** morphism. This idea is illustrated schematically by Figure 10.

A **query execution plan (QEP)** for q and a given hyperview Π is a tuple $P = (\sigma, \mathbb{B}, \mathbb{D}, \Gamma)$ consisting of (i) a substitution σ that is applied to the query graph Q, (ii) a sequence $\mathbb{B} = (b_1, \ldots, b_n)$ of binding morphisms that map subqueries B_i to the right hand sides R_i of suitable rules from Π, (iii) a sequence $\mathbb{D} = (d_1, \ldots, d_n)$ of rule dependencies that map each anchor graph A_i to the right hand side of a preceding rule p_j, $j < i$. Furthermore the query graph Q has to be completely covered by binding regions B_i and the anchor graph Q_0.

A **solution for a QEP** P is a match $m : Q\sigma \longrightarrow G$ that consists of a union of an initial match m_0 for the anchor graph Q_0 and matches m_i that have

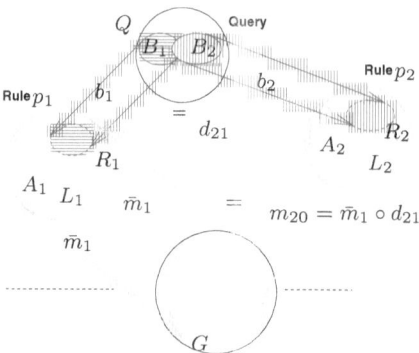

Fig. 10. Chaining of rules p_1 and p_2. Rule p_1 has already been applied, yielding a full match $\bar{m}_1$. Rule dependency morphism d_{21} maps the anchor graph of p_2 to the right hand side of p_1 and thus lifts the full match $\bar{m}_1$ for p_1 to a partial match m_{20} for p_2.

been lifted from full matches $\bar{m}_i : R_i \longrightarrow G$ for the rules p_i as illustrated in Figure 9. These matches have to coincide on overlaps of their binding areas and have to be compatible with the rule dependencies d_i. In fact, these dependencies are used to construct solutions by applying the rules $p_1, \ldots p_n$ in that order and using the d_i to obtain initial matches for the rule anchors A_i. At each step, it can be checked easily whether a new match m_i is compatible with the preceding matches m_j, $j < i$. We denote the set of all solutions for P w.r.t. Φ and a given initial match m_0 for Q_0 by $PlanOracle^{\Phi}(P|m_0)$. The set of all solutions for a query q against an initial data graph G_0 w.r.t. an oracle Φ and a hyperview Π is denoted by $Oracle^{\Phi,\Pi}(q, G_0)$. This functor iterates over all matches m_0 for Q_0 and all minimal plans for q.

The functor $Oracle^{\Phi,\Pi}$ for a hyperview Π with output schema cluster c yields an oracle competent for c. Different oracles competent for schema clusters $c_1, \ldots, c_n$ can be combined to one oracle competent for all these clusters if there are no dependencies between these clusters in the schema.

A query against $c_1, \ldots, c_n$ can be decomposed into subqueries q_i against single clusters c_i. The anchor graphs Q_{0i} may intersect because by Definition 4 need not conform to c_i. Unions of solutions m_i returned for the different q_i are solutions for q if they are compatible with each other and satisfy the constraint Γ of q.

This enables us to compose hyperviews in a way that allows information from different sources to be retrieved, restructured and combined on a higher level of abstraction as it is the case in the HyperView architecture (cf. Fig. 1). A user request for a page of a virtual HyperView Web site triggers a query against the toplevel hyperview. This query leads to a cascade of rule activations and queries resulting from these activations that finally reach the bottom level. There the WWW oracle handles the queries against the WWW by transparently loading HTML pages from the underlying sources and matching the queries against the graph representation of these pages. By means of variable bindings the data is extracted from the pages, forwarded to the toplevel and used to compute the contents of the pages returned to the user.

5   Related Work

Common to most approaches to semistructured data are data models based on labeled graphs or trees. Examples are the Object Exchange Model (OEM, [18]) used in TSIMMIS [12], and the data model of Buneman [5]. Several projects extend conventional database techniques for query processing to semistructured data by using for instance regular path expressions or automatic type coercion. Examples of query languages on semistructured data are LOREL [18], UnQL [5], StruQL [11], WebOQL [2].

Another approach is to use declarative rules for data restructuring. YAT [7] uses a tree data model and Prolog-like rules. The CGDM data model and the rule concept of HyperView are more general in supporting modularized graphs. Currently, the only other use of graph transformation rules for semistructured data is described in [3].

Several systems have been developed for integrating WWW sources, for instance TSIMMIS [12], Strudel [10], and ARANEUS [17]. The wrapper/mediator architecture of TSIMMIS encapsulates each data source with a wrapper having a uniform interface and integrates data from different wrappers using mediators. In the HyperView project, we use a similar architecture which is shown in Figure 1. Due to space limitations we have to refer to [8] for a more thorough discussion of related literature in this area.

The HyperView methodology is *novel* in its uniform use of the *same view concept at all stages*, the *demand-driven incremental view computation* mechanism, and the *use of graph transformation rules* for defining and computing these views. Demand driven incremental view computation has the advantage that response time is reduced by only loading from the sources only the information necessary to answer a user request. The uniform use of the same view mechanism reduces the training effort for view designers compared to systems such as ARANEUS or TSIMMIS that use different specification languages at different stages. Moreover, graph-transformation rules lend themselves naturally to visualization and intuitive graphical view specification. Compared to character-based data extraction (e.g. used in TSIMMIS), graph matching also provides a more robust tool for data extraction especially when combined with the virtual edges used in our HTML graph representation. This is crucial for reducing the maintenance costs of wrappers.

A design environment for migrating relational to object-oriented database schemata is described in [15]. A HyperView System does not migrate single relational databases, but incrementally computes integrated views over semistructured hyper-documents from different sources lacking an explicit schema. Hence we have developed solutions for the modularization of graphs and controlled demand-driven rule activation.

In [13], horizontal and vertical refinement of typed graph transformation systems is treated. In our approach, we also use vertical refinement to move from an abstract domain description to a concrete specification of HTML pages. However, we do not transform each level in parallel, but rather build up a graph at a higher level by matching graphs at a lower level. A typing concept similar to that of HyperView, but without modularization, is presented in [6].

6   Conclusion

In this paper we have presented the HyperView methodology for integration of information from the WWW by a hierarchy of incrementally computed views. We have focused on our view concept based on graph transformation rules which are activated in a demand-driven way. For this purpose we have introduced and formally defined the notion of an oracle against which queries can be posed. Then we have described how rule sets called hyperviews can be employed as oracles and how oracles can be composed.

The prototype of the HyperView System has been implemented in Prolog. Graphs are stored as facts and graph-transformation rules are translated into clauses of Prolog predicates. The generic Web interface uses a Java Servlet to make the HyperView System accessible from the WWW. The prototype is already in use in a digital library project [8]. A second demonstrator for the field of town and cinema information is currently under development. We expect to make this demonstrator available on the WWW soon. The prototype is also currently being enhanced to implement the rule activation mechanism in the general form presented here.

We plan to extend the formalism with a concept of inheritance for schemata and with a fusion mechanism for similar subgraphs describing the same real world entity, thus generalizing database keys.

References

1. S. Abiteboul. Querying semi-structured data. In *ICDT*, volume 6, pages 1–18, 1997. 426
2. G. Arocena and A. Mendelzon. WebOQL: Restructuring documents, databases and webs. In *Intl. Conf. on Data Engineering (ICDE 98)*, 1998. 439
3. A. Bergholz and J. C. Freitag. A three-layered approach to semistructured data: Predicate, path, and collection schemata. In *Theory and Application of Graph Transformations (TAGT'98)*, 1998. 439
4. P. Buneman. Semistructured data. In *PODS'97*, 1997. Invited Tutorial. 426
5. P. Buneman, S. Davidson, G. Hillebrand, and D. Suciu. A query language and optimization techniques for unstructured data. In *SIGMOD Conference*, pages 505–516, 1996. 439
6. I. Claßen and M. Löwe. Scheme evolution in object–oriented models. In *ICSE–17 Workshop on Formal Methods Application in Software Engineering Practice*, 1995. 433, 440

7. S. Cluet, C. Delobel, J. Siméon, and K. Smaga. Your mediators need data conversion! In *SIGMOD Conference*, 1998. 439
8. L. C. Faulstich and M. Spiliopoulou. Building HyperView wrappers for publisher web-sites. *Intl. Journal on Digital Libraries*, 1999. To appear. 439, 440
9. Lukas C. Faulstich. The formal framework of the HyperView system. Technical Report B 99-03, Inst. of Computer Science, FU Berlin, 1999.<http://www.inf.fu-berlin.de/inst/pubs/tr-b-99-03.abstract.html>. 431, 433
10. M. Fernandez, D. Florescu, J. Kang, A. Levy, and D. Suciu. Catching the boat with Strudel: experiences with a web-site management system. In *SIGMOD Conference*, 1998. 439
11. M. Fernandez, D. Florescu, A. Levy, and D. Suciu. A query language and processor for a web-site management system. In *Workshop on Management of Semistructured Data*, 1997. 439
12. Hector Garcia-Molina et al. The TSIMMIS approach to mediation: Data models and languages. *Journal of Intelligent Information Systems*, 8(2):117–132, 1997. 439
13. R. Heckel, A. Corradini, H. Ehrig, and M. Löwe. Horizontal and vertical structuring of typed graph transformation systems. *Math. Structures in Computer Science*, 6(6):613–648, 1996. 440
14. R. Heckel, J. Müller, G. Taentzer, and A. Wagner. Attributed graph transformations with controlled application of rules. In *Proc. Colloquium on Graph Transformation and its Application in Computer Science*, 1995. 432, 433
15. J. Jahnke, W. Schäfer, and A. Zündorf. A design environment for migrating relational to object-oriented database systems. In *Int. Conf. on Software Maintenance, ICSM '96*, 1996. 439
16. M. Löwe. Algebraic approach to single-pushout graph transformation. *Theoretical Computer Science*, 109:181–224, 1993. 432, 433
17. P. Merialdo P. Atzeni, G. Mecca. To weave the web. In *VLDB '97*, pages 206–215, 1997. 439
18. D. Quass, A. Rajaraman, Y. Sagiv, J. Ullman, and J. Widom. Querying semistructured heterogeneous information. In *DOOD*, number 1013 in LNCS, pages 319–344, 1995. 439

A Model Making Automation Process (MMAP) Using a Graph Grammar Formalism

Curtis E. Hrischuk

Department of Electrical and Computer Engineering,
University of Alberta, Edmonton, Canada
curtis@ee.ualberta.ca

Abstract. In a large, concurrent software system it is difficult to understand the system behavior by analyzing event traces. The difficulty arises from the need to order the large number of events and infer cause-and-effect relationships. The Model Making Automation Process (MMAP) deduces causes-and-effect event relationships so that high-level models can be generated in an automated fashion. This paper describes how MMAP uses the algorithmic attributed graph grammar formalism to specify its several model domains and the transformations between model domains. Three general approaches to developing model transformations using a graph grammar are presented as well. Experiences with using PROGRES to make these transformations are also described.

1 Introduction

A general approach to managing the complexity of software development is to use software execution models (SEMs) as engineering aids. SEMs are often used to specify the intended operation of the system (e.g., UML [20]). During the later phases of development, *realized SEMs* are constructed which characterize the software's actual behavior, for the purpose of facilitating program understanding, re-engineering, reuse, performance analysis, validation, or debugging. Often these realized SEMs are mental pictures that the analyst constructs from user requirements, design documents, source code examination, and, most importantly, experience with the system. The realized SEMs are critical for investigating differences between the specified behavior and the observed behavior. Manually constructing a realized SEM from an actual execution of a small application is possible because it is not that difficult to establish cause-and-effect relationships between events. However it is expensive, difficult, and error prone for a large, dynamic, or distributed application. The Model Making Automation Process (or MMAP) reduces the effort required to develop an accurate, realized SEM so that the software development cost is reduced.

The main contribution of this paper is to outline how MMAP uses the algorithmic graph grammar formalism to construct high-level abstractions from low-level event data. Three general approaches to developing model transformations using a graph grammar are presented. Our experience with using PROGRES as a transformation engine is also reported.

H. Ehrig et al. (Eds.): Graph Transformation, LNCS 1764, pp. 442–454, 2000.

MMAP constructs models of the execution of a distributed or concurrent software system where several scenarios occur simultaneously. A *distributed system* is composed of geographically dispersed, heterogeneous hardware with scheduled, concurrent software *objects*.[1] A *distributed scenario* is a set of coordinated interactions between shared system server objects and application specific objects. A distributed scenario is an end-to-end execution of a *distributed application*.

We assume that objects can only communicate by sending and receiving messages over communication channels that are: reliable, point-to-point, dynamically established at execution, with finite but unpredictable delay. No assumptions are made about the order of message delivery. The communication protocol may be blocking (i.e., Remote Procedure Call) as well as non-blocking (i.e., asynchronous). DCE RPC [16], CORBA [15], Java [4], and mobile agents [18] are technologies which are being used to build these types of distributed scenarios.

MMAP starts with a special application level trace that is called an ANGIOTRACE.[2] The ANGIOTRACE is a type of logical clock which is based on partial order multi-sets [19] whereas other types of logical clocks are based on a partial-ordering [24]. During the analysis, each ANGIOTRACE is converted into a graph grammar model domain called the *scenario event graph* [9]. A scenario event graph is a new coordination language [2] that characterizes all possible executions for the class of distributed systems described below. Each scenario event graph model is then converted into a more abstract scenario model description that is suited for further analysis (e.g., generating a collaboration or sequence diagram [1]). The algorithmic graph grammar formalism [14] is used as the basis for the model transformation machinery.

A prerequisite for MMAP is some executable form of the design that has instrumentation added to generate an ANGIOTRACE when it is executed. The executable design can be a very abstract executable CASE tool model, a code prototype, an event driven simulation, or an implementation.

The manner in which MMAP generates a model is described next.

2   The Automated Model Construction Strategy of MMAP

The Model Making Automation Process (MMAP) is a well-formed, traceable model construction process. It is *well-formed* because a chain of formal transformations is defined from the input model domain (an executable design) through to the target model domain (a collaboration or sequence diagram).

MMAP uses the model domains of Figure 1. They are listed in the order in which data flows in the model construction process:

a) ***Language statement***: The source code statements of an object's behavior.

---

[1] For our purpose, we consider an object to be the unit of concurrency.

[2] Angiotrace is a trademark of Angiograms for Software Analysis Inc. who can be contacted at angio@istar.ca or http://home.istar.ca/ angio/. ANGIOTRACE appears in capital letters to signify it is a trademark.

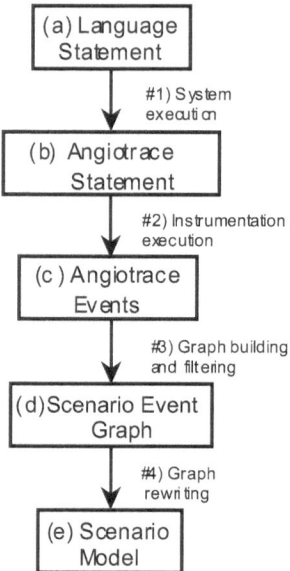

Fig. 1. MMAP Data Flow

b) *ANGIOTRACE instrumentation*: Instrumentation that atomically generates and records ANGIOTRACE events when language statements are executed.

c) *ANGIOTRACE events*: Events of a distributed scenario that are ordered according to their cause-and-effect relationships using special time stamp values.

d) *Scenario Event Graph*: A graph grammar that can characterize all of the possible executions of a distributed scenario.

e) *Scenario model*: A model that characterizes the execution of a scenario. This includes the involved objects, their individual activities, their interactions, and the communication protocol elements.

The algorithmic attributed graph grammar formalism implements transformations #3 and #4 of Figure 1. The ANGIOTRACE and scenario event graph domains are explained in more detail in the next sections. Due to space constraints the scenario model is not discussed.

3   The Scenario Event Graph (SEG)

A wide variety of specification techniques could be described as graph grammars, such as MASCOT [17], ADA [11], StateCharts [3], and actor specifications [12]. These techniques are used to explore the behavioral characteristics of a design specification. The scenario event graph (SEG) is different because it is the first graph grammar designed to characterize the execution of distributed scenarios.

The scenario event graph domain is causal, linear time, and non-interleaving. Causality constrains the scope of a scenario event graph model to a distributed scenario, segments each object's behavior into individual periods of service, and helps to identify object blocking or synchronization. Linear-time allows scenario event graphs to be derived from execution traces. Non-interleaving properly characterizes concurrency within a distributed application and captures the object architecture.

The scenario event graph is acyclic and finite, with labeled nodes that represent different event types, as well as labeled, directed edges that represent different types of causal relationships between events. It has a minimal number of node and edge types so that the semantics are clear, consistent, and non-overlapping. For example, each node type has unique properties to allow it (or prevent it) from connecting to other node types. Each edge type denotes a different type of cause-and-effect relationship between node types.

The scenario event graph is the superposition of two other types of graphs: and object event graph and a process event graph. An *object event graph* model characterizes the execution of a single object as a linear graph. A *process event graph* model characterizes the logically parallel activities of the distributed scenario, independent of the software structure or software deployment.

The next sub-sections are taken from [9]. They introduce the different node and edge types used in the object event graph and the process event graph.

3.1   The Object Event Graph

An object executes a local, cyclic algorithm as follows: (1) it receives a service request which may be an external stimulus or a message from another object, (2) it executes, and (3) it waits for the next stimulus. The first two steps are referred to as a *service period*. The execution of a service request may involve interactions with other objects.

The object event graph characterizes the cyclical execution of an object as a sequence of linear subgraphs, one for each service period. Each service period is characterized as a linear sub-graph of object events.

The object event graph is an attributed, edge labeled, directed, linear graph. It has two types of nodes and two types of edges:

- *Period Start node*: this is the first node of each object's service period.
- *Object Activity node*: represents an activity that the object performed. It is the default object node type so it is implicitly assumed.
- *Object's Next Node edge*: its target is the next node in the same object period.
- *Object's Next Period edge*: its source is the last node of an object's period and its target is the period start node of the object's next service period.

A service period ends when a node is not the source of a next node edge.

3.2 The Process Event Graph

A process event graph characterizes the potential concurrency and event sequences of a distributed scenario.

A process event graph is made of linear sub-graphs which represent potentially concurrent sequences of events. A linear sub-graph is called a *process thread*. A process event graph has special node and edge types to characterize the causal relationships of concurrent process threads.

A process event graph is a node labeled, edge labeled, binary, finite, directed, acyclic graph. The process event graph node types and edge types are:

- *External node*: it is a marker for an external stimulus.

- *Thread Begin node*: it is the beginning of a process thread.

- *Process Activity node*: it represents an activity that is performed in the process. It has an attribute to store user-defined information.

- *And-fork node*: it identifies the introduction of logical concurrency as the forking of a new process thread. It is the only node type with out-degree greater than one.

- *And-join node*: is a place holder for a synchronization between two process threads that join into a single process thread. The and-join node is the only node type with in-degree greater than one.

- *Thread End node*: ends a process thread, distinguishing between a terminated or incomplete (i.e. deadlocked) process thread.

- *Start the Process edge*: its source node is an external node and its target node is a thread begin node which is caused by the external node.

- *Process Thread's Next Node edge*: its target is the succeeding node in the same process thread.

- *Process Thread's Fork edge*: its source is an and-fork node and its target is the thread begin node of the forked process thread.

The node types are atomic.

A process event graph may have more than one external node. If so, they usually join later.

4   The ANGIOTRACE

The ANGIOTRACE is a unique, new type of logical clock based upon the partial-order multi-set [19]. A logical clock is the only reliable technique for ordering events in a distributed system [24] since it establishes a partial or total ordering on the system events that is consistent with the concurrent execution.

An ANGIOTRACE overcomes many of the difficulties associated with tracing a distributed system. The name ANGIOTRACE is derived by analogy from an angiogram. An angiogram is a visualization of an individual's blood flow that is produced by injecting a radio-opaque dye into the blood stream and taking an X ray of the dye dispersion. Similarly, an ANGIOTRACE assigns a different software dye to each distributed scenario so that each scenario's event records can be distinguished. The software dye has a special time stamp value so that a set of partial ordering relations can be used to reconstruct the cause-and-effect relationships between them. The ANGIOTRACE was introduced in [7] and has been further described in [8] and [10].

A novel aspect of the ANGIOTRACE is that it constructs two different types of graphs which order the events using cause-and-effect relationships. In comparison, a partial-ordering of events that is derived from a (classical) logical clock can only construct a single type of graph that orders events temporally. Because of this difference an ANGIOTRACE uses a different strategy for ordering the events that is based on partial order multi-sets.

Definition 1. *An ANGIOTRACE is a set of* N *recorded events that have AN-GIOTRACE time stamps assigned in accordance with an ANGIOTRACE specification.*

Definition 2. *An ANGIOTRACE specification is* $G_{Trace} = (A, R, P, O, M)$ *where:*

   A is the alphabet of event time stamps,

   R is the rules for assigning time stamp values to events at execution time,

   P is a set of event predicates,

   O is a set of partial-ordering relations, and

   $M : P \to (o_1, \ldots, o_n)$ *is the mapping of an event predicate to one, or more, valid ordering relations.*

The ordering relations are used to generate the edges between events.

An ANGIOTRACE has six event ordering relations to identify a given event's succeeding or preceding event in the object event graph or the process event graph. The ordering relations are $O = \{>^{Ob}, <^{Ob}, >^{Ps}, <^{Ps}, >^{Po}, <^{Po}\}$. Each relation is reflexive, antisymmetric and transitive. By convention, the direction of the $<$ or $>$ indicates which of the two events is being ordered: $<$ is used to look for the preceding event and $>$ is used to look for a succeeding event. For example, $<^{Ob} (e_1, e_2)$ indicates that event e_2 is provided and the preceding event in the same object event graph is bound to (e_1).

The interpretation of an ordering relation is directly related to a SEG's edge. The ordering relationships are interpreted as follows:

   $>^{Ob} (\mathbf{e_1}, \mathbf{e_2})$ orders succeeding events in the object event graph. An object's next node edge is established from the given source node e_1 to target node e_2.

   $>^{Ps} (\mathbf{e_1}, \mathbf{e_2})$ orders succeeding events in the *same process thread*. A process thread's next node edge is established from the given source node e_1 to the target node e_2.

   $>^{Po} (\mathbf{e_1}, \mathbf{e_2})$ orders succeeding process event graph events that are not in the same process thread (e.g., a fork event and its child begin event). A process thread's fork edge is established from the given source node e_1 to target node e_2.

The other edge types are established by similar graph rewrite rules.

5   Model Projections Using a Graph Grammar Formalism

MMAP is based on the novel concept of using several graph grammars to formally specify the instrumentation requirements to trace a distributed system. The graph grammars allow abstract models to be developed and analyzed.

The selection of a model transformation engine was based on the observation that all of the model domains of Figure 1 can be described as graphs with typed nodes, typed edges, and rules for connecting nodes with edges. Each model domain is a graph grammar, with its own syntax and semantics. The *syntax* identifies a properly constructed graph model by specifying the structural components (i.e., labeled nodes) and relationships between those components (i.e., labeled edges). The *semantics* identify when a graph has a valid meaning in the model domain. A graph grammar may also have *graph rewriting rules* for rewriting portions of a graph.

MMAP's general strategy for transforming a model from its input domain to its target domain is as follows. First the *application specific domain language* of the input model domain and the target domain are both formally described as separate graph grammars. Then a *projection* is developed, which is a graph transformation from the application specific language of the input domain (the input graph grammar) to the application specific language of the target domain (the target graph grammar). A projection uses graph rewriting rules to define semantic equivalences of sub-graphs in the input and output domains.

There are two benefits of using a graph grammar formalism. First, the complete syntax and semantics of a model domain can be consistently and completely defined in an application specific domain language. Second, a projection from one model domain to another can be formally defined at the domain level rather than at the much lower level of a generic programming language (e.g., C++ or Java).

5.1   Selecting a Graph Grammar Formalism

The algorithmic attributed graph grammar formalism was selected as the graph grammar formalism because it operates at the appropriate level of abstraction and tools are available. Using this formalism, a complex query is formulated as a host graph traversal and a complex graph update as a graph rewrite rule. An important and unique feature of the algorithmic graph grammar is that the selection of graph rewrite rules can be made by a complex rule selection algorithm, which may itself have nested algorithms and graph rewrite rules. This enables a model transformation to be constructed as a set of independent, declarative graph rewrite rules.

The algorithmic attributed graph grammar formalism has the three items needed to exploit a graph grammar: (1) a graph grammar specification language, (2) a graph rewrite system which implements the graph grammar specification language, and (3) a methodology for the use of both. The graph grammar specification language is used to define the model domain's graph grammar syntax, identifying the family of valid graphs that can be constructed from a given graph instance (called the *host graph*). A graph rewrite system is a tool that implements a graph grammar as a collection of rules to rewrite an input host graph. The methodology guides the development effort. These three components are available for the PROgrammed Graph REwriting System language (PROGRES) and its corresponding tools [22], [23].

A very useful feature of PROGRES is the ability to capture the application knowledge of a model domain as a *graph schema*. The graph schema is a specification of the application specific domain language. It describes the family of valid host graphs in a domain by identifying what node-types and edge types can be connected and the direction of an edge between nodes.

5.2   The Three Types of Model Projections

MMAP uses three new types of projections to transform a *host graph model* into a resulting model in the target domain.

The first projection type is a *transformation by simplification* where the input model domain language is simplified by removing terminal symbols (i.e., a node type and any corresponding edge types). This type of projection can be used to normalize a host graph before further processing.

The second type of projection is the rewriting of a host graph model, in small steps, until the rewritten graph complies with the target domain's graph grammar (*transformation by synthesis*). This strategy is shown in Figure 2. This type of translation is rule-based graph rewriting [5], [13], [6], with a set of graph rewrite rules and a control algorithm for selecting the rule(s) to be applied [21]. To accomplish this a synthesized graph syntax is developed that combines the input domain's graph syntax and the target domain's graph syntax, resulting in an application specific language that characterizes both model domains. The projection from the input domain to the target domain is constructed as a set of meaning-preserving graph rewrite rules. The graph rewrite rules use the synthesized graph syntax to convert the terminal symbols (i.e., sub-graphs of node types) from the initial domain into terminal symbols of the target domain (i.e., a replacement sub-graph). The model transformations from the ANGIOTRACE to a scenario event graph domain and from the scenario event graph to a scenario model domain, both use transformation by synthesis.

The last type of projection is to chain together several other projections into a pipeline, from the input model domain to the target model domain (*transformation by chaining*). A pipeline of projections is used by MMAP to provide repeatable, accurate, and automated construction of a high-level model from low level information.

In all three types of projections, the analyst provides a rule selection algorithm to govern the selection of the appropriate rewrite rule or projection when there is the potential for conflict.

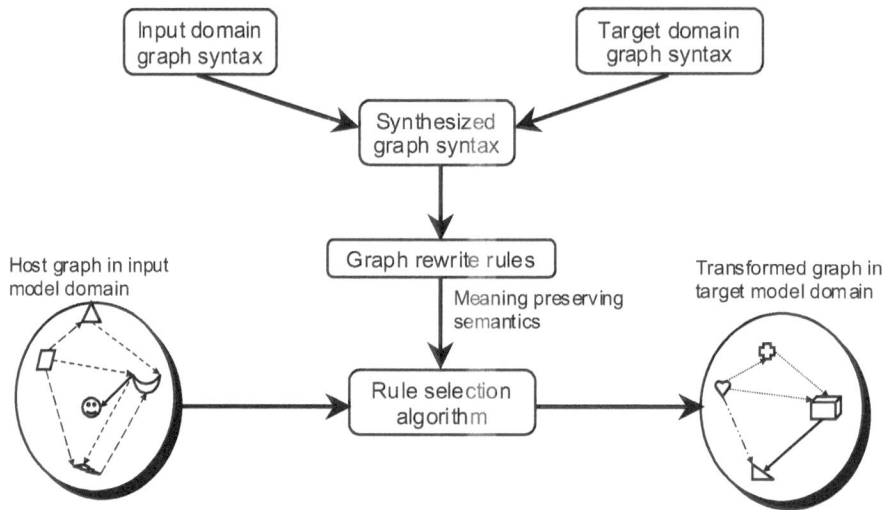

Fig. 2. Model Transformation with a Synthesized Graph Grammar

5.3   Model Transformation Correctness

A shortcoming of any transformation process which crosses a model domain is that it is very difficult to prove the correctness of the projection. A benefit of using a graph grammar is that the projection can be shown to be complete by enumeration if the input domain and target domain graph grammars are not too large. A target model domain's graph grammar can also contain assertions about the generated model to ensure that it is internally consistent. These points are discussed below.

In the graph grammar literature, the execution of graph rewrite rules have been described as *consistency preserving operations* which should not be confused with a *meaning preserving operations*. A consistency preserving graph rewrite operation produces a syntactically valid graph that is consistent with the graph schema of the input graph (i.e., use the same syntax).

A meaning preserving graph rewrite operation produces a resulting graph that preserves the model interpretation with respect to the target model domain, abstracting away details which are not important in the target domain. If a synthesized graph schema from two domains is used in the operation, then it is not (generally) possible to directly compare the meaning of the input graph to the output graph. There is then no proof system that can ensure that the rewrite operation is valid (i.e., it preserves the meaning across domains) because that meaning is provided by the analyst who constructs the rewrite operation. Therefore, preventing an incorrect model from being developed is not possible because there is no way to guard against the analyst constructing an inconsistent rewrite rule.

This identifies a limit of the automation process. It is not possible to automate a test to determine if a meaning preserving operation is correct (or incorrect). However, there are safeguards that can be taken. First, the testing and validation of each rewrite operation can be performed using a regression test suite. Next, the graph schema of the target domain can be used to perform a syntax check on the resulting model. Finally, application knowledge can be used to validate the semantics of the resulting model.

5.4 MMAP Experiences with PROGRES

PROGRES is the best tool for this research because it allowed the programming of complex graph replacement operations which were composed of many, separate graph rewrite operations. It has a language feature, called a *transaction*, that allows a graph rewrite operation to be an algorithm made up of imperative control statements and graph rewrite rules. The transaction is atomic, either succeeding in entirety or leaving the host graph in its initial state. It was found that the atomicity of the transaction is frequently used to establish consistency points during a complex graph rewrite operation. For example, a complex, recursive graph rewrite operation was developed that had several consistency points and fourteen recursive, smaller graph rewrite operations. Developing this complex transaction would not have been possible without the ability to define sub-transactions within an enclosing transaction.

Our experience with developing PROGRES specifications has led us to two design heuristics. First, the non-deterministic features of the PROGRES language were often bypassed because nondeterminism made debugging difficult. Second, we developed simple graph schemas and used in-line tests rather than developing complex graph schemas. The graph schemas were extremely useful at providing type checking at the coarse level, such as the interface points of graph rewriting rules. However, fine-grained type checking based on a precisely defined graph schema made the definition of node replacements within a graph rewrite rule difficult.

There are two features that we believe would make PROGRES programming easier. First, PROGRES would benefit from having conventional variables to store data, rather than resorting to instantiating a node in the host graph to record data for later use. An example is storing a file name for later retrieval. Another example is to have variables that are local to a section, which would be useful for enabling or disabling debugging. A compromise may be to provide write-once variables. Secondly, it would be useful if PROGRES had a non-visual, batch-mode to process several input files.

MMAP used PROGRES to transform several ANGIOTRACEs into scenario models that were used in the development of a performance model. The ANGIOTRACEs were from a simulation of a distributed database system described in [10]. The largest ANGIOTRACE was over three hundred and fifty events and required about 45 minutes of processing on a SUN Ultra 1 workstation with PROGRES version 8. Larger traces were not tested due to a limitation of the

PROGRES tools that only allows 64,000 nodes to be created during the transformation of a host graph. This not only includes the number of nodes in the host graph, but the intermediate nodes that are created during the execution of the transactions and productions. It is expected that this limitation will be lifted in future releases of PROGRES.

6   The Evolution of MMAP

MMAP has evolved over three generations with the last generation using graph grammars for automating performance model construction. A description of this evolution may be a helpful example for others who are considering using graph grammars.

The first generation of MMAP constructed performance models for a distributed application using ad-hoc rules which characterized the objects, their interactions, and the concurrency of the distributed application. However it was limited to systems which solely used a blocking RPC protocol and it assumed the availability of a global system monitor to record events in the order of occurrence [7].

The second generation of MMAP [8] overcame the limitation of pure RPC communication by characterizing asynchronous messaging. It also overcame the need for a global monitor because the ANGIOTRACE instrumentation was specified using a partial order notation based on a logical clock. An ANGIOTRACE would form an event graph that was a tree because object synchronization was not characterized and, correspondingly, no node type had an in-degree of two or more.

The third generation of MMAP [10] characterizes object synchronization and scenarios. Adding object synchronization requires a more formal, rigorous technique because the system event graph with all of the recorded events is a lattice with an embedded, directed, acyclic graph for each distributed scenario. This complex structure greatly increases the difficulty of the analysis which justifies moving to the graph grammar formalism. Using a graph grammar formalism now allows MMAP to be extended more easily.

7   Summary

The Model Making Automation Process (MMAP), which automatically generates a high-level model from event traces, has been introduced. It is in its fourth generation of development. It has overcome problems of the previous generations by using an algorithmic attributed graph grammar as its transformation formalism.

Now that MMAP uses a graph grammar it is possible to represent information from several domains in an orthogonal manner. Projections can be developed that abstract or synthesize new, accurate views of the information by transforming a host graph from an input model domain to a target domain. This suggests that MMAP will provide a uniform way to describe abstract events that include information from several domains [24].

Our experience with MMAP has shown that by describing each of the model domains as a graph grammar, the projection from one model domain to another can be formally described in the application specific language of each domain, rather than at the much lower level of a generic programming language (e.g., C or C++). Not only is this more elegant, it is the natural progression of programming: allowing domain experts to describe the problem and solution in their own language.

References

1. Grady Booch, James Rumbaugh, and Ivar Jacobson. *The Unified Modeling Language User Guide*. Addison-Wesley, Don Mills, Ontario, Canada, 1999. 443

2. D. Gelernter and N. Carriero. Coordination languages and their significance. *Communications of the ACM*, 35(2):97–107, February 1992. 443

3. Martin Glinz. An integrated formal model of scenarios based on statecharts. In Wilhelm Schäfer and Pere Botella, editors, *European Software Engineering Conference ESEC '95*, pages 254–271, September 1995. 444

4. James Gosling, Bill Joy, and Guy Steele. *The Java Language Specification*. The Java Series. Addison-Wesley, 1997. 443

5. Frederick Hayes-Roth and D.A. Waterman. Principles of pattern-directed inference systems. In D.A. Waterman and Frederick Hayes-Roth, editors, *Pattern-Directed Inference Systems*, pages 577–601. Academic Press, 1978. 449

6. L. Hess and B. Mayoh. The four musicians: Analogies and experts systems – a graphic approach. In H. Erhig, H.-J. Kreowski, and G. Rozenberg, editors, *Graph Grammars and Their Application to Computer Science*, number 532 in Lecture Notes in Computer Science, pages 430–445. Springer-Verlag, 1990. 449

7. C. Hrischuk, J. Rolia, and C. M. Woodside. Automatic generation of a software performance model using an object-oriented prototype. In *International Workshop on Modeling, Analysis, and Simulation of Computer and Telecommunication Systems (MASCOTS'95)*, pages 399–409, 1995. 447, 452

8. C. Hrischuk, C. M. Woodside, J. Rolia, and Rod Iversen. Trace-based load characterization for generating software performance models. Technical Report SCE-97-05, Dept. of Systems and Computer Engineering, Carleton University, Ottawa, Canada K1S 5B6, 1997. To appear in IEEE Trans. on Software Engineering. 447, 452

9. Curtis Hrischuk and Murray Woodside. Logical clock requirements for reverse engineering scenarios from a distributed system. Technical report, Department of Electrical and Computer Engineering, University of Alberta, Edmonton, Canada, 1999. Submitted to IEEE Trans. on Soft. Eng. 443, 445

10. Curtis E. Hrischuk. *Trace-based Load Characterization for the Automated Development of Software Performance Models*. PhD thesis, Carleton University, Ottawa, Canada, 1998. 447, 451, 452

11. Manfred Jackel. Ada-concurrency specified by graph grammars. In H. Ehrig, M. Nagl, G. Rozenberg, and A. Rosenfeld, editors, *Graph Grammars and Their Application to Computer Science*, volume 291 of *Lecture Notes in Computer Science*, pages 262–279, 1987. 444

12. Dirk Janssens and Grzegorz Rozenberg. Basic notions of actor grammars. In H. Ehrig, M. Nagl, G. Rozenberg, and A. Rosenfeld, editors, *Graph Grammars and Their Application to Computer Science*, volume 291 of *Lecture Notes in Computer Science*, pages 280–298, 1987. 444

13. Martin Korff. Application of graph grammars to rule-based systems. In H. Ehrig, H.-J. Kreowski, and G. Rozenberg, editors, *Graph Grammars and Their Application to Computer Science*, volume 532 of *Lecture Notes in Computer Science*, pages 505–519, 1991. 449

14. M. Nagl. *Building Tightly-Integrated (Software) Development Environments: The IPSEN Approach*. Number 1170 in Lecture Notes in Computer Science. Springer-Verlag, 1996. 443

15. Object Management Group. *The Common Object Request Broker: Architecture and Specification (CORBA) Revision 1.2*. Object Management Group, Framingham, Mass., revision 1.2 edition, Dec. 1993. OMG TC Document 93.12.43. 443

16. Open Software Foundation. *Introduction to OSF DCE*. Prentice-Hall, 1992. 443

17. Stephen Paynter. Structuring the semantic definitions of graphical design notations. *Software Engineering Journal*, 10(3):105–115, May 1995. 444

18. Sanjiva Prasad. Models for mobile computing agents. *ACM Computing Surveys*, 28:53, December 1996. 443

19. V. R. Pratt. Modeling concurrency with partial orders. *International Journal of Parallel Programming*, 15(1):33–71, 1986. 443, 446

20. Rational Software. *Unified Modeling Language v1.1 Semantics*, volume version 1.1. Rational Software Inc., Sept. 1997. http://www.rational.com/uml. 442

21. Stanley J. Rosenschein. The production system: Architecture and abstraction. In D.A. Waterman and Frederick Hayes-Roth, editors, *Pattern-Directed Inference Systems*, pages 525–538. Academic Press, 1978. 449

22. Andy Schürr. PROGRES: A VHL-language based on graph grammars. In H. Ehrig, H.-J. Kreowski, and G. Rozenberg, editors, *Graph Grammars and Their Application to Computer Science*, volume 532 of *Lecture Notes in Computer Science*, pages 641–659, 1991. 449

23. Andy Schürr, Andreas J. Winter, and Albert Zündorf. Graph grammar engineering with PROGRES. In W. Schäfer and P. Botella, editors, *Proceedings of the Fifth European Software Engineering Conference*, volume 989 of *Lecture Notes in Computer Science*, pages 219–234, 1995. 449

24. R. Schwarz and F. Mattern. Detecting causal relationships in distributed computations: in search of the Holy Grail. *Distributed Computing*, 7(3):149–174, 1994. 443, 446, 453

Graph-Based Models for Managing Development Processes, Resources, and Products

Carl-Arndt Krapp[1], Sven Krüppel[2], Ansgar Schleicher[3], and Bernhard Westfechtel[3]

[1] Finansys, Inc.
One World Trade Center, New York, NY 10048-0202
[2] SAP AG, Automotive Core Competence Center
Neurottstr. 16, D-69190 Walldorf, Germany
[3] Department of Computer Science III, Aachen University of Technology
D-52056 Aachen, Germany

Abstract. Management of development processes in different engineering disciplines is a challenging task. We present an integrated approach which covers not only the activities to be carried out, but also the resources required and the documents produced. Integrated management of processes, resources, and products is based on a model which is formally specified by a programmed graph rewriting system. Management tools are generated from the formal specification. In this way, we obtain a management system which assists in the coordination of developers cooperating in the development of a complex technical product.

1 Introduction

Development of products in disciplines such as mechanical, electrical, or software engineering is a challenging task. Costs have to be reduced, the time-to-market has to be shortened, and quality has to be improved. Skilled developers and sophisticated tools for performing technical work are necessary, yet not sufficient prerequisites for achieving these ambitious goals. In addition, the work of developers must be coordinated so that they cooperate smoothly. To this end, the steps of the development process have to be planned, a developer executing a task must be provided with documents and tools, the results of development activities have to be fed back to management which in turn has to adjust the plan accordingly, the documents produced in different working areas have to kept consistent with each other, etc.

Management can be defined as "all the activities and tasks undertaken by one or more persons for the purpose of planning and controlling the activities of others in order to achieve an objective or complete an activity that could not be achieved by the others acting alone" [27]. Management is concerned with processes, resources, and products:

- *Process management* covers the creation of process definitions, their instantiation to control development processes, as well as planning, enactment, and monitoring.

H. Ehrig et al. (Eds.): Graph Transformation, LNCS 1764, pp. 455–474, 2000.

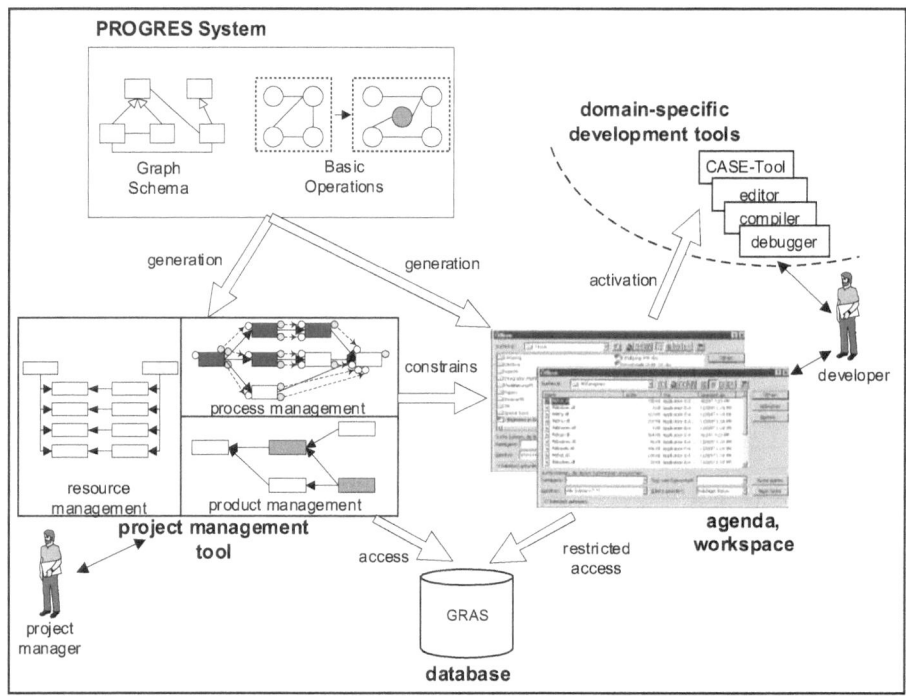

Fig. 1. A management system based on graph rewriting

- *Resource management* refers to both human and computer resources. In particular, it comprises the organization of human resources, the representation of computer resources, and the allocation of both human and computer resources to development activities.
- *Product management* deals with the documents created throughout the development life cycle, their dependencies, configurations of documents and dependencies, and versioning of both documents and configurations.

It is highly desirable to support both managers and developers through sophisticated tools assisting in the integrated management of processes, resources, and products. However, designing and implementing such tools constitutes a major challenge. In particular, the tools have to operate on complex data, and they have to offer complex commands manipulating these data.

We present a *programmed graph rewriting system* which serves as a high-level tool specification. Moreover, tools are generated from this executable specification. Our approach integrates the management of processes, resources, and products (for more detailed information on the corresponding submodels, see [12], [17], and [29], respectively). So far, we have applied our approach to three different domains, namely mechanical, chemical, and software engineering. In this paper, we will content ourselves to software engineering.

Figure 1 provides an overview of our *graph-based management system*. The management model is defined in PROGRES, which denotes both a language and an environment for the development of programmed graph rewriting systems [26]. Various kinds of tools are generated from the specification. A project management tool offers graphical views on management data and provides commands for planning and controlling development projects. Developers are supported by an agenda tool which displays a list tasks to be executed. Finally, a workspace tool is responsible for providing all documents and development tools required for executing a certain task.

The rest of this paper is structured as follows: Section 2 describes models for managing processes, resources, and products at an informal level. Section 3 presents cutouts of the corresponding PROGRES specifications. Section 4 discusses related work. Section 5 concludes the paper.

2   Management Models

2.1   Process Model

Our process management model is based on DYNAMIC Task nEts (*DYNA-MITE* [11]), which support dynamic development processes through evolving *task nets*. Editing, analysis, and execution of task nets may be interleaved seamlessly. A task is an entity which describes work to be done. The interface of a task specifies what to do (in terms of inputs, outputs, pre- and postconditions, etc.). The realization of a task describes how to perform the work. A suitable realization may be selected from multiple alternative realization types. A realization is either atomic or complex. In the latter case, there is a refining subnet (task hierarchies). In addition to decomposition relationships, tasks are connected by control flows (similar to precedence relationships in net plans), data flows, and feedback flows (representing feedback in the development process).

To illustrate these concepts, Figure 2 shows the evolution of a task net for the development of a software system. At the beginning of a software development project only little is known about the process. A Design task is introduced into the net, while the rest of the task net remains unspecified as it is dependent on the design document's internal structure (part i). As soon as the coarse design (part ii) is available, the complete structure of the task net can be specified (part iii). For each module defined in the design document, three tasks for designing the interface, for implementing the body, and for testing are inserted into the task net. Control flow dependencies are established between the tasks, partly based on the product dependencies. Implementation tasks require export and import interfaces; moreover, modules are tested in a bottom-up order.

A control flow successor may be started before its predecessor terminates (*simultaneous engineering*). For example, design of module interfaces may commence before the coarse design is fixed in its final form (part iv). During detailed design, errors may be detected, raising *feedback* to the coarse design (from DesignIntB to Design). Subsequently, a new version of the design docu-

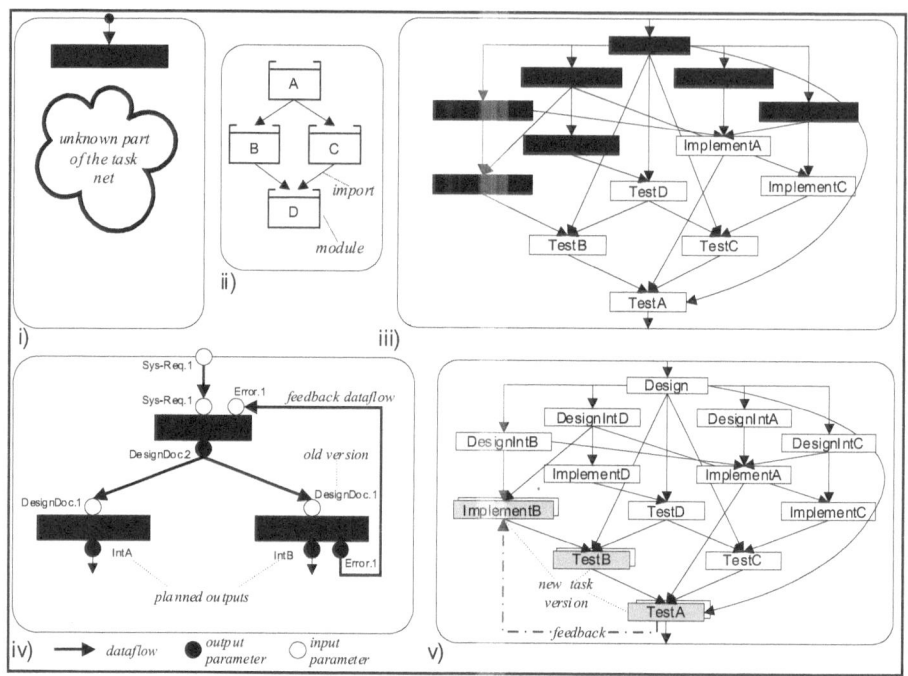

Fig. 2. Snapshot of a task net's evolution

ment (DesignDoc.2) is produced which will eventually be read by all successor tasks (so far, the successors are still using the old version).

If feedback occurs to terminated tasks, these are not reactivated. Rather, a new *task version* is derived from the old one and the old work context is reestablished (part v). This ensures traceability of the consequences of feedback. In our example, the first execution of TestA detects an error in module B. As a consequence, a new version of ImplementB is created. Version creation propagates through the task net until a new version of TestA is eventually created and executed.

2.2   Resource Model

RESMOD (*RES*ource Management *MOD*el, [17]) is concerned with the *resources* required for executing development processes. This includes human and computer resources, which are modeled in a uniform way. The notion of human resource covers all persons contributing to the development process, regardless of the functions they perform (e.g., managers as well as engineers). A similarly broad definition applies to computer resources, which cover all kinds of computer support for executing development processes, including hardware and software.

A resource is declared before it may be used in different contexts. A *resource declaration* introduces the name of a resource, its properties (attributes), and

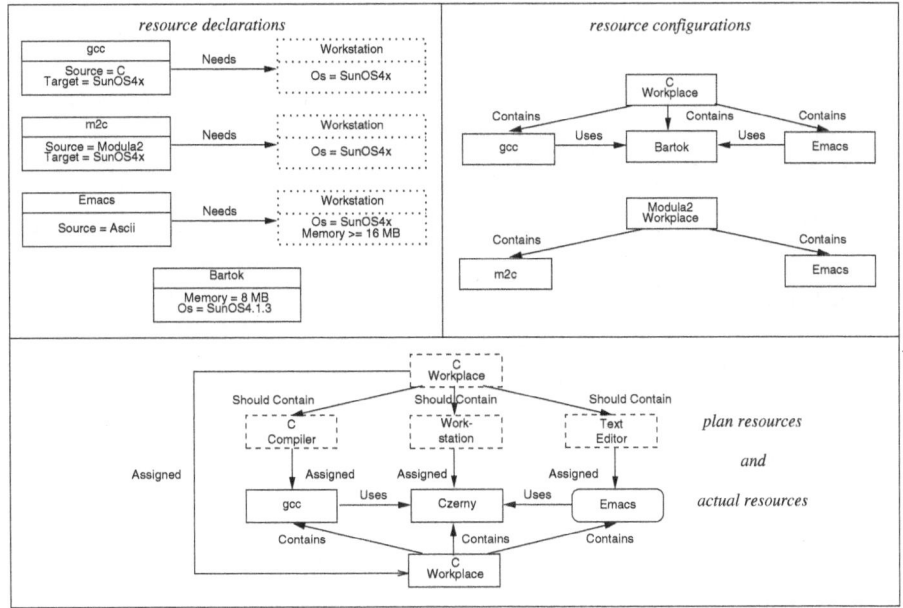

Fig. 3. Modeling of resources

(potentially) *needed resources*. For example, in Figure 3 the compiler gcc is characterized by its source language C and the target architecture SunOS4x. The compiler also needs a SunOS4x workstation, i.e., it runs on the same sort of machine for which it generates code. Note that the text editor Emacs needs at least 16 MB of main memory to execute with reasonable performance.

A *resource configuration* consists of a collection of resources. Resource configurations may be nested, i.e., a subconfiguration may occur as a component of some resource configuration. The resource hierarchy forms a directed acyclic graph. In general, a resource may be contained in multiple resource configurations. *Use relationships* connect inter-dependent components of resource configurations. These connections are established by binding formal needed resources to actual resources that have to meet the requirements stated in the resource declarations. Note that the resource configurations shown in Figure 3 are both incorrect. The first one is inconsistent because Bartok offers less main memory (8 MB) than required by Emacs. The second one is incomplete because it does not contain a workstation on which the tools may execute.

So far, we have considered *actual resources* and their relationships. In addition, RESMOD supports planning of resource requirements. A manager may specify in advance which resources he will need for a certain purpose, e.g., for running a development project. A *plan resource* serves as a placeholder for some actual resource. In the planning phase, the manager builds up configurations of plan resources. For each plan resource, he describes the requirements of a matching actual resource. Furthermore, a configuration of plan resources may

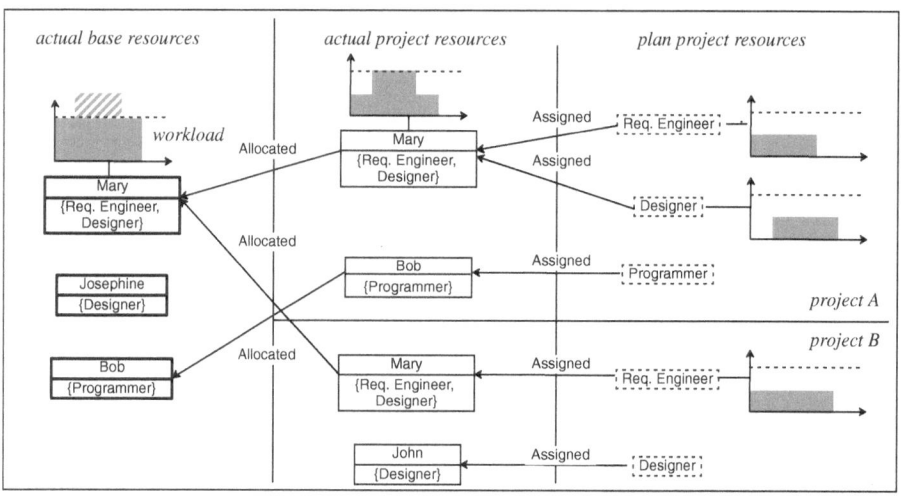

Fig. 4. Multi-project management

represent dependencies between its components. Later on, the manager assigns matching actual resources to the plan resources. In the simple example of Figure 3, C_Workplace represents a configuration of plan resources which are mapped 1:1 onto matching actual resources (in general, n:1 mappings are allowed).

Finally, RESMOD supports multi-project management (Figure 4). To this end, a distinction is made between *base resources* and *project resources*. A base resource belongs to the persistent organization of an enterprise and may be *allocated* to multiple projects. Alternatively, a resource may be acquired for a specific project only. The example given in Figure 4 demonstrates this for human resources. On the left-hand side, the available developers and their potential roles are described. For example, Mary may act as a requirements engineer or as a designer. On the right-hand side, each project manager specifies resource requirements in terms of plan resources and expected workloads. For each plan resource, a matching actual resource is either acquired specifically for this project (John), or it is drawn from the base organization (Mary). Workloads are accumulated both within a project (Mary fills multiple positions in project A) and across multiple projects (Mary is engaged in both A and B). In the example, the expected workload of Mary reveals an overload that requires rescheduling of resources.

2.3 Product Model

CoMa (*Co*nfiguration *Ma*nagement, [29]) supports version control, configuration control, and consistency control for heterogeneous documents through an integrated model based on a small number of concepts. In the course of development, documents such as requirements definitions, designs, or module bodies are created with the help of heterogeneous tools. Documents are related by man-

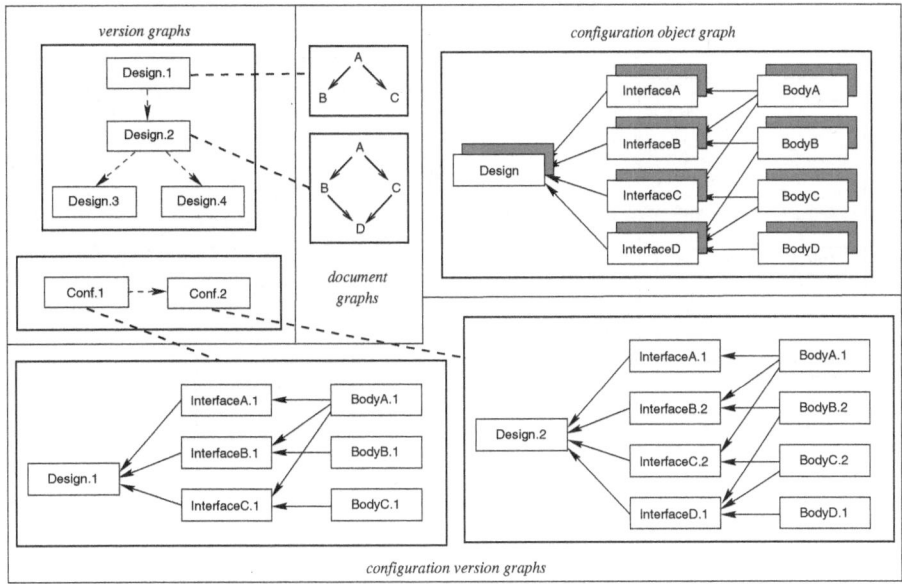

Fig. 5. Version and configuration graphs

ifold dependencies, both within one working area and across different working areas. The representation of these dependencies lays the foundations for consistency control between interdependent documents. Documents and their mutual dependencies are aggregated into configurations. Since development processes may span long periods of time, both documents and configurations evolve into multiple versions. Versions are recorded for various reasons, including reuse, backup, and provision of stable workspaces for engineers. Consistency control takes versioning into account, i.e., it is precisely recorded which versions of different documents are consistent with each other.

In order to represent objects, versions, and their relationships, the product management model distinguishes between different kinds of interrelated subgraphs (Figure 5). A *version graph* consists of versions which are connected by successor relationships. Versions are maintained for both documents and configurations. In our example, the evolution of the design of a software system is represented by a version tree. Each node refers to a *document graph* representing the contents of the respective version. Moreover, there is a version sequence representing the evolution of the overall product configuration. The contents of each version is contained in a corresponding configuration version graph.

A *configuration version graph* represents a snapshot of a set of interdependent components. Thus, it consists of component versions and their dependencies. In our example, the initial configuration version consists of initial versions of the design, module interfaces, and module bodies. In the second configuration version, the design was modified by adding a new module D. Interface and body of D were added as well; the interfaces and bodies of B and C were modified.

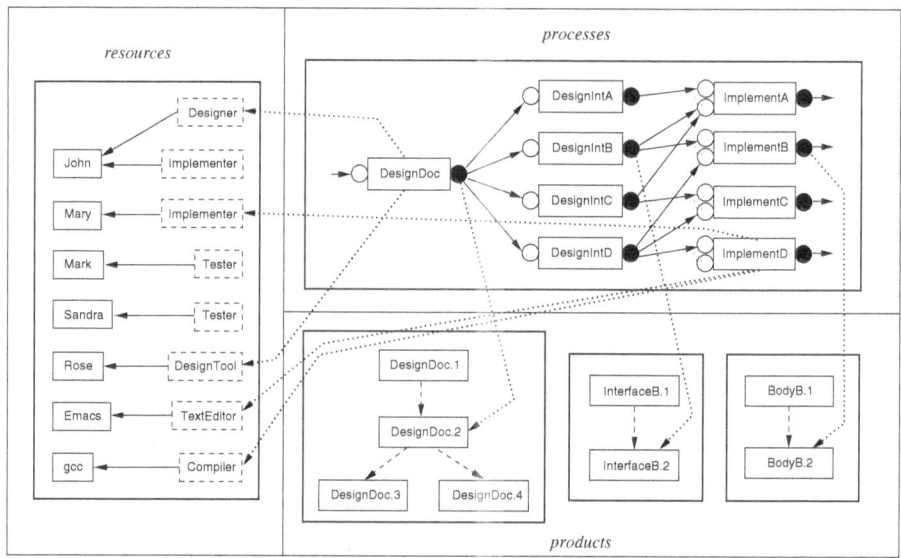

Fig. 6. Integration of processes, resources, and products

Finally, a *configuration object graph* represents version-independent structural information. It consists of documents which are connected by dependencies. For each component version (version dependency) in some configuration version graph, there must be a corresponding component object (object dependency) in the configuration object graph. Thus, the configuration object graph may be regarded as a "union" of all configuration version graphs.

2.4   Model Integration

The models for managing products, activities, and resources constitute components of an *integrated management model* (Figure 6). For example, the inputs and outputs of tasks refer to versions of documents or configurations. Furthermore, tasks are connected to both human and computer resources (responsible employees and supporting tools, respectively). Note that tasks are related to plan resources so that assignments can be made before the corresponding actual resources have been selected.

3   Formal Specification

Management models as presented in Section 2 consist of many entities which are mutually interrelated. Based on our experience in building structure-oriented environments, the data model of attributed graphs has proved suitable for the internal representation of complex data structures [19]. An *attributed graph* consists of attributed nodes which are interconnected by labeled, directed edges.

During editing, analysis, and execution of a management model, complex transformations and queries are performed on the internal data structure. We have chosen *programmed graph rewriting* in order to specify these complex operations on a high level of abstraction. In particular, we have used the specification language *PROGRES*, which is based on programmed graph rewriting [26].

Below, we will present small cutouts of the specification of the management submodels and their integration. Due to the lack of space, we will discuss only the *base models*, i.e., the model cores shared by all application domains. In order to use the models in different domains such as software, chemical, or mechanical engineering, they need to be enriched with domain-specific knowledge. For model customization, the reader is referred to [15,24].

3.1   Specification of the Process Model

After having introduced the process management model informally in Subsection 2.1, we now turn to its formal specification. Let us first present the internal data structure maintained by a process management tool as presented in Figure 7 (see Figure 2, part iv for the corresponding external representation). It consists of typed nodes and directed edges which form binary relationships between nodes. A task graph as the internal data structure of a process model instance consists of nodes representing tasks, parameters and token as references to products maintained by a corresponding product model instance. Task relations and data flows are internally represented by nodes, because they carry attributes and neither PROGRES nor the underlying graph database support attributed edges. Edges are used to connect task, parameter and token nodes.

In order to restrict the graph to meaningful structures with respect to the process model, a *graph schema* is defined which specifies a graph type. The process model's graph schema is displayed in Figure 8. Node class ITEM serves as the root class. New versions can be derived from all elements of a task net (this ensures traceability; see Figure 2, part v). A successor version is reached by following the toSuccessor edge. On the next layer of the inheritance hierarchy we mainly distinguish between process entity and process relationship types. The node class TOKEN describes nodes representing tokens that are passed along data flows. TASK nodes own PARAMETER nodes which are either INPUT or OUTPUT parameters. Tasks can **produce** tokens via output parameters and **read** tokens via input parameters. Tasks are connected by TASK RELATIONS which can be vertical or horizontal relationships. DECOMPOSITION is a vertical task relation, instances of which are used to build a task hierarchy. Horizontal relationships are CONTROL FLOW or FEEDBACK relations. Parameters in turn are connected via DATA FLOW relationships which refine task relationships.

In addition to the graph schema, various *consistency constraints* are needed to define valid task net structures. For example, it has to be enforced that the task hierarchy must form a tree and that control flow relations do not contain cycles. PROGRES offers language constructs for graphical and textual constraint definition. Due to space restrictions we will not present formal constraint definitions here (but we will show below how graph rewrite rules check constraints).

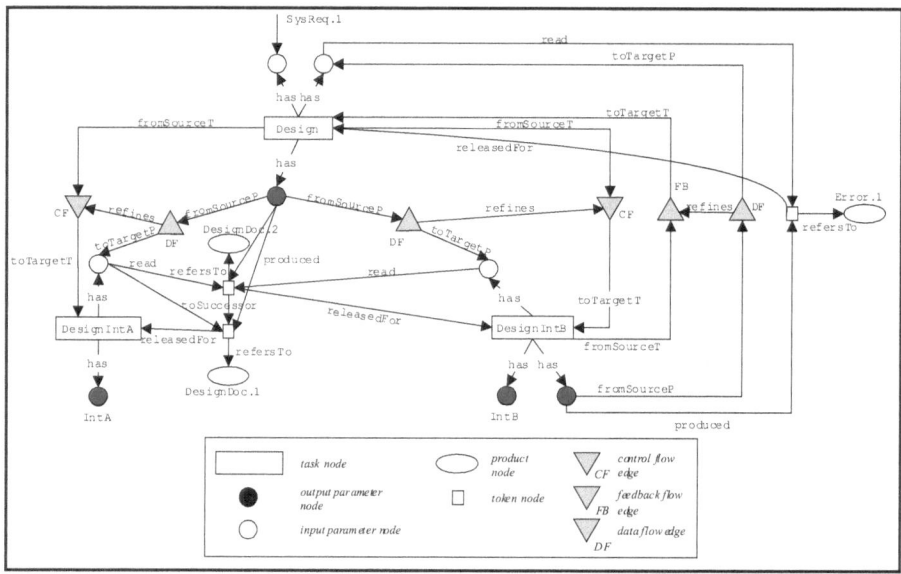

Fig. 7. Task graph

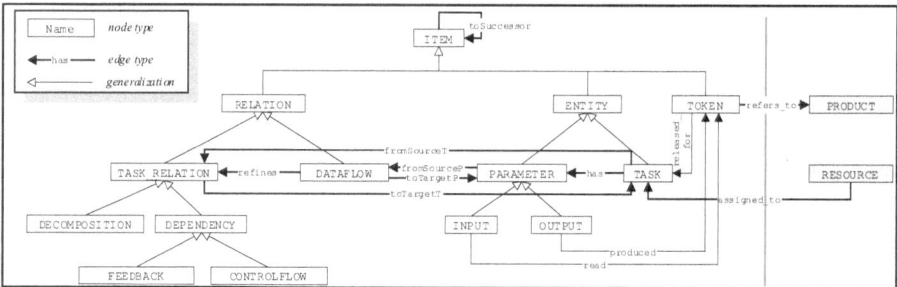

Fig. 8. The process model's graph schema

Rather, we will explain examples of the process model's operations which are divided into operations for *editing* and for *executing* a task net. Task nets can be edited by introducing new entities and relationships into the net. Equally well entities and relationships can be removed. Operations to execute a task net deal e.g. with the change of task states and the token game between tasks. Editing and execution of task nets are both described using a uniform mechanism. This allows to specify the intertwined editing and execution of task nets.

An example for an *execution operation* will be given in Subsection 3.4. As an example of an *edit operation* we chose a *graph rewrite rule* (*production* in PROGRES terminology) for feedback creation (Figure 9). The production is fed with the source and target tasks and the type of the feedback flow. In the left-hand side, a graph pattern is described that has to be found in order to create

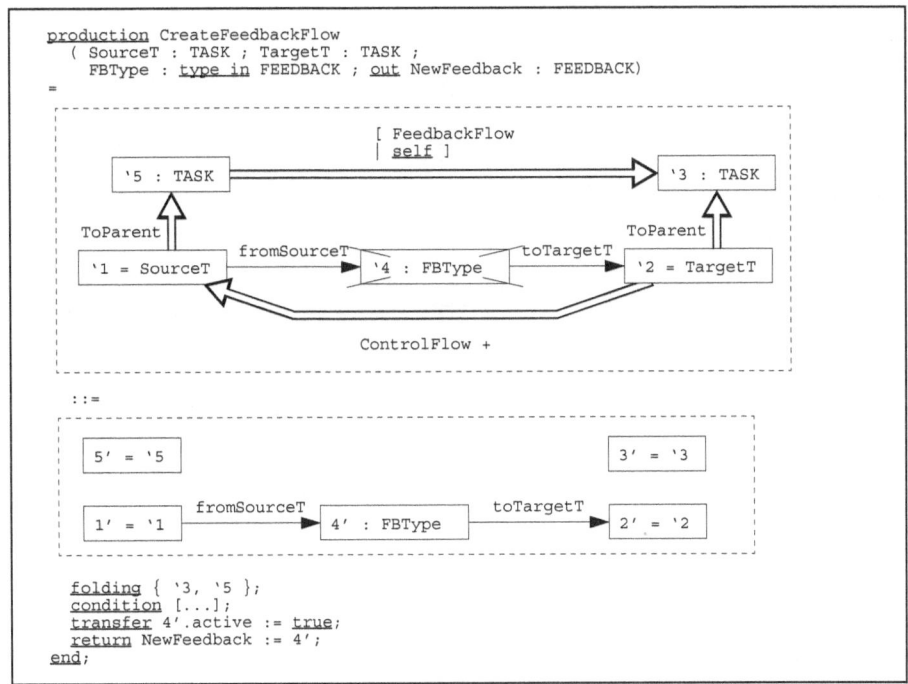

```
production CreateFeedbackFlow
  ( SourceT : TASK ; TargetT : TASK ;
    FBType : type in FEEDBACK ; out NewFeedback : FEEDBACK)
=
```

```
folding { `3, `5 };
condition [...];
transfer 4'.active := true;
return NewFeedback := 4';
end;
```

Fig. 9. Production for feedback flow creation

the feedback flow. Firstly, it has to be ensured that no feedback of the same type already exists between the two tasks. This is achieved by the negative node `4. Secondly, it has to be ensured that the feedback flow to be created is directed oppositely to control flows. When a feedback flow is created from a source to a target task (`1 and `2), there must be a (transitive) control flow path in the opposite direction. Thirdly some restrictions have to hold on the parent tasks of the feedback's source and target. If the parent tasks are unequal (`3 ≠ `5), they must be connected by a feedback flow as well. Otherwise, source and target share the same parent task (`3 = `5). This is allowed by the folding clause above the (elided) condition part. In this case, the FeedbackFlow path from `5 to `3 collapses (alternative **self** in the path definition).

If the left-hand side could be matched in the current graph, a new feedback flow is created by the right-hand side (the other nodes occurring on the right-hand side are replaced identically). The created feedback flow is set to active in the transfer part of the production (the flow becomes inactive as soon as feedback has been processed). Finally, the node representing the created feedback flow is returned to the caller.

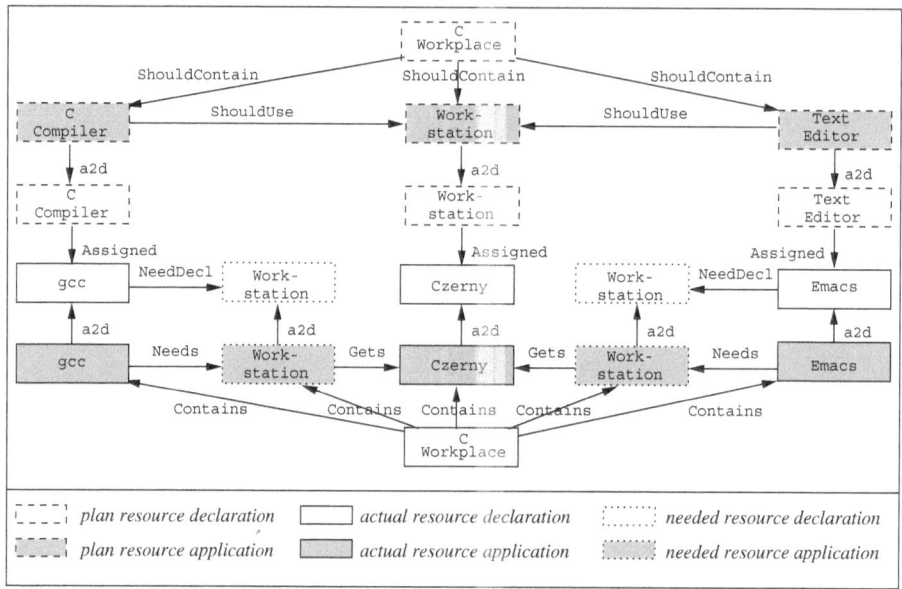

Fig. 10. Resource configurations

3.2   Specification of the Resource Model

A *resource graph* serves as the internal data structure maintained by the resource management tool. The types of nodes, edges, and attributes are defined in a graph schema. Due to the lack of space, we refrain from introducing the graph schema; instead, we briefly discuss a small example.

Figure 10 shows the internal representation of the configurations of plan and actual resources depicted earlier in Figure 3. Nodes of different classes are distinguished by means of line and fill styles (see legend at the bottom); the strings written inside the boxes are formally represented as attributes (other attributes are not shown).

The internal graph model distinguishes between *resource declarations* and *resource applications* (white and grey boxes, respectively). In this way, it is possible to model context dependent applications. As a consequence, resource hierarchies are modeled by alternating declarations and applications. For example, the text editor `Emacs` needs some workstation as host. In the `C_Workplace` shown in the figure, the workstation `Czerny` has been selected. In a different context, the editor may be run on another host. Whether the resource requirements are met, is a context dependent property which is attached to the applied occurrence of the needed resource.

Figure 11 shows operations for building up *resource hierarchies*. The base operation `CreateSubresource` creates a node that represents an applied occurrence of some resource declaration (parameter `Subres`) in some resource configuration (`Res`). The negative path on the left-hand side ensures that no cycle is

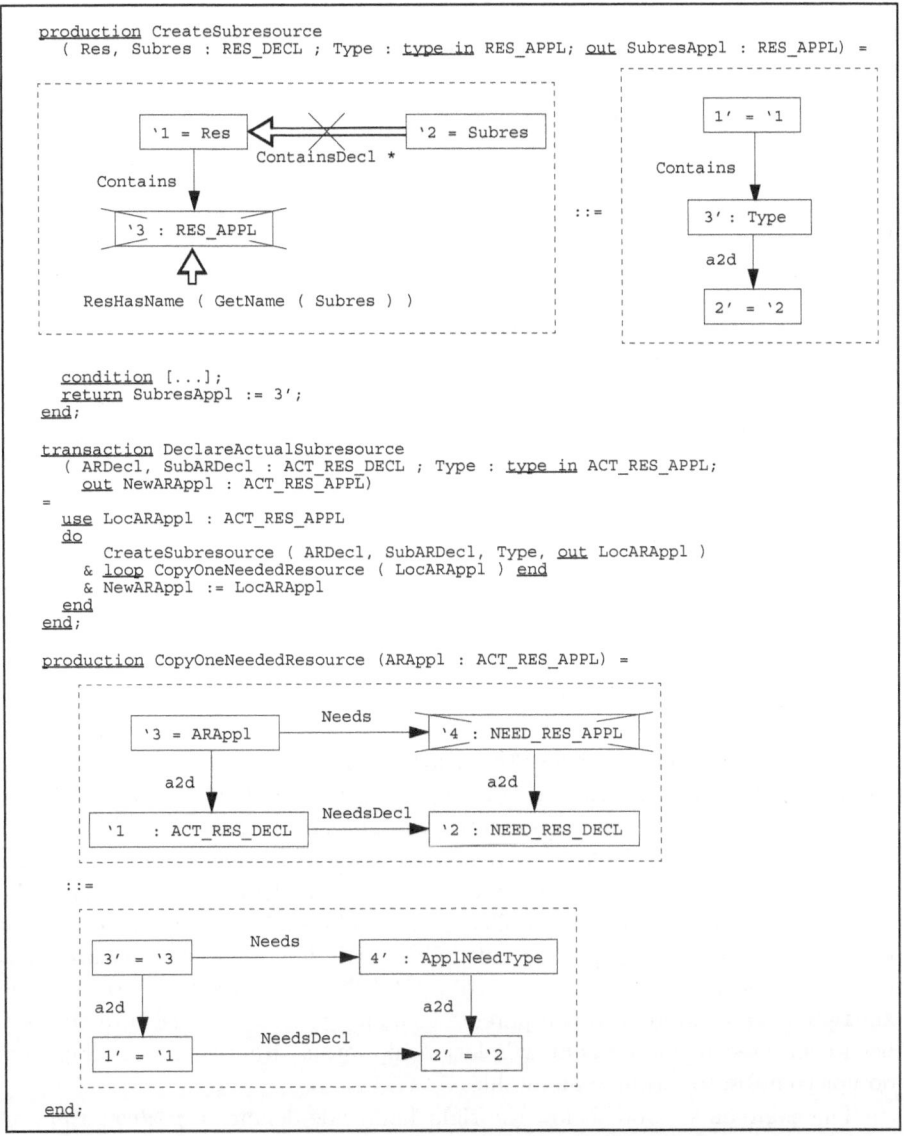

Fig. 11. Insertion of a subresource

introduced into the resource hierarchy. Furthermore, we have to check whether the child resource is already contained in the parent. This is excluded by the negative node `3: There must be no applied resource with the same name as Subres (see the restriction below the node).

The base operation does not take care of the needed resources, applied occurrences of which have to be created as well. Since this complex graph transfor-

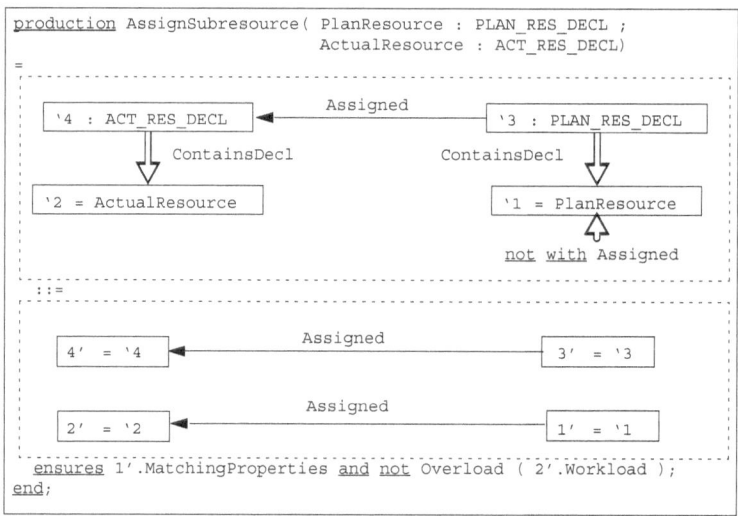

Fig. 12. Assignment of a subresource

mation cannot be expressed by a single graph rewrite rule, we compose multiple rules into a *transaction*. The transaction `DeclareActualSubresource` calls the base operation `CreateSubresource` and then performs a loop over all needed resources (the operator & and the keyword `loop` denote a sequence and a loop, respectively). `CopyOneNeededResToActualRes` creates an applied occurrence for one needed resource. The negative node on the left-hand side ensures that the applied occurrence has not been created yet, guaranteeing loop termination.

The rule shown in Figure 12 creates an *assignment relationship* between a plan resource and an actual resource (nodes '1 and '2 on the left-hand side, respectively). The rule may be applied only to subresources; assignments of root resources are handled by another rule. The restriction on node '1 ensures that no actual resource has been assigned yet to this plan resource (no outgoing `Assigned` edge). Moreover, the parents of nodes '1 and '2 must already have been connected by an assignment relationship; otherwise, resource assignments do not conform to resource hierarchies.

The **ensures** keyword below the right-hand side denotes a *postcondition* on attribute values. The rule fails if this postcondition does not hold. First, the actual resource must match the requirements attached to the plan resource. This is guaranteed when the attribute `MatchingProperties` evaluates to **true** (see below). Second, the resource assignment must not have caused an overload. This is checked by the predicate `Overload` that is applied to the attribute `Workload` of the actual resource.

How do we define the attributes used in the postcondition? The essential idea is to use *derived attributes* to this end (see [17] for further details). For example, if we have specified the resource requirements to some workstation, we want to make sure that the actual workstation meets these requirements. This can be handled as follows: Attributes `ReqMemorySize` and `ActMemorySize` are attached to the plan resource and the actual resource, respectively. The attribute `MatchingProperties` is then calculated by comparing these attributes:

```
MatchingProperties =
    self.ReqMainMemorySize <= self.Assigned.ActMainMemorySize
```

3.3 Specification of the Product Model

The specification of the product model is sketched only briefly; for further information, the reader is referred to [29].

The internal representation of version graphs, configuration version graphs, and configuration object graphs is determined according to similar design rules as in the case of process or resource management. For each subgraph of the overall *product graph*, a root node is introduced. As in RESMOD, the problem of context dependence recurs: A version may occur in different configuration versions in different contexts. Therefore, we distinguish between applied occurrences — denoted as version components — and declarations (of versions). Finally, relationships are internally represented by nodes if they are decorated with attributes or relationships between relationships have to be represented.

An example of a transformation of the product graph is given in Figure 13. The graph rewrite rule creates a version dependency (node 8') between a dependent component ('5) and a master component ('6) in a configuration version graph ('7). Several constraints have to be checked by this rule: The configuration version must not have been frozen yet (condition part). Furthermore, master and dependent component must belong to the same subgraph (edges from '5 and '6 to '7). Finally, there must be a corresponding object dependency ('4) in the configuration object graph ('1). Please recall that a configuration object graph serves as an abstraction over a set of configuration version graphs (see Subsection 2.3).

3.4 Specification of Model Integration

Due to the lack of space, we discuss model integration only briefly. We confine ourselves to the integration between process and product model. Concerning process and resource integration, we refer the reader back to Subsection 3.2 because the assignment of plan resources to tasks can be handled similarly to the assignment of actual to plan resources (Figure 12).

With respect to model integration, we favor a *loose coupling* between the submodels. That is, we connect two submodels such that the modification of one submodel has minimal impact on the other one. This is illustrated by the graph rewrite rule `Produce`, which belongs to the execution operations of DYNAMITE (Figure 14). To establish a clear separation between the process model and the product model, the output of some task does not refer to some object directly. Rather, a *token* is created which is connected to the output parameter on one hand and the produced object on the other hand. In the DYNAMITE

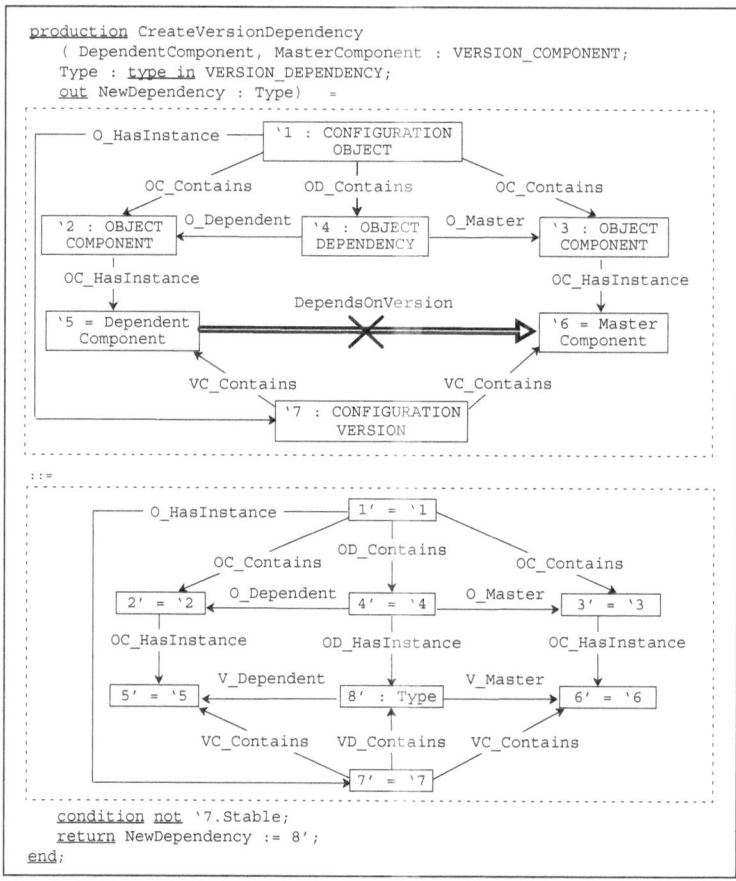

Fig. 13. Creation of a version dependency

specification, it is merely assumed that there is a node class PRODUCT from which subclasses are derived in the product model. Note that all tokens are arranged in a sequence (toSucc edges), which is required for simultaneous engineering (see Subsection 2.1). Node '2 is optional, i.e., the graph rewrite rule either creates the first token or appends a token at the end of the list.

4   Related Work

DYNAMITE is based on instance-level task nets, where tasks and relations are dynamically instantiated from types. In general, the structure of a task net is known only at run time. In contrast, process-centered software engineering environments such as Process Weaver [9] or SPADE [1] are based on populated copies, i.e., a template of a Petri net is copied, populated with tokens, and enacted. This implies that the net structure is already determined at modeling time.

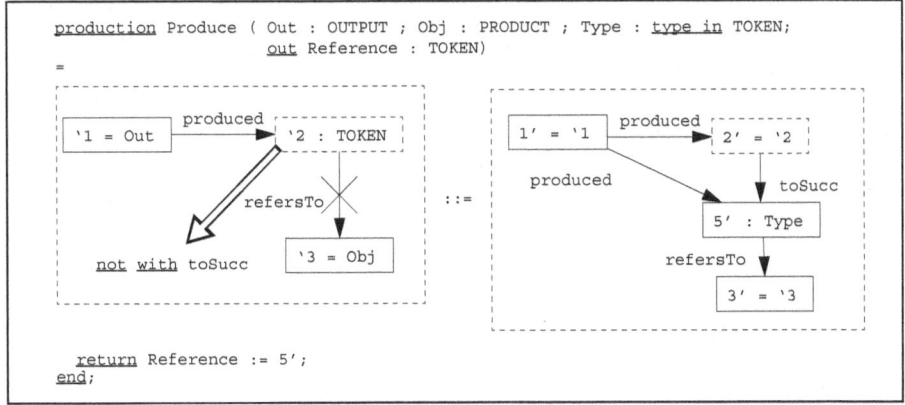

Fig. 14. Production of an output

Rule-based systems such as e.g. Marvel [14] or MERLIN [22] are more flexible in that respect. Such systems may dynamically generate plans from facts and rules, e.g. for compiling and linking program modules. While plans are maintained automatically and are normally hidden from the users, task nets are manipulated by the project manager.

RESMOD introduces fairly general concepts for resource management (resource configurations, plan and actual resources, base and project resources). These concepts can be applied to model a wide variety of organizational models. In contrast, many other systems implement a more specific model which lacks the required generality. This statement applies e.g. to the workflow management systems Leu [4], Flow Mark [13], and ORM [23]. Only MOBILE [3] is different in that it offers a very general base model for resource management. However, we believe that MOBILE is a bit too general and fails to fix anything specific to resource management. In MOBILE, a resource management model must be essentially specified from scratch, starting from general entity/relationship types.

CoMa integrates version control, configuration control, and consistency control. To this end, graphs are used to represent the complex structures encountered in product management. In particular, this distinguishes CoMa from the main stream software configuration management systems, which essentially add version control to the file system [28,18]. Concerning the relationships between versioned objects, most existing systems focus on hierarchies. This does not apply only to the file-based SCM systems mentioned above. In addition, systems relying on databases such as PCTE [21] or DAMOKLES [5] focus on versions of complex objects. In contrast, CoMa puts strong emphasis on dependencies among components of configurations. Thus, a configuration is modeled as a graph of components related by dependencies rather than just a directory.

PROGRES has been used in our group for a long time. PROGRES is a comprehensive specification language offering declarations of node types, node attributes, and edge types, derived attributes and relationships, constraints, graph tests, graph rewrite rules, control structures for combining these rules into transactions, and backtracking. In our specifications, we have exploited virtually all of these features. Moreover, we heavily rely on the PROGRES environment when building management tools. Given our requirements, the alternatives are not numerous. For example, the AGG environment [8] is based on the algebraic approach to graph rewriting, which is too restrictive for our applications. Recently, the Fujaba environment [10] has been developed which offers a graph grammar language UML [2]. Fujaba misses some essential features provided by PROGRES (in particular, derived attributes/relationships and backtracking).

5   Conclusion

We have presented an integrated model for managing development processes, resources, and products. Furthermore, we have formally specified this model by a programmed graph rewriting system. In this paper, we have presented cutouts of the formal specification of the management model. The total size of this specification is about 200 pages (counting only the base specifications).

While the examples given in this paper were taken from the software engineering domain, we have also studied applications in other engineering disciplines. A predecessor of the management model was developed in the SUKITS project, which dealt with development processes in mechanical engineering [30]. Current work is performed in the Collaborative Research Centre IMPROVE, which investigates development processes in chemical engineering [20].

References

1. S. Bandinelli, A. Fuggetta, and C. Ghezzi. Software process model evolution in the SPADE environment. *IEEE Transactions on Software Engineering*, 19(12):1128–1144, Dec. 1993. 470
2. G. Booch, J. Rumbaugh, and I. Jacobson. *The Unified Modeling Language User Guide*. Addison Wesley, 1998. 472
3. C. Bußler and S. Jablonski. An approach to integrate workflow modeling and organization modeling in an enterprise. In *Proceedings of the Third Workshop on Enabling Technologies: Infrastructure for Collaborative Enterprises*, pages 81–95, Morgantown, West Virginia, Apr. 1994. 471
4. G. Dinkhoff, V. Gruhn, A. Saalmann, and M. Zielonka. Business process modeling in the workflow management environment Leu. In P. Loucopoulos, editor, *Proceedings of the 13th International Conference on Object-Oriented and Entity-Relationship Modeling (ER '94)*, LNCS 881, pages 46–63, Manchester, UK, Dec. 1994. 471
5. K. Dittrich, W. Gotthard, and P. Lockemann. DAMOKLES, a database system for software engineering environments. In R. Conradi, T. M. Didriksen, and D. H. Wanvik, editors, *Proceedings of the International Workshop on Advanced Programming Environments*, LNCS 244, pages 353–371, Trondheim, June 1986. 471
6. H. Ehrig, G. Engels, H.-J. Kreowski, and G. Rozenberg, editors. *Handbook on Graph Grammars and Computing by Graph Transformation: Applications, Languages, and Tools*, volume 2. World Scientific, Singapore, 1999. 473, 474

7. G. Engels and G. Rozenberg, editors. *TAGT '98 — 6th International Workshop on Theory and Application of Graph Transformation*, technical report tr-ri-98-201, Paderborn, Germany, Nov. 1998. 473
8. C. Ermel, M. Rudolf, and G. Taentzer. The AGG approach: Language and environment. In Ehrig et al. [6], pages 551–602. 472
9. C. Fernström. PROCESS WEAVER: Adding process support to UNIX. In *Proceedings of the 2nd International Conference on the Software Process*, pages 12–26, Berlin, Germany, Feb. 1993. 470
10. T. Fischer, J. Niere, L. Torunski, and A. Zündorf. Story diagrams: A new graph grammar language based on the unified modeling language and Java. In Engels and Rozenberg [7], pages 112–121. 472
11. P. Heimann, C.-A. Krapp, and B. Westfechtel. An environment for managing software development processes. In *Proceedings of the 8th Conference on Software Engineering Environments*, pages 101–109, Cottbus, Germany, Apr. 1997. 457
12. P. Heimann, C.-A. Krapp, B. Westfechtel, and G. Joeris. Graph-based software process management. *International Journal of Software Engineering and Knowledge Engineering*, 7(4):431–455, Dec. 1997. 456
13. IBM, Böblingen, Germany. *IBM FlowMark: Modeling Workflow*, Mar. 1995. 471
14. G. E. Kaiser, P. H. Feiler, and S. Popovich. Intelligent assistance for software development and maintenance. *IEEE Software*, 5(3):40–49, May 1988. 471
15. C.-A. Krapp. *An Adaptable Environment for the Management of Development Processes*. Number 22 in Aachener Beiträge zur Informatik. Augustinus Buchhandlung, Aachen, Germany, 1998. 463
16. S. Krüppel. Ein Ressourcenmodell zur Unterstützung von Software-Entwicklungsprozessen. Master's thesis, RWTH Aachen, Germany, Feb. 1996.
17. S. Krüppel and B. Westfechtel. RESMOD: A resource management model for development processes. In Engels and Rozenberg [7], pages 390–397. 456, 458, 468
18. D. Leblang. The CM challenge: Configuration management that works. In W. Tichy, editor, *Configuration Management*, volume 2 of *Trends in Software*, pages 1–38. John Wiley & Sons, New York, 1994. 471
19. M. Nagl, editor. *Building Tightly-Integrated Software Development Environments: The IPSEN Approach*. LNCS 1170. Springer-Verlag, Berlin, Germany, 1996. 462
20. M. Nagl and W. Marquardt. SFB-476 IMPROVE: Informatische Unterstützung übergreifender Entwicklungsprozesse in der Verfahrenstechnik. In M. Jarke, K. Pasedach, and K. Pohl, editors, *Informatik '97: Informatik als Innovationsmotor*, Informatik aktuell, pages 143–154, Aachen, Germany, Sept. 1997. 472
21. F. Oquendo, K. Berrado, F. Gallo, R. Minot, and I. Thomas. Version management in the PACT integrated software engineering environment. In C. Ghezzi and J. A. McDermid, editors, *Proceedings of the 2nd European Software Engineering Conference*, LNCS 387, pages 222–242, Coventry, UK, Sept. 1989. 471
22. B. Peuschel, W. Schäfer, and S. Wolf. A knowledge-based software development environment supporting cooperative work. *International Journal of Software Engineering and Knowledge Engineering*, 2(1):79–106, Mar. 1992. 471
23. W. Rupietta. Organization models for cooperative office applications. In D. Karagiannis, editor, *Proceedings of the 5th International Conference on Database and Expert Systems Applications*, LNCS 856, pages 114–124, Athens, Greece, 1994. 471
24. A. Schleicher, B. Westfechtel, and D. Jäger. Modeling dynamic software processes in UML. Technical Report AIB 98-11, RWTH Aachen, Germany, 1998. 463
25. A. Schürr and A. Winter. UML packages for programmed graph rewriting systems. In Engels and Rozenberg [7], pages 132–139.

26. A. Schürr, A. Winter, and A. Zündorf. The PROGRES approach: Language and environment. In Ehrig et al. [6], pages 487–550. 457, 463

27. R. H. Thayer. Software engineering project management: A top-down view. In R. H. Thayer, editor, *Tutorial: Software Engineering Project Management*, pages 15–54. IEEE Computer Society Press, Washington, D.C., 1988. 455

28. W. F. Tichy. RCS – A system for version control. *Software–Practice and Experience*, 15(7):637–654, July 1985. 471

29. B. Westfechtel. A graph-based system for managing configurations of engineering design documents. *International Journal of Software Engineering and Knowledge Engineering*, 6(4):549–583, Dec. 1996. 456, 460, 469

30. B. Westfechtel. Graph-based product and process management in mechanical engineering. In Ehrig et al. [6], pages 321–368. 472

Deriving Software Performance Models from Architectural Patterns by Graph Transformations*

Dorina C. Petriu and Xin Wang

Carleton University
Ottawa, ON, Canada, K1S 5B6
{petriu,xinw}@sce.carleton.ca

Abstract. The paper proposes a formal approach to building software performance models for distributed and/or concurrent software systems from a description of the system's architecture by using graph transformations. The performance model is based on the *Layered Queueing Network (LQN)* formalism, an extension of the well-known Queueing Network modelling technique [16, 17, 8]. The transformation from the architectural description of a given system to its LQN model is based on PROGRES, a known visual language and environment for programming with graph rewriting systems [9-11]. The transformation result is an LQN model that can be analysed with existent solvers [5].

1   Introduction

It is generally accepted that performance characteristics, such as response time and throughput, play an important role in defining the quality of software products. In order to meet the performance requirements of such systems, the software developers should be able to assess and understand the effect of various design decisions on system performance at an early stage, when changes can be made easily and effectively. Software Performance Engineering (SPE) is a technique introduced in [15] that strongly recommends the integration of performance analysis into the software development process, starting from the earliest stages and continuing throughout the whole life cycle. According to SPE, performance must be designed and built into the system from the beginning, as opposed to the more frequent practical approach that postpones the performance concerns until the system is completely implemented, then tries to "fix" its the performance problems at this late stage. Late fixes tend to be very expensive and inefficient; in some cases, the product will never meet its original performance requirements.

The application of SPE to software development requires predictive modelling methods and tools to support performance oriented design and implementation

---

* Research partially supported by the Natural Sciences and Engineering Research Council of Canada (NSERC), and by the Communications and Information Technology Ontario (CITO).

H. Ehrig et al. (Eds.): Graph Transformation, LNCS 1764, pp. 475–488, 2000.

decisions. The *Layered Queueing Network (LQN)* is such a modelling method [16, 17, 8] and toolset [5] that was developed to capture the characteristics of concurrent and/or distributed software systems. LQN determines the delays due to contention, synchronization and serialization at both the software and hardware levels (see section 2 for a more detailed description).

LQN was applied to a number of concrete industrial systems (such as database applications, web servers, telecommunication systems [14]) and it was proven useful for providing insights into performance limitations at both software and hardware levels, for giving feedback for performance improvements to the software development team, and for supporting capacity planning under various workloads. However, the integration of performance modelling techniques, such as LQN, into the software development process as promoted by SPE is hampered by a number of factors. One is the existence of a cognitive gap between the software and the performance domains. Software developers are concerned with designing, implementing and testing the software, but they want to be spared from learning performance modelling methods and from building and solving performance models. The development teams depend usually on specialized performance groups to do the performance evaluation work, but this leads to additional communication delays, inconsistencies between design and model versions and late feedback. Also, economical pressure for "shorter time to market" leads to shorter software life cycles. There is no time left for SPE, which traditionally implies "manual" construction of the performance models.

The present paper attempts to close the gap between performance analysis and software development by proposing a formal technique based on graph transformations to build the performance model of a system from the description of the high-level software architecture (*i.e.*, the system's high-level components and their interconnections). By implementing the proposed technique in a tool, two goals can be achieved: "instant" building of the performance model shortens the time required for SPE and makes it possible to keep the model consistent with the changes in the architecture. The process of building a LQN performance model has two stages: a) generating the *model structure* from the software architecture description and b) computing the *model parameters*. These are quantitative values representing resource demands and numbers of visits, as for example "how much CPU processing time is required in average for a given software process to execute a given request" and "how many visits (in average) to various servers are necessary to complete that request". The objective of this paper is to propose a method that automates the first stage of the modelling process. Research work on the second stage is currently under way, aiming to automate the derivation of model parameters from the internal structure and behaviour of the high-level architectural components.

The paper is organized as follows: a short description of the LQN model is given in section 2, a discussion of high-level architectural patterns and the approach to translate them to LQN models is presented in section 3, the PROGRES graph schema and examples of rewriting rules are discussed in section 4, and conclusions in section 5.

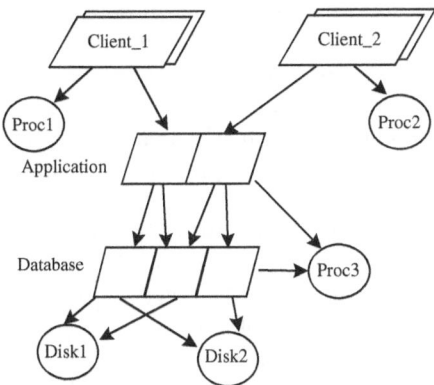

Fig. 1. Simple example of an LQN model

2   LQN Model

LQN was developed as an extension of the well-known Queueing Network (QN) model, at first independently in [16, 17] and [8], then as a joint effort [5]. The LQN toolset presented in [5] includes both simulation as well as analytical solvers that merge the best previous approaches. The main difference with respect to QN is that in LQN a server to which customer requests are arriving and queueing for service, may become a client to other servers from which it requires nested services while serving its own clients.

An LQN model is represented as an acyclic graph whose nodes are software entities and hardware devices, and whose arcs denote service requests. The software entities (named also *tasks*) are drawn as parallelograms, and the hardware devices as circles. The nodes with outgoing and no incoming arcs play the role of *clients*. The intermediate nodes with incoming and outgoing arcs play both the role of client and that of server, and represent usually software servers. The leaf nodes are pure servers, and model usually hardware servers (such as processors, I/O devices, communication network, etc.) Fig. 1 shows a simple example of an LQN model for a three-tiered client/server system: at the top there are two classes with a given number of stochastic identical clients. Each client sends requests for a specific service offered by a task named Application, which represents the business layer of the system. Each kind of service offered by an LQN task is modelled by a so-called *entry*, drawn as a parallelogram "slice". An entry has its own execution times and demands for other services (given as model parameters). In this case, each Application entry requires two different services from the Database task. Each software task is running on a processor shown as a circle; in the example, all clients of the same class share a processor, whereas Application and Database share another processor. Database uses also two disk devices, as shown in Fig. 1. It is worth mentioning that the word layered in the name LQN does not imply a strict layering of the tasks (for example, a task may call each other tasks in the same layer, or skip over layers). All the arcs used in this example represent *synchronous* requests, where the sender of a request message is blocked until it receives a reply from the provider of service. It is possible to have also *asynchronous* request messages, where the sender doesn't block after sending a request, and the server doesn't send any reply back, as shown in the following sections. Although not explicitly illustrated in the LQN notation, each software and hardware server has an implicit message queue where the incoming requests are waiting their turn to be served. Servers with more then one entry still have a single input queue, where requests for different entries wait together. The default scheduling policy of the queue is FIFO, but other policies are also supported. Typical results of an LQN model are response times, throughput, utilization of servers on behalf of different types of requests, and queueing delays. The LQN results may be used to identify the software and/or hardware components that limit the system performance under different workloads and resource allocations. Understanding the cause for performance limitations helps the development team to come up with appropriate remedies.

3   Approach for the Transformation from Software Architecture Descriptions into LQN Models

The emerging discipline of software architectures is concerned with informal and formal ways of describing the overall system structure of complex software systems. In [12] a perspective on this new discipline is presented, in [13] and [4] a number of high-level architectural patterns frequently used in today's software systems are identified and described, and in [3] a formal foundation for software architectures based on architectural connections is introduced. Graph grammars were used in [6] to describe the evolution of dynamic architectures. In this paper, we are using the informal approach of architectural patterns from [13] and [4] to explain the principles for translating an architectural description into an LQN model. We will use then some of the background on architectural connections from [3] to present our graph rewriting approach.

A relatively small number of patterns that describe the high-level architecture of a large range of software systems are identified in literature. These patterns describe the collaboration between concurrent components, which can run on a single computer or in a distributed environment. We have selected three architectural patterns as a basis for our discussion: Pipe and Filters (PF), Client-Server (CS) and BlackBoard (BB). These are frequently used to build distributed systems and present a variety of interactions between components.

PF divides the overall processing task into a number of sequential steps that are implemented as filters, while the data between filters flows through unidirectional pipes. Interesting performance problems arise in the case of active filters [4] that are running concurrently. Each filter is implemented as a process or thread that loops through the following steps: "pulls" the data (if any) from the preceding pipe, processes it and then "pushes" the results down the pipeline. The way in which the push and pull operations are implemented may also have performance consequences. Two cases are shown in the left side of Fig. 3: in (*a*)

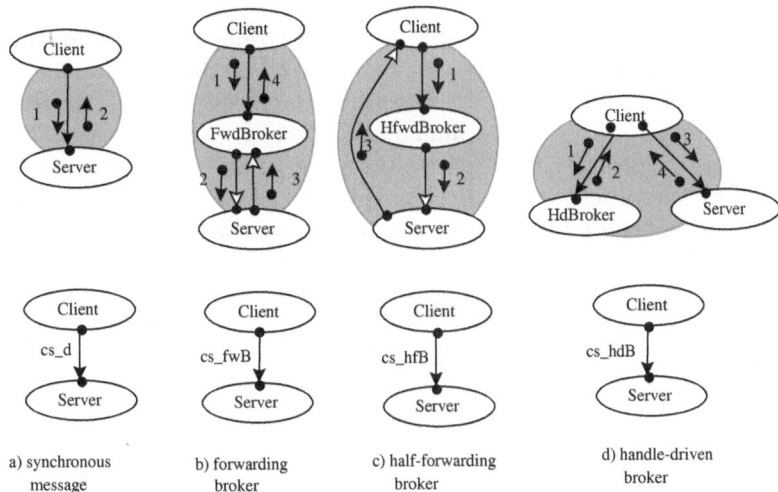

Fig. 2. Client/Server connection types

the filters communicate through asynchronous messages, whereas in (*b*) and (*c*) they communicate through a shared buffer (one pushes and the other pulls). The small arrows attached to the communication arcs indicate the data flow.

The Client-Server pattern is one of the most frequently used in today's distributed systems, especially since the introduction of new midware technology such as CORBA [7], which facilitates the connection between clients and servers running on heterogeneous platforms across local or wide-area networks. Since the communication between the architectural components has a crucial effect on performance, we will consider two different cases: direct client/server communication through a synchronous message, and a connection through a CORBA interface where the broker is the intermediary between clients and servers. In the first case shown in Fig. 2.a, the client sends a synchronous request to the server, then blocks and waits for the server's reply. Although the direction of the synchronous message is from the client to the server, the data flow is bi-directional (the request goes one way and the reply comes the other way). In the case of a CORBA interface, we distinguish several types of client/server connections [2]. In the forwarding broker pattern from Fig. 2.b, the broker relays a client's request to the relevant server, retrieves the response from the server and relays it back to the client.

The forwarding broker is at the center of all communication paths between clients and servers, and can provide load balancing or restart centrally any failed transactions. However, there is a price to pay in terms of performance: an interaction between a client and a server requires four messages, which leads to an excessive network traffic when the client, broker and server reside on different nodes. An alternative that reduces the excessive network traffic of the forwarding broker is the *half-forwarding broker* from Fig. 2.c, where the server returns the reply directly to the client. This reduces the number of messages for a client/server interaction to three, while it retains the main advantages of the forwarding broker (load balancing and centralized recovery from failure). A *handle-driven broker* (as in Fig. 2.d) returns to the client a handle containing all the information required to communicate directly with the server. The client may use this handle to talk directly to the server many times, thus reducing the potential for performance degradation. However, the client takes on additional responsibilities, such as checking if the handle is still valid after a while, and recovering from failures. Load balancing is also more difficult.

The Blackboard pattern is composed of a number of processes that share a common knowledge base stored in shared memory (see the left side of Fig. 5.) In order to insure the correctness of the common data, the access must be controlled by semaphores, locks or other similar mechanisms. The serialization brings performance effects, and must be captured in a performance model.

After the informal presentation of the chosen architectural patterns and of their performance implications, we will review briefly the formal approach to architectural connections introduced in [3], which is the basis for the graph grammar representation of software architectures proposed in the next section. According to [3], software architecture can be defined as a collection of computational *components* together with a collection of *connectors*, which describe the interactions between components. A *component type* is described as a set of *ports* and a *specification* that describes its function. Each port defines a logical point of interaction between the component and its environment. A *connector type* is defined by a set of *roles* and a *glue* specification. The roles describe the expected behaviour of the interacting parties, and the glue shows how the interactions are coordinated. The connector specification is formally described in [3] with a subset of Hoare's process algebra.

For example, in a CS pattern with CORBA interface (see Fig. 2), the connector type is defined by three roles (client, server and broker) and by the glue that shows what kind of interactions take place between participants, and in which order. Since the three kinds of brokers shown in Fig. 2.b, 2.c and 2.d behave and interact differently with the client and the server parties, each one corresponds to a different connector type. In total, we have considered four client/server connector types: one direct (Fig. 2.a) and three using the services of a broker. Another example of connector type that contains three roles (two filters and a shared buffer) is shown in the left-hand side of Fig. 3.b and 3.c. Its glue describes the "push" and "pull" operations and the constraints for correct behaviour (as for example "cannot pull data from an empty buffer", "cannot read and write to the buffer at the same time", etc.)

In our present work, we first identified the connector types associated to different architectural patterns, then defined graph transformation rules to translate each connector to an LQN submodel. The transformation approach for different connections is illustrated in Figures 3 to 5 in a more intuitive, higher-level graphical notation tailored to our problem domain, rather than in the more detailed PROGRES notation.

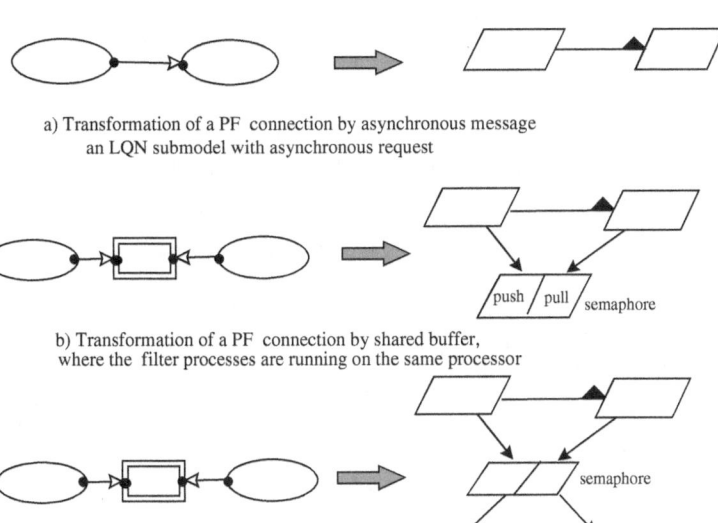

a) Transformation of a PF connection by asynchronous message
an LQN submodel with asynchronous request

b) Transformation of a PF connection by shared buffer,
where the filter processes are running on the same processor

c) Transformation of a PF connection by shared buffer,
where the filter processes are running on different processors

Fig. 3. Transformation of pipeline pattern

A software system contains many components involved in various architectural connection instances, and a component (process) may play different roles in connections of various types. Such a process owns an appropriate port for each of the roles it plays. At the same time, a port can participate in more than one connection instance of the same type.

Fig. 3.a, 3.b and 3.c show the translation of the Pipe and Filters connection types to LQN. The first is using asynchronous messages and the other two a shared buffer. The translation is quite straightforward, only the way LQN models a passive shared buffer warrants a little discussion. The pipeline connection is represented by an asynchronous LQN arc, but this does not take into account the serialization delay due to the constraint that buffer operations must be mutually exclusive. A third task is introduced, with as many entries as the number of different critical sections executed by the tasks accessing the buffer (two in this case, "push" and "pull"). It is interesting to note that, although the software architecture in Fig. 3.b and Fig. 3.c is exactly the same, the difference in the allocation of processes to processors leads to quite different LQN submodels.

Fig. 4.a, 4.b and 4.c illustrate the transformation of three Client-Server connections that have similar architectural descriptions, differentiated only by the edge type. However, their LQN models are quite different, as the connections have very different operating modes and performance characteristics. The LQN model for the direct client-server connection is quite straightforward, but those for broker connections are more interesting. The forwarding broker model uses

482   Dorina C. Petriu and Xin Wang

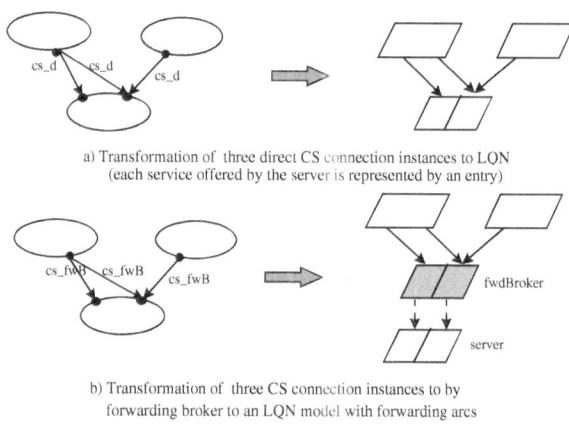

a) Transformation of three direct CS connection instances to LQN
(each service offered by the server is represented by an entry)

b) Transformation of three CS connection instances to by
forwarding broker to an LQN model with forwarding arcs

c) Transformation of three CS connection instances by
handle-driven broker

Fig. 4. Transformation of client-server pattern

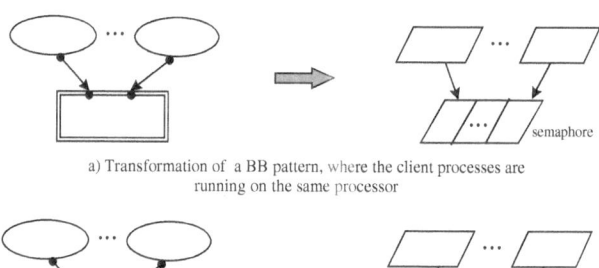

a) Transformation of a BB pattern, where the client processes are
running on the same processor

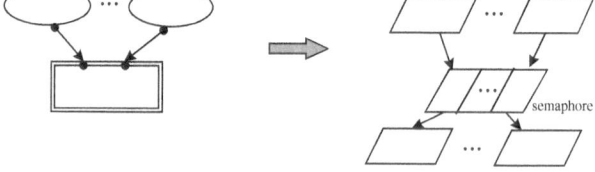

b) Transformation of a BB pattern, where the client processes are
running on different processors

Fig. 5. Transformation of blackboard pattern

LQN forwarding requests (drawn with dotted lines) with a special semantic. After accepting a request from a client, the acceptor task will do some processing, then may decide to forward the request to another task. The forwarder is free to continue its activity, while the client remains blocked, waiting for a reply. The second task that continues to serve the request may eventually complete it and send the reply to the client, or may decide to forward the request to another task. The LQN model implies that a reply will be sent to the client by the last task in the forwarding chain, but it does not represent this reply by an arrow. In Fig. 4.b, the broker is the task that receives the requests from the clients and forwards them to the appropriate entry of the server. The broker must have a separate entry for each entry it forwards to, otherwise the clients would be unable to choose the server entry they need. The LQN model for the handle-driven broker in Fig. 4.c sends two separate requests, one to the broker for getting the handle, then another to the desired server entry directly.

Fig. 5.a and 5.b show the transformation for the blackboard connection type. The performance model captures the serialization delays introduced by the constraint that the blackboard should be accessed by one client at a time. Similar to the pipeline with shared buffer, the same software architecture will generate different performance models depending on whether the clients are running on the same or on different processors. In the later case, the LQN tasks that represent the critical sections executed by the each client are co-allocated on the same processor as the respective client.

4   PROGRES Graph Transformations

The authors have followed the following principles when designing the PROGRES transformation program:

a. Each architectural component is converted to an LQN task, for which reason a common base class COMP-TASK was defined in the graph schema for components and tasks. However, the correspondence between components and tasks is not bijective, due to processes implemented in the underlying operating system or midware (such as brokers) which are not represented explicitly in the architectural view, but are explicit in the LQN view.
b. Each input port of a component is converted into an LQN entry. The correspondence between input ports and entries is not bijective either, due to broker entries.
c. The output ports do not have any correspondent in LQN. However, they play a role in the two-step translation process of server-to-server connections, as illustrated in Fig. 7.
d. Processors and devices, which are attributes in the architectural view, become full-fledged nodes in LQN (not illustrated in the paper due to space limitations). This happens because the issue of resource allocation is secondary to the software development process, but is central to performance analysis.

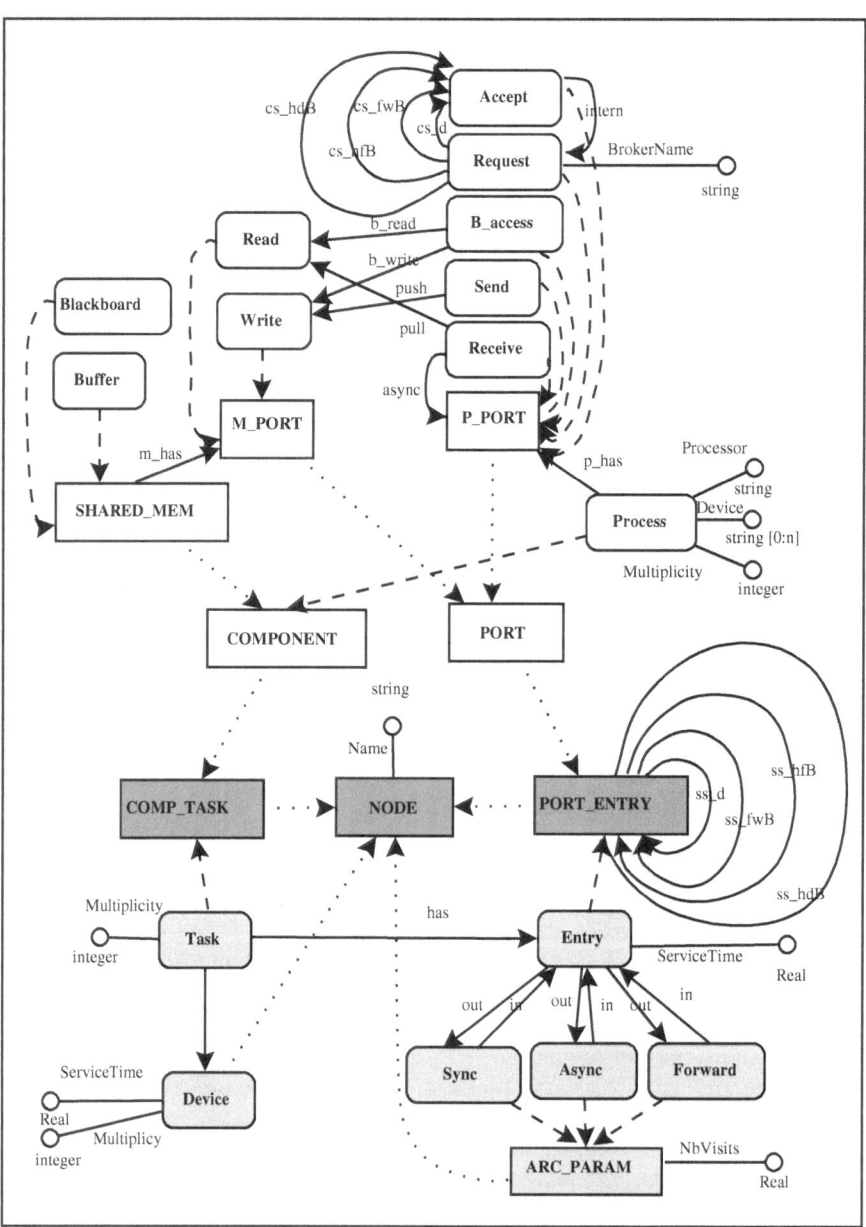

Fig. 6. The joint graph schema of the architectural description and the LQN model

·The graph schema according to the PROGRES language [9 to 11] is presented in Fig. 6. The upper part of the figure contains the *input schema* for architectural descriptions and the lower part the *output schema* for LQN models (light-gray nodes). In order to accommodate graphs in intermediary translation stages, the two schemas are joined together by three nodes shown in dark-gray at the base of the node class hierarchy (*NODE, COMP_TASK, and PORT_ENTRY*). Also, some intermediary edge types (*ss_d, ss_fwB, ss_hfB, and ss_hdB*) were found to be necessary in the process of translating server-to-server CS connections, which appear in tiered client/server systems. Such edges are illegal in both input and output schemas; they are generated and then deleted during a two-step translation process, as shown later in this section.

The input schema describes two kinds of software components and their connections: "process" (active component with its own thread of control) and "shared-memory" (passive component of either "buffer" or "blackboard" types). Each type of component has different types of ports corresponding to the roles played in various architectural connections. The edge types in the graph correspond to different connection types. An interesting example is that of the four Client Server connection types, which are differentiated in the architectural view only by their edge type (*cs_d, cs_fwB, cs_hfB and cs_hdB*, respectively). The "broker" component is not explicitly shown in the architectural view (as the broker is not actually part of the software application, but is provided by the underlying midware). However, a broker has an important impact on the system performance, and it is explicitly modelled in LQN.

The LQN graph notation has "task" and "device" and "entry" nodes, which are described by the corresponding node types in the output schema. The LQN arcs may represent three types of requests (synchronous, asynchronous and forwarding); a parameter indicates the average number of visits associated with that request. Since PROGRES edges cannot have attributes, we represent an LQN arc by three elements: an incoming edge, a node carrying the parameter and an outgoing edge.

In the case of a layered (or "tiered") client/server system, some server nodes play a dual role of server to its own clients, and of client to other servers. Such a server component owns input ports (corresponding to the server role) and output ports (corresponding to the client role). The input and output ports are linked through edges of *intern* type. Servers playing a dual role introduce an additional step in the translation of server-to-server connections, as shown in Fig.7.*a*. Firstly, the internal mapping between the input and output ports of the upper server is used to generate an appropriate number of request edges of an intermediary type (*ss_d, ss_fwB, ss_hfB, or ss_hdB*). Secondly, the translation process is applied to each intermediary edge as if it were a "normal" CS connection (*i.e.*, an *ss_d* edge is treated as a *cs_d* connection, an *ss_fwB* edge as *cs_fwB* connection, and so on.) Fig. 7.a illustrates the two-step translation process applied to a subsystem with two servers involved in several client/server connections of *cs_d* type: first the edges of *ss_d* type (represented by dotted lines) are created, then are translated into LQN like normal *cs_d* connections. Fig.7.b

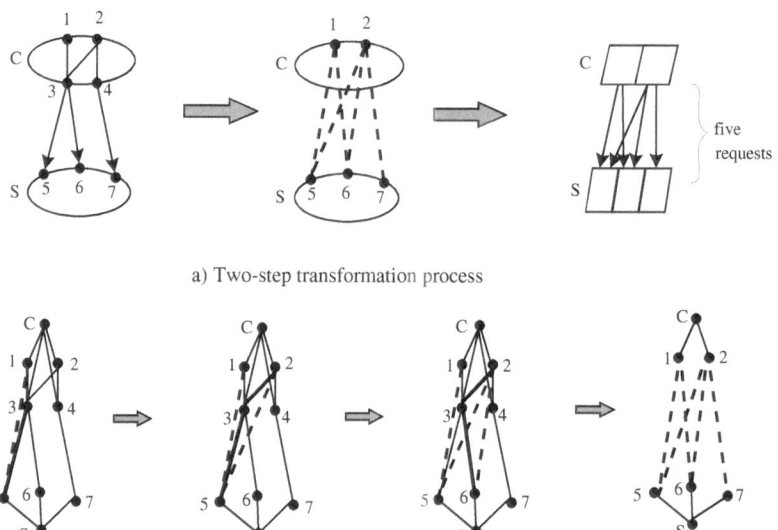

a) Two-step transformation process

b) Sequence of graph transformations for the first transformation step

Fig. 7. Illustration of the two-step transformation of server-to-server CS connections for a simple case without brokers

shows a sequence of PROGRES derivations for the first step. Each derivation deals with a two-edge link (shown in thick lines) from an input port of the upper server to an input port of the lower server. There are five such links in our example, each generating a connection edge of an intermediary type, as explained before.

The PROGRES transformation process is organized as follows: first all server-to-server CS connections are transformed into intermediary type edges as explained above, then transactions are executed for all connections found in the graph. These transactions are designed to transform any type of architectural connections into LQN sub-models. New nodes and edges of types described in the output schema are added to the graph, but not all the nodes and edges of types described in the input schema can be removed from the graph, as they may participate in more than one connection instance. Therefore, a general clean-up phase is required at the end, after all the architectural connections have been processed. For example, in the clean-up phase nodes representing output ports (which do not have an LQN equivalent) will be removed from the graph, and multiple broker tasks with the same name will be merged into a single task (in the case more than one was created by different transactions).

Fig.8 shows an example of production rule for the transformation of a CS connection with handle-driven broker. Nodes 1 and 2 represent architectural components in the left-hand side, and tasks in the right-hand side. The input

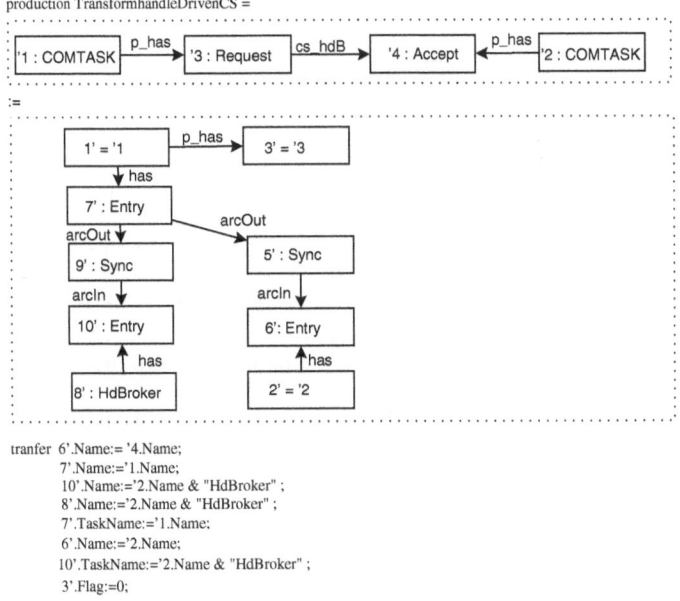

Fig. 8. Production rule for the transformation of a CS connection with handle-driven broker

port 4 is "transformed" into entry 6, and the output port 3 is kept in the graph until the clean-up stage. A broker task 8, its entry 10, client entry 7 and two request nodes 5 and 9 are also added to the graph.

5   Conclusions

The automatic generation of LQN performance models from software architecture descriptions raises a number of challenges due to the fact that each view has a different semantics, purpose and focus, which must be bridged by the translation process. The architectural view represents only the software components of the application under development, and may hide operating system and midware services. However, such services have a considerable impact on the overall system performance, and must be modelled in LQN explicitly. Another discrepancy between the two views arises from the different role played by the issue of resource allocation. Whereas the architectural view concentrates on the software under development, the performance model takes a *vertical view* of the whole system, modelling not only the application software, but also relevant characteristics of the underlying operating system, midware and hardware. The final result of the transformation process is an LQN model that can be written to a file according to a predefined format and can be solved with existing LQN solvers [5]. The analysis of an LQN model and the use of its results to identify performance trouble spots is outside of the scope of this paper, but is described in other work such as [5, 8, 14, 16, 17].

References

1. O. Adebayo, J. Neilson, D. Petriu, "A Performance Study of Client-Broker-Server Systems", in Proceedings of CASCON'97, pp 116-130, Toronto, Canada, November 1997.
2. R. Adler, "Distributed Coordination Models for Client/Server Computing", IEEE Computer, pp 14-22, April 1995.
3. R.Allen, D. Garlan, "A Formal Basis for Architectural Connection", *ACM Transactions on Software Engineering Methodology*, Vol.6, No.3, pp 213-249, July 1997.
4. F. Buchmann, R. Meunier, H. Rohnert, P. Sommerland, M. Stal, *Pattern-Oriented Software Architecture: A System of Patterns*, Wiley Computer Publishing, 1996.
5. G. Franks, A. Hubbard, S. .Majumdar, D. Petriu, J. Rolia, C.M. Woodside, "A toolset for Performance Engineering and Software Design of Client-Server Systems", *Performance Evaluation*, Vol. 24, Nb. 1-2, pp 117-135, November 1995.
6. D. Le Metayer, "Software Architecture Styles as Graph Grammars", in Proceedings of the 4-th ACM SIGSOFT Symposium on the Foundations of Software Engineering, San Francisco, USA, pp 15-23, Oct. 1996.
7. Object Management Group, *The Common Object Request Broker: Architecture and Specification*, Object Management Group and X/Open, Framingham, MA and Reading Berkshire UK, 1992.
8. J.A. Rolia, K.C. Sevcik, "The Method of Layers", *IEEE Trans. On Software Engineering*, Vol. 21, Nb. 8, pp 689-700, August 1995.
9. A. Schuerr, "PROGRES: A Visual Language and Environment for PROgramming with Graph Rewrite Systems", Technical Report AIB 94-11, RWTH Aachen, Germany, 1994.
10. A. Schuerr, "Introduction to PROGRES, an attribute graph grammar based specification language", *in Graph-Theoretic Concepts in Computer Science*, M. Nagl (ed), Vol. 411 of *Lecture Notes in Computer Science*, pp 151-165, 1990.
11. A. Schuerr, "Programmed Graph Replacement Systems", in Handbook of Graph Grammars and Computing by Graph Transformation, G. Rozenberg (ed), pp 479-546, 1997.
12. M. Shaw, D. Garlan, *Software Architectures: Perspectives on an Emerging Discipline*, Prentice Hall, 1996.
13. M. Shaw, "Some Patterns for *Software Architecture*" in *Pattern Languages of Program Design 2* (J.Vlissides, J. Coplien, and N. Kerth eds.), pp.255-269, Addison Wesley, 1996.
14. C. Shousha, D.C. Petriu, A. Jalnapurkar, K. Ngo, "Applying Performance Modelling to a Telecommunication System", Proc.of the First Int. Workshop on Software and Performance (WOSP'98), pp. 1-6, Santa Fe, New Mexico, October 1998.
15. C.U. Smith, *Performance Engineering of Software Systems*, Addison Wesley, 1990.
16. C.M. Woodside, "Throughput Calculation for Basic Stochastic Rendezvous Networks", *Performance Evaluation*, Vol. 9, Number 2, pp.143-160, April 1995.
17. C.M. Woodside, J.E. Neilson, D.C. Petriu, S. Majumdar, "The Stochastic Rendezvous Network Model for Performance of Synchronous Client-Server-like Distributed Software", *IEEE Transactions on Computers*, Vol.44, Nb.1, pp 20-34, January 1995.

Author Index

Lecture Notes in Computer Science

For information about Vols. 1–1697
please contact your bookseller or Springer-Verlag